FLUID DYNAMICS AND TRANSPORT OF DROPLETS AND SPRAYS

Second Edition

This book serves as both a graduate text and a reference for engineers and scientists exploring the theoretical and computational aspects of the fluid dynamics and transport of sprays and droplets. Attention is given to the behavior of individual droplets, including the effects of forced convection due to relative droplet–gas motion, Stefan convection due to the vaporization or condensation of the liquid, multicomponent liquids (and slurries), and internal circulation of the liquid. This second edition contains more information on droplet–droplet interactions, the use of the mass-flux potential, conserved scalar variables, spatial averaging and the formulation of the multicontinua equations, the confluence of spatial averaging for sprays and filtering for turbulence, direct numerical simulations and large-eddy simulations for turbulent sprays, and high-pressure vaporization processes. Two new chapters introduce liquid-film vaporization as an alternative to sprays for miniature applications and a review of liquid-stream distortion and breakup theory, which is relevant to spray formation.

William A. Sirignano is the Henry Samueli Professor of Mechanical and Aerospace Engineering and former Dean of the School of Engineering at the University of California, Irvine. Before that, he was the George Tallman Ladd Professor and Department Head at Carnegie-Mellon University and a professor at Princeton University. His major research and teaching interests include spray combustion, turbulent combustion and ignition, aerospace propulsion, fluid dynamics, and applied mathematics. Dr. Sirignano has written more than 450 research papers, book articles, and reports and has given more than 300 conference presentations and research seminars. He has been a formal consultant to 30 industrial organizations and federal laboratories.

Sirignano is a member of the National Academy of Engineering and a Fellow of the AIAA, ASME, AAAS, APS, and SIAM. He is the recipient of the Pendray Aerospace Literature Award, Propellants and Combustion Award, Energy Systems Award, Wyld Propulsion Award, and Sustained Service Award from the AIAA; the ASME Freeman Scholar Fluids Engineering Award; The Combustion Institute Alfred C. Egerton Gold Medal; and the Institute for the Dynamics of Explosions and Reactive Systems Oppenheim Award.

Fluid Dynamics and Transport of Droplets and Sprays

SECOND EDITION

William A. Sirignano
University of California, Irvine

CAMBRIDGE
UNIVERSITY PRESS

32 Avenue of the Americas, New York NY 10013-2473, USA

Cambridge University Press is part of the University of Cambridge.

It furthers the University's mission by disseminating knowledge in the pursuit of
education, learning and research at the highest international levels of excellence.

www.cambridge.org
Information on this title: www.cambridge.org/9781107428003

First published 2010
First paperback edition 2014

A catalogue record for this publication is available from the British Library

Library of Congress Cataloguing in Publication data
Sirignano, William A.
Fluid dynamics and transport of droplets and sprays / William A.
Sirignano. – 2nd ed.
      p.   cm.
Includes bibliographical references and index.
ISBN 978-0-521-88489-1 (hardback)
1. Spraying.   2. Atomization.   I. Title.
TP156.S6S57   2010
660'.294515 – dc21   2009027791

ISBN 978-0-521-88489-1 Hardback
ISBN 978-1-107-42800-3 Paperback

To my wife and children, Lynn Sirignano, Justin Sirignano,
Monica Sirignano, Jacqueline Vindler.

Contents

Preface

The fluid dynamics and transport of sprays comprise an exciting field of broad importance. There are many interesting applications of spray theory related to energy and power, propulsion, heat exchange, and materials processing. Spray phenomena also have natural occurrences. Spray and droplet behaviors have a strong impact on vital economic and military issues. Examples include the diesel engine and gas-turbine engine for automotive, power-generation, and aerospace applications. Manufacturing technologies including droplet-based net form processing, coating, and painting are important applications. Applications involving medication, pesticides and insecticides, and other consumer uses add to the impressive list of important industries that use spray and droplet technologies. These industries involve annual production certainly measured in tens of billions of dollars and possibly higher. Many applications are still under development. The potentials for improved performance, improved market shares, reduced costs, and new products and applications are immense. Continuing effort is needed to optimize the designs of spray and droplet applications and to develop strategies and technologies for active control of sprays in order to achieve the huge potential.

In the first edition of this book and in this second edition, I have attempted to provide some scientific foundation for movement toward the goals of optimal design and effective application of active controls. The book, however, does not focus on design and controls. Rather, I discuss the fluid mechanics and transport phenomena that govern the behavior of sprays and droplets in the many important applications. Various theoretical and computational aspects of the fluid dynamics and transport of sprays and droplets are reviewed in detail. I undertook this writing because no previous treatise exists that broadly addresses theoretical and computational issues related to both spray and droplet behavior. There are other books that address either sprays on a global scale or individual isolated droplets on the fine scale. However, no other book has attempted a true integration of these two critically related topics. My research interests have focused on the theoretical and computational aspects of the spray problem. Therefore, this book emphasizes those aspects. Major but not total attention is given to the works of my research team because we have many research publications and review papers on this subject. On the basis of these research studies over the years, a decent comprehensive portrayal of the field is achievable. I have given some emphasis to liquid-fuel droplets and to combustion applications because my experience is centered in that domain and, more important, because the high temperatures and rapid vaporization make the dynamics of the phenomenon much more interesting and general. Rapidly vaporizing sprays have a richness of the scientific phenomena and several, often disparate, time scales. The discussions are often also relevant to other important applications including materials processing, heat exchange, and coatings. Because the field of droplet and spray studies is still developing in terms of both science and technology, a critical review is undertaken here.

This book was developed largely on the basis of my lecture notes generated during several offerings of a graduate course. This treatise can serve both as a graduate-level text and as a reference book for scientists and engineers.

All of the material from the first edition is retained here, although much of it has been reorganized into different chapters. Attention is given to the behavior of individual droplets, including the effects of forced convection that are due to relative droplet–gas motion, Stefan convection that is due to the vaporization or condensation of the liquid, multicomponent liquids (and slurries), and internal circulation of the liquid. Flow-field details in the gas boundary layer and wake and in the liquid-droplet interior are examined. Also, the determinations of droplet lift and drag coefficients and Nusselt and Sherwood numbers and their relationships with Reynolds number, transfer number, Prandtl and Schmidt numbers, and spacing between neighboring droplets are extensively discussed. Results from droplet analyses are presented in a manner that makes them useful as subgrid models in spray computations. Several examples of spray computations for which these models are used are presented. The two-phase flow equations governing spray behavior are presented in various forms and thoroughly discussed. Attention is given to issues of computational accuracy and efficiency. Various configurations for spray flows are studied. Droplet interactions with vortical and turbulent fields are analyzed. Droplet behavior under near-critical and supercritical conditions is discussed.

In addition to updating and reorganizing the material from the first edition, new content has been added. This second edition is more than 50% longer than the first edition. More information has been added on the topics of droplet–droplet interactions, the use of the mass-flux potential, conserved scalar variables, spatial averaging and the formulation of the multicontinua equations, the confluence of spatial averaging for sprays and filtering for turbulence, direct numerical simulations and large-eddy simulations for turbulent sprays, and high-pressure vaporization processes. A new chapter has been included on liquid-film vaporization as an alternative to sprays for miniature applications. Another new chapter has a review of theory on liquid-stream distortion and breakup, which is very relevant to spray formation.

My interactions over more than three decades with 13 postdoctoral associates and 21 graduate students on the subject of sprays have been very productive, stimulating, and instructive. These junior (at the time) collaborators are well represented in the references. They are Boris Abramzon, Suresh K. Aggarwal, Nasser Ashgriz, Rakesh Bhatia, Jinsheng Cai, C.H. "Jeff" Chiang, Gaetano Continillo, Sadegh Dabiri, Jean-Pierre Delplanque, Amar Duvvur, Eva Gutheil, Howard Homan,

Randall Imaoka, Inchul Kim, Pedro Lara-Urbaneja, C.K. "Ed" Law, D.N. Lee, Steven Lerner, Jun Li, Mansour Masoudi, Constantine Megaridis, Carsten Mehring, Kamyar Molavi, Gopal Patnaik, Satya Prakash, M.S. Raju, Roger Rangel, David Schiller, Bartendu Seth, Simone Stanchi, Douglas Talley, Albert Tong, Guang Wu, and Jinxiang Xi. Exciting interactions with senior collaborators are also recognized: H.A. Dwyer, D. Dunn-Rankin, S.E. Elghobashi, G.J. Fix, D.D. Joseph, F. Liu, V.G. McDonell, B.R. Sanders, E. Suuberg, and S.C. Yao are identified here. Several federal funding agencies and industrial organizations have been supportive of my spray research; special recognition for continuing support goes to Julian Tishkoff of AFOSR; David Mann, Kevin McNesby, and Ralph Anthenien of ARO; and Gabriel Roy of ONR. Here I am acknowledging only those who worked with me or supported me specifically in the area of sprays. There are many others to thank for associations on other scientific problem areas. Also, I am thankful for the opportunities for intellectual exchanges and friendship with many individuals from around the world who have contributed to the disciplines of fluid dynamics, transport, and combustion and/or to technologies for energy, power, and propulsion.

Nomenclature

a	droplet acceleration; constant of curvature in stagnation-point flow; half of undisturbed sheet thickness
a, b	constants in Eq. (1.3); variable parameters in Section 3.1, Eq. (3.48); also constant parameters in equation of state (5.3)
A	constant in Section 3.1; area
$\tilde{A}$	liquid-vortex strength
b_T, b_M	corrections to transfer number
B	Spalding transfer number
B_H	energy transfer number
B_M	mass transfer number
$\mathrm{Bo} = \rho_l a R^2 / \sigma$	Bond number
c_l	liquid specific heat
c_p	specific heat at constant pressure
C_D, C_L	drag and lift coefficients
C_F	friction coefficient
d	droplet diameter; distance of vortex from droplet path
D	mass diffusivity; droplet center-to-center spacing; drag force; orifice diameter
e	thermal energy; correction in Section 8.3 for application of ambient conditions
$\vec{e}$	unit vector
E	activation energy
E_i	externally imposed electric field
f	Blasius function; droplet distribution function (probability density function)
f_H	defined by Eq. (3.72e)
f_i	fugacity of species i
F	drag force per unit mass on droplet; friction force
F_{Di}	aerodynamic forces per unit volume on droplets
F, G	functions defined in Section 3.1, Eqs. (3.44)

Fr	Froude number
G	response factor for combustion instability; also Chiu group combustion number
g	acceleration due to gravity
g_1, g_2	functions defined in Section 3.1 by Eq. (3.44)
$\mathrm{Gr} = g R^3 / \nu^2$	Grashof number
h	enthalpy; heat transfer coefficient; scale factors; film thickness
I_n	modified Bessel function of the first kind
k	nondimensional constant in Section 3.1; also turbulence wave number
k, l, m, n	wave numbers
k_E	generalized Einstein coefficient
K	constant in vaporization-rate law; constants in model equations; strain rate; cavitation number
$K(t, t - \tau)$	kernel in history integral, Section 3.4
l_c	characteristic length for liquid-stream pinch-off
L	latent heat of vaporization; differential operator; orifice length
Le	Lewis number
m	droplet mass
$\dot{m}$	droplet mass vaporization rate
m_f	mass of fluid displaced by droplet or particle
m_p	mass of particle or droplet
$\dot{M}$	mass source term (rate per unit volume); vaporization rate per unit volume for spray
M_{A1}, M_{A2}	acceleration numbers defined by Eqs. (3.65) and (3.68)
n	droplet number density; direction normal to interface; fuel mass flux
n_i	number of moles of species i
$\vec{n}$	unit normal vector
N	number of droplets; number of species in multicomponent mixture; ratio of droplet heating time to droplet lifetime
$\overline{N}$	number of droplets in array
Nu	Nusselt number
Oh	Ohnesorge number
p	pressure
Pe	Peclet number
Pr	Prandtl number
q_α	electric charge on droplet
$\dot{q}$	heat flux
Q	energy per unit mass of fuel
r	spherical radial coordinate
$\tilde{r}$	cylindrical radial coordinate
r'	nondimensional annular radius perturbation
r_f	flame radius
R	droplet radius, radius of curvature of liquid sheet
R_2	droplet-radii ratio

R	gas constant
$\mathrm{Re} = R\Delta U/\nu$	droplet Reynolds number (In discussions in which a Reynolds number based on diameter or some special properties is used, it is specified in the text.)
s	nondimensional radius; transformed variable defined by Eq. (9.14b)
S	superscalar; weighted area in numerical interpolation schemes; distribution function for radiative heat transfer; jet or droplet-surface area
Sc	Schmidt number
Sh	Sherwood number
St	Strouhal number
t	time
t_c	characteristic time for liquid-stream pinch-off
T	temperature
u, v	velocity
u_o	mean jet velocity
U	free-stream velocity; particle or droplet velocity; Bernoulli jet velocity
v	velocity in the argument of distribution function; specific volume
$\vec{v}$	velocity vector
V	volume in eight-dimensional phase space; liquid volume; characteristic velocity
$\dot{w}$	chemical-reaction rate, fractional change per unit time
W	molecular weight
$\mathrm{We} = \rho_l R(\Delta U)^2/\sigma$	Weber number
x	spatial coordinate; stream coordinate in planar case
X	mole fraction; position
y	normal coordinate in boundary layer; transverse coordinate in planar case
$\overline{y}$	liquid sheet centerline position
$\tilde{y}$	liquid sheet thickness
Y	nondimensional sheet centerline position
Y_n	mass fraction of species n
z	axial or downstream spatial coordinate in cylindrical coordinates; spanwise coordinate
Z	nondimensional temperature

Greek Symbols

α	thermal diffusivity; wave phase; growth rate
β	Shvab–Zel'dovich variables
γ	reciprocal of Lewis number; fraction of radiation absorbed at surface and in interior; ratio of specific heats
Γ	circulation

δ	distance ratio in Section 6.3; thickness of oxide layer; fraction of radiation absorbed at droplet surface; Dirac delta function; differential quantity
$\tilde{\delta}$	upstream distance for application of boundary condition
δ_{99}	the point, measured from the stagnation point, where 99% of the mass fraction variation has occurred
Δ	change in a quantity
ΔU	relative droplet velocity
ε	radiative emissivity
ε_n	mass-flux fraction for species n
ε_0	electrical permittivity of a vacuum
η	Blasius coordinate; nondimensional radius
θ	void volume fraction; porosity; azimuthal angle
κ	reciprocal of characteristic length of temperature variation
λ	thermal conductivity; eigenvalues
μ	dynamic viscosity; chemical potential
ν	stoichiometric coefficient (mass of fuel per mass of oxygen); kinematic viscosity
ρ	density
ρ_r	ratio of particle (or droplet) density to gas density
$\bar{\rho}$	bulk density
σ	surface-tension coefficient; Stefan–Boltzmann constant, nondimensional vortex-core size
τ	nondimensional time; viscous stress
τ_H	droplet-heating time
τ_L	droplet lifetime
τ^*	droplet-heating time with uniform temperature
ϕ	velocity potential; mass-flux potential; normalized stream function; entropy variable defined by Eq. (7.18)
ϕ_m	metal volume fraction
ϕ_r	ratio of acceleration numbers defined by Eq. (3.68)
φ	generic variable
Φ	fractional volume of fluid
χ	ratio of effective thermal diffusivity to thermal diffusivity; ratio of thermal diffusivities for slurry droplets
ψ	liquid volume fraction; stream function
ω	vorticity; frequency

Subscripts

0	initial condition
∞	condition at infinity
Al	aluminum property
b	boiling point Al (aluminum property); bubble
c	critical conditions
e	edge of boundary layer
eff	effective value

F	fuel vapor
g	gas phase
H	related to thermal transfer
i	index for vectorial component; index for initial value
j	index for vectorial component
k	integer index
L	lift
l	liquid phase
M	related to mass transfer
max	maximum
m	index for species; index for metal
mix	mixture
n, p	integers for numerical mesh points
N	nitrogen
ox	oxide
O	oxygen
p	particle
P	product
r, θ, z	components in cylindrical coordinates
s	droplet-surface condition
T	related to thermal transfer
wb	wet bulb
$+, -$	upper or lower liquid–gas interface

Superscripts

k	index for droplet group
.	rate or time derivative
$'$	derivative; perturbation quantity

1 Introduction

1.1 Overview

A spray is one type of two-phase flow. It involves a liquid as the dispersed or discrete phase in the form of droplets or ligaments and a gas as the continuous phase. A dusty flow is very similar to a spray except that the discrete phase is solid rather than liquid. Bubbly flow is the opposite kind of two-phase flow wherein the gas forms the discrete phase and the liquid is the continuous phase. Generally, the liquid density is considerably larger than the gas density; so bubble motion involves lower kinematic inertia, higher drag force (for a given size and relative velocity), and different behavior under gravity force than does droplet motion.

Important and intellectually challenging fluid-dynamic and -transport phenomena can occur in many different ways with sprays. On the scale of an individual droplet size in a spray, boundary layers and wakes develop because of relative motion between the droplet center and the ambient gas. Other complicated and coupled fluid-dynamic factors are abundant: shear-driven internal circulation of the liquid in the droplet, Stefan flow that is due to vaporization or condensation, flow modifications that are due to closely neighboring droplets in the spray, hydrodynamic interfacial instabilities leading to droplet-shape distortion and perhaps droplet shattering, and droplet interactions with vortical structures in the gas flow (e.g., turbulence).

On a much larger and coarser scale, we have the complexities of the integrated exchanges of mass, momentum, and energy of many droplets in some subvolume of interest with the gas flow in the same subvolume. The problem is further complicated by the strong coupling of the phenomena on the different scales; one cannot describe the mass, momentum, and energy exchanges on a large scale without detailed knowledge of the fine-scale phenomena. Note that, in some practical applications, these scales can differ by several orders of magnitude so that a challenging subgrid modelling problem results.

Detailed consideration will be given to applications in which the mass vaporization rate is so large that the physical behavior is modified. This is the most complex situation, and therefore its coverage leads to the most general formulation of the theory. In particular, as the vaporization rate increases, the coupling between the two phases becomes stronger and, as the droplet lifetime becomes as short as some of the other characteristic times, the transient or dynamic character of the problem emerges in a dominant manner.

The fast vaporization rate is especially prominent in situations in which the ambient gas is at very high temperatures (of the order of 1000 K or higher). Combustion with liquid fuels is the most notable example here. The spray combustion regime is a most interesting limiting case of the more general field of thermal and dynamic behavior of sprays. In the high-temperature domain, rapid vaporization causes droplet lifetimes to be as short as the time for a droplet to heat throughout its interior. It can be shorter than the time for liquid-phase mass diffusion to result in the mixing of various components in a multicomponent liquid. The combustion limit is inherently transient from the perspective of the droplet, richer in terms of scientific issues, and more challenging analytically and numerically than low-temperature spray problems. Vaporization might still be longer than other combustion processes such as mixing or chemical reaction; therefore it could be the rate-controlling process for energy conversion.

The spray problem can be complicated by the presence of spatial temperature and concentration gradients and internal circulation in the liquid. Interaction among droplets is another complication to be treated.

There is a great disparity in the magnitudes of the scales. Liquid-phase mass diffusion is slower than liquid-phase heat diffusion, which, in turn, is much slower than the diffusion of vorticity in the liquid. Transport in the gas is faster than transport in the liquid. Droplet diameters are typically of the order of a few tens of micrometers (μm) to a few hundreds of micrometers in diameter. Resolution of internal droplet gradients can imply resolution on the scale of micrometers or even on a submicrometer scale. Combustor or flow chamber dimensions can be 5 to 6 orders of magnitude greater than the required minimum resolution. Clearly, subgrid droplet-vaporization models are required for making progress on this problem.

Experiments have been successful primarily in resolving the global characteristics of sprays. The submillimeter scales associated with the spray problem have made detailed experimental measurements very difficult. If an attempt is made to increase droplet size, similarity is lost; the droplet Reynolds number can be kept constant by decreasing velocity but the Grashof number grows, implying that buoyancy becomes relatively more important. Also, the Weber number increases as droplet size increases; surface tension becomes relatively less important, and the droplet is more likely to acquire a nonspherical shape. Modern nonintrusive laser diagnostics have made resolution possible on a scale of less than 100 μm so that, in recent years, more experimental information has been appearing. Nevertheless, theory and computation have led experiments in terms of resolving the fluid-dynamical characteristics of spray flows.

Classical texts on droplets, including burning-fuel droplets, tend to consider an isolated spherical droplet vaporizing in a stagnant environment. In the simplified representation, the liquid has one chemical component, ambient-gas conditions are subcritical, and vaporization occurs in a quasi-steady fashion. The classical result is that the square of the droplet radius or diameter decreases linearly with time because heat diffusion and mass diffusion in the surrounding gas film are the rate-controlling (slowest) processes; this behavior is described as the d^2 law. These important phenomena are discussed in later chapters. Although most researchers are now addressing these relevant and interesting factors that cause major deviations from classical behavior, there are still some researchers who persist in the study of the classical configuration. Here we will relax these simplifications, one at a time, to gain a more accurate and more relevant understanding. Convective effects that are due to droplet motion or natural convection and subsequent internal liquid circulation are thoroughly studied. Transient heating (or cooling) and vaporization (or condensation) that are due to changing ambient conditions, unsteady liquid-phase diffusion, or unsteady gas-phase diffusion are analyzed. Multicomponent-liquid (including emulsions and slurries as well as blended liquids) droplet vaporization is studied. Near-critical and supercritical ambient conditions (and their effects on diffusion processes, phase change, solubility, and liquid-surface stripping that is due to shear) will be discussed. Interactions of droplets with other droplets and with turbulent or vortical structures are analyzed. Distortion of the spherical shape and secondary atomization of the droplets are also discussed. The effects of radiative heating of the liquid and of exothermic chemical reaction in the gas film are also studied.

Current texts do not explain in a unified fashion the various approaches to calculation of the behaviors of the many droplets present in a spray. Efficient and accurate methods for predicting the trajectories, temperatures, and vaporization rates of a large number of droplets in a spray are discussed here. Sprays in both laminar and turbulent environments are discussed.

Some comments about primary atomization and droplet-size determination are given in Section 1.2. In Chapters 2 and 3, we discuss the vaporization of individual droplets and study the phenomenon on the scale of the droplet diameter. Chapter 2 considers the case in which there is no relative motion between the droplet and the distant gas, and Chapter 3 covers the situation with a relative velocity. The theoretical models and correlations of computational results for individual droplets can be used to describe exchanges of mass, momentum, and energy between the phases in a spray flow. The vaporization of multicomponent droplets, including slurry droplets, is discussed in Chapter 4. Droplet behavior under near-critical or transcritical thermodynamic conditions is considered in Chapter 5. Secondary atomization and molecular-dynamic methods are also discussed there. Interactions among droplets and their effects on the modification of the theory are discussed in Chapter 6. The spray with its many droplets is examined first in Chapter 7. The spray equations are examined from several aspects; in particular, two-continua, multicontinua, discrete-particle, and probabilistic formulations are given. The choice of Eulerian or Lagrangian representation of the liquid-phase equations within these formulations is discussed, including important computational issues and the relationship between the Lagrangian method and the method of characteristics. Some specific computational issues are discussed in Chapter 8. Some of the theories and information in this book have already had an impact on computational codes; modification of the codes to address more recent advances should not be difficult. One shortcoming, of course, is the limited experimental verification, as just discussed.

Applications of the spray theory to special laminar-flow configurations are discussed in Chapter 9. Turbulence–spray interactions are surveyed in Chapter 10. Vorticity–droplet interactions and turbulence–droplet interactions have not yet been fully integrated into a comprehensive spray theory. These interaction studies are still active research domains, and, so far, little application to engineering practice has occurred. In Chapter 11, we discuss exchanges of mass, momentum, and energy between the phases for a type of configuration other than droplets and sprays, i.e., liquid films. In Chapter 12, current research is discussed for the distortion and disintegration of liquid streams in processes leading to the formation of sprays. Important material on the underlying governing field equations, use of conserved scalars, and a summary of droplet models are given in the Appendices.

1.2 Droplet-Size Determination

The droplet size is an important factor in its behavior. Droplet shape is another factor with profound implications. Surface tension will tend to minimize the droplet surface area, given its volume, resulting in a spherical shape for sufficiently small droplets. The size of a spherical droplet will be represented most commonly by its diameter d or radius R. In most sprays, droplets of many different sizes will exist. Vaporization, condensation, droplet coalescence, and droplet shattering will cause a temporal variation in droplet sizes. For a spray, a distribution function of the instantaneous diameter $f(d)$ is typically used to describe a spray. This function gives the number of droplets possessing a certain diameter. Often an average droplet diameter d_{mn} is taken to represent a spray. In particular,

$$d_{mn} = \frac{\int_0^\infty f(d) d^m \, \mathrm{d}d}{\int_0^\infty f(d) d^n \, \mathrm{d}d}. \tag{1.1}$$

In practice, $f(d)$ will not be a continuous function. However, for a spray with many droplets (millions can be common), the function is well approximated as a continuously varying function. One example of an average droplet is the Sauter mean diameter d_{32}, which is proportional to the ratio of the total liquid volume in a spray to the total droplet-surface area in a spray.

The aerodynamic forces on a droplet will depend on its size in a functional manner different from the dependence of droplet mass on the size. As a result, smaller droplets undergo more rapid acceleration or deceleration than larger droplets. Heating times and vaporization times will be shorter for smaller droplets. Accuracy in the initial droplet-size distribution is mandatory, therefore, if we wish to predict droplet behavior. Unfortunately, we must currently rely mostly on empirical methods to represent droplet distribution; it cannot be predicted from a first-principles approach for most liquid-injection systems.

Liquid streams injected into a gaseous environment tend to be unstable under a wide range of conditions. An important parameter is the Weber number,

$$\mathrm{We} = \frac{\rho \Delta U^2 L}{\sigma}, \tag{1.2}$$

where ρ is the gas density, ΔU is the relative gas–liquid velocity, L is the characteristic dimension of the stream, and σ is the surface-tension coefficient. The Weber number (We) is the ratio of the aerodynamic force related to dynamic pressure to the force of surface tension. Depending on the stream shape, oscillation of the stream and breakup occur above some critical value of the Weber number. These interface oscillations can occur at any wavelength, but some wavelengths will have larger rates of amplitude growth. Below the critical value of the Weber number, the surface-tension forces are large enough to overcome the aerodynamic force that tends to distort the stream. So here, the basic shape of the stream is maintained without disintegration. At higher Weber numbers, the aerodynamic force dominates, leading to distortion and disintegration. This process is called atomization.

Disintegration or atomization typically results in liquid ligaments or droplets with a characteristic dimension that is smaller than the original length scale associated with the stream. Disintegration will continue in a cascade fashion until the decreased length scale brings the Weber number for the resulting droplets below the critical value for the droplets. Other parameters will affect the critical value of the Weber number; they include the ratio of liquid density to gas density and a nondimensional representation of viscosity (e.g., Reynolds number).

Practical atomization systems use a variety of mechanisms to achieve the critical Weber numbers that are necessary. Jet atomizers use a sufficiently large pressure drop across an orifice to obtain the necessary liquid velocity. Air-assist and air-blast atomizers force air flow as well as liquid flow. The critical Weber number depends on the relative air–liquid velocity here. Some atomizers use swirl vanes for the liquid or air to create a tangential component of velocity; this can increase the relative velocity. Rotary atomizers involve spinning cups or disks upon which the liquid is flowed; the centrifugal effect creates the relative velocity. Sometimes other means are used for atomization, including acoustic or ultrasonic oscillations, electrostatic forces, and the injection of a bubbly liquid. An excellent review of practical atomization systems is given by Lefebvre (1989).

There are three general approaches to the prediction of the droplet sizes that result from atomization of a liquid stream. The most widely used approach involves the use of empirical correlations. Another approach requires the solution of the Navier–Stokes equations or of their inviscid limiting form, the Euler equations, to predict disintegration of the liquid stream. Often the linearized form of the equations is taken. The third approach assumes that, in addition to conservation of mass momentum and energy, the droplet-size distribution function satisfies a maximum-entropy principle.

In the first approach, it is common practice to fit experimental data to a number-distribution function for the droplet radius or diameter. With the current level of the theory, this is the most commonly used approach. The Rosin–Rammler distribution equation governs the volume of liquid contained in all droplets below a given diameter d. In particular, the fractional volume of liquid $\Phi(d)$ is described as

$$\frac{\int_0^d f(d')^3 \, \mathrm{d}d'}{\int_0^\infty f(d)^3 \, \mathrm{d}d} \equiv \Phi(d) = 1 - \exp\left(-\frac{d^b}{a}\right), \tag{1.3}$$

where a and b are constants to be chosen to fit the relevant experimental data. It follows that

$$f(d) = \frac{b}{a^b} d^{b-4} \exp\left[-\left(\frac{d}{a}\right)^b \right]. \tag{1.4}$$

Another correlating equation commonly chosen is the Nukiyama–Tanasawa equation, which states that

$$f(d) = a d^2 \exp(-b d^c). \tag{1.5}$$

The constant a is related to a gamma function by the condition that the integral $\int_0^\infty f(d)\,dd$ equals the total number of droplets. So two parameters remain to be adjusted to fit the experimental data.

Sometimes a Gaussian or normal distribution with the natural logarithm of the droplet diameter as the variable gives a good correlation for the experimental data. The logarithm provides a geometric rather than an arithmetic averaging. Here, the log-normal number-distribution function for droplet side is

$$f(d) = \frac{1}{\sqrt{2\pi}} \frac{1}{sd} \exp\left[-\frac{(\ln d - \ln d_{\mathrm{ng}})^2}{2s^2} \right], \tag{1.6}$$

where d_{ng} is the number geometric mean droplet diameter and s is the corresponding standard deviation. See Lefebvre (1989) and Bayvel and Orzechowski (1993) for further details on droplet-size distributions.

The second major approach to the prediction of droplet-size and -velocity distributions in a spray involves analysis guided by the first principles of hydrodynamics with an account of surface-tension forces. This approach dates back to Rayleigh (1878) but yet is still in its infancy. The theory addresses the distortion of the liquid stream that is due to hydrodynamic instability, often of the Kelvin–Helmholtz variety. The theory is limited mostly to linearized treatments, although, with modern computational capabilities, more nonlinear analysis has been occurring recently. The analyses sometimes predict the first step of disintegration of a liquid stream but generally are not able, except in the simplest configurations, to predict droplet-size distribution.

Rayleigh (1878) analyzed the temporal instability of a round liquid jet and predicted that the greatest growth rate of the instability occurs for a disturbance wavelength that is 4.508 times larger than the diameter of the undisturbed jet. With one droplet forming every wavelength, conservation of mass leads to the prediction of a droplet diameter to jet diameter ratio of approximately 1.9. Weber (1931) extended the analysis to account for spatial instability with a mean jet velocity. Hence a dependence on the now-famous Weber number was demonstrated. A useful review of the theory of instability of the round jet is given by Bogy (1979). As the Weber number of the liquid stream increases, aerodynamic effects become increasingly important and droplets of decreasing diameter result from the disintegration process. A good review of these effects with a classification of the various regimes is given by Reitz and Bracco (1982).

In many applications, the liquid stream is injected as a thin sheet to maximize the surface area and to enhance the ratio of disintegration into small droplets.

Examples of this sheet configuration are hollow cone sprays and fan sheets. The theory on this configuration is more limited than that for the round jet. Some overview discussions are given by Lefebvre (1989), Bayvel and Orzechowski (1993), and Sirignano and Mehring (2000, 2005).

The third approach in which a maximum-entropy approach is used is the youngest and least developed of the approaches to predicting droplet-size distribution. Although the concept remains controversial because all the constraints on the maximization process might not have yet been identified, it is a worthy development to follow. See, for example, Chin et al. (1995) and Archambault et al. (1998).

In summary, the current ability to predict initial droplet-size distribution in a spray is based on empirical means, but interesting and challenging theoretical developments offer promise for the future. A discussion of the relevant research is presented in Chapter 12.

2 Isolated Spherically Symmetric Droplet Vaporization and Heating

There is interest in the droplet-vaporization problem from two different aspects. First, we wish to understand the fluid-dynamic and -transport phenomena associated with the transient heating and vaporization of a droplet. Second, but just as important, we must develop models for droplet heating, vaporization, and acceleration that are sufficiently accurate and simple to use in a spray analysis involving so many droplets that each droplet's behavior cannot be distinguished; rather, an average behavior of droplets in a vicinity is described. We can meet the first goal by examining both approximate analyses and finite-difference analyses of the governing Navier–Stokes equations. The second goal can be addressed at this time with only approximate analyses because the Navier–Stokes resolution for the detailed flow field around each droplet is too costly in a practical spray problem. However, correlations from Navier–Stokes solutions provide useful inputs into approximate analyses. The models discussed herein apply to droplet vaporization, heating, and acceleration and to droplet condensation, cooling, and deceleration for a droplet isolated from other droplets. The governing partial differential equations reflecting the conservation laws are presented in Appendix A. Several coordinate systems are considered. Formulations with primitive velocity variables and formulations with stream functions are discussed. Appendix B discusses some conserved scalar variables whose analytical use can be very convenient and powerful under certain ideal conditions.

Introductory descriptions of vaporizing droplet behavior can be found in the works of Chigier (1981), Clift et al. (1978), Glassman (1987), Kanury (1975), Kuo (1986), Lefebvre (1989), and Williams (1985). Useful research reviews are given by Faeth (1983), Law (1982), and Sirignano (1983, 1993a, 1993b). The monographs by Sadhal et al. (1997) and Frohn and Roth (2000) are also noteworthy.

The vaporizing-droplet problem is a challenging, multidisciplinary issue. It can involve heat and mass transport, fluid dynamics, and chemical kinetics. In general, there is a relative motion between a droplet and its ambient gas. Here, the general aerodynamic characteristics of pressure gradients, viscous boundary layers, separated flows, and wakes can appear for the gas flow over the droplet. The Reynolds number based on the relative velocity, droplet diameter, and gas-phase properties is a very important descriptor of the gaseous-flow field. Internal liquid circulation, driven by surface-shear forces, is another important fluid-dynamic feature of the droplet problem.

These flow features have a critical impact on the exchanges of mass, momentum, and energy between the gas and the liquid phases. They are important for both vaporizing and nonvaporizing situations. The vaporizing case is complicated by regressing interfaces and boundary-layer blowing.

We will see six types of droplet-vaporization models. In order of increasing complexity, they are (i) constant-droplet-temperature model (which yields the famous d^2 law whereby the square of the droplet diameter or radius decreases linearly with time), (ii) infinite-liquid-conductivity model (uniform but time-varying droplet temperature), (iii) spherically symmetric transient droplet-heating (or conduction-limit) model, (iv) effective-conductivity model, (v) vortex model of droplet heating, and (vi) Navier–Stokes solution. There are various differences among these models, but the essential issue is the treatment of the heating of the liquid phase that is usually the rate-controlling phenomenon in droplet vaporization, especially in high-temperature gases. Some of the models will be shown to be limits of another model. These models are explained in this chapter and in the next chapter, summarized in a user-friendly manner in Appendix C, and used in spray calculations discussed in other chapters.

In the first five models a quasi-steady gas phase is often considered but that feature is not necessary; the models can be constructed to include unsteady gas phases. The first three models can be directly applied to the situation in which there is no relative motion between the droplet and the ambient gas or in which a correction based on the Reynolds number can be applied to account for convective heat transfer from the gas to the liquid. However, internal circulation is not considered in those three models. (Actually, the internal circulation has no impact on heating when liquid temperature is uniform or constant.) The effective-conductivity models account for internal circulation and internal convective heating in an ad hoc manner whereas the vortex model more directly describes the physical situation. Obviously the Navier–Stokes solutions are, in principle, exact in the continuum regime.

Generally, the thermal diffusivity in the gas phase is much larger than the liquid-phase thermal diffusivity. So transient liquid heating generally takes longer than the gas-phase transient. Often, then, a quasi-steady gas-phase behavior can be assumed, neglecting time derivatives there while keeping them for the liquid phase. An exception can occur near the thermodynamic critical point where differences between gas and liquid properties, including diffusivities, are diminished. Of course, if ambient-gas properties are undergoing a transient behavior, then the characteristic time for gas-phase diffusion (the square of characteristic length divided by gas diffusivity) must be compared with the characteristic time for the change of the ambient conditions. As gas-phase ambient conditions change more rapidly and as droplet size increases, transient behavior becomes more important; consequently, the quasi-steady gas-phase assumption is weakened. However, gas-phase quasi-steadiness is broadly applicable in practice. We will generally consider quasi-steadiness in this book but will consider exceptions. For example, we have sections that deal with situations in which the gas film surrounding the droplet has a behavior such that time derivatives become important: e.g., a droplet in a near-critical gas, the interaction of a droplet with an acoustical wave, or the interaction of a droplet with a vortical structure.

In a gas, mass diffusivity and heat diffusivity are comparable in magnitude, yielding comparable characteristic times. For a common liquid, mass diffusivity is much smaller than thermal diffusivity. So, for multicomponent liquid droplets, we shall see in Chapter 4 that mass mixing in the droplet is slower than thermal mixing.

The combustion of liquid fuels is an application in which droplet vaporization can be critical because the vaporization can be the slowest process determining the overall burning rate. Classical droplet-vaporization and -burning theory describes a fuel droplet in a spherically symmetric gas field. There, the only relative motion between liquid and gas involves radial convection that is due to vaporization. This model implies that droplets would travel through a combustion chamber at the gas velocity. Such an implication could be true, however, for only very small droplets (often much less than 30 μm in diameter). Because droplet mass is proportional to the cube of diameter and droplet drag is proportional to diameter to a power no greater than squared, we can expect acceleration to be inversely proportional to some positive power of diameter. Acceleration that is due to drag will always tend to decrease the relative velocity between gas and liquid. Of course, this tendency could be overcome in certain situations in which the gas itself is accelerating because of pressure gradients or viscous stresses in such a manner so as to increase the relative velocity. In any event, the larger droplets will have a greater kinematic inertia.

In developing the study of the gas-flow field surrounding the droplet and of the liquid flow in the droplet, certain assumptions are made. A small Mach number is considered so that kinetic energy and viscous dissipation are negligible. Gravity effects, droplet deformation, Dufour energy flux, and mass diffusion that is due to pressure and temperature gradients are all neglected. (Note, however, that thermophoresis can affect the transport rates for submicrometer particles: e.g., soot.) Radiation is commonly neglected, but it is briefly discussed. The multicomponent gas-phase mixture is assumed to behave as an ideal gas. Phase equilibrium is stated at the droplet–gas interface. Gas-phase density and thermophysical parameters are generally considered variable, unless otherwise stated.

In the limiting case in which there is no relative motion between the droplet and the gas, a spherically symmetric field exists for the gas field surrounding the droplet and for the liquid field. If a small relative velocity occurs, the droplet acceleration becomes so large that the relative velocity between the droplet and the surrounding gas immediately goes to zero. Therefore, in this limit, relative velocity remains zero even as the gas velocity varies. Here, the fluid motion is reduced to a Stefan convection in the radial direction. Many of the important issues discussed in this case also appear in the more general case with droplet motion discussed in the next chapter.

The approach here is the presentation of the zero Reynolds number (spherically symmetric) analysis in this chapter, whereas, in the next chapter, we discuss low Reynolds number analysis and analyses for intermediate to high Reynolds numbers. When these several theories are patched together, a respectable engineering model applicable over a wide Reynolds number range is developed and is discussed in Sections 3.2 and 3.3 and in Appendix C.

We defer to Chapter 3, Sections 3.4 and 3.5, the discussion of unsteady effects including transient mechanics, droplet interactions with an acoustic wave, and droplet interaction with a vortical or a turbulent structure. Multicomponent droplet vaporization, including slurry and emulsion droplets, is discussed in Chapter 4. In Chapter 5, a discussion is given for droplet behavior near and above the thermodynamic critical conditions. Interactions among droplets are addressed in Chapter 6. A summary of the useful equations for droplet models is given in Appendix C.

2.1 Theory of Spherically Symmetric Droplet Vaporization and Heating

Consider the case in which a spherical droplet vaporizes with a radial-flow field in the gas phase. The vapor from the droplet convects and diffuses away from the droplet surface. Heat conducts radially against the convection toward the droplet interface. At the droplet surface, the heat from the gas partially accommodates the phase change and the remainder conducts into the liquid interior, raising the liquid temperature at the surface and in the interior of the droplet. This type of convection is called Stefan convection. In the case of condensation, the Stefan convective flux becomes negative.

In some cases, the reactant vapor, diffusing away from the vaporizing droplet, can mix in the gas film with the other reactant, diffusing from the ambient gas, and react in an exothermic manner. The energy source enhances the droplet's heating and vaporization. This can happen with a fuel droplet's vaporizing in an oxidizing environment if the reaction time is not longer than the film-diffusion time; otherwise, a flame must envelop many droplets if it is to occur [see Chapter 6 and Sirignano (1983)].

The liquid does not move relative to the droplet center in this spherically symmetric case. Rather, the surface regresses into the liquid as vaporization occurs. Therefore heat and mass transfer in the liquid occur only because of diffusion with a moving boundary but without convection. Here, the spherically symmetric isolated droplet equations are presented. A quasi-steady assumption is often made for the gas phase because diffusion of heat and mass in the gas is usually relatively fast compared with that of the liquid; this assumption weakens as we approach the critical pressure.

Overviews of general theoretical issues related to vaporization, heating, and burning of spherically symmetric droplets can be found in the reviews of Law (1982), Sirignano (1983, 1993b), and Williams (1985). The papers by Godsave (1953) and Spalding (1951) are of historical significance in the field of fuel-droplet burning. Other noteworthy contributions are the works of Strahle (1963), Chervinsky (1969), Kotake and Okuzaki (1969), Hubbard et al. (1975), Crespo and Linan (1975), Williams (1976), Law and Sirignano (1977), Buckholz and Tapper (1978), and Bellan and Summerfield (1978a, 1978b). This list is far from complete, but a reading of these papers will provide insights into the historical development of the theoretical framework for spherically symmetric droplets vaporizing in a high-temperature environment.

2.1.1 Gas-Phase Analysis

The spherically symmetric system of equations presented in this subsection is simplified from the system of equations presented for a general multicomponent mixture in Appendix A. Only fuel vapor, oxygen, nitrogen, and product are considered here. The mass diffusivities for all species are taken to be identical and are described by D. Higher-order terms in the Mach number are neglected and pressure is taken to be uniform.

Continuity:

$$r^2 \frac{\partial \rho}{\partial t} + \frac{\partial}{\partial r}(\rho u r^2) = 0,$$

or, for the quasi-steady case,

$$\rho u r^2 = \text{constant} = (\dot{m}/4\pi). \tag{2.1}$$

Energy:

$$\frac{\partial}{\partial t}(\rho r^2 h) + \frac{\partial}{\partial r}(\rho u r^2 h) - \frac{\partial}{\partial r}\left(\lambda r^2 \frac{\partial T}{\partial r}\right) - \frac{\partial}{\partial r}\left(\sum_{m=1}^{N} \rho D h_m r^2 \frac{\partial Y_m}{\partial r}\right) = r^2 \frac{\partial p}{\partial t} - \rho r^2 Q \dot{w}_F, \tag{2.2}$$

or, when Lewis number Le $= 1$, i.e., $\lambda/c_p = \rho D$, this becomes

$$\frac{\partial}{\partial t}(\rho r^2 h) + \frac{\partial}{\partial r}(\rho u r^2 h) - \frac{\partial}{\partial r}\left(\rho D r^2 \frac{\partial h}{\partial r}\right) = r^2 \frac{\partial p}{\partial t} - \rho r^2 Q \dot{w}_F.$$

Vapor-Species Conservation:

$$L(Y_F) = \frac{\partial}{\partial t}(\rho r^2 Y_F) + \frac{\partial}{\partial r}(\rho u r^2 Y_F) - \frac{\partial}{\partial r}\left(\rho D r^2 \frac{\partial Y_F}{\partial r}\right) = \rho r^2 \dot{w}_F. \tag{2.3a}$$

Oxygen-Species Conservation:

$$L(Y_O) = \rho r^2 \dot{w}_O = \rho r^2 \dot{w}_F / \nu. \tag{2.3b}$$

Nitrogen-Species Conservation:

$$L(Y_N) = 0. \tag{2.3c}$$

Product-Species Conservation:

$$L(Y_P) = -\rho r^2 \dot{w}_F \left(\frac{\nu + 1}{\nu}\right)$$

or

$$Y_F + Y_O + Y_N + Y_P = 1. \tag{2.3d}$$

In these equations, we assume a very low Mach number (and therefore uniform pressure) and neglect Soret and Dufour effects and radiation. Also, we assume Fickian mass diffusion, Fourier heat conduction, and one-step chemical kinetics. In the limit of no chemical reaction, $Y_P = 0$ and $\dot{w}_F = 0$. Equations of state for ρ and h are also prescribed; in particular, a perfect gas is usually considered: $p = \rho RT$ and $h = \int_{T_{\text{ref}}}^{T} c_p \, dT' = \Sigma_{m=1}^{N} Y_m h_m = \int_{T_{\text{ref}}}^{T} \Sigma_{m=1}^{N} Y_m c_{p,m} \, dT'$, where m is the index for species and N is the total number of species. When the quasi-steady assumption is made, the time derivatives in Eqs. (2.1)–(2.3d) are set to zero. The preceding equations can apply to the condensation case; the Stefan velocity u would be negative and, in the interesting example, no chemical reaction occurs.

Typically, ambient conditions at $r = \infty$ are prescribed for Y_O, Y_N, Y_F, Y_P, and T. Interface conditions are also prescribed at the droplet surface. Temperature is assumed to be continuous across the interface. Phase equilibrium is assumed at the interface typically by use of the Clausius–Clapeyron relation:

$$X_{ms} = \frac{1}{p} e^{\frac{L}{RT_b}} e^{\frac{-L}{RT}} \qquad (2.4a)$$

where X_{ms} is the gas-phase mole fraction of the vaporizing species m at the interface, T_b is the boiling point at 1 atm of pressure, and L is the latent heat of vaporization for the vaporizing species. The pressure p is measured in atmospheres, and the boiling point at nonatmospheric conditions can be calculated as $\frac{L}{R}[\frac{L}{RT_b} - \ln p]^{-1}$. The mole fraction and the mass fraction of a given species are related by

$$Y_m = \frac{X_m W_m}{\sum_i X_i W_i}, \qquad (2.4b)$$

$$X_m = \frac{Y_m / W_m}{\sum_i (Y_i / W_i)}, \qquad (2.4c)$$

where the summations are taken over all species present in the gas at the droplet surface. The preceding phase-equilibrium relation is a reasonable approximation over a practical temperature range at the droplet surface. The gas mass fraction and mole fraction for the vaporizing species at the surface depend on the surface temperature, pressure, and volatility characteristics (latent heat and boiling temperature). In addition, the mass fraction depends on the molecular weight of all species present in the gas at the surface. When all molecular weights are equal, the mass fraction and the mole fraction for each species are equal.

Mass balance at the interface is imposed for each species. Stefan convection (relative to the regressing surface) and diffusion on the gas side sum to equal convection on the liquid side:

$$\frac{\dot{m}}{4\pi} Y_{i_s} - \rho D R^2 \frac{\partial Y_i}{\partial r}\bigg)_s = \frac{\dot{m}}{4\pi} \delta_{iF}, \quad i = O, F, N, P, \qquad (2.5)$$

where $\dot{m}$ is the droplet-vaporization rate and δ_{iF} is the Kronecker delta function. For nonvaporizing species, radial convection and diffusion cancel each other. The right-hand side is nonzero for only the vaporizing species.

The energy balance at the interface equates the difference in conductive fluxes to the energy required for vaporizing the liquid at the surface:

$$\lambda R^2 \frac{\partial T}{\partial r}\bigg)_s = \frac{\dot{q}_l}{4\pi} + \frac{\dot{m}}{4\pi} L = \lambda_l R^2 \frac{\partial T}{\partial r}\bigg)_{l,s} + \frac{\dot{m}}{4\pi} L = \frac{\dot{m}}{4\pi} L_{\text{eff}}, \qquad (2.6)$$

where $\dot{q}_l$ is the liquid-phase conductive heat flux at the droplet surface toward the liquid interior and L_{eff} is an effective latent heat of vaporization.

In considering Eqs. (2.1)–(2.6), it should be realized that density is immediately related to temperature through an equation of state. Therefore Eqs. (2.1) can be considered to govern the radial gas velocity. Equations (2.2)–(2.3d) are the second-order differential equations governing five quantities: temperature and the four mass fractions. Ten boundary or matching conditions would normally be required. We have a total of 12 conditions: 5 ambient conditions for the five quantities, 5 interface conditions given by Eqs. (2.5) and (2.6), a phase-equilibrium condition at the interface, and a continuous-temperature condition at the interface. Apparently, two extra conditions are presented that are needed because the problem has an eigen-value character. The vaporization rate $\dot{m}$ and the heat entering the liquid phase,

$$\dot{q}_l = 4\pi \lambda_l R^2 \frac{\partial T}{\partial r}\bigg)_{l,s},$$

are unknown a priori. This means that, if T_s were known, the gas-phase problem could readily be solved without examination of the details of the liquid phase. Another valid viewpoint is that, if the heating rate of the liquid were known, the gas-phase problem including the interface temperature could be readily determined without consideration of the liquid-phase details. However, in the general problem, neither the surface temperature nor the liquid-heating rate is known a priori. An analysis of heat diffusion in the liquid phase is required for providing an additional relationship between the liquid-heating rate and the interface temperature.

It is convenient to present the solution in terms of Shvab–Zel'dovich variables β that satisfy $L(\beta) = 0$ for either the unsteady or the quasi-steady case. See Appendix B for background information on conserved scalars. In particular, we have

$$\beta_1 = Y_F - \nu Y_O; \quad \beta_2 = h + \nu Q Y_O; \quad \beta_3 = Y_O + Y_P/(1+\nu); \quad \beta_4 = Y_N. \quad (2.7)$$

This combination of the dependent variables allows the reformulation of Eqs. (2.2)–(2.6) into a system in which only one equation has the chemical source term. In this way, some strongly nonlinear terms with the resulting analytical challenge and numerical stiffness are removed from the system. Equations (2.2)–(2.6) yield in the quasi-steady case with $\mathrm{Le} = 1$ that

$$\frac{\partial}{\partial r}(\rho u r^2 \beta) - \frac{\partial}{\partial r}\left(\rho D r^2 \frac{\partial \beta}{\partial r}\right) = 0.$$

A convenient method of solution involves the use of Eq. (2.1) and the definition of a mass-flux potential ϕ such that $\partial\phi/\partial r = \rho u = \dot{m}/4\pi r^2$. Arbitrarily setting the value of ϕ at the droplet surface to zero and integrating, we obtain $\phi = \dot{m}/4\pi R - \dot{m}/4\pi r$. Now the equation governing β can be written as

$$\frac{\partial}{\partial r}\left(r^2 \beta \frac{\partial \phi}{\partial r}\right) - \frac{\partial}{\partial r}\left(\rho D r^2 \frac{\partial \beta}{\partial r}\right) = 0.$$

Integrating once and setting the gradient to zero at infinity, we obtain

$$(\beta - \beta_\infty)\frac{\partial \phi}{\partial r} = \rho D \frac{\partial \beta}{\partial r}.$$

The next integration yields

$$\frac{\beta - \beta_\infty}{A} = \exp\left[-\int_\phi^{\phi_\infty} \frac{d\phi'}{\rho D}\right] - 1 = \exp\left[-\frac{\dot{m}}{4\pi} \int_r^\infty \frac{dr'}{\rho D(r')^2}\right] - 1, \qquad (2.8)$$

where

$$A = \frac{\beta_s - \beta_\infty}{\exp\left[-\int_{\phi_s}^{\phi_\infty} \frac{d\phi'}{\rho D}\right] - 1} = \frac{\beta_s - \beta_\infty}{\exp\left[-\frac{\dot{m}}{4\pi} \int_R^\infty \frac{dr'}{\rho D(r')^2}\right] - 1}.$$

There is no special advantage in the use of the mass-flux potential for this one-dimensional (i.e., spherically symmetric) case. However, later in the study of two or more interacting droplets, which inherently is multidimensional, the mass-flux potential will prove to be useful, especially in dealing with complex configurations. See Chapter 6 for further discussion on the use of the mass-flux potential as an independent variable.

In the limit of negligible Stefan convection, the first spatial derivative terms in Eqs. (2.2)–(2.3d) can be neglected, yielding

$$\frac{\beta - \beta_\infty}{A} = -\frac{\dot{m}}{4\pi} \int_r^\infty \frac{dr'}{\rho D(r')^2}. \qquad (2.9)$$

Note that the same result can be found by a series expansion of Eq. (2.8), whereby the first term in the expansion gives the results with neglected Stefan flow. The constants of proportionality A in Eqs. (2.8) and (2.9) are different for each of the four β functions and are determined from the interface conditions.

Equation (2.8) indicates a monotonic behavior with radius for the β functions. In the case in which no exothermic reaction occurs in the gas film, a monotonic behavior for temperature and vapor mass fraction results. With chemical reaction, we have a peak in temperature and in product concentration occurring at the flame position. An infinitesimally thin-flame assumption leads to certain simplifications in the determination of the profile in the reaction case. Without the thin-flame assumption, the reacting case requires the integration of Eq. (2.3b) coupled with Eqs. (2.7) and (2.8).

For sufficiently fast oxidation chemical kinetics, the reaction zone approaches the limit of zero thickness and a diffusion flame results. Then $\beta_1 = -\nu Y_O$ and $Y_F = 0$ outside the flame zone, whereas $\beta_1 = Y_F$ and $Y_O = 0$ inside the flame zone and $Y_F = Y_O = 0$ at the flame zone. Note that, because β_1 has a continuous first derivative, the diffusion rates of oxygen and fuel vapor into the flame will be in stoichiometric proportion. Now, determinations of Y_P and T from Eqs. (2.7) and (2.8) readily follow once the surface temperature is determined from the coupled liquid-phase solution.

In the special case of very fast chemistry, constant specific heat, and unitary Lewis number ($\rho D = \lambda/c_p$) for the gas phase, the quasi-steady gas-phase equations can be reduced to certain algebraic relations by means of the use of Eqs. (2.5), (2.7), and (2.8):

$$\dot{m} = 4\pi \Lambda(R) \ln(1 + B), \qquad (2.10)$$

where

$$\Lambda(r) \equiv \left(\int_r^\infty \frac{dr}{\rho D r^2} \right)^{-1},$$

$$B \equiv \frac{h_\infty - h_s + \nu Q Y_{O\infty}}{L_{\text{eff}}} = \frac{\nu Y_{O\infty} + Y_{Fs}}{1 - Y_{Fs}}. \tag{2.11}$$

When $\rho D = $ constant, we obtain $\Lambda(r) = \rho D r$ and $\dot{m} = 4\pi\rho D R \log(1+B)$.

The preceding analysis considers that the Lewis number is unity valued and the Spalding mass transfer number B_M and heat transfer number B_H are equal and identified simply as B. For the more general case of a nonunitary Lewis number, we have $\dot{m} = 4\pi\Lambda(R)\ln[1 + B_M]$, where $B_M = [(Y_{Fs} + \nu Y_{O_2,\infty})/(1 - Y_{Fs})]$. The more general theory for the nonunitary Lewis number can be found in Subsections 6.3.2 and 6.3.4, where it is presented for a general configuration that may include many interactive droplets. The isolated single droplet discussed here is a special case.

Note that Y_{Fs} is a function of T_s through the phase-equilibrium relation. Therefore Eqs. (2.6) and (2.11) relate interface temperature to the liquid-heating rate. When the heating rate becomes zero, $L_{\text{eff}} = L$ and Eq. (2.11) yields the constant wet-bulb temperature for the droplet. See Eq. (2.20). This yields model (i) discussed earlier in this chapter. In the case of condensation, both the Spalding transfer number B and the vaporization rate $\dot{m}$ become negative.

At the thin flame, $\beta_1 = 0$; so from Eqs. (2.7) and (2.8), we determine r_f, the position of the flame zone, as

$$\Lambda(r_f) = \Lambda(R) \frac{\ln\left(\dfrac{1 + \nu Y_{O\infty}}{1 - Y_{Fs}} \right)}{\ln(1 + \nu Y_{O\infty})} = \frac{\ln(1+B)}{\ln(1 + \nu Y_{O\infty})}. \tag{2.12a}$$

When ρD is constant with r, we have the result that

$$\frac{r_f}{R} = \frac{\ln(1+B)}{\ln(1 + \nu Y_{O\infty})}. \tag{2.12b}$$

The enthalpy at the flame h_f can be determined from Eqs. (2.7), (2.8), and (2.11). We find that

$$h_f = h_\infty + \nu Y_{O\infty} Q - \nu Y_{O\infty} \left(\frac{h_\infty - h_s + \nu Y_\infty Q}{\nu Y_{O\infty} + Y_{Fs}} \right) \tag{2.12c}$$

or, equivalently,

$$h_f = h_\infty + \nu Y_{O\infty} Q - \nu Y_{O\infty} \left(\frac{L_{\text{eff}}}{1 - Y_{Fs}} \right). \tag{2.12d}$$

The second form for the flame enthalpy agrees with the result obtained from the Superscalar, indicating that the value is independent of the particular configuration for the burning of liquid fuel. See Eqs. (B.23) and (B.24) in Appendix B. Of course, the flame temperature can be obtained directly from the enthalpy value.

For the gas inside the flame, the fuel mass fraction and the enthalpy are found from Eq. (2.8) to be

$$Y_F = -\nu Y_{O\infty}$$

$$+ \left\{ \frac{Y_{Fs} + \nu Y_{O\infty}}{\exp\left[-\frac{\dot{m}}{4\pi} \int_R^\infty \frac{dr'}{\rho D(r')^2}\right] - 1} \right\} \left\{ \exp\left[-\frac{\dot{m}}{4\pi} \int_r^\infty \frac{dr'}{\rho D(r')^2}\right] - 1 \right\}, \quad (2.13a)$$

$$h = h_\infty + \nu Q Y_{O\infty}$$

$$+ \left\{ \frac{h_s - h_\infty - \nu Q Y_{O\infty}}{\exp\left[-\frac{\dot{m}}{4\pi} \int_R^\infty \frac{dr'}{\rho D(r')^2}\right] - 1} \right\} \left\{ \exp\left[-\frac{\dot{m}}{4\pi} \int_r^\infty \frac{dr'}{\rho D(r')^2}\right] - 1 \right\}. \quad (2.13b)$$

For the gas outside the flame, the oxygen mass fraction and the enthalpy are determined from the same Eq. (2.8) as

$$Y_O = Y_{O\infty}$$

$$- \left\{ \frac{Y_{Fs}/\nu + Y_{O\infty}}{\exp\left[-\frac{\dot{m}}{4\pi} \int_R^\infty \frac{dr'}{\rho D(r')^2}\right] - 1} \right\} \left\{ \exp\left[-\frac{\dot{m}}{4\pi} \int_r^\infty \frac{dr'}{\rho D(r')^2}\right] - 1 \right\}, \quad (2.13c)$$

$$h = h_\infty$$

$$+ \left\{ \frac{h_s + Q Y_{Fs} - h_\infty}{\exp\left[-\frac{\dot{m}}{4\pi} \int_R^\infty \frac{dr'}{\rho D(r')^2}\right] - 1} \right\} \left\{ \exp\left[-\frac{\dot{m}}{4\pi} \int_r^\infty \frac{dr'}{\rho D(r')^2}\right] - 1 \right\}. \quad (2.13d)$$

The Nusselt number describes a nondimensional heat transfer rate to the droplet and is defined as

$$Nu \equiv \frac{2h'R}{\lambda}, \quad (2.14a)$$

or, equivalently,

$$Nu = \frac{2R\frac{\partial T}{\partial r}\bigg)_s}{T_\infty - T_s + \frac{\nu Q Y_{O\infty}}{c_p}}, \quad (2.14b)$$

where h' is the heat transfer coefficient for the gas-phase film surrounding the droplet. We therefore define that coefficient as

$$h' \equiv \frac{c_p \lambda \frac{\partial T}{\partial r}\bigg)_s}{h_\infty - h_s + \nu Q Y_{O\infty}} = \frac{\lambda \frac{\partial T}{\partial r}\bigg)_s}{T_\infty - T_s + \frac{\nu Q}{c_p} Y_{O\infty}}, \quad (2.14c)$$

where c_p is an average specific-heat value for the film.

The Sherwood number describes the nondimensional mass transfer rate to the droplet and is defined as

$$\text{Sh}_i = \frac{-2R}{Y_{is} - Y_{i\infty}} \left.\frac{\partial Y_i}{\partial r}\right)_s. \tag{2.14d}$$

When all binary mass diffusion coefficients are equal, the Sherwood numbers for each species are equal. Furthermore, when the Lewis number Le $= 1$, Sh $=$ Nu. The case in which Le $\neq 1$ and Sh $\neq$ Nu is discussed in Subsections 3.1.3 and 3.1.7. We can proceed by assuming constant specific heat, constant conductivity, and unitary Lewis number ($\rho D = \lambda/c_p$). Then, a combination of Eqs. (2.6), (2.10), and (2.11) yields the results that

$$\text{Nu} = \text{Sh} = 2\frac{\ln(1 + B)}{B}, \tag{2.15a}$$

$$\frac{\dot{m}}{4\pi\rho D R\text{Nu}} = \frac{B}{2} \tag{2.15b}$$

for a spherically symmetric vaporizing droplet. Note that the classical heated-sphere result (Nu $= 2$) is extracted in the nonvaporizing limit in which $B \rightarrow 0$.

In the case of a vaporizing droplet without a flame in the vicinity of the droplet, Eqs. (2.10), (2.15a), and (2.15b) still apply. However, the transfer number is

$$B = \frac{h_\infty - h_s}{L_{\text{eff}}} = \frac{Y_{Fs} - Y_{F\infty}}{1 - Y_{Fs}}. \tag{2.15c}$$

In this situation, there is no peak temperature in the surrounding gas film because of a diffusion flame. However, the ambient temperature could be driven by combustion in the far field, which occurs in the mixture of fuel vapor and oxidizer there.

Fachini et al. (1999) considered a thin spherical flame surrounding the fuel droplet with a low stoichiometric fuel–air ratio, thereby maintaining the flame sheet far from the droplet. Consequently the flame stood in the transient, developing region; the standoff distance first grew and then decreased. An asymptotic analysis, with the square root of the ratio of gas density to liquid density as the small parameter, was used. Radiative loss and nonunitary Lewis number were considered. An analytical condition for extinction was developed.

Studies of an unsteady gas-phase film can provide useful insights. Gas-phase unsteadiness is important in two situations: (a) the transient period after a droplet is introduced to the gaseous environment but before the quasi-steady gas film is established and (b) cases in which the gas-phase thermal or mass diffusivity is not an order of magnitude larger than the liquid thermal diffusivity. The two situations are related because the transient time for gas-phase unsteadiness increases as gas-phase diffusivity decreases.

In Chapter 5, near-critical and supercritical gaseous environments are considered, so the case of decreased gas-phase diffusivity is discussed there. Botros et al. (1980) considered the effect of fuel-vapor accumulation during the initial transient period; they retained time derivatives in the gas-phase equations and showed that the fuel vaporized during this transient portion is a significant fraction of the original amount of liquid fuel. However, Botros et al. did not consider droplet heating or variation of liquid temperature; the transient begins with a liquid temperature already elevated to the wet-bulb value.

A more realistic model with droplet heating would be preferred. In those cases in which gas-phase diffusivities are much larger than liquid diffusivities, the droplet heating will be rate controlling and the gas-phase will respond in a quasi-steady manner. In this type of model, the amount of fuel vapor at the surface increases as the surface temperature increases. The profile of the fuel-vapor mass fraction or concentration is coupled to the surface value in a quasi-steady fashion; therefore the profile values increase over a time scale dictated by the droplet-heating rate. In this situation the gas-phase time derivatives remain negligible compared with other terms in Eqs. (2.1) to (2.3d). Because the model of Botros et al. neglects the delay caused by droplet heating, it overestimates the rate at which fuel vapor is accumulated in the gas film.

It is noteworthy that, theoretically, diffusion to infinite lengths requires infinite time. Therefore theory indicates that the far gas field will be unsteady even after quasi-steady behavior is established over a thick gas film surrounding the droplet. In many practical situations, forced or natural convection will make diffusion over infinite length and time scales irrelevant. Rather, with convection the characteristic diffusion time is equated naturally to a finite flow residence time.

2.1.2 Liquid-Phase Analysis

Now a model of type (iii) discussed in Section 2.1 can be developed. The solution of the equation for heat diffusion through the liquid phase can be considered to provide the necessary relation between the interface temperature and the liquid-heating rate. That equation is written in spherically symmetric form:

$$\frac{\partial T_l}{\partial t} = \alpha_l \left(\frac{\partial^2 T_l}{\partial r^2} + \frac{2}{r} \frac{\partial T_l}{\partial r} \right). \tag{2.16a}$$

The boundary condition at the droplet–gas interface is obtained from a combination of Eqs. (2.6), (2.8), (2.13b), (2.15a), and (2.15b). We have

$$\left. \frac{\partial T_l}{\partial r} \right)_s = \frac{\lambda}{\lambda_l} \frac{T_\infty - T_s + \nu Q Y_{O\infty}/c_p}{R} \frac{\ln(1 + B)}{B} - \frac{\dot{m} L}{4\pi R^2 \lambda_l}$$

$$= \frac{\ln(1 + B)}{R} \frac{\lambda}{\lambda_l} \left(\frac{T_\infty - T_s + \nu Q Y_{O\infty}/c_p}{B} - \frac{L}{c_p} \right). \tag{2.16b}$$

The symmetry condition at $r = 0$ is given as

$$\left. \frac{\partial T}{\partial r} \right)_{r=0} = 0. \tag{2.16c}$$

Typically, the initial droplet temperature is assumed to be uniform at a value designated as T_0.

Equations (2.16a)–(2.16c) apply to a moving-boundary problem because the droplet surface regresses (advances) as vaporization (condensation) occurs. This requires a calculation of the boundary movement as time progresses. It is often convenient to use a coordinate transformation to a fixed-boundary problem. Such a transformation is used in formulating Eqs. (2.73)–(2.74) for a type (iv) model. We will see in Subsection 3.1.4 that the type (iii) case currently under discussion is a well-defined limit of the type (iv) case so that the same coordinate transformation to a fixed boundary readily applies.

Note that the form of Eq. (2.8) that should be used to determine the transfer number B in boundary condition (2.16b) is

$$B = (\nu Y_{O\infty} + Y_{Fs})/(1 - Y_{Fs}).$$

Here we can calculate B without knowledge of the temperature gradient at the surface. The other form of the equation for B yields an implicit relationship for the temperature gradient at the liquid side of the surface.

The time derivative is considered in the liquid phase although it is neglected in the gas phase because the liquid thermal diffusivity at subcritical conditions is much smaller than the gas-phase diffusivity. Gas-phase time derivatives are retained at near-critical and supercritical conditions. In the most general case, Eqs. (2.16a) and (2.16b) can be solved by finite-difference techniques, although approximate techniques (Law and Sirignano, 1977; Sirignano, 1983) have been used. The coupled solutions of Eq. (2.11), and Eqs. (2.12) and (2.16) yield the final results.

Sirignano (1983) provided some guidelines for estimating the ratio N of the liquid-heating time to droplet lifetime. The importance of the parameter and the resulting classification of the physical domains were identified. The challenging set of problems was shown by Sirignano to be the cases in which this ratio N is not small compared with unity. These higher values of N tend to occur in high-temperature environments. The value of N is not dependent on the initial droplet size.

The liquid-heating time τ_H, defined as the time required for a thermal diffusion wave to penetrate from the droplet surface to its center, is of the order of $R_0^2/\alpha_l = \rho_l c_l R_0^2/\lambda_l$. From Eq. (2.9) with a constant value of ρD and from a relationship between the initial liquid-droplet mass and its initial radius, the droplet lifetime $\tau_L = O\{[(1/R_0)^2 dR^2/dt]^{-1}\}$ can be estimated as $\rho_l R_0^2/[2\rho D \log(1 + B)]$. The ratio of heating time τ_H to lifetime τ_L is estimated by

$$N = \frac{\tau_H}{\tau_L} = \frac{2\rho D c_l}{\lambda_l} \ln(1 + B) = 2\frac{\lambda}{\lambda_l} \frac{c_l}{c_p} \ln(1 + B). \qquad (2.17)$$

Typical results from Law and Sirignano (1977) for the liquid-phase temperature are found in Fig. 2.1, which shows a thermal wave diffusing from the droplet surface toward its center. The surface regression is shown in Fig. 2.2. In this case, the heating time τ_H and the lifetime τ_L are of the same order of magnitude. The most interesting effect appears through the temperature dependence of the parameter B. The transient behavior is found to persist over a large portion of the droplet lifetime when the ambient temperatures are high, especially for higher-molecular-weight liquids. During the early part of the lifetime, $L_{\text{eff}} \gg L$ because of liquid heating; L_{eff} decreases through the droplet lifetime. Consequently, B increases by an order of magnitude during the droplet lifetime.

There are several special cases of interest. In the first case, which leads to a model of type (ii) as a limit of the type (iii) model, we consider that the droplet-heating time is short compared with the lifetime. This implies that the droplet

Figure 2.1. Temporal and spatial variations of liquid temperature for spherically symmetric octane droplet vaporization. Temperature versus nondimensional radial coordinate for fixed fraction of droplet lifetime (or fixed vaporized fraction of original mass). (Sirignano, 1993b, with permission of the *Journal of Fluids Engineering*.)

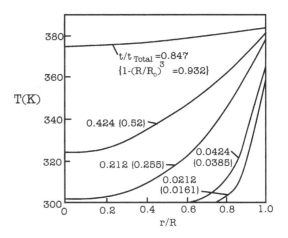

interior is quickly heated because a thermal wave diffuses from the droplet surface to its center in a short time compared with the lifetime. As a consequence, a nearly uniform liquid temperature is established quickly. Many authors have used the simplifications of this case, sometimes beyond its range of validity (see Law, 1976b, and El-Wakil et al., 1956, for example). Equation (2.17) shows that this case can occur if $\lambda \ll \lambda_l$, $c_l \ll c_p$, or B is small. Ambient temperatures of a few hundred degrees

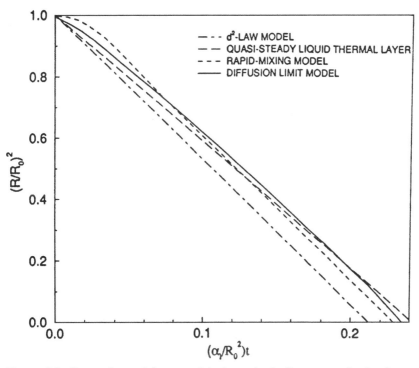

Figure 2.2. Comparison of four models for spherically symmetric droplet vaporization. Nondimensional radius squared versus nondimensional time. (Sirignano, 1993b, with permission of the *Journal of Fluids Engineering*.)

centigrade can lead to a small B value. Many heat exchanger applications are in this category.

For this case, further classification can be made with regard to the temporal behavior of the liquid temperature. Equation (2.16a) can be integrated over the droplet volume, accounting for the zero gradient at the droplet center and for the regressing surface. Furthermore, it can be assumed in this case that surface temperature is approximately equal to the volumetric average liquid temperature. As a result, we have that the heat flux entering the liquid is

$$\dot{q}_l = mc_l \frac{dT_l}{dt}, \tag{2.18}$$

where m is the instantaneous droplet mass, T_l is the average liquid temperature, and

$$\dot{q}_l = 4\pi R^2 \lambda_l \frac{\partial T}{\partial r}\bigg)_{l,s}.$$

This indicates that the liquid temperature would continue to rise until the heat flux $\dot{q}_l$ became zero, i.e., the wet-bulb temperature T_{wb} is reached. The characteristic time τ^* for this heating depends on the magnitudes of the thermal inertia mc_l and the heat flux $\dot{q}_l$. There can be shown to be two possibilities: either $\tau^* < \tau_H$ or $\tau^* > \tau_H$ for Eq. (2.18). However, if $\tau^* < \tau_H$, Eq. (2.18) is not valid because the temperature remains nonuniform on that time scale. Equation (2.18) is valid and interesting only if $\tau^* > \tau_H$.

An estimate for τ^* can be obtained by use of Eqs. (2.6) and (2.8) to yield $\dot{q}_l$, which is then substituted into Eq. (2.18). In particular, $\dot{q}_l$ is represented by a truncated Taylor series expansion about the wet-bulb temperature point. The result is

$$\frac{\tau^*}{\tau_H} = \frac{1}{3}\frac{\lambda_l}{\lambda}\left(\frac{1}{\ln\frac{1+\nu Y_{O\infty}}{1-Y_{Fs}}}\right)\left[\frac{\nu Y_{O\infty} + Y_{Fs}}{1 - Y_{Fs} + \frac{L^2}{c_p R}\frac{Y_{Fs}}{T_S^2}\left(\frac{1+\nu Y_{O\infty}}{1-Y_{Fs}}\right)}\right], \tag{2.19}$$

where Y_{Fs} is evaluated at the wet-bulb temperature. A Clausius–Clapeyron relationship has been assumed to describe phase equilibrium and gives Y_{Fs} as a function of temperature. See Eqs. (2.4a)–(2.4c). Furthermore, it can be shown from Eqs. (2.6) and (2.8) that, at the wet-bulb temperature,

$$Y_{Fs} = \frac{\dfrac{h_\infty - h_{wb} + \nu Q Y_{O\infty}}{L} - \nu Y_{O\infty}}{1 + \dfrac{h_\infty - h_{wb} + \nu Q Y_{O\infty}}{L}}, \tag{2.20}$$

which can be substituted. Here the liquid interior is not being heated and $L_{eff} = L$. Equations (2.4a)–(2.4c) and (2.20) will determine the wet-bulb temperature.

It is convenient to define the time ratio $N^* = \tau^*/\tau_L$. We always have $N^* \geq N$. It can be concluded that the characteristic time τ^* is bounded below by a quantity of the order of τ_H. Therefore this case can be divided into two subcases. In one situation, the wet-bulb temperature is reached in a time of the same order of magnitude as that of the heating time, which is very short compared with the life-time, that is, $N \leq N^* \ll O(1)$. Here, liquid temperature can be considered constant with $L_{\mathrm{eff}} = L$. Obviously this reduces the model to type (i). Then Eqs. (2.7), (2.8), and phase-equilibrium relation (2.4) for $Y_{Fs}(Ts)$ immediately yield the vaporization rate $\dot{m}$ and the droplet temperature $T_s = T_{\mathrm{wb}}$. When ρD is constant, integration of Eqs. (2.7), where $m = 4\pi R^3 \rho_l / 3$, yields the well-known d^2 law, that is,

$$(R/R_0)^2 = 1 - t/\tau_L. \tag{2.21}$$

Here the square of the radius (or diameter) decreases linearly with time. Note that the previously given estimate for τ_L is exact here, with B calculated with $L_{\mathrm{eff}} = L$.

The second subcase involves a characteristic time τ^* substantially larger than τ_H and perhaps comparable with τ_L, that is, $N \ll N^* \leq O(1)$. This requires the integration of Eq. (2.18) [coupled with Eqs. (2.4), (2.6), and (2.8) that relate $\dot{q}_l$ to $T_l = T_s$]. The results for such an integration are shown in Fig. 2.2. The use of an average temperature tends to underestimate the surface temperature during the early vaporization period because the thermal energy is artificially distributed over the droplet interior. As a result, the early vaporization rate for this subcase is less than that found in the general case by solution of diffusion equation (2.16). However, because the surface temperature is artificially low at first, more heat enters the droplet and ultimately it vaporizes faster than it does in the general case, as shown in Fig. 2.2.

Another interesting case occurs when the heating time is much longer than the lifetime: $N \gg 1$. Then a thin thermal layer is maintained in the liquid near the surface; the regression rate is too large for the thermal wave to penetrate faster than the regressing surface. This case can occur if λ_l is small or if c_l or B is large. Very large ambient temperatures of a few thousand degrees Kelvin can lead to this situation. The thin thermal layer can be considered quasi-steady so that the conductive heat flux through the surface balances the liquid convective rate toward the surface. (Convection here is measured relative to the regressing surface.) We have therefore the approximation that

$$\lambda_l \frac{\partial T_l}{\partial r}\bigg)_s = \rho_l U_l c_l (T_s - T_{l0}) \quad \text{or} \quad L_{\mathrm{eff}} = L + c_l (T_s - T_{l0}). \tag{2.22}$$

Equations (2.11) and (2.22), together with the phase-equilibrium relation, yield T_s (which here is constant with time). Then Eq. (2.10) yield $\dot{m}$ and subsequently $R(t)$. Because L_{eff} is here a constant with a larger positive value than L, another d^2 relationship results but with a lower absolute value of the slope on a time plot. Equation (2.21) can be used with τ_L, now calculated with Eqs. (2.22). One could argue that this long-heating-time model is a special subcase of the type (i) model because surface and droplet-core temperatures are constant and a d^2 law results. On the other hand, it can be seen as a subcase of type (iii) models because it does entail transient heating with spatial and temporal temperature gradients.

The two d^2-law results are also plotted in Fig. 2.2 for the purpose of comparison with the general case. The general case is the only one of the three cases discussed so far that considers an initial transient. The first (well-known) d^2 law neglects the initial thermal diffusion across the droplet interior, whereas the other d^2 law neglects the initial diffusion across the thin layer.

Note that the droplet lifetimes vary little from model to model in Fig. 2.2. However, the local slope of the curves (related to instantaneous vaporization rates) do vary more significantly. This variation implies that the vaporization rate as a function of spatial position for a droplet moving through a volume can depend significantly on the particular model.

The spherically symmetric droplet problem introduces many fundamental physical issues that remain in the problem as the flow field becomes more complex because of the relative motion between the gas and the droplet. The effects of transient heat conduction and mass diffusion, phase equilibrium at the interface, and the regressing liquid surface remain as the relative flow is introduced. The effects of relative motion are discussed in Chapter 3.

There is experimental difficulty in maintaining a near-stationary, spherical droplet. A common method has been to suspend the droplet on a fiber, assuming that the droplet behavior with the fiber still remains spherically symmetric. Dwyer and Shaw (2001) and Shaw and Harrison (2002) analyzed some of the challenges for suspended drops. Surface-tension gradients that are due to water condensation in methanol drops modify the behavior causing surface waves, instabilities, and three-dimensional behavior. Droplet shape can distort from the sphere even in reduced-gravity environments. Another modern experimental method has been to use reduced-gravity facilities for experiments (e.g., drop towers and NASA spacecraft); interesting results have been found here. Shaw et al. (2001a, 2001b) showed that radiative heating and extinction are dependent on initial droplet size. Experiments by Okai et al. (2003) indicated how the proximity of a droplet or a droplet pair to a wall causes deviation from spherical symmetry and normal behavior for burning n-heptane droplets. Okai et al. (2004) showed how an imposed electric field on a burning droplet and droplet pair causes significant variations. Dakka and Shaw (2006) and Wei and Shaw (2006) studied pressure effects. Bicomponent liquid fuels were studied by Yang et al. (1990), Shaw et al. (2001a, 2001b), Shaw and Harrison (2002), Dee and Shaw (2004), and Wei and Shaw (2006).

2.1.3 Chemical Reaction

For the case in which transport across the gas film surrounding the fuel droplet is faster than the chemical reaction in the gas, the droplet may be considered to be vaporizing in an ambiance that is chemically reacting. The amount of chemical reaction in the gas film (diffusion layer) immediately surrounding the droplet can be considered negligible. The droplet-vaporization model then neglects the chemical-reaction term; however, the gas-phase equations describing the spray flow will contain the chemical-reaction term. This approach is used in the discussions of Chapters 7 and 9. This is a situation opposite to that of the model represented by Eqs. (2.10) and (2.11), in which the chemistry is very fast.

For the fast oxidation case, it is generally not necessary to analyze the structure of the thin flame in the gas film surrounding the droplet in order to predict the vaporization rate. Heat and mass diffusion are rate controlling, and the assumption of an infinite chemical-kinetic rate that leads to Eqs. (2.10) and (2.11) is satisfactory. The structures of thin flames are relevant, however, for the prediction and the understanding of pollutant formation. Early attempts at the analysis of the thin droplet diffusion flame structure involved matched asymptotic expansions made with the reciprocal of the Damkohler number (namely, the ratio of the chemical time to the transport or residence time) as the small expansion parameter (see Kassoy and Williams, 1968; Kassoy et al., 1969; and Waldman, 1975). This approach predicted modest modifications of the vaporization rate from the infinite chemical-kinetics model; also, it could not predict extinction limits because it was inherently limited to small chemical times. More recently, Card and Williams (1992a, 1992b) and Zhang et al. (1996) used rate-ratio asymptotic methods with reduced but multistep chemical-kinetic schemes to predict extinction for methanol and *n*-heptane fuel droplets. However, sensitivity was demonstrated to the particular assumptions that are made in the model.

Bracco (1973) performed a numerical integration of a system of equations equivalent to Eqs. (2.1)–(2.6). Steady-state conditions for the gas and a uniform steady liquid temperature were considered so that the equations were reduced to ordinary differential equations. A second-order one-step chemical reaction for an ethanol droplet was used. After the temperature and major species profiles were calculated, the Zel'dovich mechanism was utilized to determine the rate of formation and the concentration profile for nitric oxide, NO. The effect of the mass diffusion of the species NO on its own rate of formation and concentration profile was found to be important. Although the overall burning rate was linear in the instantaneous droplet diameter, the rate of NO formation was proportional to the third power of the diameter. Therefore, at a fixed total mass of fuel for all droplets in an ensemble, the amount of NO formed through combustion will increase as initial average diameter increases. Of course, this assumes that the oxidation occurs in thin flames around individual droplets. A different conclusion might be reached for the droplet group combustion discussed in Chapter 6. Fuel-bound nitrogen was found to enhance NO production at low air temperatures.

Cho et al. (1991) considered a spherically symmetric time-dependent behavior for the gas film and liquid interior of a methanol fuel droplet. Detailed multicomponent transport and elementary chemical kinetics were utilized. The calculations extended over the time from ignition to extinction. Results for the burning rate, the flame-standoff distance, and the droplet diameter at extinction agreed well with experiment. Ignition-delay predictions were shorter than experimental findings. The robustness of the model was evaluated when the nitrogen in the air was replaced with helium; theory and experiment still agreed even with the significant change in transport properties.

Jackson and Avedisian (1996) considered complex chemistry for a quasi-steady spherically symmetric vaporization of *n*-heptane mixed with water. Acetylene C_2H_2 concentration was determined; it is considered a precursor of soot formation. The rate of soot formation was calculated as a function of acetylene concentration. The authors identified the needs for further improvements in the model.

Marchese and Dryer (1996) considered detailed chemical-kinetic modelling for the burning of droplets of a methanol and water mixture. Although the model is spherically symmetric, favorable comparison occurs with a microgravity drop tower experiment if they include internal liquid circulation. They argue that the circulation can be caused by droplet generation and deployment techniques or capillary action. Significant deviations from the quasi-steady d^2 law are found. Extensions to the experiment and modelling are given by Marchese et al. (1996).

2.2 Radiative Heating of Droplets

Radiation appears in several roles for droplet-heating and -vaporization problems. The most important effect is the heating of droplets by radiation from high-temperature gases. This can be important for fuel droplets in a combustion situation or for water droplets in a fire-suppression situation. Radiative losses from the droplet surface tend to be less important than conduction, except in very low-density (subatmospheric) environments, because the surface temperature is limited by the boiling point. An indirect effect is that radiative losses from a heating source for droplets (such as a flame) to the environment can reduce the source temperature and thereby decrease conductive and radiative heat transfer to the droplet.

Lage and Rangel (1993a, 1993b) studied the radiative heating for spherically single-component fuel droplets. Blackbody radiation was assumed to come from infinity (in a spherically symmetric or an axisymmetric field). Interaction between the gas film surrounding the droplet and the radiation was neglected. The radiative losses from the liquid were also neglected. The only new feature is the radiation absorbed by the liquid. Absorption of energy both at the droplet surface and in depth through the droplet were considered.

Lage and Rangel (1993a) determined the total energy absorption distribution inside a liquid droplet irradiated in an axisymmetric or spherically symmetric configuration with blackbody emission at an infinite distance. Spectral and solid-angle integration is made of the spectral absorption distribution calculated by electromagnetic theory. Volume integration and efficiency-factor spectral integration are used to obtain total absorbencies. Results were obtained for decane and water droplets in the 1–100-μm-radius range. Decane absorption was found to be less than water absorption. The geometrical-optics approximation was reasonable for droplet diameters greater than 50 μm. The spherical irradiation results were used by Lage and Rangel (1993b) together with the gas-film theory of Abramzon and Sirignano (1989) to study transient droplet heating and vaporization.

In the Lage and Rangel analysis, Eqs. (2.1)–(2.5) remain unchanged because the gas phase is passive to the radiation. Chemical-reaction rates, however, are considered to be zero in their case. The energy balance at the interface is modified to account for the radiation. Equation (2.6) is replaced with

$$\lambda R^2 \frac{\partial T}{\partial r}\bigg)_s = \lambda_l R^2 \frac{\partial T}{\partial r}\bigg)_{l,s} + \frac{\dot{m}}{4\pi} L - \delta R^2_{q_{\text{rad}}} = \frac{\dot{m}}{4\pi} L_{\text{eff}}. \tag{2.23}$$

This provides a modified definition for the effective latent heat of vaporization L_{eff}. δ is the fraction of the radiative flux $4\pi R^2_{q_{\text{rad}}}$ that is absorbed at the droplet surface. q_{rad} is the radiative flux per unit area striking the droplet surface. Of course, $0 \leq \delta \leq 1$. γ is the fraction of radiative energy absorbed both in the droplet interior

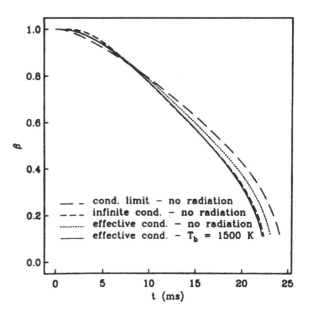

Figure 2.3. Dimensionless droplet-radius temporal variation for an n-decane droplet, initially at 300 K and 50 μm in radius, in a medium at 1000 K and 10 bars. Various heating models with and without radiation from a blackbody at 1500 K. (Lage and Rangel, 1993b, with permission of the *Journal of Thermophysics and Heat Transfer*.)

and at the surface. Therefore $\delta \leq \gamma \leq 1$. $\gamma = 1$ implies an opaque droplet, whereas $\gamma = 0$ implies a transparent droplet. $\gamma - \delta$ is the fraction absorbed in the droplet interior, and $1 - \gamma$ is the fraction of the radiative flux passing through the droplet.

Note that, in the case of a quasi-steady gas phase, Eqs. (2.7)–(2.11) remain the same in form except for the change in the definition of L_{eff}. The equation for liquid-phase heat transfer (2.16a) is replaced with

$$\frac{\partial T_l}{\partial t} = \alpha_l \left(\frac{\partial^2 T_l}{\partial r^2} + \frac{2}{r} \frac{\partial T_l}{\partial r} \right) + \frac{(\gamma - \delta) 4\pi R_{q_{\text{rad}}}^2}{\rho_l c_l} S, \qquad (2.24)$$

where the distribution function $S(r, t)$ gives the fraction of the total radiative energy absorbed per unit volume in the interior that is absorbed in the shell of volume $4\pi r^2 \, dr$. Thus,

$$\int_0^R 4\pi r^2 S(r, t) \, dr = 1.$$

The relative importance of radiative heating of the interior compared with the conductive heating is given by the value of the nondimensional grouping:

$$\frac{4\pi R^2 (\gamma - \delta) q_{\text{rad}} \displaystyle\int_0^R 4\pi r^2 S(r, t) \, dr}{\rho_l c_l \alpha_l \displaystyle\int_0^R 4\pi r^2 \left(\frac{\partial^2 T_l}{\partial r^2} + \frac{2}{r} \frac{\partial T_l}{\partial r} \right) dr} = \frac{4\pi R^2 (\gamma - \delta) q_{\text{rad}}}{4\pi \lambda_l \left(R^2 \dfrac{\partial T_l}{\partial r} \right)_{r=R}} = \frac{(\gamma - \delta) q_{\text{rad}}}{\lambda_l \dfrac{\partial T_l}{\partial r} \Big)_{r=R}}.$$

When this ratio is larger than unity, radiative heating dominates conduction. When the ratio is less than unity, conductive heating dominates in the liquid.

Lage and Rangel studied three liquid-conductive-heating models: (1) Eq. (2.24) with a finite conductivity; (2) Eq. (2.24) with an effective conductivity, following

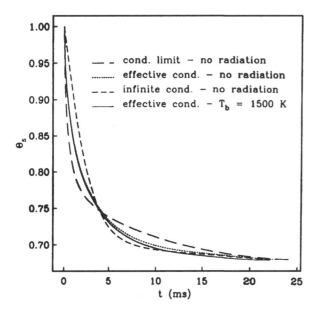

Figure 2.4. Dimensionless surface-temperature temporal variation for an *n*-decane droplet, initially at 300 K and 50 μm in radius, in a medium at 1000 K and 10 bars. Various heating models with and without radiation from a blackbody at 1500 K. (Lage and Rangel, 1993b, with permission of the *Journal of Thermophysics and Heat Transfer.*)

Abramzon and Sirignano in order to simulate the effect of internal liquid circulation; and (3) the infinite-conductivity (but finite-thermal-inertia) limit of Eq. (2.24). For the single-component droplet, they always found that conductive heating was dominant. In the spray combustion regime, the difference among the results of the three conductive-heating models was greater than the difference between the results of calculations with and without radiative heating (see Figs. 2.3 and 2.4). Figure 2.5 shows that for larger droplets most of the radiation is absorbed near the surface, whereas for smaller droplets the absorption per unit volume is more uniform. Lage and Rangel showed that the distribution of the absorption was not critical to the

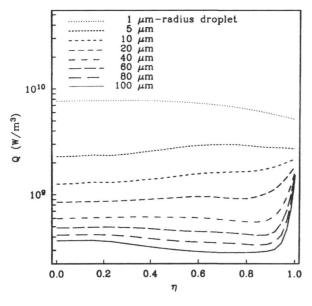

Figure 2.5. Total hemispherical absorption distribution for decane droplets irradiated by a blackbody at 1000 K under spherically symmetric conditions. (Lage and Rangel, 1993a, with permission of the *Journal of Thermophysics and Heat Transfer.*)

global results for the single-component droplet, that is, for a given value of γ, the particular value of δ made little difference.

Note that, in Chapter 4, it is explained why these conclusions do not pertain necessarily to multicomponent droplets. Also, realize that radiative heating could be greater when a flame surrounds a fuel droplet. The Lage and Rangel calculations are reasonable, however, for the common spray combustion problem in which a flame surrounds many droplets in a group (see Chapter 6).

Marchese and Dryer (1997) showed that, although radiative heat losses from droplet flames are negligible for small droplets, they can be significant for droplet diameters above 1 mm. Their radiation model indicates that, for those larger droplets, the vaporization rate is reduced because of the losses and the rate law deviates significantly from a d^2-law behavior. See also Marchese et al. (1998, 1999).

3 Convective Droplet Vaporization, Heating, and Acceleration

In this chapter, we examine the effects of droplet motion relative to the surrounding gas on the vaporization, heating, and acceleration of the droplet. The fluid velocity and scalar properties are examined for both the gas film surrounding the droplet and the liquid interior of the droplet. We use a frame of reference instantaneously travelling at the velocity of the center of the droplet; so the droplet appears stationary while the gas flows around the droplet. Still, liquid motion can occur because of internal circulation.

The relative motion between a droplet and the immediate surrounding gas results in an increase of heat and mass transfer rates in the gas film surrounding the droplets; a thin boundary layer forms over the forward section of the droplet. This boundary layer also extends over a portion of the aft section. At a sufficiently high Reynolds number (based on relative velocity, droplet radius, and gas properties), separation of the gas flow occurs at the liquid interface. Because the liquid surface moves under shear, the separation phenomenon is not identical to separation on a solid sphere; for example, the zero-stress point and the separation point are not identical on a liquid sphere, as they are on the solid sphere. The zero-stress points on the liquid sphere and the solid sphere occur at approximately the same point ($110°–130°$ measured from the forward stagnation point), but the separation point on the liquid sphere is well aft of that.

As just mentioned, the shear at the liquid surface causes a motion of the liquid at the surface. This liquid near the surface recirculates through the droplet interior so that an internal circulation results. The roughest approximation to this liquid motion is given by Hill's spherical vortex. As we will see in Subsection 3.1.2, this results in a significant decrease in the characteristic lengths and times for liquid-phase heat and mass transfer. It does not, however, justify the so-called rapid-mixing approximation, in which the characteristic times become zero.

Buoyancy effects on the gas film surrounding droplets scale with the third power of droplet diameter as compared with viscous effects; they are insignificant in many practical situations but can be important in certain laboratory experiments. Gravity effects on droplet trajectories also tend to be unimportant in high-mass-flow configurations but can be pertinent in certain situations. Surface-tension variations along the droplet surface that are due to surface-temperature variations tend not to be highly significant. There is the possibility that surface-active agents, present in the fuel by means of contamination, could cause surface-tension variations. Generally, we assume that shear is continuous across the liquid–gas interface.

At a high Reynolds number (a value in the hundreds), the inertial effects dominate the surface-tension effects (a higher Weber number situation) and droplets break into smaller droplets. These droplets will have a lower Reynolds number because of the decreased radius so that the implication is that a practical upper limit exists for the droplet Reynolds number. One can show that the maximum Reynolds number decreases as relative velocity increases because the Weber number is proportional to radius times the square of the relative velocity.

Gas-phase density and thermophysical parameters are generally considered variable, unless otherwise stated. Liquid-phase viscosity is generally taken as variable in the finite-difference calculations, but density and other properties are typically taken to be constant. The basic equations governing the gas flow are the Navier–Stokes equations for a viscous, variable-density, and variable-properties multicomponent mixture. The liquid-phase primitive equations are the incompressible Navier–Stokes equations; however, the stream-function–vorticity axisymmetric formulation is typically used (see Appendix A). These equations are also described in general form by Williams (1985) and in boundary-layer form by Chung (1965). Useful descriptions of the Navier–Stokes equations for single-component (variable-density and constant-density) flows can be found in many reference books; those of Landau and Lifshitz (1987), Howarth (1964), Sherman (1990), and White (1991) are recommended. The formulation for the gas and the liquid phases of the axisymmetric droplet problem complete with boundary, interface, and initial conditions can be found in Chiang (1990).

3.1 Convective Droplet Vaporization

In most applications, droplets in a spray will be moving at some relative velocity to the surrounding gas. The Reynolds number, based on the relative velocity, droplet radius, and gas properties, can be as large as the order of 100. Convective boundary layers and separated near wakes can therefore surround the droplet. For a liquid–gas (or liquid–liquid) interface, shear stress and tangential velocity do not generally become zero at the same point; the separation point is defined as the zero-velocity point where the streamline actually leaves the surface. The boundary layer enhances the heat and mass transport rates over the values for the spherically symmetric droplet. Furthermore, the shear force on the liquid surface causes an internal circulation that enhances the heating of the liquid. As a result, the vaporization rate increases with increasing Reynolds number. Most of this section addresses forced convection, whereby relative motion is caused by viscous or pressure stresses imposed on the gas or, before atomization, on the liquid or by body forces acting on the droplets. The situations wherein relative droplet–gas motion is caused by gravity forces acting on a gas of varying density (i.e., free or natural convection) are discussed in Subsection 3.1.8.

One commonly used empirical result is the Ranz and Marshall (1952) correlation that corrects the spherically symmetric vaporization rate $\dot{m}_{ss}$ as follows:

$$\dot{m} = \dot{m}_{ss}[1 + 0.3\text{Pr}^{1/3}(2\text{Re})^{1/2}]. \qquad (3.1)$$

For the Frossling (1938) correlation, the preceding factor 0.3 is replaced with 0.276. Here Eq. (2.10) would be used to provide $\dot{m}_{ss}$. This formula is based on certain quasi-steady, constant-radius, porous-wetted-sphere experiments. It does not account for transient heating, regressing interface, and internal circulation. Here the Reynolds number (Re) is based on the radius and the relative velocity of the droplet and gas density and viscosity; the Prandtl number (Pr) is based on gas properties. It is seen that an increase in the Reynolds number will lead to increased rates of heat transfer, mass transfer, and vaporization, and therefore to a decreased droplet lifetime. The use of these types of corrections on models of type (i), (ii), or (iii) is common in the literature. It does have some serious flaws, however.

For Reynolds numbers that are very large compared with unity, the second term in the preceding factor is dominant, and we can expect that the Nusselt number and vaporization rate will increase with the square root of the Reynolds number. Because the vaporization rate for spherically symmetric flow is proportional to droplet radius (from which the d^2 law follows), the vaporization rate with the convective correction is proportional to the radius to the three-halves power. From this statement, it follows that the droplet lifetime is proportional to the initial droplet diameter to the three-halves power for large Reynolds numbers.

The wetted-porous-sphere experiment, however, simulates only a droplet that has a uniform, steady droplet temperature at the wet-bulb temperature. The wetted sphere does not reproduce the real droplet effect of liquid-phase heat or mass transport and internal circulation. Those effects are not important for slowly vaporizing droplets for which the droplet lifetime is long compared with a characteristic time for transport in the liquid. In such a case, liquid diffusivity may be considered infinite, and the details of the internal flow field are not important because the internal gradients of the scalar properties (such as temperature or concentration) go to zero.

For that reason, a porous-wetted-sphere experiment cannot simulate the situation for fuel droplets in combustors, in which the vaporization rates are sufficiently high that internal transport and therefore internal circulation are interesting. The experiments of El-Wakil et al. (1956) also did not approach the range of high ambient temperatures and high vaporization rates so that their results are in agreement with those of Ranz and Marshall. For the preceding reasons, it is not recommended that one use results from porous-wetted-sphere experiments or from droplets vaporizing in low-temperature environments for combustor analysis.

In addition, there are some noteworthy modifications that should be made to the results of Ranz and Marshall and El-Wakil et al. First, it was shown by Lerner et al. (1980) that the extension of the empirical 1/3 averaging rule for the Prandtl and the Reynolds numbers to the convective correction is necessary when large temperature differences exist across the gas film surrounding the droplet. Again, the Reynolds number based on the average properties can be more than five times the number based on ambient properties. The original experiments of Ranz and Marshall and El-Wakil et al. did not have such large temperature differences.

Second, Sirignano (1979) has shown that the natural-logarithm functional dependence on the transfer number B is not correct and can, especially in an unsteady situation, produce serious error. The $\ln(1 + B)$ dependence results from the analysis of the spherically symmetric, quiescent vaporizing droplet. In a quasi-steady situation, the value of B does not vary greatly over a range of typical hydrocarbon fuels in a combustion situation. Therefore this dependence has never really been well tested experimentally for convecting droplets (high Reynolds number) in a quasi-steady combustion environment because $\ln(1 + B)$ has relatively small variation. However, as shown by Sirignano and Law (1978), the effective transfer number can vary by an order of magnitude during the droplet lifetime when transient heating effects are properly considered. This implies that, for the convective situation, the proper functional dependence on the effective transfer number becomes important. The $\ln(1 + B)$ dependence simply cannot apply when a convective boundary layer exists over the droplet, as demonstrated by the theoretical developments of Prakash and Sirignano (1978, 1980), Law et al. (1977), Sirignano (1979), and Lara-Urbaneja and Sirignano (1981). Its erroneous use in high Reynolds number situations becomes serious as the unsteady effects (to be discussed shortly in Subsection 3.1.3 and later) become important when heating rates and vaporization rates are high.

For these reasons, a new approach to droplet vaporization with slip in a high-temperature environment became necessary. The problem cannot be viewed as a perturbation to the spherically symmetric case when the droplet Reynolds number is high compared with unity. At best, we have an axisymmetric situation. Models of types (iv), (v), or (vi) are better suited to account for the basic phenomena present in axisymmetric configurations.

Generally, we can safely assume that the gas phase is quasi-steady even if the situation is near the critical thermodynamic conditions or the extinction domain (if a flame exists about the droplet). This quasi-steady behavior near the critical conditions or the extinction domain is not present in the spherically symmetric case. The reason for the different behaviors is due to the large difference in the length scales for gas-phase diffusion. In the spherically symmetric case, the gas-film thickness is an order of magnitude greater than the droplet radius, whereas, in the large Reynolds number situation, the film thickness (or the boundary-layer thickness) is much smaller than the droplet radius. Note that (i) the characteristic time for diffusion is proportional to the diffusion length squared so that the difference between the preceding two cases becomes even more significant; (ii) the gas Prandtl and Schmidt numbers are of the order of unity so that the gas viscous, thermal, and mass diffusion layers are all of the same approximate thickness and characteristic time; and (iii) the residence time in the boundary layer equals the diffusion time across the boundary layer.

3.1.1 Evaluation of Reynolds Number Magnitude

Usually the pressures and gas temperatures in a combustor are high, leading to a large droplet Reynolds number. Consider, for example, the ambient-gas pressure, density, temperature, and viscosity to be $p = 20$ atm, $\rho_\infty = 0.7 \times 10^2$ gm/cm^3, $T_\infty = 1000$ K, and $\mu = 0.4 \times 10^3$ P, respectively. With an initial droplet radius R_0 of 25 μm or greater and an initial relative gas-droplet velocity of 25 m/s or greater, we may expect a Reynolds number (based on radius rather than diameter) greater than 100. The implication is that a laminar large Reynolds number boundary layer will exist over the droplet surface. Obviously, with atmospheric combustors, the density and therefore the Reynolds number would be much smaller, but this major conclusion about high Reynolds number is true for a wide range of interest. The Reynolds number would increase as the ambient temperature decreases.

It can be shown that the Reynolds number remains high for most of the droplet lifetime in many typical situations. Yuen and Chen (1976) estimated that the drag data for vaporizing droplets can be correlated with the drag data for solid spheres if a certain average viscosity in the boundary layer is used for calculation of the Reynolds number. Improvements on their analysis will be discussed in Subsections 3.1.5 and 3.1.7. Let us consider, for the moment, that the drag coefficient $C_D = 12/\mathrm{Re}$. (Note that many investigators would use $C_D = 24/\mathrm{Re}$, where their Reynolds number is based on diameter, not radius.) This relationship for the drag coefficient is quite accurate as applied to nonvaporizing spheres for $\mathrm{Re} = 1$ or less. It underestimates by a factor of 3 or so at $\mathrm{Re} = 100$ but is still useful for the order-of-magnitude estimate that follows. The drag law is given by

$$m\frac{\mathrm{d}u_l}{\mathrm{d}t} = \frac{C_D}{2}\rho_g \pi R^2 |u_r|(-u_r), \tag{3.2}$$

where the mass of the droplet is

$$m = (4/3)\pi \rho_l R^3,$$

u_r is the relative droplet velocity, u_l is the absolute droplet velocity, ρ_l is the liquid density, ρ_g is the gas density, and R is the droplet radius.

It follows that

$$\frac{\mathrm{d}u_l}{u_r} = \frac{9}{2}\frac{\rho_g}{\rho_l}\frac{|u_r|R}{\mathrm{Re}}\frac{\mathrm{d}t}{R^2}.$$

The quantity $|u_r|R/\mathrm{Re}$ will not change with time as velocity and radius change. For the representative values just cited, that constant will be of the order of 10^{-5} m^2/s. If the density ratio ρ_g/ρ_l is of the order of 10^{-2}, it follows that the characteristic time for decrease of relative velocity is

$$\tau = R^2/K, \tag{3.3}$$

where $K = 0\,(10^{-7}\ \mathrm{m}^2)$ in our case. Now R^2 may be taken as an average or $R_0^2/2$. It may be considered that τ is ~ 1 ms.

The lifetime τ_L of the size droplet in a high-temperature environment with the relative velocity indicated will be ~ 1 ms also. Note further that, as the initial radius increases, the lifetime increases with a power somewhat weaker than the square because of a convective correction; that is, empirical evidence inferred from wetted porous spheres indicates that

$$\tau_L \sim R^2[1 + 0.3\mathrm{Pr}^{1/3}(2\mathrm{Re})^{1/2}]. \tag{3.4}$$

Note again that we base the Reynolds number on radius here in contrast to the original formulation of the correlation. So, for very low Reynolds numbers compared

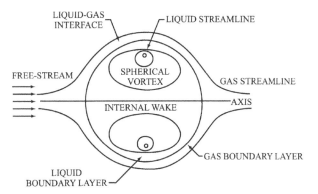

Figure 3.1. Vaporizing droplet with relative gas–droplet motion and internal circulation. (Sirignano, 1993b, with permission of the *Journal of Fluids Engineering.*)

with unity, $\tau_L \sim R^2$, whereas for high Reynolds numbers, $\tau_L \sim R^{3/2}$. The velocity-relaxation time τ_v will increase with R^2 as long as the drag law is obeyed. (The data of Yuen and Chen indicate that C_D decreases as the Reynolds number increases up to a Reynolds number of $\sim$1000.) We may conclude that, for initial droplet radii of 25 μm and greater, the time for velocity relaxation that is due to drag is comparable with or greater than the droplet lifetime.

The change in droplet radius during the major portion of the droplet lifetime is not significant with regard to the determination of the order of magnitude of the Reynolds number. For example, as droplet mass decreases by a factor of 8 (or $\sim$1 order of magnitude), the radius decreases by only a factor of 2. On account of the time scales for change in radius and relative velocity, the Reynolds number decreases by less than 1 order of magnitude during the period when most of the droplet mass vaporizes. Therefore, in many practical combustion situations, we may expect a laminar, high Reynolds number (of the order of 100) boundary layer to remain on the droplet for most of its lifetime. Because gas-phase Prandtl numbers are close to unity, the same conclusion can be made about the value of the gas Peclet number and the gas-phase thermal boundary layer. It should be understood that practical situations will also occur wherein the Reynolds number will be smaller, perhaps of the order of 10 or less.

The velocities in the liquid phase associated with the internal circulation have been found by Prakash and Sirignano (1978, 1980) to be more than 1 order of magnitude lower than the relative gas-droplet velocity. However, the liquid density is 2 to 3 orders of magnitude greater than the gas density (depending on the pressure level) so that, as a result, the liquid-phase Reynolds number, based on droplet radius, liquid surface velocity, liquid density, and liquid viscosity is as large, or larger, than the droplet Reynolds number. Because the Prandtl numbers for ordinary liquids are of the order of 10, the liquid-phase Peclet numbers are very large compared with unity (of the order of 1000). We can expect a thin viscous layer and a thinner thermal layer to exist in the liquid at the surface.

3.1.2 Physical Description

The configuration to be studied for an isolated droplet is an axisymmetric flow, as depicted in Fig. 3.1. Droplets can turn in a flow or the gas flow can change direction, but typically the characteristic time for a change of vectorial direction of the relative velocity is long compared with the residence time for an element of gas to flow past the droplet. Therefore axisymmetry is a good approximation. The droplet still retains a spherical shape, provided that the Weber number remains of the order of unity or less.

The axisymmetric flow field within and around the droplet is shown in Fig. 3.1 as including four major regions. Region I is the inviscid free stream of gas flowing over the droplet and outside its wake, as seen from a frame of reference moving with the droplet. Region II includes the gas-phase viscous boundary layer and near wake. Region III contains the liquid viscous layer and the liquid internal wake, which is the zone along the axis of symmetry through which the liquid circulating out of the boundary layer passes. Region IV is the toroidal core region within the droplet that is rotational but nearly shear free and can be approximated as a Hill's spherical vortex.

Let us discuss the various assumptions implied by this model of a slipping droplet. First, it is assumed that the droplet remains spherical; this has been shown by Lerner et al. (1980) to be reasonable for hydrocarbon droplets with diameters of 400 μm or less and relative velocities of tens of meters per second. Obviously the Weber number is a key parameter. Operation in the near-critical region, as discussed by Faeth (1977), would lead to a failure of the spherical assumption.

Another important assumption involves the axisymmetry of the flow field and the absence of droplet rotation. Rotation of the droplet would be caused by a torque on the droplet. Once the droplet is formed, the major cause of a torque would be a net moment that is due to friction forces on the droplet surface. Some asymmetry in the flow is necessary here to accomplish this because velocity gradients must exist in the approaching gas flow. This turbulent scale can be of the order of the droplet diameter ($\sim$100 μm or so) in practical combustor situations. Estimates of the net torque and the perturbation in the friction force at the surface that is due to velocity gradients of this maximum amount indicate that (i) the characteristic time for rotation of a solid particle in the flow would be very long compared with the time of transit for an element of gas moving past the particle, and (ii) the fractional change in frictional force on the particle or droplet surface would be of the order of the reciprocal of the droplet Reynolds number. The process of estimation is now presented. We expect a balance between the inertial terms and the friction along the droplet surface so that

$$\frac{\mathrm{d}(\rho u^2)}{\mathrm{d}x} \sim \frac{\partial \tau}{\partial y} \sim \frac{\tau}{\delta}, \tag{3.5}$$

where δ is the gas boundary-layer thickness and is of the order of $R\mathrm{Re}^{1/2}$; note that the approximation sign again implies that quantities are of the same order of magnitude. All velocities in this discussion are relative droplet–gas velocities. Because the relevant arc length along the surface is of the order of the droplet radius, we can say that

$$\rho u^2 \sim \tau \frac{R}{\delta} \sim \tau \, \mathrm{Re}^{1/2}. \tag{3.6}$$

Whenever a fluctuation occurs, the axisymmetry can be lost, and a difference in the velocity and the stress is found across the droplet. It is expected from approximation (3.6) that these differences Δu and $\Delta \tau$ must be related by

$$2\rho u \Delta u \sim \mathrm{Re}^{1/2} \Delta \tau. \tag{3.7}$$

If Δu is dictated by a fluctuating turbulent eddy at the Kolmogorov scale, it follows that

$$\Delta u \sim v/\eta \sim v/R, \tag{3.8}$$

where it has been assumed that the eddy size η is of the order of the droplet size. Furthermore, the friction force F will be given by

$$F \sim \tau R^2, \tag{3.9}$$

so that the net friction force ΔF that is due to the fluctuation in the velocity is

$$\Delta F \sim R^2 \Delta \tau, \tag{3.10}$$

and the torque on the droplet is

$$T \sim R \Delta F \sim R^3 \Delta \tau. \tag{3.11}$$

Combining approximations (3.7), (3.8), and (3.11) yields

$$T \sim \rho u v \, \mathrm{Re}^{-1/2} R^2. \tag{3.12}$$

Now the moment of inertia of the droplet I will be of the order of $\rho_l R^5$ so that the acceleration

$$w \sim \frac{T}{I} \sim \frac{\rho}{\rho_l} uv \, \mathrm{Re}^{-1/2} R^{-3} \tag{3.13}$$

or

$$w \sim \frac{\rho}{\rho_l} \frac{u^2}{R^2} \mathrm{Re}^{-3/2} \sim \frac{\rho}{\rho_l \, \mathrm{Re}^{-3/2}} \frac{1}{t_{\mathrm{res}}^2}, \tag{3.14}$$

where t_{res} is the residence time for an element of gas flowing past the droplet. If we consider the angular acceleration w to be roughly constant with time and if we define the time for one rotation as t_{rot}, we may obtain from approximation (3.14) that

$$\frac{t_{\mathrm{rot}}}{t_{\mathrm{res}}} \sim \left(\frac{\rho_l}{\rho}\right) \mathrm{Re}^{3/4} > 1. \tag{3.15}$$

Approximations (3.5)–(3.9) will yield

$$\frac{\Delta F}{F} \sim \frac{2}{\mathrm{Re}}. \tag{3.16}$$

Approximations (3.15) and (3.16) support the conclusions concerning the large magnitude of the droplet rotational time and the small magnitude of the fractional change in the friction force.

If a net torque were applied to a liquid, it is intuitively accepted by this author that the droplet would be induced to undergo an asymmetric internal circulation because of the shearing-force imbalance rather than to rotate as a solid body. The basis for this belief is that the moment of inertia of the liquid in the circulatory motion is significantly less than that associated with a solid-body-like rotation. The solid-body rotation model is a worst case, in some senses. In any event, conclusions (i) and (ii) at the beginning of this subsection indicate that rotation or asymmetric circulation is not significant for droplets with the large Reynolds number under consideration here.

The same type of analysis can be used to consider the effects of initial droplet rotation established during the formation or atomization process. The indication is that the net torque that is due to surface friction would oppose the rotation, which would be relaxed in the time required for the droplet to travel tens of diameters (but not hundreds of diameters). Again, droplet rotation does not seem very important.

In Section 3.3 and Section 3.4, a discussion of the effects of the temporal and spatial changes in the velocity field is given. Currently we neglect the effect of smaller-scale turbulence in causing a deviation of the external and the internal droplet-flow fields from axisymmetry. For turbulent length scales larger than the droplet and for turbulent times scales longer than the flow residence time, axisymmetry and gas-phase quasi-steadiness are reasonable assumptions.

The Prandtl number and the Schmidt number for the typical hydrocarbon fuel are of the order of 10 and of the order of 100, respectively. Because of the high liquid density, the Reynolds number and the Peclet numbers for the droplet interior can also be orders of magnitude larger than unity. Two conclusions may be drawn from this order-of-magnitude analysis. First, the transient time for heating and mass diffusion in the liquid phase can be quite long compared with the relaxation time for the internal circulation to be established because of the large Prandtl and Schmidt numbers. Indications are that the heating time is comparable to and the mass diffusion time is much longer than the droplet lifetime. Second, the large Peclet numbers demonstrate that thin boundary layers exist near the liquid surface.

At large liquid Peclet numbers, we expect that an axisymmetric thermal boundary layer exists near the liquid–gas interface; as the boundary layer flows away from the rear stagnation point, an internal wake forms. The thermal layer and thermal internal wake are thinner than the viscous layer and the viscous internal wake of Region III in Fig. 3.1, but otherwise are qualitatively similar. The toroidal core will be unsteady because heating is slow at a large Prandtl number but in the limit may be assumed to be one dimensional when cast in an optimal coordinate system, as demonstrated by Prakash and Sirignano (1978, 1980). The time required for circulating an element of liquid once around a stream surface is very small compared with other characteristic times (a consequence of the high Peclet number). Within the core, therefore, the convection will tend to make the temperature (and also concentration) uniform along the stream surface, although variations will exist across the stream surfaces. Diffusion will then tend to be normal to the stream surfaces in the limit. Temperature variations and heat diffusion may in the limit be neglected along the stream surface. Therefore the problem is recast in new orthogonal coordinates whereby one coordinate is constant along a stream surface; the two-dimensional unsteady problem is transformed to a one-dimensional unsteady problem. The unsteady core is qualitatively similar but slightly larger than the Hill's vortex core (Region IV of Fig. 3.1).

Another assumption used in the approximate models that should be discussed relates to the effect of surface regression on the flow field. Basically the reasonable argument is advanced that residence time for a gas element flowing past the droplet and the circulation time for a liquid element in the droplet are orders of magnitude less than the droplet lifetime. The surface regression on those time scales is miniscule and can result in only negligibly small perturbations to the flow field. The smallest length scales that we will see in this problem will be the thicknesses of the heat and mass diffusion layers at the liquid surface. As long as a liquid element at the surface circulates many times $[O(10)$ or greater$]$ before the surface regresses that distance, our approximate model will be reasonable. The assumption is best for the viscous layer, next best for the thermal layer, but weakest for the mass diffusion layer because the liquid Prandtl number is of the order of tens and the liquid Schmidt number is of the order of hundreds. In high-temperature environments, such as combustors, the vaporization rate becomes so high that a fluid element in the boundary layer and internal wake (Region III) will not circulate many times before the completion of vaporization; this assumption is rather weak therefore, and its relaxation should be studied in the future.

Continuity at the liquid–gas interface would indicate that the surface velocity in the aft region of the droplet could be reversed on account of the existence of the near wake. Results of numerical calculations indicate that, above a certain Reynolds number, a small secondary vortex will exist at the aft region of the liquid sphere. The Hill's vortex model is oversimplified, therefore, in that range where a secondary vortex must exist. Numerical results by Rivkind and Ryskin (1976) indicate that whenever the internal fluid (liquid) density is much larger than the external fluid (gas) density, the secondary vortex is quite small compared with the primary vortex. This can be explained by the fact that, as the shear stress goes to zero at a certain point on the droplet surface, the tangential velocity there is still positive. Aft of that point, the shear stress reverses sign and the heavier liquid drags the lighter gas for some distance before stagnation (zero velocity) occurs. The tangential velocity at the surface is therefore in the same direction as the primary external flow for most of the surface area. It follows that, for our situation, the assumption of a single spherical vortex is reasonable as a first approximation.

Another important assumption is that the fluid motion in the droplet interior and in the gas-phase boundary layer is considered to be quasi-steady. The transient period for the establishment of the motion is quite small compared with the droplet lifetime. That characteristic time is the vorticity diffusion time given by the radius squared divided by the kinematic viscosity. Note that the time for heat diffusion across the droplet is much longer than the vorticity diffusion time by the factor of the Prandtl number whereas the mass diffusion time is longer by the factor of the Schmidt number. For a droplet vaporizing in a combusting environment, the droplet temperature and concentrations vary spacially through the droplet and behave in an unsteady manner.

In addition to the initial transient, some unsteadiness could result from vortex shedding in the gas from the droplet surface. Based on Strouhal number estimates, the frequency of shedding would be high compared with other phenomena in the droplet. Also, the vortex shedding, if it does occur, would affect only the aft side of the droplet and should not perturb the gas boundary layer on the droplet through which most of the heat and mass transfer occurs. It is assumed therefore that any effect of vortex shedding averages is negligible.

The evaluations of the vaporization rate and heating rate have been performed by both approximate analyses and exact solutions of the coupled gas- and liquid-flow equations by finite-difference calculations. These analyses and their results are discussed in Subsections 3.1.6 and 3.1.7.

3.1.3 Approximate Analyses for Gas-Phase Boundary Layer

There are various analyses that describe the behavior of a quasi-steady, laminar, gas-phase boundary layer that exists at the interface of a droplet with Reynolds number (based on relative velocity) that is larger than unity. The quasi-steady boundary-layer equations are obtained from the axisymmetric form of the equations in Appendix A after a transformation to coordinates better suited for the boundary-layer study. See White (1991) for a general exposition of the boundary-layer theory. The earliest studies (type 1) were performed by Prakash and Sirignano (1978, 1980) for single-component liquids and by Lara-Urbaneja and Sirignano (1981) and Law et al. (1977) for multicomponent liquids. Prakash and Sirignano (1978, 1980) and Lara-Urbaneja and Sirignano (1981) used integral boundary-layer techniques to analyze the gaseous flow over a vaporizing-droplet surface. These authors were the first to identify the essential physics of the convective droplet-vaporization problem. They resolved the dependencies on two spacial coordinates for both the gas-phase and the liquid-phase boundary layers. The analyses were made earlier than those described in this subsection and also were more complex without offering clear advantages. They were not yielding as complete and as detailed information as the finite-difference solution of the Navier–Stokes elliptic-flow equations. At the same time, the integral techniques were too complex to incorporate into extensive spray computations. Later analyses that are examined more closely here were developed by Sirignano (1979), Tong and Sirignano (1982a, 1982b, 1983, 1986a, 1986b) (type 2), and Abramzon and Sirignano (1989) (type 3). They offer useful simplifications to the studies of Prakash and Sirignano and Lara-Urbaneja and Sirignano. In particular, type 2 and type 3 models reduce the dimensionality of the problem by averaging over the coordinate in the main flow direction (film models).

The Prakash and Sirignano analysis and the Lara-Urbaneja and Sirignano analysis (type 1) considered two-dimensional quasi-steady behavior in the boundary layers on the two sides of the droplet–gas interface. This made the amount of calculation in a multidrop spray analysis impractical. So those analyses were never used in a spray analysis. The Tong and Sirignano analysis (type 2) used a locally self-similar form of the gas-phase boundary layer and an approximation of the liquid-phase boundary layer that reduced the dimensionality significantly. This droplet-heating model is often identified as the vortex model because it explicitly identifies a spherical vortex in the liquid. Subsequently, Tong and Sirignano's droplet analysis was used extensively in spray computations. See, for example, Aggarwal et al. (1984), Raju and Sirignano (1989, 1990a), and Rangel and Sirignano (1988b).

All of the previously mentioned models applied to high Reynolds number domains in which a Prandtl-type gas-phase boundary layer existed on the droplet. Abramzon and Sirignano (1989) developed a more robust model (type 3) that applied to zero or very low Reynolds number cases as well as to boundary-layer situations. It also retained the reduced dimensionality of the Tong and Sirignano models. The liquid-phase portion of this model is often identified as an effective-conductivity model since it implicitly represents the effect of the liquid vortex through an effective-conductivity term that becomes larger than the actual conductivity. It has been applied to many spray computations as the subgrid droplet model. See, for example, Continillo and Sirignano (1988), Bhatia and Sirignano (1991), and Delplanque and Sirignano (1995, 1996).

All of these approximate analyses are limited strictly to boundary layers and do not predict the behavior in the near wake. Some of the analyses can predict the point of separation on the droplet surface. They use either integral techniques or local similarity assumptions for the analysis. Because most of the heat and mass transport for the Reynolds number range of interest (of the order of 100 and less) occurs before the point of separation, vaporization and droplet-heating rates can be well predicted. However, because form drag is competitive in magnitude with friction drag, the drag coefficient cannot even be approximated by these analyses; input is needed for these values.

TYPE 2 MODEL. Now some details are given for the type 2 gas-phase boundary-layer analysis. Under the assumption that the droplet Reynolds number is large compared with unity (but not so large that instability or turbulence occurs), a thin, gas-phase boundary layer exists on the surface of the droplet. In our application, we can reasonably neglect kinetic energy and viscous dissipation. Also, $\mathrm{Pr} = \mathrm{Sc} = 1$, and one-step chemistry can be assumed, following Tong and Sirignano. The x, y coordinates are tangent and normal to the droplet surface, respectively, with r representing the distance from the axis of symmetry. The governing equations in quasi-steady boundary-layer form follow from the equations in Appendix A:

Continuity:

$$\frac{\partial(\rho u r)}{\partial x} + \frac{\partial(\rho v r)}{\partial y} = 0. \tag{3.17}$$

x Momentum:

$$L\left(\frac{u}{u_e}\right) = \rho u \frac{\partial}{\partial x}\left(\frac{u}{u_e}\right) + \rho v \frac{\partial}{\partial y}\left(\frac{u}{u_e}\right) - \frac{\partial}{\partial y}\left[\mu \frac{\partial}{\partial y}\left(\frac{u}{u_e}\right)\right]$$

$$= \left[\frac{\rho_e}{\rho} - \left(\frac{u}{u_e}\right)^2\right]\rho\frac{\mathrm{d}u_e}{\mathrm{d}x}, \tag{3.18}$$

where

$$\frac{\mathrm{d}p_e}{\mathrm{d}x} = \rho_e u_e \frac{\mathrm{d}u_e}{\mathrm{d}x}.$$

Note that $u_e(x)$ is the potential-flow velocity immediately external to the boundary layer. The y-momentum equation states that pressure gradients in the y direction are negligible.

Energy:

$$L(h) = -Q\dot{w}_F. \tag{3.19}$$

Species Conservation

$$L(Y_i) = \dot{w}_i, \qquad i = O, F, P, N. \tag{3.20}$$

The x-momentum equation can be further simplified as the ambient temperature is much higher than the surface temperature and the ambient velocity is much higher than the tangential surface velocity. The right-hand side of Eq. (3.18) goes to zero at the outer edge of the boundary layer; also, it becomes negligibly small at the droplet surface. The ad hoc assumption can be made, following Lees (1956), that the right-hand side is everywhere zero because the quantity in square brackets is negligibly small. Note that the pressure gradient can still be nonzero; it is the transverse variation of the dynamic pressure that is small.

Two interesting cases are readily studied: the stagnation-point flow ($r = x$ and $u_e = ax$) and the shoulder region [$\theta = \pi/2$, $r = R$, $u_e = (3/2)U_\infty$], where the pressure gradient is zero and the flow locally behaves like a flat-plate flow, that is, because the right-hand side of Eq. (3.18) is negligible, local similarity is believed to be a very good approximation. The well-known similarity solution,

$$u = u_e(x)\frac{\mathrm{d}f}{\mathrm{d}\eta}(\eta) \equiv u_e(x)f'(\eta),$$

is found, where $f(\eta)$ satisfies the Blasius equation resulting from Eqs. (3.17) and (3.18) under the previously mentioned approximations and identities:

$$\frac{\mathrm{d}^3 f}{\mathrm{d}\eta^3} + f\frac{\mathrm{d}^2 f}{\mathrm{d}\eta^2} = 0, \tag{3.21}$$

where

$$\eta = \frac{ru_e \int_0^y \rho\,\mathrm{d}y'}{\left(\int_0^x \rho_e\mu_e u_e r^2\,\mathrm{d}x'\right)^{1/2}}.$$

The vaporization rate per unit area is given by

$$(\rho v)_s = -Af(0). \tag{3.22}$$

Note that $A = (2K\rho_e\mu_e)^{1/2}$ for the stagnation point and $\rho_e\mu_e(u_e/2\int_0^x \rho_e\mu_e\,\mathrm{d}x')^{1/2}$ for the shoulder region. K is the strain rate near the stagnation point.

Two of the three boundary conditions for third-order, nonlinear ordinary differential equation (3.21) governing $f(\eta)$ are

$$\frac{\mathrm{d}f}{\mathrm{d}\eta}(0) = u_s/u_e, \qquad \frac{\mathrm{d}f}{\mathrm{d}\eta}(\infty) = 1.$$

These conditions reflect the prescribed free-stream velocity and the continuity of tangential velocity at the droplet interface. The third boundary condition is developed from Eq. (3.22) and requires a coupling with the solution of the energy equation.

The definitions given in Eqs. (2.7) can be used to solve Eqs. (3.19) and (3.20) for the case of rapid chemical kinetics. With the simplified version of Eq. (3.18) previously discussed, it can be shown that for all of the β functions

$$\beta = \beta_s + \frac{u - u_s}{u_e - u_s}(\beta_e - \beta_s) = \beta_s + \frac{f'(\eta) - f'(0)}{1 - f'(0)}(\beta_e - \beta_s). \tag{3.23}$$

Equations (3.19) and (3.20) are second-order partial differential equations normally associated with five boundary conditions at the outer edge of the boundary layer and five boundary conditions at the droplet surface. The ambient conditions for temperature (or enthalpy) and mass fractions are provided. Five boundary conditions at the droplet surface are given by conservation of energy and species mass flux as follows:

$$\left(\lambda \frac{\partial T}{\partial y}\right)_{g,s} = \left(\lambda_l \frac{\partial T}{\partial y}\right)_{l,s} + (\rho v)_s L \equiv (\rho v)_s L_{\text{eff}}, \tag{3.24}$$

$$(\rho v)_s Y_{Fs} - \left(\rho D \frac{\partial Y_F}{\partial y}\right)_s = (\rho v)_s, \tag{3.25}$$

$$(\rho v)_s Y_{i,s} - \left(\rho D \frac{\partial Y_i}{\partial y}\right)_s = 0, \quad i = O, P, N. \tag{3.26}$$

These conditions are analogous to Eqs. (2.5) and (2.6) for the spherically symmetric case. In addition, we have a phase-equilibrium relation, (2.4), and a continuity condition on the temperature at the interface. The extra two conditions are required because the vaporization rate (per unit area) $(\rho v)_s$ and the liquid-heating rate (per unit area) $(\lambda_l \frac{\partial T}{\partial y})_s$ are not given a priori. They are eigenvalues of the problem. Complete determination requires the coupling with the heat transport problem in the liquid phase.

From Eqs. (3.22) through (3.26) and the definition in Eq. (2.11), we can develop the third boundary condition for the Blasius differential equation. We find that

$$\frac{f''(0)}{[-f(0)][1 - f'(0)]} = \frac{1}{B}. \tag{3.27}$$

Actually, the Blasius function depends on η, B, and u_s/u_e because B and u_s/u_e appear in boundary conditions (3.22) and (3.27). So $f(0)$ implies $f(0, B, u_s/u_e)$. Furthermore, $f(0)$ is negative for the vaporization conditions. It can also be shown that

$$\lambda_l \frac{\partial T}{\partial y}\bigg)_{l,s} = -Af(0)\left(\frac{h_e - h_s + QvY_{O\infty}}{\frac{1 + vY_{O\infty}}{1 - Y_{Fs}} - 1} - L\right). \tag{3.28}$$

An interesting comparison can be made between this convective case and the spherically symmetric case. Equations (3.22), (3.24), and (3.28) can be combined to reproduce an equation exactly like the second equality of Eq. (2.11) except that h_e appears instead of h_∞. This demonstrates a great similarity in the physics of the two cases. Furthermore, the same formula for the wet-bulb temperature applies in the spherically symmetric case, in the stagnation-point region, and in the shoulder region under similarity conditions.

The surface flow is expected to reach zero-normal-temperature gradient shortly after the point of zero-pressure gradient. Therefore most of the heat and mass transport occur on the forward side or the shoulders of the droplet; little occurs on the downstream side. The preceding analyses for the stagnation region and for the shoulder region can be used to construct a reasonable estimate for the global liquid-heating rate and vaporization rate.

Use of the definition of Nusselt number given by Eq. (2.14) together with Eqs. (2.11) and (3.28) yields the following relationship:

$$\mathrm{Nu} = \frac{k[-f(0)]}{B}\mathrm{Re}^{1/2}, \qquad (3.29)$$

where k is a positive nondimensional coefficient of the order of unity that is determined when the heat flux is averaged over the droplet surface in an approximate manner (based on the two local solutions previously discussed). $k = 0.552\sqrt{2}$ or $k = 0.6\sqrt{2}$ can be justified by comparison with Frossling (1938) or Ranz and Marshall (1952) correlations. Again, $f(0)$ is negative with vaporization (see Sirignano, 1979). Also, we have the global vaporization rate by means of a similar averaging process:

$$\dot{m} = 2\pi k\mu_e R[-f(0)]\,\mathrm{Re}^{1/2}. \qquad (3.30)$$

Under the Le $=1$ assumption and the surface-averaging procedure previously described, the Sherwood number defined in Eq. (2.14) becomes identical to the Nusselt number given by Eq. (3.29).

The results of Eqs. (3.29) and (3.30) are based on an analysis limited to a high droplet Reynolds number. There is an important domain between the spherically symmetric case (zero Reynolds number case) and the thin laminar boundary-layer case (high Reynolds number case). A more robust vaporization model that covers a wide range of Reynolds number is needed. This is especially important because a given droplet can experience a range of Reynolds numbers during its lifetime. Typically the droplet Reynolds number will decrease with time as the droplet diameter and the relative velocity decrease. There are exceptions to the monotonic behavior, e.g., oscillatory ambient flow in which large fluctuations of relative velocity (including change in direction) can occur.

TYPE 3 MODEL. One ad hoc method for developing a more robust model (type 3) was presented by Abramzon and Sirignano (1989). Now some of the details of that model are presented. The Nusselt number and the vaporization rate are each given by a composite of two asymptotes (zero Reynolds number limit and large Reynolds number limit). We have that

$$\mathrm{Nu} = \mathrm{Sh} = \frac{2\ln(1+B)}{B}\left\{1 + \frac{k}{2}\frac{[-f(0)]}{\ln(1+B)}\mathrm{Re}^{1/2}\right\}, \tag{3.31}$$

$$\dot{m} = 4\pi\rho D R\ln(1+B)\left\{1 + \frac{k}{2}\frac{[-f(0)]}{\ln(1+B)}\mathrm{Re}^{1/2}\right\}. \tag{3.32}$$

Note that $f(0) = f(0, u_s/u_e, B)$ reaches a finite limit as B becomes large. The preceding results therefore disagree strongly with the Ranz–Marshall or the Frossling correlations except in a limit of small B values. The problem with those correlations is that they were developed for only a very narrow range of B values that varied by less than 1 order of magnitude.

Under the assumption that $\mathrm{Sc} = 1$, the ratio $\dot{m}/\mathrm{Nu}$ is identical for the three pairs of denominator and numerator given by Eqs. (2.15a) and (2.15b), (3.29) and (3.30), and (3.31) and (3.32). See Eq. (2.15b).

Abramzon and Sirignano actually made two further extensions. First, they considered general values for Schmidt, Prandtl, and Lewis numbers, relaxing the unitary conditions and allowing for variable properties. Second, they considered a range of Falkner–Skan solutions (White, 1991) to develop the average transport rates across the gas boundary layer on the droplet surface. Thereby, Eqs. (3.17)–(3.20) were solved for cases with the nonzero-pressure gradients in order to simulate the flow around spherical droplet. With the definitions

$$B_H = \frac{h_e - h_s}{L_{\mathrm{eff}}}, \qquad B_M = \frac{Y_{Fs} - Y_{F\infty}}{1 - Y_{Fs}}, \tag{3.33}$$

they demonstrated that

$$\mathrm{Nu} = 2\frac{\ln(1+B_H)}{B_H}\left[1 + \frac{k}{2}\frac{\mathrm{Pr}^{1/3}\mathrm{Re}^{1/2}}{F(B_H)}\right] \tag{3.34a}$$

$$\mathrm{Sh} = 2\frac{\ln(1+B_M)}{B_M}\left[1 + \frac{k}{2}\frac{\mathrm{Sc}^{1/3}\mathrm{Re}^{1/2}}{F(B_M)}\right], \tag{3.34b}$$

$$\dot{m} = 4\pi\frac{\lambda R}{c_{pF}}\ln(1+B_H)\left[1 + \frac{k}{2}\frac{\mathrm{Pr}^{1/3}\mathrm{Re}^{1/2}}{F(B_H)}\right]$$

$$= 4\pi\rho D R\ln(1+B_M)\left[1 + \frac{k}{2}\frac{\mathrm{Sc}^{1/3}\mathrm{Re}^{1/2}}{F(B_M)}\right]. \tag{3.34c}$$

Note that when $\mathrm{Pr} = \mathrm{Sc}$, we have $B_M = B_H$. Otherwise,

$$B_H = (1+B_M)^a - 1, \tag{3.35a}$$

where

$$a \equiv \frac{c_{pF}}{c_p}\frac{1}{\mathrm{Le}}\frac{1 + \dfrac{k}{2}\dfrac{\mathrm{Sc}^{1/3}\mathrm{Re}^{1/2}}{F(B_M)}}{1 + \dfrac{k}{2}\dfrac{\mathrm{Pr}^{1/3}\mathrm{Re}^{1/2}}{F(B_H)}}. \tag{3.35b}$$

A correlation of numerical results for the Falkner–Skan solutions by Abramzon and Sirignano (1989) shows that

$$F(B) = (1+B)^{0.7}\frac{\ln(1+B)}{B} \quad \text{for } 0 \le B_H, \; B_M \le 20, \; 1 \le \text{Pr}, \text{Sc} \le 3. \tag{3.36}$$

Comparison with the Frossling (1938) correlation for small values of B yields $k = 0.552\sqrt{2} = 0.781$ whereas the comparison with the Ranz–Marshall (1952) correlation yields $k = 0.6\sqrt{2} = 0.848$.

Following Clift et al. (1978), Abramzon and Sirignano suggest that, for $\text{Re} \le 5$, the Reynolds-number-dependent terms in Eqs. (3.34) and (3.35) should be replaced. The factor

$$\left[1 + \frac{k}{2}\frac{\text{Pr}^{1/3}\text{Re}^{1/2}}{F(B_H)}\right]$$

should be replaced with

$$\left[1 + \frac{(1+2\text{RePr})^{1/3}\max[1,(2\text{Re})^{0.077}] - 1}{2F(B_H)}\right]$$

for $\text{Re} \le 5$. Similarly, the factor

$$\left[1 + \frac{k}{2}\frac{\text{Sc}^{1/3}\text{Re}^{1/2}}{F(B_M)}\right]$$

should be replaced with

$$\left[1 + \frac{(1+2\text{ReSc})^{1/3}\max[1,(2\text{Re})^{0.077}] - 1}{2F(B_M)}\right]$$

for that same range of Re.

When $\text{Le} = 1$ ($\text{Pr} = \text{Sc}$), then $\text{Nu} = \text{Sh}$ and Eq. (2.15b) applies. When $\text{Re} = 0$, we have an extension of the spherically symmetric nonburning case for $\text{Le} \ne 1$.

The extension of the Abramzon and Sirignano model to the case with a thin diffusion flame in the boundary layer is straightforward in the $\text{Pr} = \text{Sc} = 1$ limit. Then the transfer number B_H ($= B_M$ here) can be defined by Eq. (2.11) instead of by the first of Eqs. (3.33). Now Eqs. (3.34) and (3.36) can be used.

Other approximate or asymptotic techniques have been utilized to solve related problems. Hadamard (1911) and Rybczynski (1911) solved for the creeping flow around a nonvaporizing liquid droplet. Acrivos and Taylor (1962) and Acrivos and Goddard (1965) analyzed the heat transfer for this type of flow and estimated the Nusselt number. Harper and Moore (1968) and Harper (1970) solved the high Reynolds number nonvaporizing droplet; boundary layers on both sides of the droplet interface were analyzed. Rangel and Fernandez-Pello (1984) studied the high Reynolds number vaporizing droplet with an isothermal liquid assumption. Chung et al. (1984a, 1984b) and Sundararajan and Ayyaswamy (1984) studied condensing droplets by perturbation and numerical methods. In Section 3.2, vaporizing droplets in low Reynolds number flows are discussed. An extension of Eqs. (3.34) is developed to give a more robust form; see Eqs. (3.62a, 3.62b).

In the case in which many components exist in the liquid phase, it is necessary to track the vapor components individually as the species advect and diffuse through the gas phase. If there are n vaporizing components in the liquid, additional $n - 1$ field equations of the form of Eq. (3.20) and additional $n - 1$ boundary conditions of the form of Eq. (3.25) must be added to the system. Also, a phase-equilibrium relationship is required for each component and the chemical source term in energy equation (3.19) must be modified.

The gas-phase equations are strongly coupled to the liquid-phase equations through the continuity of velocity and temperature and the balance of mass, force, and energy. See, for example, Eqs. (3.22), (3.24), (3.25), and (3.27).

Lozinski and Matalon (1992) performed an interesting study of a vaporizing spinning spherical droplet. The quasi-steady vaporization was enhanced by the spinning because of the creation of an induced secondary convective current inward toward the droplet poles and outward from its equator. The augmentation was found to be proportional to the fourth power of the angular velocity. The d^2 law was no longer followed.

3.1.4 Approximate Analyses for Liquid-Phase Flows

The jump in shear stress across a liquid–gas interface equals the gradient of surface tension (Levich, 1962). Sirignano (1983) argues that generally the temperature and composition variations along the surface of a droplet are too small to cause a significant gradient of surface tension. Therefore we consider a continuity of shear stress across the interface. However, see the discussion at the end of Subsection 3.1.7.

The liquid-phase responds to the viscous shear force at the interface by circulating within the droplet. Toroidal stream surfaces result with low velocities. Because of the larger liquid density, the Reynolds number for the internal circulation (based on droplet diameter, maximum liquid velocity, and liquid properties) can be of the same order of magnitude as or higher than the Reynolds number for the gas flow over the droplet. Peclet numbers are even higher in the liquid because of the large Prandtl and Schmidt numbers. Heat and mass transport in the liquid will therefore behave in a fashion highly dissimilar to the transport of momentum or vorticity. The analysis of the hydrodynamics in the approximate models differs substantially from the analyses of heat and mass transport on account of the large Prandtl and Schmidt numbers.

It is convenient to address the incompressible liquid-phase hydrodynamics by use of the vorticity-stream-function formulation. The vorticity equation can be developed by taking the curl of the momentum equation. For an incompressible fluid, we have

$$\frac{D\vec{\omega}}{Dt} = \vec{\omega} \cdot \nabla\vec{u} + \nu\nabla^2\vec{\omega}. \tag{3.37}$$

In a planar flow, $\vec{\omega}$ and $\vec{u}$ are orthogonal. So the first term on the right-hand side becomes zero in that case. Therefore, in the inviscid limit, the vorticity vector $\vec{\omega}$ is constant along a particle path. In the case of a nonswirling axisymmetric flow, $\vec{\omega}$ is always directed in the local θ direction. Then it can be shown that

$$\vec{\omega} \cdot \nabla \vec{u} = \frac{\omega v}{\tilde{r}}\vec{e}_{\theta} = \frac{\omega}{\tilde{r}}\vec{e}_{\theta}\frac{D\tilde{r}}{Dt}, \tag{3.38}$$

where v is the radial component of velocity and $\tilde{r}$ is the radial coordinate in cylindrical coordinates. The cylindrical coordinate forms of the axisymmetric vorticity and stream-function equations are presented in Appendix A. In the inviscid limit, it follows from the preceding equations that $\vec{\omega}/\tilde{r}$ is constant along a particle path with a magnitude of $\omega/\tilde{r}$. With axisymmetry, the vorticity vector always points in the azimuthal direction. In the steady (or quasi-steady) liquid flow, the particle path is a closed streamline so that $\omega/\tilde{r}$ is a function of the stream function ψ. Note that the Prandtl number and the Schmidt number for a typical liquid are large compared with unity. Diffusions of heat and mass are therefore slow compared with diffusion of vorticity in the liquid. A quasi-steady hydrodynamic behavior is established in a short time compared with the transient time for heating or mixing. It is also assumed that the transient time is sufficiently short so that the hydrodynamics instantaneously adjusts to droplet-diameter changes resulting from vaporization. In our simplified models, quasi-steady hydrodynamic behavior is assumed and transient heating is allowed.

VORTEX MODEL. The Hill's spherical vortex (Lamb, 1945; Batchelor, 1990) is a well-known solution of Eqs. (3.37) and (3.38) in the inviscid limit that also satisfies matching interface conditions with an external potential flow. In this special case, $\omega/\tilde{r}$ has the same constant value for all values of the stream function. The spherical vortex solution provides a basis for the vortex models [type (v)] of droplet heating and vaporization. It has been shown that

$$\omega = 5\tilde{A}r\sin\theta = 5\tilde{A}\tilde{r}, \tag{3.39}$$

$$\psi = -\frac{1}{2}\tilde{A}r^2(R^2 - r^2)\sin^2\theta = -\frac{1}{2}\tilde{A}\tilde{r}^2[R^2 - (\tilde{r}^2 + z^2)]$$

$$= -\frac{3}{4}U\frac{\tilde{r}^2}{R^2}(R^2 - \tilde{r}^2 - z^2). \tag{3.40}$$

Note that U is the instantaneous relative velocity between the droplet and the ambient gas. The maximum potential-flow velocity at the interface is 3/2 times that value. We will not assume in our analysis that a potential flow exists immediately near the surface; rather, allowance is made for a viscous boundary layer. So we consider $\tilde{A}$ to be the liquid-vortex strength and relate it simply to the maximum velocity at the liquid surface, which can be an order of magnitude more or less (depending on density and viscosity) than the extrapolated potential-flow velocity value. In particular

$$\tilde{A} = U_{\text{max}}/R. \tag{3.41}$$

Therefore Eq. (3.40b) is neglected when the viscous boundary layer is considered. Rather, Eqs. (3.40a) and (3.42) are used.

It is convenient to define the nondimensional variables and $\phi = 8\psi/\tilde{A}R^4$ and $s = r/R$. Then

$$\phi = 1 - 4s^2(1 - s^2)\sin^2\theta. \tag{3.42}$$

Note that $\phi = 1$ at the interface ($s = 1$) and at the center of the internal wake ($\theta = 0$, $\theta = \pi$) and that $\phi = 0$ at the vortex center ($s = 1/\sqrt{2}$ and $\theta = \pi/2$).

The preceding solution has been established effectively as a high Reynolds number limit behavior, that is, the viscous term has been neglected in the derivation so that an inviscid solution for the internal liquid flow results. The low Reynolds number quasi-steady limiting behavior is given by the Hadamard–Rybczynski solution (Hadamard, 1911; Rybczynski, 1911; Lamb, 1945; Batchelor, 1990). There, the inertial term in both the surrounding gas and the interior liquid is neglected. The remaining linear system is solved by separation of variables. Their solution in the external phase differs significantly from the high Reynolds number solution. The most interesting result is that the Hill's spherical vortex solution given by Eqs. (3.39), (3.40a) or (3.42), and (3.41) applies for the Hadamard–Rybczynski low Reynolds number solution. This can be explained by the fact that the vector $\vec{\omega} = \omega\vec{e}_\theta = 5\tilde{A}r\vec{e}_\theta$ actually satisfies Laplace's equation, so that the viscous term in Eq. (3.37) goes to zero without assuming zero viscosity. (Note that the scalar ω is not harmonic, however.) Viscosity affects only the value of the constant $\tilde{A}$ through the interface matching process. Hill's spherical vortex solution actually applies over a wide range of Reynolds numbers; in the vorticity equation, both the inertial (nonlinear) terms and the diffusion (viscous) term are individually equal to zero. In the original momentum equation, the viscous term balances exactly the pressure-gradient term and the inertial (nonlinear) terms are identically zero. This fortuitous character is shared with Couette and Poiseuille flows. The low Reynolds number solution can provide the basis for an interesting perturbation on the spherically symmetric vaporization case, as discussed in Section 3.2. Other analytical papers in which droplet internal flows are discussed are those by Chao (1962), Harper and Moore (1968), and Prakash and Sirignano (1978, 1980).

The particular value of the vortex strength will not be important provided that (i) the circulation time in the droplet is shorter than other characteristic times, and (ii) the surface velocity is small compared with the relative gas–droplet velocity. (Both of these are reasonable assumptions in the practical high Reynolds number situations.) It is fortunate that the vortex strength value is of secondary interest because the error in its calculation can be significant because of the variable viscosity that is due to large temperature differences through the boundary layers in the liquid and in the gas film.

The axisymmetric form of the energy equation is presented with cylindrical coordinates in Appendix A. In a large Peclet number situation, heat and mass transport within the droplet will involve a strong convective transfer along the streamline with conduction primarily normal to the stream surface. In the limit of zero Peclet number, only conduction occurs. Over the full range of Peclet numbers, the heat and mass transport problems are axisymmetric and unsteady. With a certain coordinate transformation, the large Peclet number problem can be cast as a one-dimensional, unsteady problem.

In place of the cylindrical coordinates $\tilde{r}$ and z, we can use the stream function ψ (a measure of distance normal to the stream surface) and ξ (a measure of distance in the local-flow direction). The azimuthal coordinate η is maintained in the transformation. There is no variation in the η direction because of the axisymmetry and, for rapid circulation (high Peclet number), the variation in the ξ direction is negligible. The only variation therefore comes because of conduction in the ψ direction. Transforming the coordinates and averaging temperature along the stream surface, we find that the liquid-phase energy equation in the vortex models becomes

$$F(\psi)\frac{\partial T_l}{\partial t} = \alpha_l \frac{\partial}{\partial \psi}\left[G(\psi)\frac{\partial T_l}{\partial \psi}\right], \tag{3.43}$$

where the following definitions have been made for integrals over the closed fluid path:

$$F(\psi) = \oint \frac{h_\xi\, d\xi}{u_l} = \frac{8}{\tilde{A}R}g_1(\phi), \quad G(\psi) = \oint \frac{h_\eta h_\xi}{h_\psi}\, d\xi = \frac{\tilde{A}R^5}{8}g_2(\phi). \tag{3.44}$$

These relationships can also be regarded as definitions of $g_1(\phi)$ and $g_2(\phi)$. Note that h_η, h_ξ, and h_ψ are scale factors of the transformation and $u_l(\psi, \xi)$ is the local velocity that is tangential to the stream surface (Prakash and Sirignano, 1978, 1980). In particular, $h_\eta = r\sin\theta$, $h_\xi = \tilde{A}(r^5\cos^5\theta)/2u_l$, and $h_\psi = 1/(r\,u_l\sin\theta)$. The differential $h_\xi\, d\xi$ is an element of length along the streamline and $(h_\xi/u_l)d\xi$ is a differential Lagrangian time; the cyclic integral $F(\psi)$ is therefore the circulation time. It can be shown that

$$V(\psi) = 2\pi \int_0^\psi F(\psi')\, d\psi', \quad V(\phi, t) = 2\pi R^3(t)\int_0^\phi g_1(\phi')\, d\phi' \tag{3.45}$$

give the volume enclosed by the stream surface ψ. Note that here ψ is referenced to the vortex center where its value is set to zero, with the maximum value at the droplet surface. Using ϕ as defined by Eq. (3.43), we can state that

$$V(\psi) = V[R(t), \phi] = V(t, \phi), \quad \left.\frac{\partial \phi}{\partial t}\right)_\psi = -\frac{\left.\frac{\partial V}{\partial t}\right)_\phi}{\left.\frac{\partial V}{\partial \phi}\right)_t}. \tag{3.46}$$

It follows that

$$\left.\frac{\partial T_l}{\partial t}\right)_\psi = \left.\frac{\partial T_l}{\partial t}\right)_\phi - \frac{\left.\frac{\partial V}{\partial t}\right)_\phi}{\left.\frac{\partial V}{\partial \phi}\right)_t}\frac{\partial T_l}{\partial \phi}. \tag{3.47}$$

A combination of Eqs. (3.44), (3.45), and (3.47) leads to the following one-dimensional form of the diffusion equation:

$$\frac{\partial T_l}{\partial \tau} = a(\phi, \tau)\frac{\partial^2 T_l}{\partial \phi^2} + b(\phi, \tau)\frac{\partial T_l}{\partial \phi}, \tag{3.48a}$$

where $\tau = \alpha_l t/R_0^2$, $a(\phi, \tau) = (R_0/R)^2\ g_2(\phi)/g_1(\phi)$, and $b(\phi, \tau) = (R_0/R)^2 g_2'(\phi)/g_1(\phi) + (3/R)(dR/dt)\int_0^\phi g_1(\phi')d\phi'/g_1(\phi)$, where $g_2'(\phi)$ is the derivative and ϕ' is a dummy variable. Tong and Sirignano (1982b) demonstrated that convenient approximations to $g_1(\phi)$, $g_2(\phi)$, $g_2'(\phi)$ and therefore to $a(\phi, \tau)$ and $b(\phi, \tau)$ can be found. See Eqs. (8.1) and (8.2), but do note that the nondimensional time has been defined differently in those equations.

The boundary condition at the edge of the liquid core ($\phi = 1$) is based on a thin quasi-steady boundary layer existing in the liquid adjacent to the droplet interface. The heat flux at the droplet interface then equals the heat flux instantaneously at the edge of the liquid core. From Eqs. (3.24), (3.28), and (3.29) and from the rules of transformation and nondimensionalization, the boundary condition at $\phi = 1$ becomes

$$\left.\frac{\partial T_l}{\partial \phi}\right)_s = \frac{R}{4}\left.\frac{\partial T_l}{\partial r}\right)_s = \frac{\lambda}{\lambda_l}\frac{(T_\infty - T_s + \nu Q Y_{O\infty}/c_p)\mathrm{Nu}}{8} - \frac{\dot{m}L}{16\pi R\lambda_l}.$$

Here, Nusselt number Nu and vaporization rate $\dot{m}$ can be determined through either of equation pairs (3.29) and (3.30) or (3.31) and (3.32). Using Eqs. (3.29) and (3.30), we have, with $\mathrm{Pr} = 1$,

$$\left.\frac{\partial T_l}{\partial \phi}\right)_s = k[-f(0)]\frac{\mathrm{Re}^{1/2}}{8}\frac{\lambda}{\lambda_l}\left[\frac{(T_\infty - T_s + \nu Q Y_{O\infty}/c_p)}{B} - \frac{L}{c_p}\right]. \qquad (3.48\mathrm{b})$$

See the comment following Eqs. (2.16) about the determination of the transfer number in this boundary condition.

The boundary condition at the vortical center $\phi = 0$ is that a regular solution exists in spite of the fact that the coefficient a in Eq. (3.48a) is singular there. This boundary condition is used to discard any singular solution, keeping the first term on the right-hand side of Eq. (3.48a) at the value zero. The boundary condition on the finite-difference computation is that

$$\frac{\partial T_l}{\partial \tau} = b(0, \tau)\frac{\partial T_l}{\partial \phi} \qquad (3.48\mathrm{c})$$

at $\phi = 0$. Finally we specify typically the initial condition that

$$T_l(\phi, 0) = T_{l0}, \qquad (3.48\mathrm{d})$$

that is, a uniform initial temperature exists.

Certain interesting observations can be made even before the solution of the equation: (i) Eq. (3.48a) is a diffusion equation with variable coefficients. With no vaporization, the coefficients depend on only ϕ, but in general they vary with both τ and ϕ. (ii) The equation appears to be linear but, in actuality, a certain hidden nonlinearity is present. Namely, the regression rate $\mathrm{d}R/\mathrm{d}t$ will depend on surface temperature. Other nonlinearities that are due to the temperature dependence of transport and thermodynamic properties could appear in a more exact analysis. One can, however, still take advantage of the apparent linearity in the analysis. (iii) Because g_1 and g_2 are each proportional to vortex strength and because they or their derivatives appear in only ratio forms, the coefficients a and b are independent of vortex strength. This implies that there is no such phenomenon as a rapid-mixing limit. Diffusion equation (3.48a) does not change form as vortex strength (or, equivalently, circulation) increases and no limiting condition results. (iv) The characteristic length for diffusion is related to the shortest distance from the vortex center to the stream surface coincident with the edge of the liquid boundary layer. If the edge of the layer were approximated as the surface of the droplet (an infinitesimally thin layer), this distance would become $R(1 - 1/\sqrt{2}) \sim 0.3R$. The characteristic length is $\sim$30% of the characteristic length for liquid heating in the spherically symmetric case. Because characteristic time is proportional to the square of the distance, the heating time is less than 10% of the corresponding time for the spherically symmetric case. (v) In the limit of rapid regression rate (large dR/dt), the coefficient b becomes large, resulting in the formation of a quasi-steady layer at the edge of the core, whereby there is an approximate balance between the second-derivative term and the pseudoconvective term in Eq. (3.48a). The portion of the last term in that equation that contains the product of dR/dt and $(\partial T_l/\partial \phi)$ is named the pseudoconvective term because it becomes a convective term in a frame of reference fixed to the regressing droplet surface. The direction of the convection here is normal to the droplet surface. This boundary layer in this limit would amount to a relaxation zone that regresses into the interior of the droplet as rapid vaporization occurs; the core temperature would remain intact at the initial temperature until the relaxation zone arrives. This limit can occur only when the droplet lifetime is much shorter than the droplet-heating time and has not been observed in practical situations at typical combustor temperatures and with typical fuels. The analog of this situation will be possible in the case of liquid-phase mass diffusion in which a characteristic diffusion time can be much larger than the droplet lifetime. This point is discussed in Chapter 4. We would expect this relaxation zone appearing in this limit to overlay and to incorporate the boundary layer previously discussed as the relaxation zone between the two-dimensional gas-phase solution and the one-dimensional core solution.

When droplet slip is present, the droplet Nusselt number increases with the square root of the Reynolds number so that heat flux increases and droplet lifetime decreases. As mentioned in point (iv) above, the characteristic heating time is less than that value for the nonslipping case but the heating time is independent of vortex strength (and implicitly Reynolds number for a fixed viscosity) at large vortex strength (or large Reynolds number). We already know that, in the non-slipping, noncirculating case in a typical combustor environment with typical fuels, the droplet-heating time and droplet lifetime are comparable. Note that here both scale with initial droplet radius squared. Coincidentally, for Reynolds numbers of the order of 100, both times are again comparable, although reduced from their values at zero Reynolds number. The heating time still scales with initial radius squared but the lifetime no longer does; in a quasi-steady situation, the lifetime can be shown to be proportional to initial radius squared divided by the square root of Reynolds number (effectively a radius to the three-halves power dependence). At lower Reynolds numbers, we may expect the heating time to be shorter than the lifetime so that a uniform temperature approximation could become reasonable. At higher Reynolds numbers, lifetimes would become shorter and the rapid regression limit of preceding point (v) would be approached; however, in practice, the Weber number tends to become large at high Reynolds number and droplets disintegrate into smaller droplets with decreased Reynolds numbers.

Then Eq. (3.48a) can be solved with proper matching conditions at the interface, i.e., Eqs. (3.25), (3.26), and (3.28) plus the phase-equilibrium condition and continuity of temperature. A boundary condition at $\phi = 0$, the vortex center, prescribes that the heat flux goes to zero there.

The liquid-transient-heating phenomenon with internal circulation then involves unsteady heat conduction from $\phi = 1$ (the warm droplet surface and the warm axis of symmetry) toward $\phi = 0$ (the relatively cool vortex center). Temperature is a monotonically increasing function of ϕ with the gradient diminishing with time. The limit of uniform but time-varying temperature results as the liquid thermal diffusivity goes to infinity. Contrary to earlier beliefs by some investigators, the uniform-temperature limit does not result from infinitely rapid internal circulation. As previously shown, infinitely fast circulation or an infinite liquid Peclet number has the result that the finite temperature gradient becomes normal to the stream surfaces. Note that the averaging of the temperature over the stream surface eliminated the convection term from Eq. (3.48a); only conduction is represented therein. Furthermore, the transformation from ψ to ϕ modified the vaporizing-droplet problem from a moving-boundary problem to a fixed-boundary problem. The effect of the regressing interface appears in the coefficient of that diffusion equation.

It was shown by Tong and Sirignano (1982a, 1982b, 1983, 1986a, 1986b) and Sirignano (1993a) that liquid-phase heat diffusion equation (3.48a) for the vortex model and its counterpart mass diffusion equation can be simplified when the change in droplet radius that is due to vaporization occurs slowly compared with changes in liquid temperature. Under that assumption, the nonlinearities introduced by the coefficients in Eq. (3.48a) can be modified to give an approximate piecewise linear behavior for the equation. Equation (3.48a) is approximated by Eq. (8.1) with the simplified coefficients. A Green's function analysis reduces the equation to an integral form whereby a quadrature gives the liquid temperature at any point as a function of the surface heat flux. An integral equation results that relates surface temperature to surface heat flux. The Green's function (which is the kernel function in the integral equation) is obtained as an eigenvalue expansion. Tong and Sirignano (1986b) showed that the problem could be reduced to a system of ordinary differential equations, thereby improving computational efficiency at a given accuracy. See Section 8.1 for details.

EFFECTIVE-CONDUCTIVITY MODEL. An alternative approach [type (iv)] to the analysis of the liquid-phase heat diffusion was proposed by Abramzon and Sirignano (1989). Whereas the type (v) model reduces the dimensionality of the problem in an asymptotically correct manner (isotherms become identical with stream functions as the Peclet number becomes infinite), the type (iv) model reduces dimensionality in an ad hoc manner. In particular, an effective thermal diffusivity $\alpha_{l\,\text{eff}}$ was used wherein

$$\alpha_{l\,\text{eff}} = \chi \alpha_l, \qquad \chi = 1.86 + 0.86 \tanh[2.225 \log_{10}(\text{Pe}_l/30)]. \tag{3.49}$$

Equation (3.49) results from the fitting of numerical results by Johns and Beckmann (1966) for mass transfer between a droplet with internal circulation and a moving external immiscible liquid. The liquid-phase Peclet number Pe_l depends on liquid properties, droplet radius, and the maximum liquid velocity. Here, a spherically symmetric pseudotemperature field is solved with the diffusion equation,

$$\frac{\partial T_l}{\partial t} = \frac{\alpha_{l\,\text{eff}}}{r^2}\frac{\partial}{\partial r}\left(r^2\frac{\partial T_l}{\partial r}\right) = \alpha_{l\,\text{eff}}\left(\frac{\partial^2 T_l}{\partial r^2} + \frac{2}{r}\frac{\partial T_l}{\partial r}\right). \tag{3.50}$$

The use of the effective thermal diffusivity presents an accurate description of the characteristic heating time and thermal inertia of the liquid. By means of Eq. (3.49), the effective diffusivity monotonically increases with the maximum liquid velocity. Therefore it can vary with time. It is bounded below by the molecular diffusivity and above by 2.72 times that finite value. It is expected therefore that the surface-temperature history will be given with only a small error by this ad hoc analysis. The details of the internal temperature field can be grossly in error. For example, the pseudotemperature will be a minimum at the center of the droplet whereas the actual temperature is a minimum at the vortical center.

Abramzon and Sirignano recast Eq. (3.50) in terms of certain nondimensional variables. Defining

$$\zeta = \frac{r}{R(t)}, \qquad \tau = \frac{\alpha_l t}{R_0^2}, \qquad r_s(\tau) = \frac{R(t)}{R_0},$$

$$z = \frac{T_l - T_{l0}}{T_{l0}}, \qquad \beta = \frac{1}{2}\frac{d}{d\tau}\left(r_s^2\right),$$

we obtain

$$r_s^2\frac{\partial z}{\partial \tau} - \beta\zeta\frac{\partial z}{\partial \zeta} = \frac{\chi}{\zeta^2}\frac{\partial}{\partial \zeta}\left(\zeta^2\frac{\partial z}{\partial \zeta}\right). \tag{3.51}$$

This form of the equation offers the advantage of a fixed boundary even as the actual droplet surface regresses. Note of course that χ will vary with time according to Eq. (3.49). The boundary condition at the droplet–gas interface for partial differential equation (3.51) is obtained from a combination of Eqs. (2.6) with the definitions of Eqs. (2.14a), (2.14b), (2.14c), and (2.15) and with the just mentioned coordinate transformation. In particular, we obtain at $\zeta = 1$

$$\frac{\partial z}{\partial \zeta}(1, \tau) = \frac{\lambda}{\lambda_l}\left(\frac{T_\infty - T_s}{2T_{l0}}\right)\text{Nu} - \frac{\dot{m}L}{4\pi R\lambda_l T_{l0}}.$$

Using Eqs. (3.34), we have

$$\frac{\partial z}{\partial \zeta}(1, \tau) = \ln(1 + B_H)\left[1 + \frac{k}{2}\frac{\text{Pr}^{1/3}\text{Re}^{1/2}}{F(B_H)}\right]\frac{\lambda}{\lambda_l}\left(\frac{T_\infty - T_s}{T_0 B_H} - \frac{L}{c_p T_0}\right). \tag{3.52a}$$

To avoid an implicit relationship for the temperature gradient at the liquid side of the surface, do not calculate B_H directly from Eqs. (3.33). Rather, calculate B_M from Eqs. (3.33) and then calculate B_H from Eqs. (3.35). See the discussion following Eq. (3.36) about the modification of the Reynolds-number-dependent term in Eq. (3.52a) for low values of that parameter. The condition at $\zeta = 0$ reflects the symmetry in the profile of temperature; namely,

$$\frac{\partial z}{\partial \zeta}(0, \tau) = 0. \tag{3.52b}$$

The determination of the coefficient $\beta(\tau)$ in Eq. (3.51) uses the droplet-volume–mass relationship $m = 4\pi \rho_l R^3/3$ and the previously mentioned nondimensionalized scheme. We obtain

$$\beta = -\dot{m}/4\pi \rho_l \alpha_l R,$$

where $\dot{m}$ is positive during vaporization and β is negative. Typically, droplet initial temperature is taken to be uniform at the value T_0. Use of Eqs. (3.34c) yields

$$\beta = -\frac{\lambda}{\lambda_l}\frac{c_l}{c_{pF}}\ln(1 + B_H)\left[1 + \frac{k}{2}\frac{\text{Pr}^{1/3}\text{Re}^{1/2}}{F(B_H)}\right]. \tag{3.52c}$$

The Peclet number in Eq. (3.49) is based on the maximum velocity $U_{\max}$ at the droplet surface, the droplet diameter, and liquid properties; that is,

$$\text{Pe}_l = \frac{2U_{\max}R\rho_l c_l}{\lambda_l} \tag{3.52d}$$

Hill's spherical vortex solution (3.40) can be used to determine the maximum surface velocity. The tangential velocity is continuous across the gas–droplet interface. If the external flow were a potential flow, the maximum surface velocity would be 3/2 times the relative droplet–gas velocity. Moreover, the viscous effects that reduce the gas velocity at the interface have been used in this derivation. The friction on the gas side of the surface is assumed to act only on the front half of the sphere. The total friction on the gas side is equated to the total friction on the entire surface for the liquid side. Then with a friction force F at the interface, we use the definition for the friction coefficient C_F that

$$F = \left(2\pi R^2\right)\frac{1}{2}C_F\rho(\Delta U)^2, \tag{3.52e}$$

where ΔU is the relative droplet–gas velocity. We are considering that

$$F = 2\pi R^2 \int_0^{\pi/2} \tau_{g,s}\sin^2\theta \, d\theta.$$

Also

$$F = 4\pi R^2 \int_0^{\pi/2} \tau_{l,s}\sin^2\theta \, d\theta$$

$$= 4\pi R^2 \int_0^{\pi/2} \mu_l\frac{\partial u_l}{\partial r}\sin^2\theta \, d\theta. \tag{3.52f}$$

Note that we are not stating that the shear stress $\tau_{g,s} = \tau_{l,s}$. There is actually a thin liquid boundary layer between the liquid-vortex core and the gas boundary layer. The force on the gas–liquid interface is set equal to the force on the edge of the liquid core. The thin liquid boundary layer is approximated to be subject to the friction force from the gas over a surface area $2\pi R^2$ but to transmit the force to the liquid core over a surface area $4\pi R^2$.

One sine term appears in the integral because of surface-area weighting, and the other multiplying sine term appears because only the axial component of the local surface friction force contributes to the global force. The velocity gradient $(\partial u_l/\partial r)_s$

is determined from Eqs. (3.40a) and (3.41). Combining the result with Eqs. (3.52e) and (3.52f), we obtain that the maximum velocity

$$U_{\max} = \frac{1}{6\pi} \frac{\mu_g}{\mu_l} C_F \mathrm{Re} \Delta U. \qquad (3.52\mathrm{g})$$

Here the Reynolds number is based on the droplet diameter. Based on the reports of Clift et al. (1978) and Renksizbulut and Yuen (1983), the following friction coefficient correlation for a vaporizing droplet is recommended:

$$C_F = 12.69 \mathrm{Re}^{-2/3} (1 + B_M)^{-1}. \qquad (3.52\mathrm{h})$$

Now the time-dependent effective diffusivity described by Eq. (3.49) can be determined. Note that Eq. (3.52g) contains an improvement in the calculation of the coefficient compared with that of Abramzon and Sirignano, who used 1/32 instead of $1/6\pi$.

Abramzon and Sirignano demonstrated that their simplified model that reduced the dimensionality of the problem still gave an accurate prediction of the droplet vaporization when compared with the axisymmetric calculations.

In the case of a multicomponent liquid fuel, mass diffusion in the liquid phase becomes important. As the more volatile substance is vaporized faster from the surface, more of that substance will diffuse from the interior of the droplet to the surface to vaporize. For n liquid components, $n-1$ liquid-phase mass diffusion equations must be solved for $n-1$ mass fractions; the other mass fraction can immediately be deduced because the mass fractions sum to unity. The mass diffusion equations can be placed in a form equivalent to that of Eq. (3.48a). These equations have been solved by approximate and exact methods (Lara-Urbaneja and Sirignano, 1981; Tong and Sirignano, 1986a, 1986b; Continillo and Sirignano, 1988, 1991; and Megaridis and Sirignano, 1991, 1992a). The problem is especially interesting and challenging because the liquid mass diffusivity is typically an order of magnitude smaller than the liquid thermal diffusivity so that a new time scale and a greater degree of stiffness are created. See Chapter 4 for more discussion on this topic.

3.1.5 Droplet Drag Coefficients

The drag coefficient C_D for the droplet is determined from the aerodynamic force and a dynamic pressure based on a relative velocity. The drag force D is given by $D = C_D \frac{1}{2} \rho (\Delta U)^2 \pi R^2$.

Yuen and Chen (1976) found that the droplet drag coefficient is close to that for a solid sphere of the same diameter. An approximate balance appears to exist between the decrease on friction drag and the increase in pressure drag that is due to vaporization. They show, however, that the appropriate Reynolds number (to be used in extracting the drag coefficient from the standard curve for a sphere) is based on a certain 1/3 averaging rule for weighting the free stream and the surface viscosity. In particular,

$$\mu_r = \mu_s + \frac{1}{3} (\mu_\infty - \mu_s), \qquad (3.53)$$

where r, s, and ∞ indicate reference (averaged), surface, and ambient values, respectively. Note that if the most intense vaporization were on the aft surface of the droplet it is plausible that pressure drag would decrease and the conclusions of Yuen and Chen would not apply. This could occur with droplets that have wake flames.

Equation (3.53) weighs the surface viscosity heavier than the free-stream viscosity. For a droplet in a hot-gas environment, the weighted Reynolds number is higher and the weighted drag coefficient is lower than the quantities based on the free-stream values. For a hot droplet (e.g., molten metal) in a cold gas, the weighted Reynolds number is lower and the weighted drag coefficient is higher.

The general results of Yuen and Chen were confirmed in the experiments of Lerner et al. (1980), who used the averaging principle for both density and viscosity. They also showed that the drag law could be applied to ellipsoidal droplets if the effective diameter D is taken as $(a^2 b)^{1/3}$, where a and b are the major and the minor axes, respectively, and their values are within 10% or so of each other. Note that the Reynolds number based on the averaged reference values can be several times greater than a Reynolds number based on free-stream properties, increasing the likelihood of a high Reynolds number boundary-layer phenomenon.

The more recent experimental results are in disagreement with the results of Ingebo (1967), which were taken from spray measurements and therefore were marred by entrainment effects. Namely, an average droplet in a group of droplets will penetrate further into the gas than a single isolated droplet of the same size and initial velocity; the group of droplets will cause a significant motion of the gas, affecting the relative gas–droplet velocity. It follows that the increased penetration of the droplets in the spray would give the appearance of a reduced drag coefficient when actually it is the reduction in the relative velocity that decreases the drag. The Yuen and Chen drag coefficients are favored over the Ingebo results because it is desirable to have a drag coefficient that relates to an individual droplet rather than to the global spray character. The entrainment effects would automatically be considered if the appropriate momentum balance were performed on both the gas and the liquid phases. It must be cautioned, however, that, for dense spray regions, the effect of droplet spacing on the drag coefficient should be taken into account.

Renksizbulut and Yuen (1983) concluded that the Stefan convection reduces the drag coefficient by a factor of $1 + B_M$. At a low Reynolds number,

$$C_D = \frac{24}{\mathrm{Re}_m (1 + B_M)} \tag{3.54}$$

is recommended. Here the Reynolds number Re_m should be based on the instantaneous values of the average gas-film viscosity given by Eq. (3.53), the free-stream density, the relative velocity, and the droplet diameter. For a Reynolds number equal to 30 and above, see the correlations obtained from Navier–Stokes solutions and given by Eqs. (3.55) in Subsection 3.1.7.

3.1.6 Results from Approximate Analyses

The solution to Eqs. (3.48) [and to Eq. (3.2), which is similar] was obtained by finite-difference methods (Lara-Urbaneja and Sirignano, 1981), and more recently

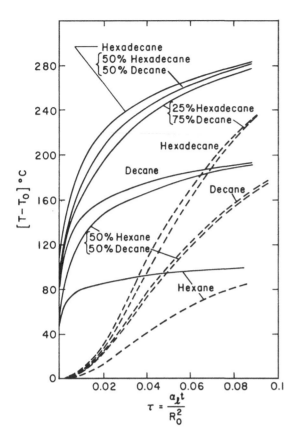

Figure 3.2. Liquid temperature versus time for axisymmetric vaporization and a variety of fuels: $T_0 = 300$ K, 10 atm. The solid curves give surface temperatures at $\theta = 90°$; the dashed curves give vortex-center temperatures. (Sirignano, 1983, with permission of *Progress in Energy and Combustion Science.*)

approximate analytical solutions (Tong and Sirignano, 1982b) were developed. The liquid-phase solutions are matched at every instant of time with the gas-phase solution. In addition to the integral methods (Prakash and Sirignano, 1978, 1980; Lara-Urbaneja and Sirignano, 1981) for the gas-phase solution, other more expedient approximate solutions (Tong and Sirignano, 1983) have been suggested.

Figure 3.2 from Lara-Urbaneja and Sirignano (1981) shows the behavior of the surface temperature at 90° off the axis of symmetry and of the vortex-center temperature for three different fuels of varying volatility. A nondimensional time of 0.1 is approximately the droplet lifetime. It is seen that the transient period covers the droplet lifetime (especially for the least volatile fuels) because the temperature is continually rising. The surface temperature exceeds the center temperature so that spatial gradients are important. The boiling points are never reached; the maximum temperatures occur at the surface near the end of the lifetime and still are 80 to 100 °C below the boiling point (in these cases with 10 atm of pressure). Absolutely no justification can be found for a quasi-steady assumption for the liquid core.

Prakash and Sirignano resolved the gas-phase boundary layer on the droplet and the liquid boundary layer by a two-dimensional quasi-steady treatment. The liquid vortical core was treated by the one-dimensional, unsteady model of Eq. (3.48a). Tong and Sirignano showed that similarity analysis for the gas-phase boundary layer could be used and that the temperature change through the liquid-phase boundary layer was not significant so that the liquid layer structure need not be resolved.

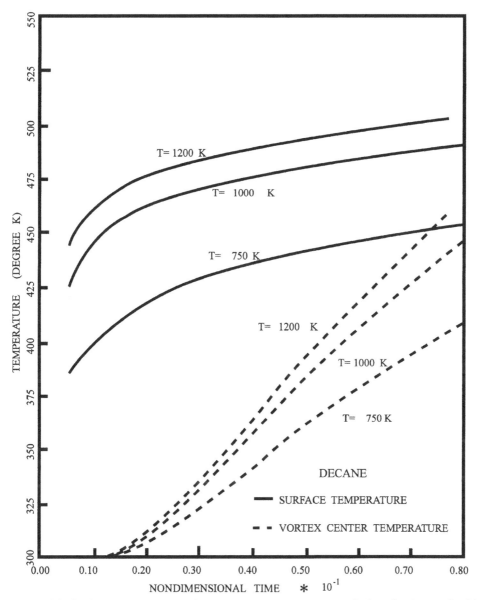

Figure 3.3. Surface-temperature and vortex-center-temperature variations for decane liquid droplets at droplet Re = 100 and various ambient temperatures.

The high Reynolds number, quasi-steady gas-phase boundary-layer analysis coupled with the model of heat diffusion in the internal vortical flow gives a reasonable representation of the quantitative behavior of the velocity and thermal fields in the gas boundary layer and in the liquid. Figure 3.3 shows results of Tong and Sirignano for surface temperature versus time and for temperature at the vortical center versus time. Spatial variations in the liquid temperature are seen to occur. The surface temperature gradually increases with time toward the wet-bulb temperature whereas the lower vortical-center temperature increases at a faster rate, thereby decreasing the spacial variance.

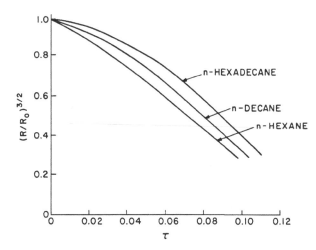

Figure 3.4. Size variation versus time for three fuels with axisymmetric vaporization. (Prakash and Sirignano, 1980, with permission of the *International Journal of Heat and Mass Transfer.*)

Figure 3.4 from Prakash and Sirignano (1980) shows radius to the three-halves power plotted versus time for three different fuels. The curvature or deviation from a linear relationship is a result of the transient behavior. With time, the behavior becomes closer to linear for all three fuels, especially the more volatile *n*-hexane. Figure 3.4 demonstrates that the less volatile fuels will undergo the more pronounced transient behavior but even the normal hexane fuel, which is extremely volatile, will have a transient behavior for much of its lifetime. Figure 3.5 from Tong and Sirignano shows that the rate of change of the radius (which is related to the vaporization) is slower at first and later increases, giving a clear indication of important transient effects.

The gas-film model coupled with the effective liquid conductivity allows for coverage of a much wider range of Reynolds number. This ability makes the model of Abramzon and Sirignano more practical for use in a spray calculation, in which even the droplets that begin with a large relative velocity and large Reynolds number witness a deceleration that eventually gives them lower values of the relative velocity and Reynolds number. Figures 3.6 and 3.7 show typical variations of the gas-phase Reynolds number, liquid-phase Reynolds number, and liquid-phase Peclet number with time. The Peclet number strongly influences the effective conductivity through Eq. (3.49); it is seen that it can vary over several orders of magnitude in the droplet lifetime. The high Peclet number assumption that leads to the establishment of Eqs. (3.44) and (3.48a) and the high gas-phase Reynolds number assumption that allows for thin boundary layers are not valid over the complete lifetime of the droplet.

Figures 3.8–3.10 compare droplet radius, surface temperature, and vaporization-rate variations during the droplet lifetime for several liquid-phase models. The most accurate model is the extended model, which yields the solution for the liquid-phase axisymmetric energy equation with the velocity field determined by Hill's spherical vortex solution. See Abramzon and Sirignano (1989) for details. The effective-conductivity model is seen to agree very well in terms of these results, which are sufficient to give a useful coupling with the gas phase in spray calculations. The Abramzon and Sirignano model does provide a uniformly good

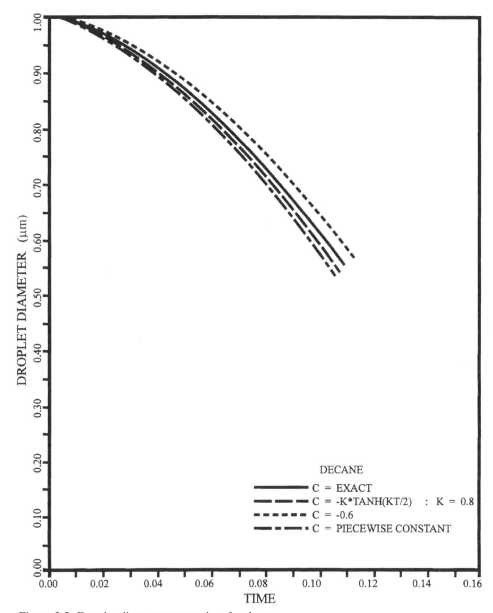

Figure 3.5. Droplet diameter versus time for decane.

description of the internal liquid temperature; it is better at very small Peclet numbers at which its spherically symmetric structure is closer to the real situation. Figures 3.8–3.10 show that the effective-thermal-conductivity model results and the actual thermal-conductivity model results are significantly different, implying that internal liquid convection is very important. The infinite-conductivity model also produces very different results. It is noteworthy that small differences in the droplet-radius and the surface-temperature curves of Figs 3.8 and 3.9 relate to much larger differences in vaporization rate, as shown in Fig. 3.10. Talley and Yao (1986) developed a semiempirical effective-conductivity model that produced good agreement

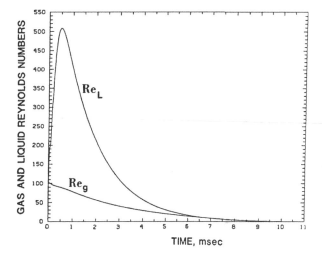

Figure 3.6. Gas-phase Reynolds number Re_g and liquid-phase Reynolds number Re_L versus time. (Abramzon and Sirignano, 1989, with permission of the *International Journal of Heat and Mass Transfer.*)

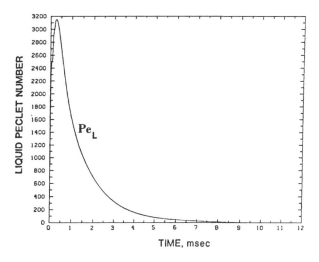

Figure 3.7. Liquid-phase Peclet number versus time. (Abramzon and Sirignano, 1989, with permission of the *International Journal of Heat and Mass Transfer.*)

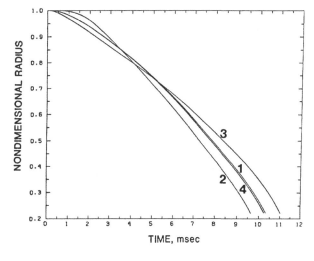

Figure 3.8. Nondimensional droplet radius versus time: extended model (curve 1), infinite-conductivity model (curve 2), finite-conductivity model (curve 3), and effective-conductivity model (curve 4). (Abramzon and Sirignano, 1989, with permission of the *International Journal of Heat and Mass Transfer.*)

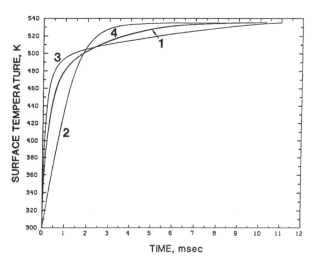

Figure 3.9. Surface temperature (in degrees Kelvin) versus time: various models. (Abramzon and Sirignano, 1989, with permission of the *International Journal of Heat and Mass Transfer*.)

with experiments. Furthermore, it predicted faster liquid-phase mixing than the Tong and Sirignano model.

None of these approximate models addresses the separated elliptic-flow region. Typically, heat and mass transfer are substantially reduced in that region for the Reynolds number range of interest, so that the droplet-heating rate and the droplet-vaporization rate can still be predicted satisfactorily. The pressure or form drag is a major portion of the total droplet drag at velocities above the creeping flow range. Therefore drag coefficients cannot be predicted well by these models. Either correlations for the drag coefficients obtained from experiment or correlations obtained from finite-difference computations can be utilized in principle as inputs to these approximate models. Experiments have generally yielded only correlations that do not account for vaporization (blowing); it is known that solid-sphere data, for example, result in overprediction of the drag if used for vaporizing droplets. One exception is the experimental work of Renksizbulut and Yuen (1983), in which the influence of the transfer number was determined. In Subsection 3.1.7, computational solutions

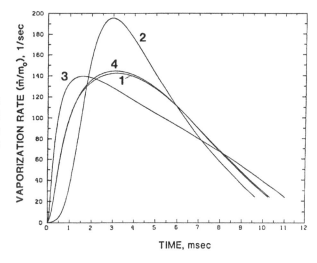

Figure 3.10. Vaporization rate versus time: various models. (Abramzon and Sirignano, 1989, with permission of the *International Journal of Heat and Mass Transfer*.)

of the Navier–Stokes equation and derived correlations are discussed. The correlation for the drag coefficient can be used in the simplified model, although the first uses of these models preceded the availability of the correlation.

Some final comments about the validity of the rapid-mixing approximation are in order. The rapid-mixing concept is somewhat flawed; internal circulation does not reduce mixing time to a negligible amount. Through a change in characteristic length scale, internal circulation can reduce the mixing time by 1 order of magnitude at large vortex strength. However, that limit remains fixed as vortex strength increases so that the time never goes to zero even as strength becomes infinite. Because vortex strength is related to the degree of slip, Nusselt number and vortex strength increase together for a given droplet. So lifetime and mixing (or heating) time decrease together. Except for a small (but not zero or very small) Reynolds number, uniform liquid temperatures will not result for droplets vaporizing in a combustor. Conceptually, it is not correct to relate the uniform temperature limit to a rapid mixing that is due to internal circulation; it would be correct conceptually (but not practical for combustion applications) to relate the uniform-temperature limit to either the limit of infinite diffusivity or the limit of very slow vaporization (perhaps infinite latent heat of vaporization).

3.1.7 Exact Analyses for Gas-Phase and Liquid-Phase Flows

Exact analyses for gaseous flows over liquid droplets and internal liquid flows have been performed by several investigators. These Navier–Stokes computations resolve the inviscid flow outside of the boundary layer and wake, as well as the viscous, thermal, and diffusive layers, the recirculating near wake, and part of the far wake. The shear-driven internal liquid circulation and transient heating are also resolved. Therefore elliptic-flow regions as well as the hyperbolic and the parabolic regions are analyzed in detail. In addition to the global and the local heat and mass exchange between the gas and the liquid (which is primarily determined by the solution of the parabolic-flow regions), these computations yield the drag force on the droplet (which requires resolution of the elliptic near wake as well as the other flow regions).

Exact calculations of the flow around and within vaporizing droplets serve several purposes. First, they provide detailed insight into the phenomena of heat, mass, and momentum transport for the droplet field. Second, these calculations provide a basis for comparison and verification of the simplified models that can be used in spray calculations. Finally, the calculations can be made for a range of parameters, yielding correlations for lift coefficient, drag coefficient, Nusselt number, Sherwood number, and other similarity parameters that can be used in the simplified models.

The calculations for a vaporizing droplet that is moving through a gas require the solution of a complex set of nonlinear coupled partial differential equations. The problem is inherently unsteady because droplet size is continually changing because of vaporization, relative velocity is also changing because of droplet drag, and temperatures are varying on account of droplet heating. The problem can be considered to be axisymmetric for a spherical droplet. The primitive forms of the axisymmetric equations are provided in Appendix A. Because the boundary is moving because of droplet vaporization, adaptive gridding is required for finite-difference computations.

Generally, implicit finite-difference techniques are used and the gas-phase primitive variables (velocity components, temperature, pressure, and mass fractions) are calculated directly, without transformation to other variables. The axisymmetric, unsteady form of the governing equations are solved with stiff upstream boundary conditions and zero-derivative downstream boundary conditions. The liquid and the gas flows are coupled at the spherical droplet surface by conditions of continuity on temperature, species, and global normal mass fluxes and tangential shear force and balance of normal momentum and normal heat flux. The stream-function-vorticity method is typically used for the incompressible liquid. Chiang (1990) provides details of the most recent numerical methodology with adaptive nonuniform numerical grids.

Conner and Elghobashi (1987) considered the Navier–Stokes solution for laminar flow past a solid sphere with surface mass transfer. Dwyer and Sanders (1984a, 1984b, 1984c) performed finite-difference calculations, assuming constant properties and constant density. Patnaik et al. (1986) relaxed the density assumption in their calculations but considered other properties to be constant. These calculations were made for a hydrocarbon fuel droplet vaporizing in high-temperature air so that the heating and the vaporization were highly transient. Haywood and Renksizbulut (1986), Renksizbulut and Haywood (1988), and Haywood et al. (1989) solved the problem of a fuel droplet's vaporizing into a fuel-vapor environment at moderate temperature and the problem of a fuel droplet's vaporizing in air at 800 K and 1 atm of pressure. They considered variable properties and variable density. Chiang et al. (1992) extended the computational theory to high-temperature and high-pressure air environments with fuel droplets, allowing for variable properties and variable density with multicomponent gaseous mixtures. They showed by comparison that the constant-property calculations of Raju and Sirignano (1990b) for two droplets in tandem and Patnaik et al. (1986) could overpredict drag coefficients by as much as 20%. Of course, appropriate averaging of the properties (between free-stream values and droplet-interface values) to determine the constant property for the calculation could reduce the error. They also calculated the deceleration of the droplet, accounting for the noninertial frame of reference. Dandy and Dwyer (1990) considered a nonuniform three-dimensional flow past a sphere with generation of a lift force.

The results of Chiang (1990) and Chiang et al. (1992) are shown in Figs. 3.11–3.17. Figures 3.11 and 3.12 show the instantaneous gas-phase velocity field and the liquid-phase streamlines, respectively, at a time when the Reynolds number is 23.88, based on droplet radius. The decrease in relative droplet–gas velocity that is due to drag and the decrease in droplet radius that is due to vaporization imply that the droplet Reynolds number is decreasing with time. Features such as the near-wake separation and recirculation and the internal liquid circulation are clearly seen. Figure 3.13 shows the liquid-phase isotherms at three points in time as the droplet decelerates from an initial Reynolds number of 100. High Peclet number behavior dominates in the early period, but, later, conduction in the streamwise direction competes with convection. Figures 3.14 and 3.15 provide typical results for the drag

GAS-PHASE VELOCITY VECTORS

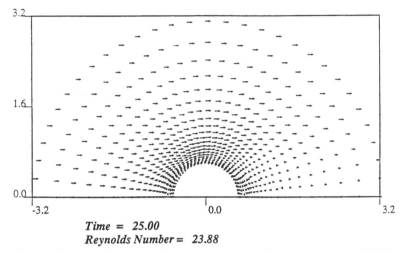

Time = 25.00
Reynolds Number = 23.88

Figure 3.11. Gas-phase velocity vectors at a nondimensional time of 25.0 and a Reynolds number of 23.88. (Chiang et al., 1992, with permission of the *International Journal of Heat and Mass Transfer.*)

coefficient and the Nusselt number. Figure 3.16 shows the drag coefficient as a function of instantaneous Reynolds number, which is decreasing with time. The drag coefficient is not monotonically decreasing with increasing Reynolds number because of the dependence on the transfer number. There is a weak sensitivity to the initial liquid temperature but a strong sensitivity to the model for droplet heating and vaporization. These nondimensional numbers are reduced substantially below the values for nonvaporizing spheres because of the blowing effect in the boundary

LIQUID-PHASE STREAM FUNCTION

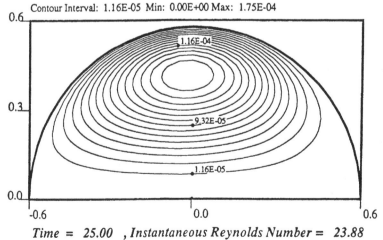

Contour Interval: 1.16E-05 Min: 0.00E+00 Max: 1.75E-04

Time = 25.00 , Instantaneous Reynolds Number = 23.88

Figure 3.12. Liquid-phase stream-function contours at a nondimensional time of 25.0 and a Reynolds number of 23.88. (Chiang et al., 1992, with permission of the *International Journal of Heat and Mass Transfer.*)

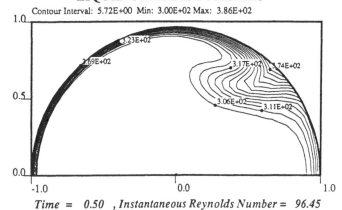

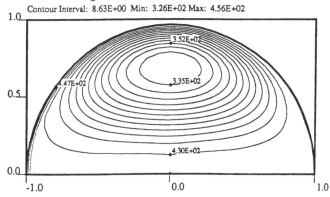

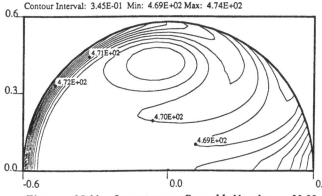

Figure 3.13. Transient history of droplet heating. Liquid-phase isotherms at three nondimensional times and corresponding instantaneous Reynolds numbers: (a) time = 0.50 and Re = 96.45, (b) time = 5.00 and Re = 76.06, (c) time = 25.00 and Re = 23.88. (Chiang et al., 1992, with permission of the *International Journal of Heat and Mass Transfer.*)

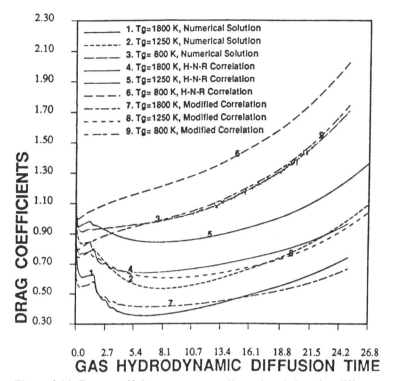

Figure 3.14. Drag coefficients versus nondimensional time for different ambient temperatures: numerical results and correlations. (Chiang et al., 1992, with permission of the *International Journal of Heat and Mass Transfer.*)

layer. The contributions of the friction drag, pressure drag, and thrust (which is due to vaporization) are compared with each other in Figure 3.17. Clearly friction drag is affected the most by ambient temperature and heating rate through the modification of the vaporization (blowing) rate. The impact of blowing rate (through the ambient-temperature variation), displayed by Figs. 3.14 and 3.15, is observed to be very strong.

Correlations of the numerical results were obtained by Chiang and Sirignano (1993a), relating these nondimensional groupings to the instantaneous Reynolds number and transfer number B. They are

$$C_D = (1 + B_H)^{-0.27} \frac{24.432}{\text{Re}_m^{0.721}}, \qquad \text{Nu} = 1.275(1 + B_H)^{-0.678}\text{Re}_m^{0.438}\text{Pr}_m^{0.619},$$

$$\text{Sh} = 1.224(1 + B_M)^{-0.568}\text{Re}_m^{0.385}\text{Sc}_m^{0.492}, \tag{3.55}$$

where Re_m, Pr_m, and Sc_m are based on average gas-film values (except for the use of free-stream density in Re_m) and varied from 30 to 200, 0.7 to 1.0, and 0.4 to 2.2, respectively. Also, B_H and B_M are given by Eqs. (3.33) and cover the ranges from 0.4 to 13 and from 0.2 to 6.5, respectively. These correlations are shown by Figs. 3.14 and 3.15 to fit the high-temperature multicomponent-gas situation much better than the previously developed (H–N–R) correlations of Haywood et al. (1989). The correlations of Chiang et al. can be used for gas-film model development without

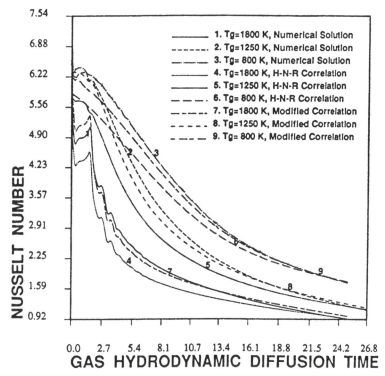

Figure 3.15. Nusselt number versus nondimensional time for different ambient temperatures: numerical results and correlations. (Chiang et al., 1992, with permission of the *International Journal of Heat and Mass Transfer*.)

use of any of the models previously discussed. Note that vaporization rate can be readily related to the Sherwood number. Model development for the liquid phase is still necessary because many parameters in the correlations introduce surface values. An alternative approach would be to use only the drag coefficient correlation with the approximate model. Under the implied condition of quasi-steady behavior in the gas film, the correlations describe the exchanges of mass, momentum, and energy between the two phases. The transient heating of the liquid phase must be analyzed in the simplified model to obtain the surface temperature and the liquid-heating rate for input to the transfer number that appears in the correlations.

From Eqs. (2.14), (3.25), and (3.55), the following relationship can be constructed:

$$\frac{\dot{m}}{4\pi\rho DR} = \frac{\text{Sh}\,B_M}{2} = 0.612 B_M (1 + B_M)^{-0.568} \text{Re}_m^{0.385} \text{Sc}_m^{0.492}. \tag{3.56}$$

Correlations (3.55) and (3.56) can be used for a gas-film model. They can, for example, be combined with a type (ii), (iv), or (v) liquid-core-heating model to predict droplet heating, vaporization, and motion. At this point, the correlations have not been used for this purpose. A comparison of Eqs. (3.55) and (3.56) with the model given by Eqs. (3.34) and (3.36) shows some qualitative similarities. A quantitative comparison has not been made.

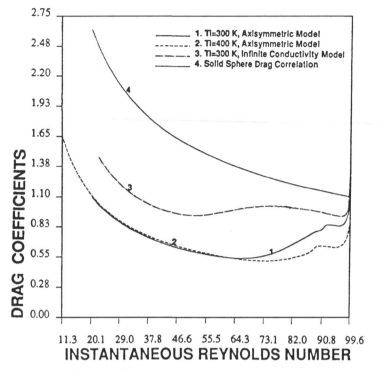

Figure 3.16. Drag coefficient versus instantaneous Reynolds number: different models and initial liquid temperature. (Chiang et al., 1992, with permission of the *International Journal of Heat and Mass Transfer*.)

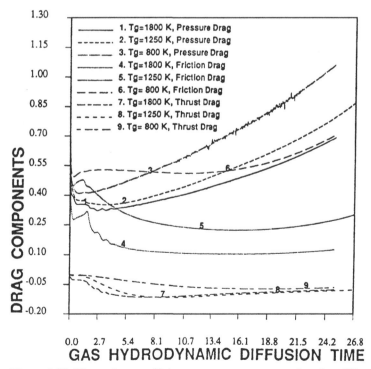

Figure 3.17. Three drag coefficient components versus time for different ambient temperatures. (Chiang et al., 1992, with permission of the *International Journal of Heat and Mass Transfer*.)

   In some recent calculations for droplets near the critical temperature, liquid-density variations are taken into account. Specifically, Chiang and Sirignano (1991) replaced the liquid-phase stream function and vorticity calculation with a velocity and pressure calculation. In current research, a cubic equation of state is applied for the liquid and the gas. Also, dissolving and diffusion of the ambient-gas components in the liquid are considered near the critical point.

SURFACE-TENSION EFFECTS. A number of investigators have examined the effect of surface-tension variation along the droplet interface. The temperature dependence of the surface tension results in an augmentation of the shear stress experienced by the liquid. A general matching condition at the liquid–gas interface is given by

$$\mu_g \frac{\partial u}{\partial y}\bigg)_{g,s} = \mu_l \frac{\partial u}{\partial y}\bigg)_{l,s} - \frac{\partial \sigma}{\partial x},$$

where the y direction is normal locally to the interface and positive in the outward direction whereas u and x are locally tangent to the interface. In previous discussions, the surface tension σ did not vary with the position x along the surface so that the shear stress was continuous across the interface. Niazmand et al. (1994), Shih and Megaridis (1996), Dwyer et al. (1996), Masoudi and Sirignano (1998), and Dwyer et al. (1998) considered the variation of surface tension and its effect. Axisymmetric and three-dimensional calculations were performed. Both pure vaporization and burning cases were considered. The calculations without a flame endured too short a portion of the droplet lifetime to show any conclusive global effects of surface tension. However, Dwyer et al. (1998) found profound effects when a flame was established in the near wake of the droplet. Surface tension became lower on the aft side so that the liquid could be pulled forward near the surface. Also, the expansion of gases from the flame could cause a reversal in the direction of the droplet drag so that it could become a thrust.

3.1.8 Free Convection

In most applications involving droplet heating and vaporization, the sizes of the droplets are sufficiently small that buoyancy effects in the gas film surrounding the droplets are not significant. The relative importance of buoyancy on the vaporization rate increases as $R^{3/4}$, compared with the gravity-free spherically symmetric case, and as $R^{1/4}$, compared with the large Reynolds number forced-flow case. Therefore situations can occur in which buoyancy is important and must be considered. Later in this subsection, these situations are depicted by large values of the Grashof number. One set of examples is laboratory studies with very large droplets that allow enhanced resolution.

   Some noteworthy studies were performed by Gollahalli (1977), Potter and Riley (1980), Fernandez-Pello and Law (1982a,1982b), Wu et al. (1982), and Fernandez–Pello (1983). Gollahalli evaluated buoyancy effects on the wake-flame structure of a burning droplet in a buoyant condition. Potter and Wiley performed a numerical analysis of the buoyant gas film surrounding a vaporizing-fuel sphere. They considered a thin-flame sheet, unitary Lewis number, a constant temperature for the condensed phase, no internal circulation, phase equilibrium governed by the Clausius–Clapeyron relation [Eq. (2.4a)], a high Reynolds number gas boundary layer, and constant density with the Boussinesq approximation for the gravity force. The Grashof number was identified as an important nondimensional grouping:

$$\mathrm{Gr} = \frac{g R^3}{\nu^2}.$$

Note that for an $O(100\text{-}\mu\mathrm{m})$ radius droplet in air at earth gravity, the Grashof number will vary between $O(10^{-1})$ and $O(10)$, depending on the air pressure and temperature. In this range, buoyancy is typically not of great significance. However, a droplet radius that is $O(1 \text{ mm})$ results in a 3-orders-of-magnitude increase in the Grashof number; then significant buoyancy effects can occur.

The Potter and Riley numerical solution of the boundary-layer equations did not result in a separated flow and a large wake. The mass transfer rate was proportional to the factor $R^{7/4}g^{1/4}$. This can be shown from mass conservation of the liquid fuel to imply that

$$\frac{\mathrm{d}R^2}{\mathrm{d}t} \sim -R^{3/4}g^{1/4}.$$

Fernandez-Pello and Law (1982a) used locally similar high Reynolds number boundary-layer analysis to predict the behavior of a burning droplet with natural convection. The unitary Lewis number, thin-flame sheet, Clausius–Clapeyron phase equilibrium, and constant-liquid temperature without internal circulation are assumed. Variable density is considered. The solutions are obtained as a two-term series expansion, with the leading term giving the stagnation-point solution. Separated flow is predicted; a pool-burning model is used to predict the small vaporization rates in the aft region that lies within the separated flow domain. They also predict a result consistent with the $R(t)$ determined by Potter and Riley, although the Fernandez-Pello and Law representation of the proportionality constant is more general and more detailed.

Wu et al. (1982) did a stagnation-point analysis on a burning spherical fuel particle to determine the extinction criterion in a natural convective mode.

Fernandez-Pello (1983) considered mixed (forced- and free-) convective heating, vaporization, and burning for a spherical fuel particle. He examined ignition, steady burning, and extinction for the stagnation-point region and introduced the mixed convection ratio,

$$\phi = 4(\mathrm{Gr}/9\mathrm{Re}^2)\frac{|T_s - T_\infty|}{T_\infty} = [4g R/9(\Delta U)^2]\frac{|T_s - T_\infty|}{T_\infty}, \qquad (3.57)$$

which is essentially the reciprocal of a Froude number. Fernandez-Pello found that the forced-flow stagnation-point vaporization rate should be multiplied by the factor

$$(1 + \phi^2)^{1/8}$$

to obtain the vaporization rate for the mixed-mode convective burning. He presents a simplified governing equation for $R(t)$ that is equivalent to the result that

$$\dot{m} = \sqrt{6}\pi \frac{\lambda R}{c_p}\mathrm{Re}^{1/2}(1 + \phi^2)^{1/8}\frac{B}{(1 + B)0.8}. \qquad (3.58)$$

Fernandez-Pello's simplified equation masks a weaker dependence on ϕ from the solution of the differential equations that he shows graphically. To obtain the limit of a purely forced-convective problem, $\phi \to 0$. If we then compare the result with Eq. (3.34b) in its high Reynolds limit, we find two differences: (a) there is a minor difference in the exponents of $(1 + B)$ obtained from curve fits and (b) the Fernandez-Pello coefficient is larger by a factor of 1.44. That difference results from a different averaging process over the droplet surface. If the Fernandez-Pello vaporization rate is reduced by a factor of 1.44, it will yield the Ranz–Marshall correlation in the limit of large Reynolds number and small B.

The Fernandez-Pello (1983) result is limited to high Reynolds numbers so that comparisons cannot be made in the general Reynolds number range. Fernandez-Pello and Law (1982b) performed a mixed convective study of the stagnation-point behavior on a vaporizing sphere by using an approach similar to that of Fernandez-Pello (1983). The weighting factors for the forced and the free modes are not consistent with those of Fernandez-Pello (1983), which was not explained by the authors. It is difficult to compare the limit of infinite ϕ (which keeps $\mathrm{Re}^{1/2}\phi^{1/4}$ finite but $\mathrm{Re} \to 0$) with the results of Fernandez-Pello and Law (1982a) for pure natural convection. Differences in the normalization scheme plus other differences have prevented an easy comparison.

Rangel and Fernandez-Pello (1984) extended the work of Fernandez-Pello (1983) to include the effects of internal circulation. Rangel and Fernandez-Pello (1985) examined droplet ignition in the mixed convective mode.

A general conclusion of these studies is that either forced or free convection results in a significant deviation from the d^2 vaporization law. As already noted in Subsection 3.1.1, quasi-steady vaporization with forced convection results in a linear decrease in $R^{3/2}$ in time for a high Reynolds number. Natural convection leads to the high Grashof number quasi-steady result that $R^{5/4}$ decreases linearly with time.

3.2 Low Reynolds Number Behavior

To this point, we have considered only spherically symmetric vaporization (droplet Reynolds number $\mathrm{Re} = 0$) and high Reynolds number ($\mathrm{Re} \gg 1$) vaporization. Following Abramzon and Sirignano (1989), a composite law was presented from these two limiting behaviors that is intended to cover the fall-droplet Reynolds number range of practical interest. Some limited research has been done on the low Reynolds number case and is worthy of discussion here.

Fendell et al. (1966) considered a burning droplet in a forced-convective flow. They developed their theory assuming a Stokes flow and neglecting inertial terms in the near-gas film. See also Fendell (1968). Gogos et al. (1986) extended their work to account for practical situations in which the vaporization can produce a Stefan (radial) flow that yields a Reynolds number that is not small compared with unity. Gogos et al. retained the inertial term in the analysis but did still take advantage of the smallness of the Reynolds number based on the free-stream relative velocity.

We discuss here the results of an inner and outer matched asymptotic expansion analysis by Gogos et al. (1986). Their outer expansion followed the work of Fendell et al. (1968). Other aspects of their analysis, including transient liquid heating, followed the work of Chung et al. (1984a, 1984b), Sadhal and Ayyaswamy (1983), and Sundararajan and Ayyaswamy (1984). Gogos et al. considered a constant-density case and a variable-density case.

Gogos et al. perturbed the quasi-steady gas-phase equations, using the expansion parameter ε that equals the droplet Reynolds number based on radius. They also defined a Reynolds number A_{00} based on the droplet radius, the gas viscosity, and the radial surface-gas velocity in the limiting case of zero free-stream velocity. $\varepsilon \ll 1$ was considered, but $A_{00} = O(1)$ was allowed. In the analysis they considered the Lewis number equal to unity, constructed Shvab–Zel'dovich variables, and assumed that a thin flame exists. In the limit as $\varepsilon \to 0$, they recovered the spherically symmetric behavior discussed in Section 2.1. Here we refer to the vaporization rate for the spherically symmetric case as $\dot{m}_0$; it is given by Eq. (2.15b).

Gogos et al. obtain

$$\frac{\dot{m}}{\dot{m}_0} = 1 + \varepsilon \frac{A_{01}}{A_{00}}, \tag{3.59}$$

plus higher-order terms that are not resolved. Recasting their results for the constant-density case in our nomenclature, we obtain

$$A_{00} = \frac{1}{\mathrm{Sc}} \ln(1 + B_{00}), \tag{3.60a}$$

$$\frac{A_{01}}{A_{00}} = \frac{\mathrm{Sc}}{2} + \frac{B_{01}}{(1 + B_{00}) \ln(1 + B_{00})}, \tag{3.60b}$$

where B_{00} is the transfer number B in the spherically symmetric case and εB_{01} [resulting from the substitution of Eq. (3.60b) into Eq. (3.59)] is the perturbation to the transfer number that is due to the low Reynolds number free stream. The first term on the right-hand side of Eq. (3.60b) can be viewed as a corrective coefficient for the characteristic diffusive length across the gas film. The second term is a corrective factor for the function of the transfer number based on a correction of the droplet surface temperature that is due to the convective effect.

B_{00} and B_{01} are determined by coupling of the gas phase with a transient liquid-heating analysis. Gogos et al. determined a surface temperature T_{00} for the spherically symmetric case and an average temperature perturbation over the droplet surface εT_{01} that is due to the low Reynolds number.

The approach of Gogos et al. requires that the perturbation analysis also be performed on the liquid-phase analysis to identify T_{00} and T_{01} distinctly. An abbreviation of that process is possible without any loss of accuracy. Equations (3.59) and (3.60) can be organized as follows:

$$\dot{m} = 4\pi \rho D R \left(1 + \frac{\mathrm{Re\,Sc}}{2} \right) \ln(1 + B) \tag{3.61a}$$

plus higher-order terms in the Reynolds number. This form is based on the knowledge that the second term on the right-hand side of Eq. (3.60b) for A_{01} follows from a perturbation of the A_{00} term directly. The difference between our form here and the original form of Gogos et al. is higher order in the Reynolds number. Note that $\mathrm{Pr} = \mathrm{Sc}$ in the formulation. Now we do not require the formal perturbation analysis for the transient liquid heating. It is possible now to use a liquid-heating model, such as the effective-conductivity model, that directly yields the instantaneous average surface temperature for the droplet.

Gogos and Ayyaswamy (1988) used methodology similar to the approach of Gogos et al. (1986) to analyze the vaporization (without combustion) of a slowly moving droplet. Their results can be placed in the form of Eq. (3.61a) except that the transfer number B should now be defined as $(Y_s - Y_\infty)/(1 - Y_s)$, where Y is the mass fraction of the vapor of the droplet liquid.

Note that Gogos et al. (1986) in their constant-density analysis did allow for heat to conduct into the interior of the droplet. Gogos and Ayyaswamy (1988) also considered that flux of heat into the interior. Surprisingly, Gogos et al. (1986) in their variable-density analysis did not consider that possibility; they assumed that all the heat conducted from the gas for the surface was required for overcoming the latent heat of vaporization. They could not obtain the appropriate transient thermal behavior with that limitation. Also, it was not necessary to make that limitation; the latent heat of vaporization in their analysis could be readily replaced with the effective latent heat given by Eq. (2.6). Then the variable-density formulation of Gogos et al. could also be manipulated into the form of Eq. (3.61a).

The Nusselt number for low Reynolds number vaporizing droplets can be obtained as

$$Nu = \frac{2\ln(1+B)}{B}\left(1 + \frac{Re\,Sc}{2}\right). \tag{3.61b}$$

Clearly, as $Re \to 0$, the spherically symmetric form given by Eq. (2.15a) is yielded. In the nonvaporizing limit whereby $B \to 0$, the result of Brunn (1982) is obtained, that is, $Nu = 2(1 + Pe/4)$, where the Peclet number Pe is based on the droplet diameter.

Note that a new robust model of droplet vaporization over a broad Re range can be formulated to replace the Abramzon–Sirignano model when Eqs. (3.61a) and (3.61b) are used. Instead of using the spherically symmetric results to form a composite solution, we can combine the low Reynolds number solutions with the high Reynolds number model to achieve a model that better fits the asymptote for small Re. We could replace Eqs. (3.34a) and (3.34b) with

$$Nu = Sh = \frac{2\ln(1+B)}{B}\left[1 + \frac{Re\,Sc}{2}e^{-cRe} + \frac{k}{2}\frac{Sc^{1/3}Re^{1/2}}{F(B)}(1 - e^{-cRe})\right], \tag{3.62a}$$

$$\dot{m} = 4\pi\rho DR\ln(1+B)\left[1 + \frac{Re\,Sc}{2}e^{-c\,Re} + \frac{k}{2}\frac{Sc^{1/3}Re^{1/2}}{F(B)}(1 - e^{-c\,Re})\right], \tag{3.62b}$$

where c is a constant to be adjusted. Equations (3.62a) and (3.62b) have not been tested and should be considered as a conjecture at this point. Obvious modifications in Eqs. (3.52a) and (3.52c) would result from the use of Eqs. (3.62a) and (3.62b). These equations could replace the equations in Table C.3 of Appendix C.

There are some other noteworthy works on slowly moving vaporizing droplets: Wichman and Baum (1993), Jog et al. (1996), Ackerman and Williams (2005), and Del Alamo and Williams (2007). Ackerman and Williams used asymptotic analysis based on a small value of the ratio of far-field convective velocity to sur-face Stefan velocity. The burning-rate constant was shown to increase linearly with this ratio value. Del Alamo and Williams developed matched expansions for the low Reynolds number case. They considered acceleration, deceleration, and oscillation of the droplet motion. In the inner region near the drop surface, a quasi-steadiness was found except for short durations under abrupt changes in acceleration. The outer region was fully transient with the droplet appearing as a point source at large distances. With the use of a Green's function, the outer region is represented as a history integral.

3.3 Droplet Vaporization in an Oscillating Gas

The interaction of a vaporizing droplet with a surrounding oscillating field has many applications. One application involves fuel-droplet vaporization in an oscillating or pulsating combustion chamber. Acoustical fields, often of large enough amplitude to constitute nonlinear oscillations, occur in unstable combustors. The heating rate, vaporization rate, and trajectory of a fuel droplet in the combustor can be dramati-cally affected by the oscillatory state; the average vaporization rate or equivalently the droplet lifetime can be modified under the oscillations. The gas oscillations in the combustor are driven by (receive energy from) the oscillating combustion pro-cess. In combustors in which the combustion rate is determined by the vaporization rate, it is important to understand the feedback and response associated with the vaporizing droplet in the oscillating field.

Combustion instability has been observed in many liquid-fueled combustors, including furnaces, rockets, turbojets, afterburners, and ramjets. The combustion oscillations increase heat transfer rates and structural vibrations, thereby providing considerable incentive to seek an understanding of this undesirable phenomenon. Generally speaking, combustion instability results from the coupling between the combustion and the fluid dynamics of the system. Through this coupling, oscillatory energy is supplied by the combustion of the liquid fuel to sustain the oscillation.

As stated by Rayleigh (see Harrje and Reardon, 1972), the general criterion for wave growth or decay is, in the simplest terms, that a wave will grow if sufficient energy or mass is added in phase with the pressure. Conversely, the wave will damp if the addition is out of phase. When applied to a system in which several mecha-nisms are releasing energy or mass at once, the growth or the decay of the wave is determined by the net in-phase or out-of-phase energy or mass addition.

During the Apollo rocket program, vaporization was analytically studied as a rate-controlling mechanism for instability. Strahle (1964, 1965a, 1965b, 1966) de-termined the gas-film response to oscillatory gas pressure and velocity without ac-counting for a thermal wave in the liquid interior. Essentially, liquid thermal inertia in this model was infinite, and the liquid did not respond to ambient fluctuations. Priem and Heidmann (1960) and Heidmann and Wieber (1966) examined the effect of the liquid thermal inertia but assumed that the liquid temperature was uniform throughout the droplet interior. This model overestimated the thermal inertia of the liquid. Neither of these models provided convincing evidence that vaporization was the driving mechanism for the instability. According to these studies, the gain from the oscillatory vaporization process was not sufficient to satisfy the well-known Rayleigh criterion.

It was shown in Sections 2.1 and 3.1 that the thermal wave through the liquid and the related spatial variations of temperature are important, that is, in the high-temperature environment of a combustor, a droplet vaporizes sufficiently fast that the liquid-heating time is of the same order as that of the droplet lifetime. Transient heating and liquid thermal inertia are therefore important in steady combustor operation. Consequently we expect droplet thermal inertia to be important in oscillatory combustion.

Tong and Sirignano (1989) examined the effect of the oscillating gas pressure and velocity on the droplet-vaporization rate calculated from the authors' model for both standing and travelling waves. The vaporization rates of an array of repetitively injected droplets in a combustor were obtained from the summation of individual droplet histories. The response factor approach, which is basically an application of the Rayleigh criterion, was used. Because travelling shock waves commonly form in longitudinal-mode instability (Sirignano and Crocco, 1964), both sinusoidal and shock-wave (sawtooth) oscillations were considered. They are both of practical relevance, and each yields a different vaporization response.

A rough model of a one-dimensional combustor was created with vaporization-controlled combustion occurring in a concentrated fashion at one location in the combustor. The pressure and the temperature oscillations were assumed to follow an isentropic relationship. Hexane fuel droplets were continually introduced at initial temperatures of 300 K. The transient droplet heating and vaporization were prescribed by the model of Tong and Sirignano (1982a, 1982b, 1983) discussed in Section 3.2. Sinusoidal variations of pressure or gas velocity were imposed in a range up to 800 Hz.

The Rayleigh criterion for feedback of energy to the oscillation can be evaluated by determination of the factor

$$G \equiv \frac{\int_0^{2\pi} M' p' \, d\theta}{\int_0^{2\pi} (p')^2 \, d\theta},$$

where M' and p' are the longitudinal-mode oscillatory perturbations in vaporization rate per unit volume and pressure, each normalized by their steady-state (or unperturbed) values. θ is the phase in the oscillation. New droplets are continually introduced during the cycle of oscillation. It has been shown by Crocco and Cheng (1956) that (for small-amplitude sinusoidal pressure, temperature, and velocity oscillations) G must exceed some minimum value, dependent on the location of the combustion zone. The smallest threshold occurs with combustion concentrated at a pressure antinode and a near-velocity-node behavior at the combustor exit; its value is $(\gamma + 1)/2\gamma$, where γ is the ratio of the specific heats. In addition to the study of sinusoidal oscillations, Tong and Sirignano also examined oscillations with shock waves; sawtooth pressure oscillations simulating nonlinear longitudinal-mode oscillations with shock waves were imposed.

The calculations of Tong and Sirignano (1989) indicated that, under certain circumstances, the vaporization-rate response function was large enough to drive the

Table 3.1. *Vaporization-rate responses for four types of sinusoidal oscillations*

Oscillation	Frequency (Hz)							
	100	200	300	400	500	600	700	800
Pressure oscillation only	0.338	0.517	0.331	0.422	0.370	0.368	0.330	0.321
Velocity oscillation only	0.891	1.481	1.373	1.462	1.411	1.460	1.419	1.444
Travelling wave (pressure and velocity fluctuations in-phase)	1.229	1.998	1.795	1.884	1.781	1.828	1.748	1.765
Standing wave (pressure and velocity fluctuations out-of-phase)	1.008	0.569	0.400	0.332	0.260	0.248	0.184	0.213

instability. The frequency of the oscillation did affect the magnitude of the response function because a finite vaporization time is involved; the peak in the response for a droplet with an initial diameter of $\sim$100 μm occurs at a few hundred cycles per second. In particular, the ratio of the period of oscillation to the droplet-heating time is important. At high frequencies, the response was not sufficient to drive the instability. Table 3.1 presents some typical results. For a sinusoidal oscillation, the pressure and velocity are 90° out-of-phase. Velocity sensitivity for the vaporization rate is greater than pressure sensitivity so that optimal response can occur at some distance from the pressure antinode (equivalently the velocity node). However, the response function would be zero if combustion occurred at a pressure node, even if the vaporization rate maximized at that point.

For the sawtooth wave, velocity and pressure fluctuations tend to be in-phase. A sufficiently large response can occur in certain situations so that an instability can be sustained.

Duvvur et al. (1996) performed a numerical solution of the Navier–Stokes equation for the oscillating field surrounding a single droplet coupled with the internal liquid flow. They extended the variable-property computational model of Chiang et al. (1992) to account for the case in which an acoustical wave of much longer wavelength than the droplet diameter passed the droplet. So, like Tong and Sirignano (1989), they considered a uniform but time-varying free stream for the relative velocity and ambient pressure and temperature experienced by the droplet. Unlike Tong and Sirignano, they accounted for unsteady effects in the gas film surrounding the droplet and for the change in relative velocity that is due to drag forces.

Duvvur et al. generalized the preceding integral formula for the response factor to involve double integrals over space and time. In their case, the vaporization of each droplet occurred as the droplet moved through space. Tong and Sirignano considered that the vaporization occurred in a sufficiently small volume to be considered as a fixed location.

Two different configurations were studied. In both cases, the mean gas flow and the droplet moved in the positive longitudinal direction. In one case, the gas moved faster than the droplet so that the wake was ahead of the droplet. In the other case, the droplet moved faster than the gas so that the wake was behind the droplet. In the first case, a positive (negative) perturbation in the gas velocity that is due to an acoustic wave resulted in a positive (negative) perturbation to the relative velocity. For the second case, the perturbations to the gas velocity and to the relative velocity had opposite signs. The response factor was positive in the former case and negative in the latter case, indicating different effects on combustion instability in vaporization-controlled combustion. In a ramjet combustor, we expect the gas to be faster, whereas in a rocket engine, the droplets are faster in the upstream portion of the combustor and slower in the downstream portion.

The response factor for the "gas faster" case achieved a value of 5.81 when the oscillation period was 0.6 times the characteristic droplet-heating time. This value is considered sufficient to drive combustion instability in the longitudinal mode with a distributed energy release that is due to combustion (see Crocco and Cheng, 1956). At ratios of the oscillation period to heating time equal to 0.3 or 0.9, the response factor was too low, so a strong frequency dependence was found. Some modest dependence on the amplitude of the oscillation was also found. When the pressure amplitude was decreased by a factor of 2, the response factor changed from 5.81 to 5.02 for the "gas faster" case previously cited. The effects of volatility were studied when the latent heat of vaporization was varied; a decrease in the response factor from the value of 5.81 to 4.97 was found as the latent heat increased from the value of octane to the value of decane.

The two studies (those of Tong and Sirignano and Duvvur et al.) conclude that, for certain frequency ranges and initial droplet sizes, instability in a vaporization-rate-controlled combustor can occur. In Section 9.6, an analysis of a ramjet combustor also supports that conclusion.

Okai et al. (2000) performed experiments for a single droplet and for a droplet pair, vaporizing in an acoustic field. The octane droplet-vaporization rate increased with acoustic intensity at small or moderate intensity and low frequency. The burning rate was proportional to the frequency. Droplets in pairs vaporized more slowly than isolated droplets, which is consistent with the theories of Chapter 6.

3.4 Individual Droplet Behavior in an Unsteady Flow

The dispersion of particles or droplets in a turbulent field has been a subject of major research interest over the past few decades. Hinze (1972, 1975) discusses the rudimentary aspects of fluid and particle dispersion. According to Hinze (1975), the first analytical work on particle motion in a turbulent field was performed by Tchen, who took the equation for particle motion derived by Basset (1888), Boussinesq (1903), and Oseen (1927) and applied it to a particle in a homogeneous turbulent field. Stokes drag, pressure-gradient force, added mass, and the Basset history integral correction were considered in the analysis. The equation of particle motion is discussed later in this section. Implicitly, the analysis is limited to low droplet Reynolds numbers and to cases in which the turbulent eddy is large compared with the particle size. From the works discussed in Section 3.1, we know how to extend the prediction of quasi-steady droplet or particle motion to higher Reynolds numbers. The second restriction is a major one; there are practical situations in which the droplet or particle size is comparable to the smallest (Kolmogorov) scale of turbulence. Therefore the droplet experiences not only a temporal variation in its local free stream but nonuniformity as well.

Relatively little research has been performed on the interaction of turbulent eddies with individual droplets or particles. An important parameter is the ratio of the turbulent length scale to the droplet diameter. Existing theories typically assume that the droplet is much smaller than the turbulent length scale, i.e., $Rk \ll 1$ for all values of the wave number k in the turbulence spectrum. In this limit, a quasi-uniform free stream is experienced by the droplet. Of course, temporal changes can still occur in the free stream. However, it can be shown that, for many flow devices that operate at high Reynolds numbers, the Kolmogorov scale actually becomes of the order of 100 μm and compares to the droplet size.

A second important parameter is the ratio of the characteristic time for change in the velocity fluctuation to the residence time for the gas (continuous-phase) flow past the droplet. If U is the mean relative droplet velocity and u' is the velocity fluctuation of the gas, this ratio is given by $U/u' Rk$. For values of the order of unity or smaller, the flow over the droplet is unsteady. Only when the ratio is large compared with unity can the quasi-steady-flow assumption be made. In that case and with $Rk \ll 1$, the values of the drag and lift coefficients and the Nusselt and Sherwood numbers are not affected by the turbulent flow. (Note that the aerodynamic forces and the transport rates would still be affected by the turbulent fluctuations.)

The conventional practice is to use certain corrections on the drag force for unsteadiness in the relative droplet velocity. The earliest works on this subject are described by Basset (1888), Boussinesq (1903), and Oseen (1927). We can write, following Maxey and Riley (1983),

$$\frac{4\pi}{3}\rho_l R^3 \frac{du_{li}}{dt} = 6\pi\mu R(u_i - u_{li}) + \frac{2}{3}\pi\rho R^3 \left(\frac{du_i}{dt} - \frac{du_{li}}{dt}\right)$$

$$+ 6R^2\sqrt{\pi\rho\mu}\int_0^t \frac{\left(\frac{du_i}{dt} - \frac{du_{li}}{dt}\right)}{\sqrt{t - t'}}\,dt'$$

$$+ \frac{4}{3}\pi\rho R^3 \frac{Du_i}{Dt} + \frac{4}{3}\pi R^3(\rho_l - \rho)g_i. \qquad (3.63)$$

This equation may be viewed as an elaboration of Eq. (7.17) or Eq. (7.25) that details the aerodynamic forces. Note the distinction between the two time derivatives: D/Dt indicates a Lagrangian derivative following an element of gas at the neighborhood of the droplet whereas d/dt represents the Lagrangian derivative following the moving particle or droplet. With the two-continua concept, the Lagrangian derivative following the element of gas describes the behavior of the gas properties averaged over the droplet neighborhood. A simple interpretation for an isolated droplet is that the Lagrangian gas-property derivative pertains to the local ambient conditions for the droplet. In the case in which spatial gradients of the gas-velocity field are negligible (uniform flow), the distinction between the two time derivatives of the gas velocity disappears. The first term is the steady-state drag determined by Stokes for a low Reynolds number. The account of internal circulation by means of the Hadamard–Rybczynski result (Hadamard, 1911; Rybczynski, 1911) will reduce the coefficient of this term from 6 to 4. The second term is the apparent or added-mass term that accounts for the inertia of the gas in the boundary layer and wake of the droplet. It is negligible when the gas density is much smaller than the liquid density (which is usually the case). The third term is the Basset history integral correction force; within the context of a low Reynolds number flow, it corrects for temporal variations in the relative velocity. It can be shown by an order-of-magnitude argument that this correction is negligible if the second (time-ratio) parameter previously mentioned is large compared with unity. The fourth term accounts for the forces on the droplet created by pressure gradients and spatial variations in the viscous stress. See Chapter 7. The last term represents the net effect of gravity on the droplet, i.e., gravity force less the buoyancy force. (Note that the hydrostatic pressure gradient is automatically cancelled from the fourth term as written.)

The terms in Eqs. (7.14), (7.17), and (7.25) are given on a per-unit-volume basis whereas Eq. (3.63) and later Eqs. (3.64), (3.72a), and (3.73) are written for a given droplet, particle, or bubble. The differing factor is the droplet number density. The first three terms on the right-hand side of Eq. (3.63) multiplied by the number density equal the aerodynamic force per unit volume F_{Di} discussed in Chapter 7.

The effect of an electric charge on the droplet motion can be represented by the addition of the terms

$$q_\alpha E_i + \frac{1}{4\pi\varepsilon_0} \sum_{\alpha \neq \beta} \frac{q_\alpha q_\beta}{r_{\alpha\beta^3}} r_{\alpha\beta,i}$$

on the right-hand side of Eq. (3.63) or of Eqs. (3.64), (3.72a), and (3.73). q_α is the electric charge on the droplet that is identified by the index integer α whereas q_β are the charges on the neighboring droplets that are indexed by other integer values; E_i is the electric-field vector imposed externally; ε_0 is the permittivity of a vacuum; $r_{\alpha\beta}$ is the scalar distance between charged droplets; and $r_{\alpha\beta,i}$ is the vector distance between those droplets. The second term in the equation represents the results of the space charge.

Some exploratory research was done on the use of electrostatic charges to influence the atomization of sprays and the trajectories of the droplets. Within a given liquid element, the electric charges are mutually repelling and tend to move to the surface of the liquid. As liquid streams and ligaments disintegrate into droplets, the charges do not distribute in proportion to volume but to some lower power of the length scale; smaller droplets tend to have more charge per unit mass than larger droplets. Although all droplets are mutually repelling, the effect is greater on the smaller droplets. See Kelly (1984), Bankston et al. (1988), Ganan-Calvo et al. (1994), and Gomez and Chen (1997) for some further details.

Maxey (1993) included the effect of the initial velocity difference between the sphere and the carrier fluid in the particle-motion equation of Maxey and Riley (1983). The additional term is $6R^2\sqrt{\pi\mu\rho}\,[u(0) - u_l(0)]/\sqrt{t}$. Maxey and Riley did include the Faxen (1922) correction for nonuniform flows, which is not included in Eq. (3.63). That correction adds $R^2\nabla^2 u_i/6$ to the relative velocity in the Stokes steady-state term and in the history term. $R^2\nabla^2 u_i/10$ is added in the apparent mass term.

Equation (3.63) is seriously flawed for application to sprays in turbulent flows, in spite of its extensive use. It does not account for situations with high wave numbers in which Rk is of the order of unity or larger. It also requires correction for higher Reynolds numbers, correction for effects of vaporization and Stefan flow, and corrections for the proximity of other droplets. In the quasi-uniform limit, these corrections are discussed in Chapters 3, 4, and 6. If values of both the first and the second parameters (length and time ratios) identified in the early part of this section are large, a nonuniform but quasi-steady flow is obtained. As discussed in Chapter 6, some studies of these three-dimensional flows have been made. However, in general, more research is required here.

Formula (3.63) applies for the motion of a sphere in a nonuniform creeping (low Reynolds number) flow. We obtain the widely used equation (e.g., Berlemont et al., 1990) for a noncreeping flow by empirically modifying the first right-hand-side term in Eq. (3.63) to be

$$\frac{1}{2} C_{D\text{std}} \pi R^2 \rho |u_i - u_{li}|(u_i - u_{li}),$$

where $C_{D\text{std}}$ is the drag coefficient from the (steady) standard drag curve. A simpler form of Eq. (3.63) is obtained when the second, third, and fourth terms on the right-hand side are neglected; that form is used in many practical engineering calculations, assuming particles with large response time relative to the time scale of the flow. Odar and Hamilton (1964) and Odar (1966) studied experimentally the force on a guided sphere rectilinearly oscillating in an otherwise stagnant fluid for $0 \leq \text{Re} \leq 62$. Re is based on diameter in this chapter. They proposed an equation for the motion of a sphere with a finite Reynolds number based on their experimental study as

$$\frac{4\pi}{3} \rho_l R^3 \frac{du_{li}}{dt} = -\frac{1}{2} C_{D\text{std}} \pi R^2 \rho |u_l| u_{li} - C_a \frac{2\pi}{3} \rho R^3 \frac{du_{li}}{dt}$$

$$- C_h 6 R^2 \sqrt{\pi \mu \rho} \int_0^t \frac{du_{li}/d\tau}{\sqrt{t - \tau}} d\tau, \tag{3.64}$$

with C_a and C_h obtained experimentally and given by

$$C_a = 2.1 - 0.132 M_{A1}^2 / (1 + 0.12 M_{A1}^2),$$
$$C_h = 0.48 + 0.52 M_{A1}^3 / (1 + M_{A1})^3.$$

M_{A1} is the dimensionless relative acceleration number defined by

$$M_{A1} = \frac{2R}{|u - u_l|^2} \left| \frac{d|u - u_l|}{dt} \right|. \tag{3.65}$$

Note that $C_a \to 1$ and $C_h \to 1$ as $M_{A1} \to \infty$.

More recently, the computations by Rivero et al. (1991), Mei et al., (1991), and Chang and Maxey (1994, 1995) showed that the added-mass term for finite Reynolds number flows is the same as predicted for the low Reynolds number flow and potential-flow theory over a wide range of the acceleration number (i.e., C_a is unity).

Mei et al. (1991) studied an unsteady flow over a stationary sphere with small fluctuations in the free-stream velocity at finite Reynolds numbers ($0.1 \leq \text{Re} \leq 40$)

by using a finite-difference method and found that the Basset force term in the equation of particle motion should have a kernel that decays much faster than $1/\sqrt{t-\tau}$ at large times. Mei and Adrian (1992) and Mei (1994) considered the same problem as Mei et al. (1991) but for $\mathrm{Sl}_w \ll \mathrm{Re} \ll 1$ by using a matched asymptotic expansion, where the Strouhal number Sl_w is based on the angular frequency of the free stream and the sphere radius. They proposed a modified Basset force on the bases of the analytical result at small Reynolds numbers for low frequency, the numerical result at finite Reynolds numbers for low frequency, and the unsteady Stokes result for high frequency. Their proposed Basset force term is

$$6\pi\mu R \int_{-\infty}^{t} K(t-\tau,\tau)\frac{\mathrm{d}(u_i-u_{li})}{\mathrm{d}\tau}\mathrm{d}\tau, \tag{3.66}$$

with the broad-frequency-range approximation for the integral kernel given by

$$K(t-\tau,\tau) = \left\{ \left[\frac{\pi(t-\tau)\nu}{R^2} \right]^{0.25} + \left[\frac{\pi|u(\tau)-u_l(\tau)|^3}{2R\nu f_H^3(\mathrm{Re}_t)}(t-\tau)^2 \right]^{0.5} \right\}^{-2}, \tag{3.67}$$

where $f_H(\mathrm{Re}_t) = 0.75 + 0.105\,\mathrm{Re}_t(\tau)$ and $\mathrm{Re}_t = |u(t)-u_l(t)|2R/\nu$. Now Eq. (3.67) shows that the history kernel decays initially as $t^{-1/2}$ but as t^{-2} at large time.

Mei and Adrian (1992) found that the original Basset integral kernel is valid for only high frequencies at low and finite Reynolds numbers. They modified the kernel by interpolating the term representing the original Basset kernel and the term representing low frequencies [the second right-hand-side of Eq. (3.67)]. However, they developed the term representing low frequencies under the assumption of small-amplitude oscillation of the free stream. Therefore their modified kernel of Eq. (3.67) does not correctly predict the behavior of a spherical particle undergoing large acceleration or deceleration at low frequency. For example, expression (3.66) produces a less accurate solution for the sphere drag with a lower density ratio than for the case with a high density ratio, or, in other words, expression (3.66) produces a less accurate solution with higher deceleration than with lower deceleration. This suggests that the term representing low frequencies should be weighted by the acceleration magnitude. Kim et al. (1998) proposed a weighting function that contains the time derivative of the relative velocity M_{A1} and the ratio ϕ_r of M_{A2} to M_{A1}. M_{A1} was defined by Eq. (3.65), and M_{A2} and ϕ_r are defined as

$$M_{A2}(t) = \frac{(2R)^2}{|u-u_l|^3}\left|\frac{\mathrm{d}^2|u-u_l|}{\mathrm{d}t^2}\right|, \qquad \phi_r(t) = \frac{M_{A2}}{M_{A1}}. \tag{3.68}$$

The derivation of the history force is based on the condition that the particle be present at all times and the lower limit of integration should be negative infinity. Now we show that the term associated with the initial velocity difference can be derived from the history integral with a $-\infty$ lower limit:

$$\int_{-\infty}^{t} K(t-\tau,\tau)\frac{\mathrm{d}}{\mathrm{d}t}[u_i(\tau)-u_{li}(\tau)]\mathrm{d}\tau. \tag{3.69}$$

A particle instantaneously appearing in a fluid at $t=0$ is the same as a particle being in a stagnant fluid for $-\infty < t < 0$ and then having the bulk fluid velocity make a step change at $t = 0$. Thus the relative velocity term in history integral (3.69) can be rewritten as

$$[u_i(\tau) - u_{li}(\tau)]_{-\infty < \tau \le t} = [u_i(0^+) - u_{li}(0^+)]H(\tau) - \{[u_i(0^+) - u_{li}(0^+)]$$

$$- [u_i(\tau) - u_{li}(\tau)]\}_{0 < \tau \le t}, \tag{3.70}$$

where $H(\tau)$ is the Heaviside step function and the term in the braces is assumed to be zero for $t < 0$. Equation (3.70) is differentiated with respect to τ to obtain

$$\left\{ \frac{d}{d\tau}[u_i(\tau) - u_{li}(\tau)] \right\}_{-\infty < \tau \le t}$$

$$= [u_i(0^+) - u_{li}(0^+)]\delta(\tau) + \left\{ \frac{d}{d\tau}[u_i(\tau) - u_{li}(\tau)] \right\}_{0 < \tau \le t}. \tag{3.71}$$

Let m_p be the mass of the sphere and m_f be the mass of the fluid in a volume equal to the sphere volume. Define $K_1(t) \equiv K(t, 0)$. For a steady flow before $t = 0$ with an impulse at $t = 0$, Eq. (3.63) is replaced with

$$m_p \frac{du_{li}}{dt} = \frac{1}{2} C_{Dstd} \pi R^2 \rho |u - u_l|(u_i - u_{li}) + \frac{1}{2} m_f \left(\frac{Du_i}{Dt} - \frac{du_{li}}{dt} \right) + m_f \frac{Du_i}{Dt}$$

$$+ 6\pi \mu R \int_{0^+}^{t} K(t - \tau, \tau) \frac{d(u_i - u_{li})}{d\tau} d\tau + (m_p - m_f)g$$

$$+ 6\pi \mu R K_1(t)[u_i(0^+) - u_{li}(0^+) - u_i(0^-) + u_{li}(0^-)]. \tag{3.72a}$$

Note that this equation was derived for a uniform gas flow so that the distinction between the two Lagrangian time derivatives for the gas velocity has disappeared.

Kim et al. (1998) developed a modified form of the kernel for a wide range of frequencies and Reynolds numbers. Their correlation for the kernel matches the low Reynolds number asymptotes at low and high frequencies. It also matches their Navier–Stokes equation solutions over wide ranges of frequency and Reynolds number. In particular, we have

$$K(t - \tau, \tau) = \left\{ \left[\frac{\pi(t-\tau)\nu}{R^2} \right]^{0.5/c_1} + G1 \left[\frac{\pi}{2} \frac{|u(\tau) - u_l(\tau)|^3}{R \nu f_H^3(\text{Re}_t)}(t - \tau)^2 \right]^{1/c_1} \right\}^{-c_1},$$

$$\tag{3.72b}$$

$$G(\tau) = \frac{1}{1 + \beta \sqrt{M_{A1}}(\tau)}, \tag{3.72c}$$

$$\beta = \frac{c_2}{1 + \phi_r \phi_r^{c_4}/[c_3(\phi_r + \phi_r^{c_4})]}, \tag{3.72d}$$

$$f_H = 0.75 + c_5 \text{Re}_t(\tau), \tag{3.72e}$$

where $\text{Re}_t = |u(\tau) - u_l(\tau)|2R/\nu$.

The low Reynolds number asymptotes are correctly obtained with no dependence on the choice of c_1 through c_5. Equation (3.72c) shows that, as M_{A1} is reduced, $G(\tau)$ approaches unity and the present kernel [Eq. (3.72b)] becomes similar in form (at both the high and low frequency limits) to the kernel of Eq. (3.67). On the other hand, as M_{A1} becomes large, $G(\tau)$ approaches zero, and the present kernel becomes of the same form as that of Basset (1888). The β in the expression of $G(\tau)$ is not a constant but a function of ϕ_r, the ratio of M_{A2} to M_{A1}. This function behaves as follows: $\beta \cong c_2 c_3 / \phi_\tau^{c_4}$ for $\phi_\tau \gg 1$. The kernel becomes of the same form as that of Basset (1888) as the dimensionless frequency becomes large. Kim et al. showed that the function $K_1(t)$ can be approximated as

$$K_1(t) = \left\{ \left[\frac{\pi t \nu}{R^2} \right]^{0.5/c_1} + G_1 \left[\frac{\pi}{2} \frac{|u(0) - u_l(0)|^3}{R \nu f_H^3 (\mathrm{Re}_{t0})} t^2 \right]^{1/c_1} \right\}^{-c_1}, \qquad (3.72f)$$

$$G_1 = \frac{1}{1 + c_6 \mathrm{Re}_{t0}^{-0.25} (\rho_r + 0.5)^{-0.5}}, \qquad (3.72g)$$

where $\mathrm{Re}_{t0} = |u(0) - u_l(0)| 2R/\nu$ and ρ_r is the ratio of particle density to fluid density.

Equations (3.72f) and (3.72g) show that, as the density ratio ρ_r becomes large, G_1 approaches unity, and the function $K_1(t)$ becomes proportional to $1/t^2$ at large time. This indicates that, when ρ_r is large, the drag that is due to the initial velocity difference decays rapidly with time and becomes negligible compared with the quasi-steady drag C_{Dstd}. Conversely, when ρ_r and the initial Reynolds number are both small, G_1 is small, and the $1/\sqrt{t}$ term in the function $K_1(t)$ remains important, even at large time. Thus, when the density ratio is small in low Reynolds number flows, the drag that is due to the initial velocity difference decays slowly with time and is not negligible compared with the quasi-steady drag C_{Dstd}.

For a fixed particle, G_1 becomes unity, and thus the hydrodynamic force does not depend on the density of the particle. This fixed particle limit can be viewed as an infinite particle-density limit. For a fixed particle, the left-hand side of Eq. (3.72a) should be replaced with the negative of the applied force required for fixing the particle under the action of the hydrodynamic forces.

The six constants were determined to be

$$c_1 = 2.5, \qquad c_2 = 22.0, \qquad c_3 = 0.07,$$

$$c_4 = 0.25, \qquad c_5 = 0.126, \qquad c_6 = 17.8. \qquad (3.72h)$$

Equation (3.72a) has not been tested for nonuniform free streams. It might fail when the characteristic length for variation in the free stream becomes as small as or smaller than the particle dimension.

To evaluate the contributions to the total drag from each term on the right-hand side of Eq. (3.72a), we may rewrite that equation as

$$m_p \frac{du_l}{dt} = \frac{1}{2} \pi R^2 \rho |u - u_l| (u - u_l) C_{Dtot}$$

$$= \frac{1}{2} \pi R^2 \rho |u - u_l| (u - u_l) (C_{Dstd} + C_{Dadd} + C_{Dfld} + C_{Dhis} + C_{Dini} + C_{Dgrv}), \qquad (3.73)$$

where C_{Dstd} represents the quasi-steady drag coefficient from the (steady) standard drag curve, C_{Dadd} is the drag coefficient that is due to the added-mass force, C_{Dfld}

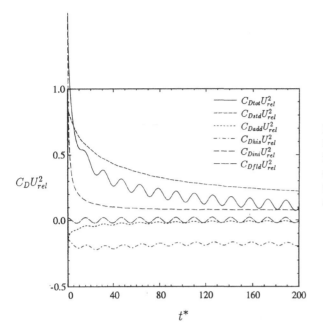

Figure 3.18. C_{Dtot}, C_{Dstd}, C_{Dadd}, C_{Dhis}, C_{Dini}, and C_{Dfld} multiplied by U_{rel}^2 as a function of time obtained from the new equation for $\mathrm{Re}_{t0} = 150$ and $\rho_r = 5$ with the base flow oscillating with $\mathrm{Sl}_w = 0.4$ and $\alpha_1 = 0.02$ (Kim et al., 1998).

equals the drag coefficient that is due to the carrier-fluid acceleration or the gradient of the pressure and the shear stress at the position of the sphere, C_{Dhis} represents the drag coefficient that is due to the (unsteady) history force that is the integral of the past relative acceleration of the sphere weighted by the kernel K, C_{Dini} is the drag coefficient that is due to the initial velocity difference between the carrier fluid and the sphere, and C_{Dgrv} equals the force coefficient that is due to the net gravity force.

Kim et al. (1998) show that the numerical results of the Eq. (3.72a) compare more favorably with solutions of the full Navier–Stokes equations for unsteady, axisymmetric flow around a freely moving sphere injected into a quiescent fluid or a fluid oscillating with time in the x direction as

$$u_x(t) = \alpha_1 |u_l| \sin \omega t,$$

where α_1 is a constant controlling the amplitude and ω is the angular frequency.

Figure 3.18, from Kim et al. (1998), shows the component drag coefficients and the total drag coefficient of the sphere multiplied by the square of the relative velocity as a function of time ($0 \le t^* \le 200$) for $\mathrm{Re}_{t0} = 150$ and $\rho_r = 5$ with the base flow oscillating with Strouhal number $\mathrm{Sl}_w = 0.4$ and $\alpha_1 = 0.02$. This figure presents the relative magnitudes of the component drag forces. The quasi-steady drag term and the history drag term yield the largest contributions in this case.

C_{Dfld} is strongly affected by and roughly proportional to the magnitude of the dimensionless frequency. For low values of the density ratio (e.g., high gas pressure and density) the history term becomes important. The total drag estimated by Eq. (3.72a) is within a few percent of the Navier–Stokes solution for this oscillating-flow case.

It should be understood that some velocity fluctuations in the wave-number range in which Rk is of the order of unity will be generated at sufficiently high droplet Reynolds numbers. We can expect that vortex shedding of the flow over a droplet can occur, producing eddies of a size of the same order of magnitude as the droplet.

Coimbra and Rangel (1998) obtained an analytical solution to Eq. (3.63) in the unsteady Stokes regime. A fractional differential operator was used to solve the integrodifferential equation. The new approach was applied to three configurations: (i) a particle falling under gravity, (ii) the ambient fluid accelerating linearly in time, and (iii) impulsive start of the fluid. They confirmed the findings of previous researchers that the history integral is important unless the particle density is much greater than the fluid density.

Michaelides and Feng (1994) extended the approach of Maxey and Riley to the droplet energy equation. Some discussion is given in Chapter 7 following Eq. (7.22). Coimbra et al. (1998) extended the approach of Coimbra and Rangel (1998) to the solution of Eq. (7.22) in the limit of a small Biot number and a negligible Peclet number. A linearized radiation law was assumed to facilitate the exact solution. For typical spray conditions, the solution was essentially quasi-steady without an important contribution from the history integral.

Little research has been done on convective heat and mass transport within the gas film surrounding a droplet in a turbulent spray flow. On account of its relevance to combustion instability in liquid-fueled and liquid-propellant systems, a substantial amount of research has been performed on the impact of long-wavelength ($Rk \ll 1$) fluctuations on a spray (Priem and Heidmann, 1960; Priem, 1963; Strahle, 1964, 1965a, 1965b, 1966; Heidmann and Wieber, 1965, 1966; Harrje and Reardon, 1972; Tong and Sirignano, 1989; and others). See Section 3.3 for details. The fluctuations of transport rates in this case with uniform but temporally varying free streams approaching the droplet have been found to be significant. Extension of these analyses to the high-wave-number domain is necessary for understanding fully the behavior of sprays involved in heat and mass exchange with a turbulent flow. Some limited work in that direction has been performed by Masoudi and Sirignano (1997, 1998, 2000).

Masoudi and Sirignano (1997, 1998, 2000) extended the work of Kim et al. (1995, 1997) to consider heat transfer to a sphere that is interacting with a vortex. The parameter range of interest involved comparable characteristic dimensions for the spherical particle and the initially cylindrical vortex. Incompressible three-dimensional Navier–Stokes equations were solved by finite-difference equations. Both uniform-free-stream-temperature and stratified-free-stream-temperature fields were considered, with stratification having a profound effect when the characteristic length scale for the stratification was comparable to the vortex dimension. The transient response of the Nusselt number is sensitive to the geometric configuration, e.g., vortex initial size and initial position. Time-averaged Nusselt numbers were compared with the steady axisymmetric flow value for the Nusselt number. They are nearly the same for the case in which there is no temperature stratification and the vortex approaches the sphere along the line of symmetry (for the steady

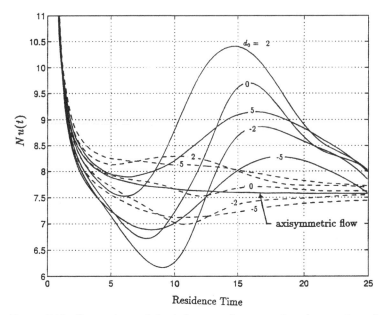

Figure 3.19. Comparison of the influence of vortex advection on the sphere transient Nusselt number $Nu(t)$ in a uniform-temperature (dashed curves) versus a stratified-temperature (solid curves) field [$v_{max0} = 0.4$, $\sigma_0 = 1$, Re $= 100$, with $k = 0$ (dashed curves) and 0.3 (solid curves)]. (Masoudi and Sirignano, 1998, with permission of the *International Journal of Heat and Mass Transfer.*)

base flow). The unsteady case is inherently asymmetric, and the time-averaged Nusselt number varies monotonically with vortex initial distance from the symmetry line, base-flow Reynolds number, and vortex circulation. Asymptotic values are reached. A correlation in self-similar form is found for the sphere in the uniform free stream. Kim et al. (1995) calculated lift and torque.

For the stratified-free-stream-temperature case, the transient phenomena are different, resulting in a threefold increase in the impact of the vortex on the change in the Nusselt number compared with the uniform-stream-temperature case. Correlations are also obtained for this nonuniform case. When the product of the vortex circulation and the gas-field temperature gradient is positive (negative), the time-averaged Nusselt number for the sphere increases monotonically with an increase (decrease) in the vortex circulation and with an increase (decrease) in the vortex initial distance from the symmetry axis. Again, asymptotic values are reached. The stratified-temperature-case correlation is presented as the sum of two correlations, one for the uniform stream and the other dependent on the temperature values. The uniform-temperature-case correlation then is a special case of the correlation for the stratified case. The correlations are shown to be valid both for solid spheres and for liquid spheres with internal flow.

The free-stream temperature is represented as

$$T(x, y, z) = \frac{T_{max} + T_{min}}{2} + \left(\frac{T_{max} - T_{min}}{2} \right) \tanh(\kappa x), \qquad (3.74)$$

where the maximum and the minimum free-stream temperatures are $T_{\max}$ and $T_{\min}$, respectively. The reciprocal of κ is the characteristic length over which the temperature varies in the free stream.

The flow direction is given by z, and the axis of the initial cylindrical vortex is oriented in the y direction. The steady axisymmetric correlation is given as

$$\mathrm{Nu} = 1 + (1 + \mathrm{Pr}\,\mathrm{Re})^{1/3}\mathrm{Re}^{0.077}, \tag{3.75}$$

where $\mathrm{Re} \leq 400$ (see Clift et al., 1978).

For the uniform-free-stream temperature the averaged Nusselt number $\overline{\mathrm{Nu}}$ is given as

$$\frac{\overline{\mathrm{Nu}}_{\mathrm{uniform}}}{\overline{\mathrm{Nu}}_{\mathrm{axisymmetric}}} = 1 + 0.019\frac{\Gamma_0}{2\pi}\mathrm{Re}^{0.40}\tanh\left(\frac{d_0}{2\sigma_0^{0.6}}\right), \tag{3.76}$$

where Γ_0 is the initial vortex circulation, Re is the base-flow Reynolds number, d_0 is the initial distance of the vortex from the symmetry axis, and σ_0 is the core radius of the initial vortex. Of course, the hyperbolic tangent function reaches asymptotes of ± 1 as the argument goes to $\pm\infty$.

For the nonuniform-free-stream temperature, the correlation for the Nusselt number becomes

$$\frac{\overline{\mathrm{Nu}} - \overline{\mathrm{Nu}}_{\mathrm{uniform}}}{\overline{\mathrm{Nu}}_{\mathrm{axisymmetric}}} = \pm 2.04 \left[\frac{T_{\max} - T_{\min}}{\dfrac{T_{\max} + T_{\min}}{2} - T_s}\right]\left(\frac{\Gamma_0}{2\pi}\right)^{1.1}\frac{\kappa^{0.8}}{d_0}\tanh\left(\frac{0.35d_0}{\sigma_0^{0.6}}\right). \tag{3.77}$$

The positive sign is taken when the product $\Gamma_0\kappa$ is positive, and the negative sign is taken when that product is negative. T_s is the surface temperature of the sphere. Note that the Reynolds number dependence in Eq. (3.77) is implied through the use of Eqs. (3.75) and (3.76) for substitution.

Figure 3.19 shows the results for $\mathrm{Nu}(t)$ in various cases. A few residence units of time are required for the start transient so the time averages in Eqs. (3.76) and (3.77) are taken over the nondimensional time from 2 to 25. The temporal deviation of the Nusselt number from the axisymmetric value is seen to be much greater for the stratified free stream than for the uniform free stream.

Although the configurations studied by Masoudi and Sirignano are highly idealized, they do indicate that the interaction of small-scale vortical fluctuations with a spherical particle or droplet causes variations in the heating (or cooling) rates that are worthy of attention in turbulent flows. Masoudi and Sirignano (2000) show for vaporizing droplets in a variable-density, variable-composition gas that the Stefan flow reduces the perturbation in the Nusselt number because of the vortical interaction. However, the Sherwood number perturbation is found to be significant.

4 *Multicomponent-Liquid Droplets*

There are various complications that occur when a multicomponent liquid is considered (Landis and Mills, 1974, and Sirignano and Law, 1978). Different components vaporize at different rates, creating concentration gradients in the liquid phase and causing liquid-phase mass diffusion. The theory requires the coupled solutions of liquid-phase species-continuity equations, multicomponent phase-equilibrium relations (typically Raoult's Law), and gas-phase multicomponent energy and species-continuity equations. Liquid-phase mass diffusion is commonly much slower than liquid-phase heat diffusion so that thin diffusion layers can occur near the surface, especially at high ambient temperatures at which the surface-regression rate is large. The more volatile substances tend to vaporize faster at first until their surface-concentration values are diminished and further vaporization of those quantities becomes liquid-phase mass diffusion controlled.

Mass diffusion in the liquid phase is very slow compared with heat diffusion in the liquid and extremely slow compared with momentum, heat, or mass diffusion in the gas film or compared with momentum diffusion in the liquid. In fact, the characteristic time for the liquid-phase mass diffusion based on droplet radius is typically longer than the droplet lifetime. Nevertheless, this mass diffusion is of primary importance in the vaporization process for a multicomponent fuel. At first, early in the droplet lifetime, the more volatile substances in the fuel at the droplet surface will vaporize, leaving only the less volatile material that vaporizes more slowly. More volatile material still exists in the droplet interior and will tend to diffuse toward the surface because of concentration gradients created by the prior vaporization. This diffusion is balanced by the counterdiffusion of the less volatile fuel components toward the droplet interior.

The result is that different components will have different vaporization rates, and those rates can vary significantly during the droplet lifetime. Two limits exist in theory: (a) the rapid regression or zero-diffusivity limit, and (b) the uniform concentration or infinite-diffusivity limit. In practical combustion situations, the diffusion rate is finite but tending to be slow. Limit (a) could be interesting in some cases but limit (b) would be interesting only in noncombusting situations in which the ambient-gas temperature is low and the vaporization rate is low. Limit (b) is the distillation limit and is the analog of the droplet-heating condition described by a type (ii) model in Chapter 2.

The rapid regression limit is one in which a diffusion layer (of zero thickness in this limit) exists at the surface and regresses in a quasi-steady manner with the surface. There is a balance between normal (to the surface) convection and normal diffusion in this thin layer, with the concentrations outside the layer (in the droplet interior) remaining uniform and constant with time at the initial values. A conservation-of-mass principle immediately yields the result that the vaporization rate for each component is proportional to its initial composition; e.g., for a 50% decane and 50% octane liquid-fuel mixture, 50% of the mass vaporized per unit time is decane and 50% is octane in this limiting condition. The value of the concentration of each liquid component at the surface will be self-adjusting to whatever value is necessary to assume this vaporization rate. Obviously, the more volatile component (octane in this case) will assume a lower concentration value at the surface than its interior value whereas the less volatile substance (decane) will assume a higher value than its interior value. If the gas diffusivities of each of the components were identical, the gas-phase concentrations at the surface would be proportional to the vaporization flux of each of the components, which in turn would be proportional to the initial concentration of the fuel.

In a low-temperature environment, the droplet lifetime might be so long that diffusion can be considered to be fast and concentrations in the liquid may be assumed to be uniform but time varying. Note that the required droplet lifetime for uniform concentration would be an order of magnitude greater than the necessary lifetime for the uniform-temperature condition because of the large difference between diffusivities for heat and mass. Within this limit, another limiting condition of sequential vaporization of the components exists when the volatilities of the fuel components are different from each other. In this distillation limit, the most volatile component will first vaporize completely from the droplet with negligible vaporization of the other components; then the next most volatile component will completely vaporize without the vaporization of the other remaining components. The distillation process continues until the least volatile substance vaporizes in sequence. The required combination of low ambient temperature and widely varying volatility (say, as measured by the boiling points) of the components makes this limit not very interesting in combustion. Furthermore, this argument indicates the danger of applying low-temperature experimental results to predict behavior under combustion conditions.

The gas-phase Lewis number and Prandtl number are close to unity so that mass transfer, as well as heat and momentum transport, may be typified as quasi-steady. The individual fuel-vapor components may diffuse at somewhat different rates and may react chemically at very different rates, but no one has carefully investigated these effects for droplet vaporization in a combusting environment. Some works (Landis and Mills, 1974; Law et al., 1977; Law and Sirignano, 1977) tend to consider only a uniform diffusivity and a thin flame (or none at all).

It is convenient to replace the boundary conditions given by Eq. (2.5) or Eq. (3.25) in the multicomponent case. The fractional vaporization rate ε_m for species m is also the mass-flux fraction for species m in a quasi-steady gas phase and is given by

$$\rho v \varepsilon_m = \rho v Y_m - \rho D \frac{\partial Y_m}{\partial n},$$
(4.1a)

and the species-vaporization rate is

$$\dot{m}_m = \varepsilon_m \dot{m},$$
(4.1b)

where density, mass fraction, normal component of velocity, and normal component of mass fraction gradient are evaluated at the gas side of the droplet surface with outward toward the gas side taken as positive. Clearly, $\sum \varepsilon_m = 1$. The latent heat of the vaporizing fuel becomes

$$L = \sum_m \varepsilon_m L_m.$$
(4.2a)

where L_m is the latent heat for species m. We also will later use that

$$L_{\text{eff}} = \sum \varepsilon_m \left[L_m + \frac{4\pi R^2}{\dot{m}} \lambda \frac{\partial T}{\partial n} \right] = \left(\sum \varepsilon_m L_m \right) + \frac{4\pi R^2}{\dot{m}} \lambda \frac{\partial T}{\partial n}.$$
(4.2b)

Raoult's Law becomes the phase-equilibrium statement at the surface and replaces the Clausius–Clapeyron relation that was used for single-component mixtures. The ratio of the mole fraction on the gas side of the surface X_{mgs} to the liquid-side mole fraction X_{mls} is

$$\frac{X_{mgs}}{X_{mls}} = \frac{1}{p} \exp \left[\frac{L_m}{R_m} \left(\frac{1}{T_{b,m}} - \frac{1}{T_s} \right) \right].$$
(4.3a)

See the discussion following Eq. (2.4a) to understand the dependence of the boiling point at nonatmospheric pressure on the pressure p and the atmospheric boiling temperature $T_{b,m}$. For a single-component liquid, the surface temperature cannot exceed the boiling point because the mole fraction cannot exceed the value of one. However, it is clear from Raoult's Law that the surface temperature for a multicomponent liquid can exceed the nominal boiling point of the most volatile component (designated as that component with the lowest boiling temperature in the pure state). So, for that component, boiling in the mixture occurs at a higher temperature than for the pure state because the mole fraction is not maximized at the nominal boiling point.

The mass fractions Y_m and the mole fractions X_m are related by

$$Y_m = \frac{X_m W_m}{\sum_i X_i W_i},$$
(4.3b)

$$X_m = \frac{Y_m / W_m}{\sum_i (Y_i / W_i)},$$
(4.3c)

where W_m is the molecular weight of species m. These relations between mole fractions and mass fractions apply for both gas and liquid species. Therefore, given T_s

and Y_{mls}, X_{mls} can be obtained that yields X_{mgs} from Eq. (4.3a), and then Y_{mgs} can be determined. Consequently,

$$Y_{m,s} = \frac{W_m X_{mls} \sum_{i=1}^{N}(Y_{i,s}/W_i)}{p} e^{L_m/RT_{b,m}} e^{-L_m/RT_s}. \qquad (4.3d)$$

Raoult's Law implies a linear relation between vapor pressure and liquid-phase composition measured in moles or mole fraction. Some practical multicomponent liquids will deviate from the ideal linear relation. Sometimes maxima or minima can occur in the plots of the partial pressure of a component to the liquid-phase mole fraction. In cases of an extreme deviation from Raoult's Law, empirical modifications of Eq. (4.3a) should be sought in the physical chemistry literature.

The mass transfer number in Eqs. (3.33) is replaced with

$$B_M = B_m = \frac{\left(\sum_{i,F} Y_i\right)_s - \left(\sum_{i,F} Y_i\right)_\infty}{1 - \left(\sum_{i,F} Y_i\right)_s} = \frac{Y_{m,F,s} - Y_{m,F,\infty}}{\varepsilon_{m,F} - Y_{m,F,s}}, \qquad (4.4)$$

where the summations of mass fraction at infinity and at the surface include only those components present in the liquid phase. Here, in our approximate analyses, we assume that all binary diffusion coefficients are equal. The transfer numbers for each of the species are therefore identical because transport for each species becomes identical. From Eq. (4.4), it is seen that, if each species' mass fraction is known at the droplet surface and at infinity, then each species-vaporization rate fraction (or species mass-flux fraction) is determined by $\dot{m}_m = \varepsilon_m \dot{m}$. The species mass fraction in the gas at the droplet surface can be found from Eqs. (4.3a), (4.3b), and (4.3c) if the surface temperature and the mass fractions on the liquid side of the surface are known.

Sirignano (2002) developed the Super Scalar for multicomponent-liquid vaporization that remains uniform valued over the gas field under certain conditions. The scalar S_{mv} is given by $S_{mv} = h + \frac{L_{eff} Y_m}{\varepsilon_m - Y_{m,s}}$. The same scalar value holds for any species m. See Appendix B for more information.

The augmentations of the droplet-heating and -vaporization models of Chapters 2 and 3 to account for multicomponent liquids are discussed in Sections 4.1 and 4.2 for the spherically symmetric case and the convective case, respectively. Vaporization of an array of multicomponent-liquid droplets is discussed in Subsection 6.3.5. A summary of the useful equations is given in Appendix C. The heating and the vaporization of slurry droplets are analyzed in Section 4.3, and the analysis for emulsified-liquid droplets is discussed in Section 4.4.

4.1 Spherically Symmetric Diffusion

The gas-phase equations are modified from those presented in Section 2.2. Vapor-species-conservation equation (2.3a) is replaced with a set of partial differential species-conservation equations, one for each species vaporizing from the liquid. Boundary conditions (2.5) are modified as previously mentioned. The source terms in Eqs. (2.2), (2.3a), and (2.3b) must be modified to account for the reaction of each of the fuel-vapor species. In the limit of fast chemical kinetics, a thin flame results. With the assumption of Le = 1, results for vaporization rate and Nusselt number will still be provided by Eqs. (2.10) and (2.15). However, the energy per unit mass released in chemical reaction is the sum over all fuel species, i.e., $Q = \sum_i \varepsilon_i Q_i$. Also, the latent heat of vaporization is given by Eq. (4.2) and the phase equilibrium is now described by Eqs. (4.3). The determination of the fractional vaporization rate (or, equivalently, the fractional mass flux) ε_m for each species from Eqs. (4.3) is now necessary to close the problem. To obtain those fractions, an analysis of the liquid-phase composition is necessary. Note that the Sherwood number as defined by Eq. (2.14d) is identical for all species and equal to the Nusselt number because we will consider all binary diffusion coefficients to be equal and Le = 1.

The concentration or mass fraction $Y_{l,m}$ of the mth species in the composition of the spherical liquid droplet obeys a diffusion equation of spherically symmetric form:

$$\frac{\partial Y_{l,m}}{\partial t} = D_l \left(\frac{\partial^2 Y_{l,m}}{\partial r^2} + \frac{2}{r} \frac{\partial Y_{l,m}}{\partial r} \right). \tag{4.5a}$$

This form follows from the species-conservation equations presented in Appendix A and is identical to that of heat diffusion equation (2.16a), except that the mass diffusivity D_l is much smaller than the heat diffusivity α_l. Here, we have the analog to a type (iii) droplet-heating model discussed in Chapter 2. This equation applies in the domain $t \geq 0, 0 \leq r \leq R(t)$, so that we have a moving-boundary problem. It may be transformed to a fixed-boundary problem with the consequence that a convective term appears because of the surface regression. On account of the low diffusivity value, this convective term becomes much more important here than the corresponding term in the heat equation. The importance of this term is highlighted in the previous discussion of the rapid regression limit.

The boundary condition at the liquid side of the droplet surface is developed from Eqs. (2.10), (2.14), (2.15), (4.1), and (4.4):

$$\left. \frac{\partial Y_{l,m}}{\partial r} \right)_s = \frac{\rho D}{\rho_l D_l} \frac{\ln(1 + B)}{R} (Y_{l,ms} - \varepsilon_m). \tag{4.5b}$$

At the center of the droplet, symmetry yields

$$\left. \frac{\partial Y_{l,m}}{\partial r} \right)_{r=0} = 0. \tag{4.5c}$$

Typically, initial conditions for each species state a uniform composition when the droplet is first introduced:

$$Y_{l,m}(0, r) = Y_{l,m0}. \tag{4.5d}$$

This equation and its boundary conditions must be applied concurrently with phase-equilibrium conditions (4.3) and energy equations (2.16) to obtain the complete solution.

Landis and Mills (1974) solved the coupled heat and mass transfer problem for vaporizing bicomponent droplets. Pentane, hexane, and heptane were respectively mixed with octane. Their results indicate that batch distillation (both heat and mass diffusivities are taken to be infinite in this model) or infinite mass diffusivity models are highly inaccurate in predicting the results for vaporization rates of the individual components. The agreement between the more exact diffusion-controlled model and these oversimplified models appears to be slightly better at 600 K ambient temperature (low overall vaporization rate) than at 2300 K (high overall vaporization rate), but it is still quite poor. One might expect that the batch distillation model would improve slightly as the difference in volatilities of the two components increased. However, their paper does not address this point. For the heptane/octane mixture with 2300 K ambient temperature, the rapid regression limit appears to have been reached; a diffusion layer exists, and each individual component vaporization rate is approximately proportional to its initial concentration. The authors indicate that diffusion will not be as dominantly controlling for lower vaporization rates but will still be important and the batch-distillation model would not apply.

Aggarwal and Mongia (2002) examined droplet-vaporization models relevant to high-pressure combustion in gas-turbine engines, comparing bicomponent behavior with that of a single-component surrogate fuel. They found large differences in the predictions between the batch-distillation and diffusion-limit models and recommended that the diffusion-limit model be used at high pressures.

In the opinion of this author, the batch-distillation model (or an infinite mass diffusivity model) should never be applied to a combustion situation because vaporization rates are too high. On the other hand, only at extremely high vaporization rates will the rapid regression limit apply. In general, it is advised to use the finite-diffusivity model whereby the diffusion equation is solved. However, care must be taken to resolve the fine structure near the surface that results from the small value of liquid mass diffusivity.

Landis and Mills (1974) also showed that disruptive boiling or microexplosions are also possible because, for certain regions of the droplet interior, the equilibrium vapor pressure of the more volatile component can exceed the ambient pressure. This is an interesting phenomenon that may be important in multicomponent or emulsified spray combustion. The interested reader should seek the review by Law (1982). Landis and Mills further suggest, in passing, that internal circulation would decrease the differences between the batch-distillation model and the diffusion-controlled model because effectively the diffusivity is increased. In Section 4.2 it is demonstrated that their conjecture is not correct because the convective mechanism that causes the internal circulation also increases the transport rates through the gas film and thereby increases the vaporization rate.

Zhang and Law (2008) addressed the interesting question of the dependence of liquid-phase mass diffusivity on temperature and pressure. Substantial decrease in the diffusivity was found at elevated pressure and temperature. Accordingly, in the droplet context, the lower-molecular-weight fuels might not have a higher diffusivity than higher-weight fuels because they have lower boiling points and thereby face temperature limitation in the liquid phase. The thermal diffusivity also shows variation but significantly less variation than the mass diffusivity. The calculations indicate that gradients of the liquid-phase mole fraction across the droplet decrease by roughly a factor of 2 as pressure is increased from 1 to 15 atm.

Liquid-phase mass diffusion also becomes important at pressures near or above the critical pressure of the liquid, even if it initially is a pure component. Ambient gases dissolve in the liquid to a significant extent as the critical pressure is approached; then mass diffusion occurs in the liquid phase. It is noteworthy that the actual critical pressure and temperature vary spatially with composition (see Chapter 5). Typically, the dependence on composition is very nonlinear, and the critical pressure for a mixture can be greater than the critical pressure of any component. Therefore subcritical conditions can exist at an interface with a distinct discontinuity between liquid and gas even if the pressure is above the critical pressure of the original pure component in the droplet. See Shuen et al. (1992) and Delplanque and Sirignano (1993) for detailed analyses of these problems.

The importance of the Soret and Dufour effects and the validity of the binary diffusion approximation with regard to liquid-phase mass diffusion in the droplet-vaporization problem remain to be determined.

Experiments in reduced-gravity (NASA microgravity) facilities allowed for improved understanding through the study of droplets as large as 5.4 mm. With the negligible gravity levels found with drop towers and the Space Shuttle, the spherical shape can be maintained for these very large droplets. See Shaw et al. (2001a) who, guided by theory, determined the effective liquid-phase diffusivity for alcohol mixtures from their data. Bubble nucleation, growth, and disruption was observed with some occurrence of microexplosions. Some indication of water absorption in the droplet was found.

The concepts discussed in this section have some applications to the vaporization of emulsified fuels, which are discussed in Section 4.4. The reader is also recommended to read the review by Law (1982). The phenomenon of microexplosions is an interesting possibility in miscible-fuel mixtures and emulsified-fuel mixtures. It has been observed in laboratory situations and could have practical implications (see Law, 1982). Although no evidence exists, it seems conceivable that slurries made from coal with volatile contents could also exhibit microexplosion behavior.

Lage et al. (1995) showed the effect of radiative heating on a bicomponent vaporizing droplet. This analysis builds on the studies of Lage and Rangel (1993a, 1993b) discussed in Section 2.3. Radiation absorption does influence vaporization in high-temperature environments. The in-depth radiation absorption can move the position of maximum liquid temperature from the droplet surface to the droplet center. This temperature increase at the droplet center might lead to heterogeneous nucleation and microexplosion. Without large volatility differences between the two liquid components, flash vaporization occurs before the superheating limit is reached. It is conjectured by Lage et al. that large volatility differences might lead to reaching the superheating limit with resulting homogeneous nucleation and microexplosion. A variable liquid-density model, allowing for thermal expansion of the liquid, was included.

Schiller et al. (1996a, 1998) considered the vaporization of energetic fuels that undergo exothermic liquid-phase decomposition. The model included bubble formation and growth in the droplet due to the decomposition. Unsteady temperature and composition profiles and gasification rates were determined for cases with and without oxidation in the surrounding gas film. Very high liquid temperatures and very rapid gasification rates were found compared with those of conventional liquids. The results scaled very differently from the results for conventional fuels. The decomposition was modelled as a nonequilibrium process so that no limiting liquid temperature such as a boiling point was allowed in the model. Improved modelling in the future should provide a more realistic prediction of the peak temperature. Nevertheless, some interesting new physics are explored in these papers.

Law (1998) gives a useful overview of the combustion of high-energy liquid and slurry fuels. He discusses liquid-phase reactions, especially of organic azides; shell formation from the vaporizing droplet and its effect on burning and vaporization rates; and soot formation, especially as related to the burning of fuels with highly strained molecules and high carbon-to-hydrogen ratios, e.g., cubanes and benzvalenes.

4.1.1 Continuous-Thermodynamics Models

Some interesting work has been performed with a distribution function to represent the composition of a multicomponent liquid. The model assumes that the molecular weight of the components form a continuous function rather than the discrete values chosen when a finite number of species are present. The mass fraction, mole fraction, partial density, and concentration of the liquid components are given for each value of the molecular weight. A few moments of the distribution are considered in the analysis, and evolution equations are developed for these moments. So, for cases in which many species are present, large savings of computational resources can be achieved by use of a few moment-evolution equations rather than by use of a species-conservation equation for each component.

The early papers on the concept were presented by Bowman and Edmister (1951), Edmister and Bowman (1952), Gal-Or et al. (1975), Briano and Glandt (1983), Rätzsch and Kehlen (1983), Cotterman et al. (1985), and Chou and Prausnitz (1986). Applications to problems with droplets were made by Tamim and Hallett (1995), Lippert and Reitz (1997), Hallett (2000), Harstad et al. (2003), and Harstad and Bellan (2004a,b). The concept has been used for direct numerical simulations (DNSs) of two-phase flows undergoing transition to turbulence by LeClercq and Bellan (2004, 2005a, 2005b) and by Selle and Bellan (2007a, 2007b, 2007c, 2009). See the discussion on turbulent spray flows in Chapter 10.

Gal-Or et al. (1975) considered the distribution function $\rho(M)$ for the density, with the total density given as $\overline{\rho} = \int_0^\infty \rho(M) dM$. Although M is continuous, each value of M is construed as a distinct species. The authors defined normalized moments as $m_n = \int_0^\infty \frac{\rho(M)}{\overline{\rho}} M^n dM$; so $m_0 = 1$; m_1 is the expected value; and $m_2 - m_1^2$ is the variance. The velocity of each species is given as v_M with the center-of-mass velocity for the mixture given by $\overline{v} = \int_0^\infty \frac{\rho(M)}{\overline{\rho}} v_M dM$. Then, the flux for each species is given as $J_M = \rho(M)[v_M - \overline{v}]$. A few evolution (i.e., conservation) equations can track the change in the moments, representing the changes in the distribution function. Thereby, the tracking of a very large number of individual species is avoided and a huge savings in computational costs can result.

The gamma probability distribution function (Γ-PDF) has commonly been used in these analysis; specifically, $f_\Gamma(M) = \frac{(M-\gamma)^{\alpha-1}}{\beta^\alpha \Gamma(\alpha)} \exp\left(-\frac{M-\gamma}{\beta}\right)$. Γ is the classical gamma function, and α, β, and γ are constants that adjust the probability distribution function. Distribution functions are constructed for both the liquid and vapor phases, with γ having the same value for both phases. See, for example, Tamim and Hallett (1995), Lippert and Reitz (1997), and Hallett (2000). Certain shortcomings were noted with that single-mode distribution function. Consequently, Bellan and co-workers introduced the two-mode double-Γ-PDF: $P(M, \varepsilon) = (1 - \varepsilon) f_\Gamma(M, \alpha_1, \beta_1) + \varepsilon f_\Gamma(M, \alpha_2, \beta_2)$. This produced better results than the single Γ-PDF when compared with solutions obtained by tracking up to 32 species with traditional species-conservation equations. See Harstad et al. (2003) and Harstad and Bellan (2004a, 2004b).

The method of continuous thermodynamics has not yet been developed for convecting droplets. Droplets with spherical symmetry have been considered, sometimes with empirical correlations (e.g., Ranz–Marshall corrections) for relative drop-gas motion. Resolution of spatial variations of composition and temperature in the droplet interior has also not yet been made.

4.2 Liquid-Phase Mass Diffusion with Convective Transport

4.2.1 Approximate Analyses

Whenever droplet slip causes internal circulation, geometrical details of the situation will be modified but the same types of limiting behavior will exist. The discussion of the internal liquid flow is given in Subsection 3.1.4. For a liquid Peclet number (based on mass diffusivity) of the order of unity, the convection will cause the diffusion problem to become axisymmetric and no longer spherically symmetric as it was at the zero Peclet number. However, at a very high Peclet number, the convection is so fast that concentration becomes uniform along a stream surface. Again, a one-dimensional problem, but now in a stream-surface coordinate system, will result. The solution for the liquid-phase mass fraction or concentration can be developed from the analysis for liquid temperature given in Subsection 3.1.4. The axisymmetric form of the liquid-species-conservation equation can be reduced by one dimension in analogous fashion to the temperature.

The mass fraction $Y_{l,m}$ of species m in the liquid phase will be governed by

$$\frac{1}{\gamma} \frac{\partial Y_{l,m}}{\partial \tau} = a(\phi, \tau) \frac{\partial^2 Y_{l,m}}{\partial \phi^2} + b(\phi, \tau) \frac{\partial Y_{l,m}}{\partial \phi}, \qquad (4.6a)$$

where $a(\phi, \tau)$ and $b(\phi, \tau)$ are defined as in Eq. (3.48a). The parameter γ is the ratio of the mass diffusivity to the thermal diffusivity (a reciprocal of the Lewis number) for the liquid and is typically quite small compared with unity. This equation has the exact same form as heat diffusion equation (3.48a) except for the coefficient of the time-derivative term, which reflects the very important difference between the characteristic times for heat diffusion and for mass diffusion. Otherwise, the five points made following Eq. (3.48d) would still apply to Eq. (4.6a) and are not repeated here. Equation (4.6a) describes the concentration behavior with a vortex model so that we have the analog to a type (v) droplet-heating model.

Using definition (2.14), general boundary conditions (3.1), and the results given by Eqs. (3.29), (3.30), and (4.4), together with the knowledge that Sh = Nu for Le = 1, we have, at the edge of the liquid core,

$$\left.\frac{\partial Y_{l,m}}{\partial \phi}\right)_s = \frac{k[-f(0)]\text{Re}^{1/2}}{8}\frac{\rho D}{\rho_l D_l}\left(Y_{l,ms} - \varepsilon_m\right). \tag{4.6b}$$

At the vortical center, we have the boundary condition that results from the elimination of any singular solution:

$$\frac{\partial Y_{l,m}}{\partial \tau} = \gamma b(0,\tau)\frac{\partial Y_{l,m}}{\partial \phi} \tag{4.6c}$$

at $\phi = 0$. The usual initial condition is that the composition is uniform, namely,

$$Y_{l,m}(\phi, 0) = Y_{l,m0}. \tag{4.6d}$$

Lara-Urbaneja and Sirignano (1981) solved this equation with the appropriate boundary, matching, and initial equations for multicomponent fuel-droplet vaporization. They considered quasi-steady boundary layers in the gas phase and in the liquid phase adjacent to the droplet interface. Integral methods were used in the boundary-layer analysis. Figure 3.2 shows the results for liquid temperature versus time at a position on the surface compared with a position at the vortex center; various fuel combinations of varying volatility are considered, and an ambient pressure of 10 atm (and therefore elevated boiling points) is considered. It is seen that the behavior is transient and spatial gradients of temperature exist over the droplet lifetime for all mixtures considered. The degree of transient behavior decreases as the volatility of the mixture increases, but even hexane demonstrates a very transient behavior. The boiling points of the fuels are never reached during their lifetime. In bicomponent mixtures, the heavier component seems to dominate the thermal behavior, i.e., the 25%–75% mixture of hexadecane and decane (or the 50%–50% decane and hexane) behaves much more like pure hexadecane (or pure decane) than like decane (or hexane). The apparent reason for this behavior is that the more volatile component vaporizes quickly, and most of the material near the surface consists of the heavier, less volatile component.

Figure 4.1 shows how the liquid mass fraction for the more volatile species in a bicomponent mixture typically varies spatially and temporally. The species is depleted quickly near the surface and then diffuses from the interior to the surface. The characteristic diffusion time is clearly much longer than the droplet lifetime (approximately $\tau = 0.1$ here); the mass fraction is very nonuniform throughout the lifetime. The ambient-gas temperature for this calculation is 1000 K, the pressure is 10 atm, and the initial mixture consists of 50% decane and 50% hexadecane. It is seen that the mass diffusion layer near the surface $\phi = 1$ thickens with time and does not remain as a thin layer near the surface, so this solution is not close to the rapid regression limit. It is conjectured that if 2300 K were taken as the ambient-gas temperature and the mixture were heptane and octane, as studied by Landis and Mills for spherical symmetry, there would be qualitative agreement with their results, namely, the higher ambient temperature (higher heat flux) and more volatile

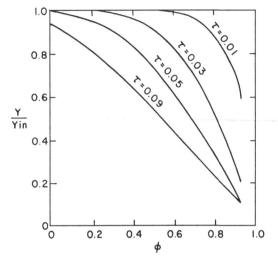

Figure 4.1. Mass fraction of liquid decane in an initial 50% decane–50% hexadecane mixture. (Lara-Urbaneja and Sirignano, 1981, with permission of The Combustion Institute.)

fuel would result in a sufficiently high vaporization rate to obtain the rapid regression limit.

It can be seen from Fig. 4.2 that the more volatile substance (decane) vaporizes early whereas the heavier substance (hexadecane) vaporizes later during the lifetime. Because a fuel droplet moves through a combustor, the implication is that certain regions will be relatively rich in the more volatile fuel vapors whereas other regions will be relatively rich in the less volatile substance. Stratification of this type would have significant impact on the characteristics of ignition, flame stability, and pollutant formation. In fact, these influences on those important combustion features are what provide the impetus for multicomponent vaporization studies.

Tong and Sirignano (1986a) showed that the vaporization rate of a multicomponent droplet in a high-temperature gas with relative velocity (and therefore internal circulation) would not be accurately predicted by assuming either a stagnant fluid (with only internal diffusion but noadvection) or infinitely rapid mixing in the

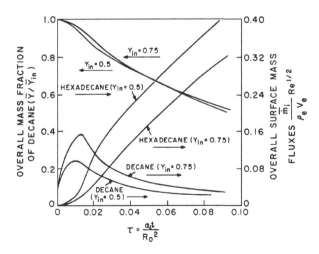

Figure 4.2. Overall decane mass fraction and vaporization rates in decane–hexadecane mixtures. Y_{in}, initial decane mass fraction; τ, nondimensional time. (Lara-Urbaneja and Sirignano, 1981, with permission of The Combustion Institute.)

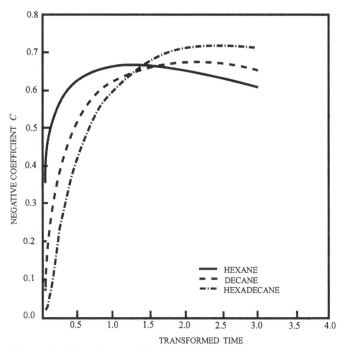

Figure 4.3. Negative coefficient *C* versus time.

limit. They showed that the liquid-mixing rate was enhanced by the internal advection but still remained finite and had a controlling effect on the vaporization rate. Their results were consistent with those of Lara-Urbaneja and Sirignano, but they introduced many useful simplifications related to the gas-phase and liquid-phase boundary layers. In particular, Tong and Sirignano were able to reduce the dimensionality of the droplet boundary-layer analysis by using locally similar solutions. This reduced computational time substantially and made the model feasible for use in a spray calculation in which many droplets are present.

Some interesting results were obtained by Tong and Sirignano, following the analysis described in Section 8.1, by using the simplified Eq. (8.1) and more efficient algorithms. Figure 4.3 shows the variation of the coefficient *C* with time. Note that, during the earlier part of the droplet lifetime, the assumption of slowly varying *C* appears to be poor. Nevertheless, results for temperature, concentration, vaporization rate, and diameter indicate essentially perfect agreement between our approximate solution and direct solution of the equations by finite-difference methods.

Some important differences exist between the results for different models of the liquid-phase heating. The most common approach in the literature is to assume uniform but time-varying temperature in the liquid. Effectively, this means infinite liquid conductivity. Another model neglects the internal vortex and liquid circulation and assumes spherically symmetric conduction. Figure 4.4 compares the simplified vortex model results with results from the other two models. The infinite-conductivity model clearly is very poor. The spherical diffusion model predicts higher initial surface temperatures because it does not account for heat

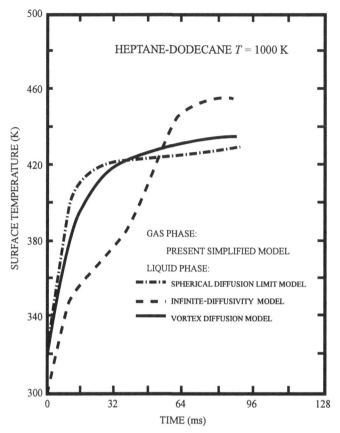

Figure 4.4. Surface temperature versus time comparisons for three liquid-phase heating models. (Sirignano, 1993a, with permission of Taylor & Francis, Inc.)

convected to the droplet interior from the surface. For the same reason, it under-predicts surface temperature later in the droplet lifetime.

Results for a hexane–decane bicomponent droplet vaporizing at an ambient temperature of 1000 K were also obtained. The conditions are the same as in the previous single-component droplet case. Although the value of 30 is more representative of typical liquid fuels, Lewis numbers of 1, 10, and 30 were used. The results are shown in Figs. 4.5–4.8.

Figure 4.5 shows the timewise variations of droplet-surface temperature and nondimensionalized droplet diameter. As in the single-component droplet case, the curves rise rapidly initially and then level off to approach asymptotically the wet-bulb temperature. The results are similar to those in the single-component droplet case.

The timewise variation of the surface liquid-phase hexane mass fraction is shown in Fig. 4.6 along with the timewise variation of the relative vaporization mass flux of hexane. The curves for the surface liquid-phase hexane mass fraction fall rapidly initially and then level off at an almost constant value. This is mostly due to the large resistance of mass diffusion within the droplet. All the curves, with the exception of the $C = 0$ case, rise slightly later in the droplet lifetime. Initially, the

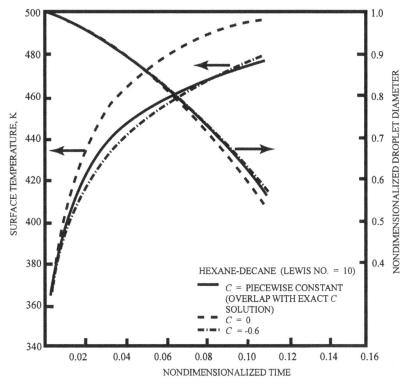

Figure 4.5. Surface temperature and droplet diameter versus time (Le = 10). (Sirignano, 1993a, with permission of Taylor & Francis, Inc.)

liquid-phase hexane concentration at the droplet surface decreases because the liquid diffusion and convection are slow to offset the high fractional vaporization rate of hexane. Later, the fractional vaporization rate of hexane becomes small because of the increase in surface temperature and decrease in hexane concentration at the liquid surface. At the same time, the liquid convection becomes more important as vaporization becomes more efficient. Eventually, the fast liquid convection and the small fractional vaporization rate of hexane result in a residue of hexane at the liquid surface.

The curves for the relative vaporization mass flux are consistent with the curves for the surface liquid-phase hexane mass fraction. Initially, the rapid drop in the surface liquid-phase hexane mass fraction results in a rapid drop in the relative vaporization rate of hexane. Later, the curve levels off. In the limit of infinite vaporization rate, each individual vaporization rate will be approximately proportional to the corresponding initial concentration. As surface temperature increases, the fuel-vapor mass fraction at the droplet surface also increases. This leads to the timewise increase in the total vaporization mass flux shown in Fig. 4.7. Results for Le = 1 and 30 were also obtained and are given in Fig. 4.8 along with results for Le = 10 for comparison. An exact value of C was used in these calculations.

The model of Abramzon and Sirignano (1989) provides another method for the evaluation of droplet behavior (see the discussion in Chapter 3). That model offers the advantage of applying for zero or very low Reynolds numbers as well as high

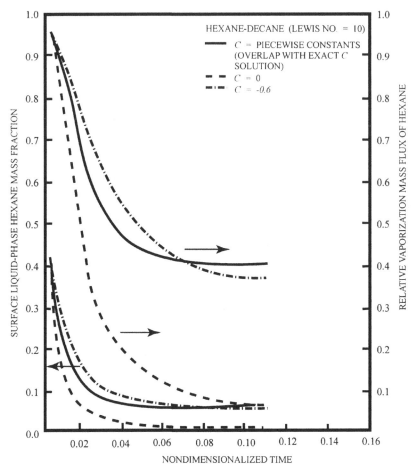

Figure 4.6. Surface liquid-phase mass fraction and relative vaporization mass flux of hexane versus time (Le = 10). (Sirignano, 1993a, with permission of Taylor & Francis, Inc.)

Reynolds numbers; the Tong and Sirignano model was limited to Reynolds numbers that yield a viscous Prandtl-type boundary layer on the droplet. Continillo and Sirignano (1988, 1991) showed that the Abramzon and Sirignano model could be readily extended to multicomponent liquid vaporization. Neglecting internal circulation, they used Eqs. (4.1)–(4.5) plus Eqs. (3.31)–(3.36). This modification therefore uses a boundary-layer description of the gas phase and a spherically symmetric liquid-phase model. Delplanque et al. (1991) did extend the full Abramzon–Sirignano model to a multicomponent case, accounting for internal circulation by means of an effective diffusivity model. The liquid-phase mass diffusion equation is written as

$$\frac{\partial Y_{l,m}}{\partial t} = D_{l\,\mathrm{eff}}\left(\frac{\partial^2 Y_{l,m}}{\partial r^2} + \frac{2}{r}\frac{\partial Y_{l,m}}{\partial r}\right),$$

where

$$D_{l\,\mathrm{eff}} = \chi_d\, D_l,$$

$$\chi_d = 1.86 + 0.86\tanh[2.225\log_{10}(\mathrm{Re}_l\mathrm{Sc}_l/30)]. \tag{4.7}$$

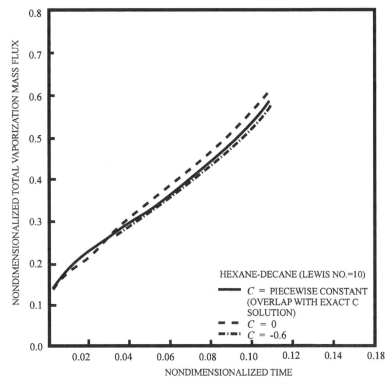

Figure 4.7. Total vaporization mass flux versus time (Le $= 10$). (Sirignano, 1993a, with permission of Taylor & Francis, Inc.)

Now we use the coordinate transformation that was utilized to obtain nondimensional equation (3.51) and conditions (3.52). In particular, we obtain

$$\frac{1}{\gamma}\left(r_s^2 \frac{\partial Y_{l,m}}{\partial \tau} - \beta \zeta \frac{\partial Y_{l,m}}{\partial \zeta}\right) = \frac{\chi_d}{\zeta^2} \frac{\partial}{\partial \zeta}\left(\zeta^2 \frac{\partial Y_{l,m}}{\partial \zeta}\right). \qquad (4.8a)$$

The boundary condition at the outer surface follows from Eqs. (4.1), (2.14), and (3.34):

$$\left.\frac{\partial Y_{l,m}}{\partial \zeta}\right)_{\zeta=1} = \frac{\rho D}{\rho_l D_l}\ln(1 + B_M)\left[1 + \frac{k}{2}\frac{\mathrm{Sc}^{1/3}\mathrm{Re}^{1/2}}{F(B_M)}\right](Y_{l,ms} - \varepsilon_m), \qquad (4.8b)$$

where B_M and $F(B_M)$ are given by Eqs. (3.33) and (3.36), respectively. See the discussion following Eq. (3.36) about the replacement of the Reynolds-number-dependent term in Eq. (4.8b) for low values of that parameter. The boundary condition at the droplet center follows from symmetry:

$$\left.\frac{\partial Y_{l,m}}{\partial \zeta}\right)_{\zeta=0} = 0. \qquad (4.8c)$$

Normally, uniform initial composition is chosen so that

$$Y_{l,m}(0, \zeta) = Y_{l,m0}. \qquad (4.8d)$$

We will discuss later the use of this model in a spray application (see Chapter 9).

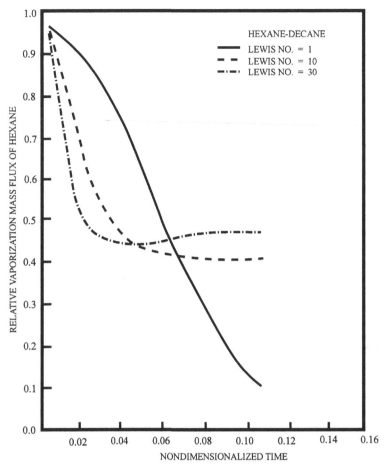

Figure 4.8. Relative vaporization mass flux of hexane versus time. (Sirignano, 1993a, with permission of Taylor & Francis, Inc.)

4.2.2 Exact Analyses

Megaridis and Sirignano (1991,1992a) considered axisymmetric, unsteady multi-component-droplet vaporization by using a Navier–Stokes solver that was an extension of the analysis of Chiang et al. (1992) for single-component droplets. Variable thermophysical properties and non-Fickian gas-phase mass diffusion were considered. The potential for microexplosion was examined. The difference between Fickian and non-Fickian gas-phase diffusion did not result in significant changes in drag coefficients, heating rate, vaporization rate, or droplet-surface property values. Changes in the value of the liquid-phase Lewis number produced some significant differences on the droplet-heating and -vaporization rates but did not substantially affect the drag coefficient. For the benzene/n-octane bicomponent liquid droplet studied, superheat liquid conditions were not achieved, indicating little probability for a microexplosion. Benzene/n-decane mixtures were also studied, and it was concluded that an increase in the difference between the volatilities of the two components increases the probability of a microexplosion. The interesting phenomenon was also found that an increase in concentration of the less volatile component

Contour Interval: 2.72E-02 Min: 4.94E-01 Max: 7.67E-01

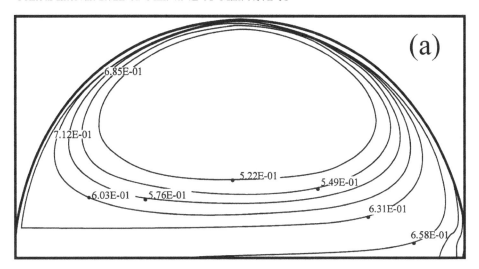

Contour Interval: 2.93E-02 Min: 6.95E-01 Max:9.88E-01

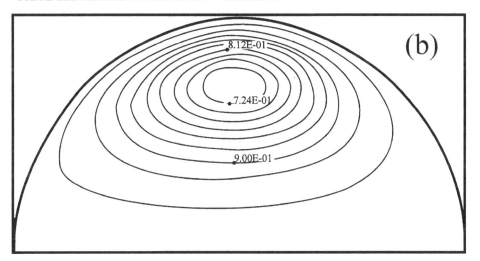

Figure 4.9. Liquid-phase *n*-octane mass fraction contours in an *n*-octane/benzene mixture at two different nondimensional times: (a) $\tau = 5$, (b) $\tau = 25$. The ambient-hot-gas flow is from left to right. The outer semicircle represents the droplet surface.

can sometimes result in a higher vaporization rate because of higher liquid-heating rates.

Figure 4.9 shows the instantaneous contours for liquid *n*-octane in an *n*-octane/benzene mixture. The less volatile component actually has a higher concentration near the surface and will diffuse toward the vortex center.

4.3 Metal-Slurry Droplet Vaporization and Combustion

There is an interest within the combustion community in developing slurry fuels with high volumetric energy-release rates for potential application in combustion and propulsion systems. A variety of solid constituents such as coal, boron, aluminum, and beryllium in the slurries were studied (see Szekeley and Faeth, 1982; Antaki, 1986; Takahashi et al., 1986, 1989; Antaki and Williams, 1987; Wong and Turns, 1987). The fundamental research of metal-slurry fuels is justified because metal particles, within high-temperature oxidative environments, burn and release substantial amounts of energy, thus having important advantages over conventional all-liquid fuels. The liquid components of slurries serve both as a carrier and a fuel; therefore slurry fuels have the advantage of the liquid-fuel properties (spray injectibility and pumpabililty) that make their utilization in aerospace applications very attractive. Even though slurry fuels are widely used today, their vaporization, ignition, and combustion characteristics are not fully understood. In this section those characteristics are discussed. The primary examples follow from the works of Megaridis and Sirignano (1992b) and Bhatia and Sirignano (1992, 1993a, 1993b). They studied aluminum particles in hydrocarbon fuels.

Two extreme models of slurry droplets are considered. In one model, only one large spherical particle resides in a spherical hydrocarbon fuel droplet. In the other model, many very fine-metal particles reside in a spherical fuel droplet. As liquid-fuel vaporization occurs, the fine-metal particles agglomerate. In both models, the remaining metal heats and liquefies after the fuel is vaporized so that a spherical metal droplet results. The metal droplet vaporizes on further heating. The metal vapors mix with oxygen and react exothermically, driving the heating and vaporization of the metal droplet. A major product of the metal-oxidation reaction is condensed metal oxide. In the next three subsections, we consider the vaporization and the burning of a hydrocarbon fuel droplet with a single metal particle, of a hydrocarbon fuel droplet with many fine-metal particles, and of a metal droplet, respectively. In Section 9.5, a flame propagating through a slurry spray is analyzed by use of the fine-particle model and the metal-droplet model.

4.3.1 Burning of a Fuel Droplet Containing a Single Metal Particle

Here we consider two submodels. In one case, there is no relative motion between the gas and the droplet, and the metal particle remains in the center of the droplet so that a spherically symmetric situation results. In the other case, a relative motion with axisymmetric convective heating occurs; more rapid heating results and the metal particle can move under hydrodynamic forces.

In the combustion of metal-slurry droplets, the liquid carrier vaporizes and burns first, exposing the dry metal to a hot oxidizing environment, and a film of metal oxide forms on the metal surface. This oxide coating acts as a barrier to further surface oxidation of the metal and must be retracted before the metal can ignite. The dry metal heats to the melting point of the aluminum (933.1 K) followed by the phase change. Subsequent heating causes further temperature rise and slow oxidation of the metal until the oxide coating melts (2315 K). From Glassman (1960) and Wong and Turns (1987), ignition is presumed to occur as soon as the oxide coat melts and allows the metal vaporization to begin.

SPHERICALLY SYMMETRIC CASE. The effect of unsteady gas-phase heat and mass transfer on the metal ignition is important. After liquid burnout, the flame will collapse inward onto the metal surface, thus affecting the temperature rise of the solid agglomerate. First, we solve the unsteady spherically symmetric gas-phase conservation equations, coupled with a transient description for the liquid–metal-slurry heating. The key assumptions are that a stagnant flow field occurs around the droplet, the gas-phase Lewis number is unity, and a single-step, infinitely fast chemical reaction occurs in the gas phase.

The gas-phase analysis considers the energy equation, oxidizer and fuel-species equations, continuity equation, and perfect-gas equation of state in the gas phase. The equations are identical to Eqs. (2.1)–(2.6). Shvab–Zel'dovich variables and mass coordinates are used. The liquid-phase and the solid-phase energy equations are also considered together with the balance of heat and mass fluxes, phase equilibrium, and the continuity of temperature at the interface.

The metal and the liquid components of the slurry are separated, and the local thermophysical properties are readily determined. The energy equation for transient, spherically symmetric droplet heating can be obtained from Eq. (2.16a) in the following nondimensional form:

$$r_s^2 \frac{\partial Z}{\partial \tau} - \frac{1}{2} \eta \frac{dr_s^2}{d\tau} \frac{\partial Z}{\partial \eta} = \frac{1}{\eta^2} \frac{\partial}{\partial \eta} \left\{ \chi \eta^2 \frac{\partial Z}{\partial \eta} \right\}, \qquad (4.9a)$$

where the following definitions are used:

$$r_s = R/R_0, \qquad Z = T/T_i, \qquad \tau = \alpha_l t / R_0^2, \qquad \eta = r/R,$$

and χ is given by (a) $\chi = 1$ for an all-liquid droplet or the liquid region ($\eta > r_m/R$) of the slurry droplet or by (b) $\chi = \alpha_m/\alpha_l$ for the metal region ($\eta > r_m/R$) of the slurry droplet. The initial and boundary conditions are

$$\text{at } r = 0, \quad Z = 1, \qquad (4.9b)$$

$$\text{at } \eta = 0, \quad \left. \frac{\partial Z}{\partial \eta} \right|_{\eta=0} = 0, \qquad (4.9c)$$

$$\text{at } \eta = 1, \quad \left. \frac{\partial Z}{\partial \eta} \right|_{\eta=1} = \frac{q_d r_s r_{li}}{4\pi \lambda_l T_0 R^2}. \qquad (4.9d)$$

At the liquid–metal interface the heat flux is continuous, and this condition is incorporated by use of a harmonic averaging of the liquid and the metal thermal conductivities (Patankar, 1980). Also, liquid mass conservation implies that

$$\dot{m}_F = -4\pi \rho_l R^2 \dot{R}. \qquad (4.10)$$

The quasi-steady theory of droplet vaporization and combustion results in a constant ratio of flame radius to droplet radius. Thus the theory will be extremely limited in describing the physics near the end of the liquid-vaporization process. Furthermore, with the quasi-steady assumption it is not possible to model the collapse of the flame and the subsequent conductive heating of the dry metal. In this stage of burning of slurry droplets, the dry metal heats rapidly until the phase change of metal occurs. As the metal thermal diffusivity is large (~3 orders of magnitude larger than the liquid diffusivity, and Biot number $\approx 7 \times 10^{-4} \ll 1$), we assume that the metal is isothermal. Note that the fuel vapor is still burning in the gas phase. The metal heat rise may be given by

$$\frac{4}{3}\pi r_m^3 \rho_m c_{pm} \frac{dT_s}{dt} = \dot{q}_m. \tag{4.11}$$

The heat flux $\dot{q}_m$ consists of $\dot{q}_d$ and also the fluxes that are due to radiation and chemical reaction, i.e.,

$$\dot{q}_m = \dot{q}_d + \dot{q}_{\text{rad}} + \dot{q}_{\text{ch}}. \tag{4.12}$$

In the study by Wong and Turns (1987), a model for the rate of oxidation of an aluminum agglomerate, based on the work of Alymore et al. (1960), is given as

$$\dot{q}_{\text{ch}} = A_{\text{sp}} C X_O^{0.45} \exp(-E/RT),$$

where

$$A_{\text{ch}} \propto A_{\text{sp}}(d/d_p)(1-\theta), \qquad C \propto Q_{\text{Al}} K_0(1-\theta)/d_p, \tag{4.13}$$

$\dot{q}_{\text{ch}}$ represents the heat generated by the chemical reaction $A_{\text{sp}} = \pi d^2/4$, A_{ch} is the reactive surface area, X_O is the ambient-oxygen mole fraction, the activation energy $E = 51.9$ kcal/mole (from Alymore et al.), d and d_p are the agglomerate and the particle diameters, respectively, θ is the porosity, K_0 is the frequency factor, and Q_{Al} is the heat of reaction of aluminum per unit mass. The preceding relation is critical in explaining the agglomerate ignition times, and the factor C is determined from their own experimental results. The main conclusion from the Alymore study is that they were unable to propose a quantitative analysis of aluminum oxidation. Also, the Alymore study is for single aluminum strips, rather than an agglomerate. Further, the cited study is limited to aluminum oxidation in the temperature range 400–650 °C, whereas Eq. (4.13) gives significant fluxes for temperatures above 1000 °C. Thus the model proposed by Wong and Turns may not withstand the generalizations. Therefore Bhatia and Sirignano (1992) proposed another approach.

To model the heating process of the dry particle, consider a spherical particle of initial radius r_{Al}, oxidizing in an environment consisting of the remnant flame of the vaporized liquid fuel. (This approach is a modification of the suggestion by Mills, 1990). The particle radius may increase because of surface oxidation of metal, even though the density of the oxide is larger than that of the metal ($\rho_{\text{Al}} = 2.7$ gm/cc, $\rho_{\text{ox}} = 3.97$ gm/cc). Balancing heat at the particle surface, we have

$$\frac{d}{dt}\left[\frac{4}{3}\pi r_{\text{Al}}^3(\rho c_p)_{\text{Al}} T + \frac{4}{3}\pi \left(R^3 - r_{\text{Al}}^3\right)(\rho c_p)_{\text{ox}} T\right]$$

$$= \dot{m}_0 \Delta H_c + 4\pi R^2 \lambda_g \left.\frac{\partial T}{\partial r}\right|_{r=R} - 4\pi R^2 \sigma \varepsilon \left(T^4 - T_\infty^4\right), \tag{4.14}$$

where the particle temperature is taken to be uniform. We are making the assumption of fast aluminum oxidation and slow oxidizer diffusion and neglecting any equilibrium laws at the oxide–gas interface that may determine the rate of oxidizer flow. Conduction from the flame is, initially, the prime heating mechanism of the dry particle. The radiation heat transfer effects are not significant during the preignition process (Friedman and Macek, 1962, 1963). An estimate of the relative importance of radiation to convection can be made as

$$\dot{q}_r/\dot{q}_c = \frac{\varepsilon \sigma r_m \left(T^4 - T_\infty^4 \right)}{h \left(T - T_\infty \right)}, \qquad (4.15)$$

where h is the convection heat transfer coefficient and $\varepsilon \approx 0.3$, say. Further, for $\mathrm{Nu} \approx 2$, $h = \lambda_g / r_m$. Then

$$\dot{q}_r/\dot{q}_c = \frac{\varepsilon \sigma \left(T^4 - T_\infty^4 \right)}{h \left(T - T_\infty \right)} \approx \frac{\varepsilon \sigma r_m \left(T^4 - T_\infty^4 \right)}{\lambda_g \left(T - T_\infty \right)} \approx \begin{cases} \dfrac{4\varepsilon \sigma r_m T^3}{\lambda_g} & \text{for} \quad T \approx T_\infty \\[3mm] \dfrac{\varepsilon \sigma r_m T^3}{\lambda_g} & \text{for} \quad T \gg T_\infty \end{cases}. \qquad (4.16)$$

For $r_m = 10~\mu\mathrm{m}$, $\lambda_g = 0.15~\mathrm{W/(mK)}$, $T = 2300~\mathrm{K}$, we get $\dot{q}_r/\dot{q}_c = 0.055$. This represents the maximum relative value of the radiant flux. Thus the contribution of the radiant flux is small. The calculations neglect radiation heat transfer. We need an additional relation for the oxygen transfer rate to the particle surface. Balancing mass at the particle surface, we have

$$\dot{m}_{os} = 4\pi R^2 (\rho D)_g \frac{\partial Y_O}{\partial r}\bigg|_{r=r_s} = 4\pi R^2 (\rho D)_{ox} \frac{\partial Y_{O,ox}}{\partial r}\bigg|_{r=R}, \qquad (4.17)$$

where $\dot{m}_{os}$ is the oxidizer mass flux at the oxide–gas interface and, for simplicity, we have assumed Fickian diffusion of the oxygen through the aluminum oxide, governed by a transient, spherically symmetric equation as

$$\rho_{ox} \frac{\partial Y_{O,ox}}{\partial t} = \frac{1}{r^2} \frac{\partial}{\partial r} \left[(\rho D)_{ox} r^2 \frac{\partial Y_{O,ox}}{\partial r} \right]; \qquad r_{Al}(t) \leq r \leq R(t), \qquad (4.18a)$$

with the initial and the boundary conditions

$$\text{at } t = 0, \quad Y_{O,ox} = 0, \qquad (4.18b)$$

$$\text{at } r = r_{Al}(t), \quad Y_{O,ox} = 0, \qquad (4.18c)$$

$$\text{at } r = R(t), \quad Y_{O,ox} = Y_{O,os}. \qquad (4.18d)$$

A model for the coefficient of diffusion for the oxygen through the solid aluminum oxide is given by Bhatia and Sirignano (1992). When the transformation

$$\eta = \frac{r - r_{Al}(t)}{R(t) - r_{Al}(t)}, \quad 0 \leq \eta \leq 1, \qquad (4.19)$$

is used, Eq. (4.19) becomes

$$\frac{\partial Y_{O,ox}}{\partial t} = \frac{1}{R - r_{Al}} \left[(1 - \eta)\dot{r}_{Al} + \eta \dot{R} + \frac{2D_{ox}}{r} \right] \frac{\partial Y_{O,ox}}{\partial \eta} + \frac{D_{ox}}{(R - r_{Al})^2} \frac{\partial^2 Y_{O,ox}}{\partial \eta^2}. \qquad (4.20)$$

The rate of growth of the oxide film can be obtained by stoichiometry of the reaction $2\mathrm{Al} + 3\mathrm{O} \rightarrow \mathrm{Al}_2\mathrm{O}_3$, requiring that 9 kg of aluminum be consumed for each 8 kg of oxygen, such that

$$\frac{\mathrm{d}}{\mathrm{d}t} \left\{ \frac{4}{3} \pi \rho_{Al} r_{Al}^3 \right\} = -\frac{9}{8} \dot{m}_0. \qquad (4.21)$$

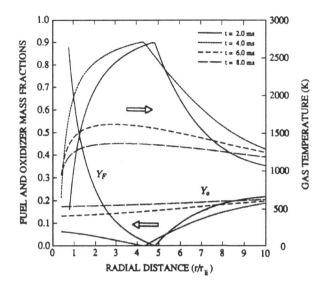

Figure 4.10. Gas-phase temperature and fuel and oxidizer mass fraction profiles for a slurry droplet: $r_{li} = 50$ μm, $r_m = 20$ μm, $p_\infty = 10$ atm, and $T_\infty = 1000$ K. (Bhatia and Sirignano, 1992, with permission of *Combustion Science and Technology*.)

Similarly, as 17 kg of oxide shell are formed for each 8 kg of oxygen consumed, the outer radius of the particle is given by

$$\frac{d}{dt}\left[\frac{4}{3}\pi\rho_{ox}\left(R^3 - r_{Al}^3\right)\right] = \frac{17}{8}\dot{m}_0. \tag{4.22}$$

It is well known that a continuous film of oxide grows over a nascent aluminum surface that is exposed to an oxidizing atmosphere (see, e.g., Hatch, 1984, and Merzhanov et al., 1977). At room temperature, the limiting thickness δ_i is approximately 25–30 Å. Here δ_i varies between 10 and 30 Å.

The oxygen flow rate at the Al–Al$_2$O$_3$ interface is given by

$$\dot{m}_0 = 4\pi r_{Al}^2 (\rho D)_{ox} \left.\frac{\partial Y_{O,ox}}{\partial r}\right|_{r=r_{Al}}. \tag{4.23}$$

Equations (4.20)–(4.23), along with the unsteady gas-phase conservation equations and heat balance equation (4.14), complete the problem definition. Continuous heating of aluminum particles causes the phase change, which occurs at 933.1 K, and, during this change, the metal temperature remains constant. After the phase change, the metal heating continues to be described by Eq. (4.14). The droplet keeps heating until the melting point of the oxide is reached at 2300 K. At this point, the ignition is assumed to occur.

A second-order, implicit, finite-difference scheme was used by Bhatia and Sirignano to solve the system of gas, liquid, and metal equations.

The results shown in Fig. 4.10 are obtained with air at 10 atm and 1000 K, and with a droplet initial radius equal to 50 μm and a liquid initial temperature equal to 300 K. The profiles of the gas-phase fuel and oxidizer fractions and the temperature for a 20-μm metal particle are shown. They are similar to those observed for the corresponding all-liquid droplet, until the completion of vaporization. After this time, the oxygen continues to diffuse toward the metal surface, as noted by the nonzero gradient in its profile. The flame collapses onto the metal surface, and the

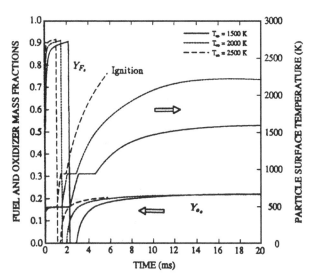

Figure 4.11. Fuel and oxidizer mass fraction and surface temperature versus time for slurry droplet: $r_{li} = 50$ μm, $r_m = 20$ μm, $p_\infty = 10$ atm, and $\delta_i = 30$ Å. (Bhatia and Sirignano, 1992, with permission of *Combustion Science and Technology*.)

droplet-surface temperature rises rapidly. For both the 10- and the 20-μm particle sizes, the metal temperature never reaches the ignition temperature of 2300 K because the conduction from the flame is the dominating mechanism of heat transfer. It is conceivable that much smaller particles with subsequently lower thermal content might reach the melting point.

An initial oxide layer thickness of 30 Å is assumed. The diffusivity of the oxygen through the oxide layer is sufficiently low so that a time delay exists between the flame collapse and the release of heat from the metal oxidation. Figure 4.11 shows the slurry-droplet history for various ambient temperatures. Note that, without extremely high ambient temperatures or forced convection, ignition of the metal particle does not occur. The energy from the burning of the hydrocarbon fuel in the slurry droplet is not sufficient to ignite the metal.

Next, we address the effects of relative droplet–gas motion and gas-phase convection on the vaporization rates and its subsequent burning process. Earlier studies used a lumped-capacity formulation that assumes that the slurry droplet has a spatially uniform temperature, thus implying infinite thermal conductivity of the liquid and the metal. Although this may be a reasonable limit for the metal, it is certainly inappropriate for the liquid fuel, as shown in Chapter 2. This analysis includes the effects associated with gas-phase convection, internal circulation, and finite conductivity of the liquid as a first step in developing a comprehensive model that can be used for spray combustion calculations.

AXISYMMETRIC CASE. Droplet motion and deceleration of the relative droplet-gas velocity that are due to drag are considered. From the work of Abramzon and Sirignano (1989), the gas-phase energy transfer and species mass transfer account for the boundary-layer motion over the droplet and the liquid-phase energy equation accounts for the circulatory motion. The same balance and equilibrium conditions are used at the gas–liquid interface and the metal–liquid interface as were used in the spherically symmetric analysis. In addition, velocity and shear stress

contour interval: 4.31E-02 from: -1.98E-01 to: 1.31E+00

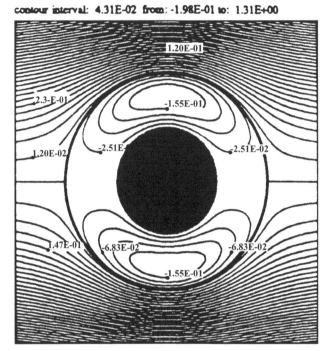

Figure 4.12. Nondimensional liquid-phase stream function for a slurry droplet. (Bhatia and Sirignano, 1993a, with permission of *Combustion Science and Technology*.)

are matched at the surface. Variable thermophysical properties, nonunitary Lewis number, and an envelope flame in the gas-phase boundary layer are considered. Details of the model can be found in Bhatia and Sirignano (1993a).

A typical stream-function plot for the slurry droplet is found in Fig. 4.12. Note that a modified spherical vortex occurs. For large liquid Peclet numbers, the internal convection determines the isotherm pattern, and the isotherms and stream functions are approximately the same surfaces. Because of drag and deceleration, the Peclet number decreases with time and the isotherms approach spherical surfaces.

The results for axisymmetric calculations are found to agree qualitatively with those for the spherically symmetric calculations. The vaporization rates approximate those rates for all-liquid droplets. As the metal-particle size is increased, the droplets are seen to accelerate slower because of the added mass. Ignition-time results are shown in Figure 4.13.

Values of the equivalence ratio of less than 0.3 did not result in ignition of the metal particles considered in this study. Note that the equivalence ratio is used to deduce only the ambient temperature experienced by the droplet. This temperature results from the combustion of the liquid fuel in many droplets. Increasing the amount of the liquid fuel results in a higher gas temperature. This implies faster heating of the dry particle and hence shorter ignition times. Depending on (among other parameters) the gas temperatures and the solid loading in the slurry, the metal ignition times can be several times larger than the liquid burnout times. An increase of the initial Reynolds number results in a small decrease in the ignition times

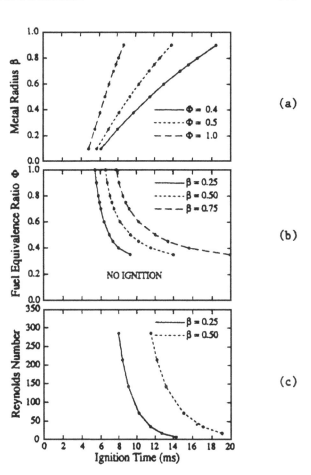

Figure 4.13. Ignition times for slurry droplets burning in air at $T_\infty = 1500$ K and $p_\infty = 10$ atm: (a) effect of initial nondimensional metal-particle radius, (b) effect of fuel-equivalence ratio based on initial droplet-fuel mass, (c) effect of Reynolds number based on initial relative droplet–gas velocity. (Bhatia and Sirignano, 1993a, with permission of *Combustion and Flame*.)

(because the metal heating is initially convection dominated). The metal ignition times increase rapidly with rising metal-particle size and can be several times larger than the liquid-fuel burnout time.

Megaridis and Sirignano (1992b) have solved the Navier–Stokes equations for this vaporization problem. Particle motion can be important according to their results so neglect of that motion can be challenged. Figure 4.14 shows the calculated eccentricity of the metal particle as an increasing function of time and Reynolds number. The eccentricity is defined as the nondimensional distance between the metal-particle center and the droplet center. The liquid in the droplet decelerates faster because of drag (caused by the relative motion with the gas) than the metal particle. Therefore the metal particle moves forward relative to the center of the droplet. This raises the potential of secondary atomization caused by the particle's breaking through the liquid surface. Calculations, however, lost numerical resolution and were stopped as the metal particle approached the gas–liquid interface.

In summary, simple models have been constructed to study combustion of a metal-slurry droplet with one metal particle. When the gas-phase quasi-steady assumption is relaxed for a slurry droplet burning in a quiescent atmosphere, the vapor burnout time can be significantly larger than the liquid-vaporization time. In

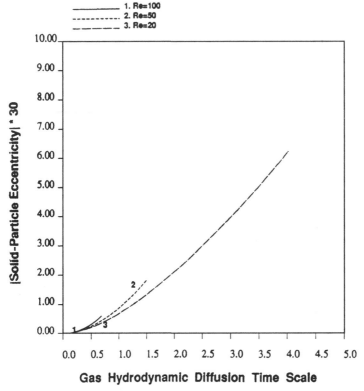

Figure 4.14. Metal-particle eccentricity versus time for different initial Reynolds numbers (Megaridis and Sirignano, 1992b).

the absence of forced convection, the flame does not have sufficient energy to ignite the aluminum particles. Forced-convective effects are considered by use of the film theory to study the gas phase. As the metal-particle size is increased, the droplets accelerate slower because of the added mass. The liquid Peclet number rises, resulting in greater heat flux into the droplet interior. When combustion of the isolated slurry droplet is considered, the heat diffusion from the hot gases is the primary heating mechanism for the dry metal for temperatures up to 2000 K, above which the heat flux from the surface oxidation of the aluminum also becomes important. A Navier–Stokes solution predicts the movement of the particle within the liquid.

4.3.2 Liquid Vaporization from Fine-Metal-Slurry Droplets

This subsection addresses vaporization of a liquid fuel from a slurry droplet composed of a large number of solid particles (aluminum) dispersed and initially suspended in a liquid hydrocarbon (*n*-octane). Previous experimental studies have established that, after an initial decrease that depends on solid loading, the droplet radius becomes fixed. A rigid porous shell is established on the droplet surface, and the agglomerates formed are hollow. This aspect was addressed by Lee and Law (1991), who postulated the formation of a bubble inside the slurry droplet. However, the dynamic equilibrium conditions within the droplet interior and the gas phase were not considered. The slurry droplet of initially known metal and liquid mass fractions is treated as a spherical droplet of spatially varying and time-varying average properties.

LIQUID-PHASE ANALYSIS. Here, the metal slurry is conceptualized as consisting of a suspension of small metal particles in a spherical liquid droplet. In this case, the property variations (in space and time) result from varying temperature and composition of the slurry droplet as it heats and liquid fuel vaporizes. Several studies are available to predict thermal conductivity of a solid–liquid mixture. We use the general mixture rule from Nielsen (1978):

$$\frac{\lambda_{\text{mix}}}{\lambda_l} = \frac{1 + AB\phi_m}{1 - B\psi\phi_m}, \tag{4.24}$$

where A is a constant that depends on the shape of the solid particles, state of agglomeration, and nature of the interface. ϕ_m is the metal volume fraction. It is related to the generalized Einstein coefficient k_E by

$$A = k_E - 1. \tag{4.25}$$

In our case, we take $k_E = 3 (A = 2)$, following Nielsen. The constant B is given by

$$B = \frac{\lambda_m/\lambda_l - 1}{\lambda_m/\lambda_l + A}. \tag{4.26}$$

The reduced concentration term ψ is approximated by

$$\psi \approx 1 + \frac{1 - \phi_{\text{max}}}{\phi_{\text{max}}^2}\phi_m, \tag{4.27}$$

where ϕ_{max} is the maximum metal volume fraction. Geometrical considerations show that, for spherical particles, $0.5236 \leq \phi_{\text{max}} \leq 0.7405$, depending on the type of the packing. There is some experimental evidence that $\phi_{\text{max}} \approx 0.54$ (Cho et al., 1989; Lee and Law, 1991), which corresponds to the most open packing. The theoretical value of $\pi/6$ is used here.

Figure 4.15 shows the vaporization sequence of the slurry droplet. Stage I is characterized by surface regression; here, the vaporization proceeds in a manner similar to that for pure liquid droplets (with modified droplet properties). The onset of shell formation (Stage II) is given by the appearance of a thin shell of solid particles forming on the droplet surface. The manner of further vaporization depends on the permeability and the strength of the shell. This is discussed in some detail under Stage II. Further vaporization causes an increase in the shell thickness and formation of a bubble in the interior of droplet, whose diameter increases with time. When the shell thickness is at the maximum value, the drying of the pores starts (Stage III). Stage IV involves the dry agglomerate heating, which begins when all the liquid has evaporated. Stage V is the metal-coalescence stage and is characterized by the coalescence of the individual solid particles into a single molten droplet.

Mass conservation of liquid implies that an interior bubble of saturated vapor (or vapor plus gas) must form, for a constant-radius liquid depletion to occur from the interior core through the porous shell. Earlier studies, e.g., Antaki and Williams (1986, 1987) and Antaki (1986), did not take the bubble formation into account, thus

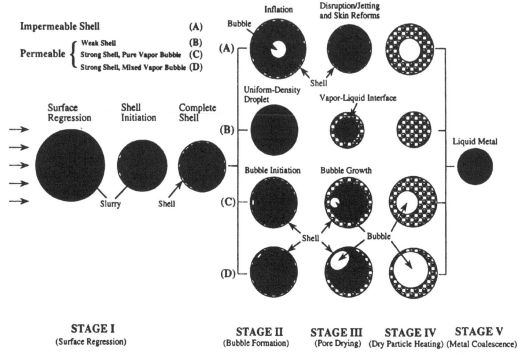

Figure 4.15. Schematic of the history of a slurry droplet with many particles (Bhatia and Sirignano, 1993b).

implicitly assuming a uniform metal-particle density inside the slurry droplet. The shell–bubble formulation is proposed in the work of Lee and Law. However, the static and thermodynamic equilibrium at the liquid–bubble interface requires some further consideration. The temperature in the liquid core is lower than at the gas–liquid interface. Hence the saturated vapor pressure in the bubble must be lower than the environmental pressure. Thus the static equilibrium cannot be maintained, and the vapor in the bubble will have to condense back into the liquid phase.

Investigating the conditions under which a vapor bubble might exist in the core, we have to examine the structure of the shell. The hydrocarbon is a wetting liquid. As previously noted, the shell contains a known ($\phi_{\max} = \pi/6$) fraction of the metal. Hence ~48% of the volume is void. The characteristic radius associated with these micropores is approximately of the same order as the radius of the solid particles forming the shell [Marcus (1972) suggests that the radius of curvature $r_c \approx 0.41r_m$ for randomly packed beds.] Consider a wick structure in the shell, as shown in Fig. 4.16. Developing a rough argument, we assume that the radius of curvature of the meniscus at the gas–liquid interface is approximately the same as the pore radius. In the case of quasi-static equilibrium,

$$p_l = p_\infty - \frac{2\sigma_o}{R_c},\tag{4.28}$$

$$p_b = p_l + \frac{2\sigma_i}{R_b} = p_\infty - \left\{\frac{2\sigma_o}{R_c} - \frac{2\sigma_i}{R_b}\right\},\tag{4.29}$$

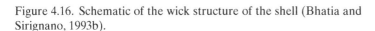

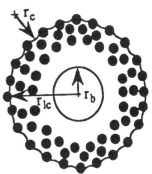

Figure 4.16. Schematic of the wick structure of the shell (Bhatia and Sirignano, 1993b).

where σ_o and σ_i denote the surface-tension coefficient at the outer and the inner gas–liquid interfaces, respectively, and p_l, p_b, and p_∞ are the pressures in the liquid, bubble, and ambient, respectively.

The bubble radius R_b is at least an order of magnitude larger than the pore radius, and hence the second term in the braces may be neglected in comparison with the first. Hence, if the stress (defined as $2\sigma_o / R_c$) is large, the capillarity might cause the pressure in the vapor bubble to be maintained at a higher value than the saturation pressure. In general, the pressure difference may not satisfy Eq. (4.29), therefore causing an instability of the bubble in the center of the droplet.

STAGE I: SURFACE REGRESSION. The regression history of the slurry diameter, because of vaporization, is dependent on the initial solid loading. For higher initial solid loadings (with solid volume fractions close to $\phi_{\max}$), the slurry diameter may not regress at all. We consider the more general case of low initial solid loading.

For simplicity, we assume (from Lee and Law) that the porous shell contains the solid particles in the shell thickness and the volume is swept away because of vaporization. The shell thickness δ may be found by use of solid-mass conservation:

$$\frac{4}{3}\pi R_0^3 \phi_{m0} = \frac{4}{3}\pi \left[R^3 - (R - \delta)^3 \right] \phi_{\max} + \frac{4}{3}\pi (R - \delta)^3 \phi_{m0}, \tag{4.30}$$

where ϕ_{m0} and $\phi_{\max}$ denote the initial and the maximum solid volume fractions, respectively.

To model the heat transfer in the liquid phase, we take the configuration to be spherically symmetric. In this case, the nondimensional form of the energy equation can be extended from Eq. (4.7) for the slurry material [with $\eta = r/R(t)$, $\tau = \alpha_{\mathrm{mix}0} t / R_0$, $r_s = R(t)/R_0$]:

$$r_s^2 \frac{\partial Z}{\partial \tau} - \left\{ \frac{1}{2}\eta \frac{dr_s^2}{d\tau} - \frac{c_{pm}}{c_{p\mathrm{mix}}} \frac{v_m R Y_m}{\alpha_{\mathrm{mix}0}} \right\} \frac{\partial Z}{\partial \eta} = \frac{(\rho c_p)_{\mathrm{mix}0}}{(\rho c_p)_{\mathrm{mix}}} \frac{1}{\eta^2} \frac{\partial}{\partial \eta} \left\{ \frac{\lambda_{\mathrm{mix}}}{\lambda_{\mathrm{mix}0}} \eta^2 \frac{\partial Z}{\partial \eta} \right\}, \tag{4.31a}$$

where the average density and the specific heat at any radial location of the slurry droplet can be obtained by mass averaging:

$$\rho_{\mathrm{mix}} = \left\{ \frac{Y_m}{\rho_m} + \frac{1 - Y_m}{\rho_l} \right\}^{-1}, \qquad c_{p\mathrm{mix}} = Y_m c_{pm} + (1 - Y_m) c_{pl}. \tag{4.31b}$$

The velocity of the metal particles v_m is zero in the droplet core and, in the shell, is given by $-\mathrm{d}\delta/\mathrm{d}t$. The initial and the boundary conditions are

$$at\ r = 0, \quad Z = 0, \tag{4.31c}$$

$$at\ \eta = 0, \quad \frac{\partial Z}{\partial \eta} = 0, \tag{4.31d}$$

$$at\ \eta = 1, \quad \frac{\partial Z}{\partial \eta} = \frac{\dot{Q}_l}{4\pi \lambda_{\mathrm{mix}} R T_0}. \tag{4.31e}$$

The droplet-radius reduction is governed by

$$\frac{\mathrm{d}R}{\mathrm{d}t} = -\dot{m}_l / (4\pi \rho R^2). \tag{4.32}$$

The vaporization progresses according to Eqs. (4.31), and to the equations given later in the subsection on gas-phase analysis, until a solid shell forms on the droplet surface. There is experimental evidence from Charlesworth and Marshall (1960) that the shell formation begins from the front stagnation point and proceeds rapidly toward the rear stagnation point. A critical stage is reached when the solids in the outermost layer cannot be compacted anymore. On physical grounds, the shell formation should begin as soon as the shell porosity reaches a critical value. Lee and Law characterize one onset of this stage by a critical packing and a critical shell thickness δ_c, which is taken to be three times the solid-particle diameter. The effect of the internal circulation on the shell formation of a vaporizing slurry droplet is not well understood. The study of Chung (1982) concludes that, for high Reynolds number (Re_r or $\mathrm{Re}_\theta \sim 100$) cases, the Basset plus curvature forces are stronger than the pressure force and the particle moves toward the droplet surface. Some similar observations are reported in the experimental work of Charlesworth and Marshall. The characteristic time for stabilization of the internal vortex, t_{hydr}, depends on the liquid Reynolds number. At $\mathrm{Re} \leq 1$ (viscous regime), this may be estimated as $t_{\mathrm{hydr}} \sim R^2/v_l$. At high liquid Reynolds number, $\mathrm{Re} \gg 1$, the vorticity disturbance transfer from the surface to the interior is convection and $t_{\mathrm{hydr}} \approx R/U_s \approx R^2/v_l \mathrm{Re}_l$. This is approximately 1–2 orders of magnitude less than the droplet lifetime. The time taken for shell formation is comparable to the lifetime. Hence the internal circulation may play a dominant role in the onset of the shell. This is especially true for smaller droplets. However, this effect is neglected in the present study. We assume that the shell is formed by the radially inward-swept solids as the liquid vaporizes. Following Lee and Law, we assume that the critical solid volume fraction corresponds to that for the most open packing (i.e., $\phi_{\mathrm{max}} = \pi/6$) and the corresponding shell thickness is three times the solid-particle diameter.

STAGE II: LIQUID VAPORIZATION THROUGH THE SHELL. The phenomena during this stage are dependent on the ambient pressure and temperature, the type of the liquid fuel, and the particle size. The manner of liquid vaporization depends on the permeability, the effective pore radius, the strength of the shell, and, depending on these, there are several models (refer to Fig. 4.15).

(A) **Impervious Shell, Any Ambient Pressure:** High-molecular-weight organic dispersing agents are often added to liquid fuel to stabilize the slurries. These additives may pyrolize as the dry outer surface of the shell gets heated to temperatures higher than the liquid-fuel boiling point, thus making the shell impervious (Wong and Turns, 1987; Cho et al., 1989). Also, the pore spaces between the metal particles could be physically filled by addition of submicrometer particles, e.g., $\sim$0.35-μm carbon particles in a slurry droplet consisting of 4-μm aluminum particles, as in the work of Wong and Turns (1987). If the shell is impervious to the gases and the liquid flow, then the droplet heating will result in nucleation inside the shell. Further heating will cause additional internal vaporization, raising the pressure inside the droplet. This situation is likely to result in the following cases:

  (i) **Swelling and Jetting:** If the shell is plastic, it is going to inflate because of internal vaporization and pressure buildup. This stretching of the shell may increase the permeability such that jetting of the liquid fuel (and solid particles) occurs to relieve the internal pressure. The phenomenon of swelling and jetting has been experimentally observed in several situations (Antaki and Williams, 1987; Lee and Law, 1991). Further, Wong and Turns report formation of secondary spheres, attached to the primary slurry droplet.

  (ii) **Disruption/Fracture:** If the shell remains impervious, then disruption is going to occur. This case has been observed by many researchers (Takahashi et al., 1986, 1989; Wong and Turns, 1987; Cho et al., 1989; Lee and Law, 1991).

(B) **Permeable Shell, Low Buckling Resistance:** The shell formed on the droplet surface has a certain buckling resistance, which would depend on its mechanical properties. If the buckling resistance is lower than the stress, the shell will collapse, and the result will be uniform density vaporization without any bubble. In such a case, the surface regression occurs until the metal packing inside the droplet reaches the critical value. The remaining liquid will get vaporized by the vapor–liquid interface retreating into the porous-sphere interior. This case was used by Antaki and Williams (1986). Of the numerous experimental studies available, this case does not seem to be generally observed. It is remarked here that Lee and Law (1991) and Takahashi et al. (1989) present evidence of a recessed pit on the outer surface of the shell. Lee and Law speculate that this phenomenon is coupled to the flow field because the flat portion of the droplet is always situated in the rearward stagnation region. However, this is probably a result of stress, as just explained, and the shell wall caves in near the rear stagnation region as the shell is weakest and rigidizes last there. Usually, the shell buckling resistance is larger than the stress, and this situation is considered in the next two cases.

(C) **Permeable Shell, Low Ambient Pressure:** In this model, the ambient pressure and the stress that is due to the surface tension on the gas–liquid interface are of the same order of magnitude. So a liquid-fuel-vapor bubble can be supported inside the slurry droplet, and this enables conservation of mass in the event of vaporization from a sphere of constant radius. This case is termed the shell–bubble case. This nucleation will start on the inside of the shell from the front stagnation point, where heat transfer rate is maximum. Marangoni convection, i.e., thermocapillary force, can cause the vapor bubble to be attracted to the warmer temperature (Young et al., 1959). Thus the bubble may remain attached to the forward stagnation point. Subsequent vaporization will cause the bubble to grow. Note that, in such a situation, the stress value could be constantly adjusting with the growth in the bubble radius. As previously noted, this situation is generally unstable and unlikely to exist.

(D) **Permeable Shell, High Ambient Pressure:** The largest value of the stress is limited by the largest available statistical pore size and the liquid-surface tension. If the stress exceeds bubble point, then blow-through occurs and stress gets fixed at the bubble-point value. In such a case, the surrounding gas (mostly fuel vapor and nitrogen) enters the shell from the largest pore, and the resulting bubble consists of fuel vapor and the gas. The bubble will be located asymmetric to the droplet center, as shown in Fig. 4.15. Scanning electron microscope pictures made by Lee and Law and Cho et al. support this observation. This situation is likely to occur when the ambient pressure is relatively higher than the bubble point, as would be expected in realistic combustors.

Depending on the slurry contents, at various times it is quite possible that a combination of the preceding situations exists. In this study, we use the shell–bubble model to describe the slurry-droplet history. Researchers do not agree about whether the porous shell formed can be dry or wet with the liquid carrier, and that is a topic of active research. Lee and Law consider both limiting cases, and, from their analytical procedure and experimental results, they reach the conclusion that wet-shell gasification is more realistic. The dry-shell case is shown to cause the shell maximum porosity to be 0.68, implying that $\phi_{max} = 0.32$, which is well outside the range $0.5236 \leq \phi_{max} \leq 0.7405$, and thus leading to physically unrealistic (very loose) packing of the solid particles. Note that their formulation has three potential sources of error:

1. The mass flux of liquid vaporizing internally to the bubble is neglected.
2. The burning rate is given by the classical expression

$$\dot{m}_l = \frac{\pi}{2}\rho_l R_c K, \tag{4.33}$$

   where K is the burning constant. The use of R_c in Eq. (4.33) is justified in the wet-shell case. However, in the dry-shell case, vaporization occurs at the liquid radius $R < R_c$, which should be used to determine the burning rate. This is also true in the model of Cho et al. (1989).
3. The effect of capillarity and the pressure balance has not been considered.

In this study, we consider the more general permeable-shell case [(D) in the preceding list] with high ambient pressure, with the gas bubble consisting of fuel vapor and ambient gas attached to the droplet surface. In such a case, to maintain quasi-static equilibrium, it follows from Eq. (4.29) that

$$p_b = p_v + p_g = p_\infty - \left\{ \frac{2\sigma_o}{R_c} - \frac{2\sigma_i}{R_b} \right\}, \tag{4.34}$$

where p_v and p_g denote the partial pressures of the fuel vapor and the ambient gas (inert) in the bubble. If the bubble is not attached to the surface, that implies that $p_g = 0$, which is case (C) in the preceding list.

To develop a simplified model, we assume that, only for the purpose of heat transport calculations, the bubble is placed spherically symmetric with respect to the droplet center. Thus a one-dimensional calculation is made inside the droplet to determine the temperature profile.

To obtain the shell thickness δ, solid-mass conservation yields

$$\frac{4}{3}\pi R_0^3 \phi_{m0} = \frac{4}{3}\pi [R^3 - (R - \delta)^3]\phi_{max} + \frac{4}{3}\pi \left[(R - \delta)^3 - R_b^3 \right] \phi_{m0}, \tag{4.35}$$

where R is the outer radius of the slurry droplet.

In the one-dimensional heat diffusion equation, with the following transformations,

$$\tau = \frac{\alpha_{mix\,0} t}{R_0^2}, \quad r \geq 0, \tag{4.36}$$

$$\eta = \frac{r - R_b(t)}{R(t) - R_b(t)}, \quad 0 \leq \eta \leq 1, \tag{4.37}$$

$$Z = \frac{T - T_0}{T_0}, \quad r_l(t) = R(t)/R_0,$$

$$r_b(t) = R_b(t)/R_0, \quad \xi(t) = \frac{R_b(t)}{R(t) - R_b(t)}, \tag{4.38}$$

we get the transformed energy equation with fixed boundaries:

$$(r_s - r_b)^2 \frac{\partial Z}{\partial \tau} + \left[\frac{a(r)(R - R_b)}{\alpha_{mix\,0}} - \eta(r_s - r_b)\frac{dr_s}{d\tau} - (1 - \eta)(r_s - r_b)\frac{dr_b}{d\tau} \right]\frac{\partial Z}{\partial \eta}$$

$$= \frac{(\rho c_p)_{mix\,0}}{(\rho c_p)_{mix}} \frac{1}{(\eta + \xi)^2} \frac{\partial}{\partial \eta}\left[\frac{\lambda_{mix}}{\lambda_{mix\,0}}(\eta + \xi)^2 \frac{\partial Z}{\partial \eta} \right]. \tag{4.39}$$

In Eq. (4.39) the thermal diffusion resulting from unequal species velocities v_m and v_l (i.e., interdiffusion or Dufour effect) is neglected. The function $a(r)$ is given by

$$a(r) = \begin{cases} \left(\dfrac{R_b}{r}\right)^2 \dfrac{dR_b}{dt} & \text{for } R_b < r < (R - \delta) \\[2ex] \left(\dfrac{R_b}{r}\right)^2 \dfrac{(\rho c_p)_l}{(\rho c_p)_{mix}} \dfrac{1 - Y_m}{1 - \phi_{max}} \dfrac{dR_b}{dt} & \text{for } (R - \delta) < r < R \end{cases}. \tag{4.40}$$

Equation (4.39) with $r_b(t) = 0$ and $dr_b/d\tau = 0$ recovers the energy equation for the surface regression, Stage I, Eq. (4.31a). During Stage II, the droplet outer radius remains constant, $r_s(t) = R/R_0$, and $dR/d\tau = 0$. The initial and the boundary conditions for Eq. (4.39) are given by

$$\text{at } r = 0, \qquad Z = 0, \tag{4.41}$$

$$\text{at } \eta = 0, \qquad Z = T_b(t)/T_i, \tag{4.42}$$

$$\text{at } \eta = 1, \qquad \frac{\partial Z}{\partial \eta} = \frac{[R - R_b(t)]\dot{Q}_l}{4\pi \lambda_{\text{mix}} \tau_{lc}^2 T_0}. \tag{4.43}$$

We assume a perfectly mixed bubble, i.e., the incoming mixture mixes perfectly to a uniform fuel mass fraction Y_{F_b} and the temperature T_b. The gas thermal diffusivity is $\sim$2 orders of magnitude smaller than the liquid thermal diffusivity. The incoming mass flow rate into the bubble ($\dot{m}_e$), the bubble-vaporization–condensation rate ($\dot{m}_b$), and the bubble volume (V_b) are determined by expressing the bubble-conservation equations as

$$\dot{m}_b + Y_{Fs}\dot{m}_e = \frac{d}{dt}(\rho_b Y_{Fb} V_b), \tag{4.44}$$

$$\dot{m}_b + \dot{m}_e = \frac{d}{dt}(\rho_b V_b), \tag{4.45}$$

$$\dot{m}_l + \dot{m}_b = \rho_l \frac{dV_b}{dt}. \tag{4.46}$$

The preceding three equations, along with the net mass flux [Eq. (4.61)] are solved for the four unknowns: $\dot{m}_l, \dot{m}_e, \dot{m}_b$, and V_b. The bubble density ρ_b is determined from an equation of state:

$$\rho_b = p_b \bar{M}/R_u T_b, \tag{4.47}$$

and the fuel mass fraction is determined as a function of bubble temperature T_b from the Clausius–Clapeyron relation.

The bubble temperature is found by making an energy balance:

$$(mc_v)_b \frac{dT_b}{dt} = \dot{m}_e c_{pe}(T_s - T_b) + 4\pi r_b^2 \lambda_{\text{mix}} \left.\frac{dT}{dr}\right|_{r=r_b} - \dot{m}_b L(T_b). \tag{4.48}$$

In the Lee and Law model, $\dot{m}_b$ and $\dot{m}_e$ are not considered. In the earlier work by Antaki (1986) and Antaki and Williams (1986), the bubble formation is not considered, which is contrary to experimental evidence because hollow shells have been found after vaporization for slurry droplets without heavy loading. To model impervious case (A) and collapsed-shell case (B), we can put the external and the incoming flux rates to zero, respectively.

STAGE III: PORES DRYING. This stage begins when the shell thickness reaches a maximum value (and therefore the bubble radius equals the inner-shell radius) and all the solid particles in the drop are at maximum compactness. Depending on the thickness of the shell, in general there is still a substantial mass of liquid trapped in the pores of the agglomerate. The liquid vaporization continues to be described by the equations under Stage III. When the bubble radius equals the inner-shell radius, $R_{bc} = R - \delta_{\text{max}}$, the surface-tension effects on the liquid–gas interface on the droplet surface and the liquid–bubble interface cancel. Hence the bubble pressure is assumed to be the ambient pressure and the external gas-inflow rate $\dot{m}_e$ is assumed to be equal to zero. In this subsection, we use an expanding-bubble–regressing-liquid model. The bubble is allowed to expand into the shell because of the effects of temperature increase, which change gas density and the fuel mass fraction equilibrium inside the bubble. The conservation equations are given by

$$\dot{m}_b = \frac{\mathrm{d}}{\mathrm{d}t}(\rho_b Y_{Fb} V_b) = \frac{\mathrm{d}}{\mathrm{d}t}(\rho_b V_b), \qquad (4.49)$$

where the bubble volume is given by

$$V_b = \frac{4}{3}\pi R_{bc}^3 + \frac{4}{3}\pi (R_b^3 - R_{bc}^3)(1 - \phi_{\max}). \qquad (4.50)$$

The second part of Eq. (4.49) implies that the mass of the nonfuel species remains constant during the pore-drying stage. During this stage, we distinguish between the outer-shell radius R and the outer-liquid radius R_l. Because the liquid surface is regressing, the external vaporization rate and the liquid radius are given by

$$\dot{m}_l = 2\pi \bar{\rho}_g \bar{D}_g R_l \mathrm{Sh}^* \ln(1 + B_M) \quad \text{[see Eq. (3.34c)]}, \qquad (4.51)$$

$$\dot{m}_l + \dot{m}_b = -\rho_l \frac{\mathrm{d}V_l}{\mathrm{d}t} = -\rho_l \frac{\mathrm{d}}{\mathrm{d}t}\left[\frac{4}{3}\pi (R_l^3 - R_b^3)(1 - \phi_{\max})\right], \qquad (4.52)$$

and the volume of the dry pores on the outside of the wet pores is assumed to be equal to the volume of the dry pores on the inside of the wet pores. The thermal conductivity of the agglomerate is high; hence the energy conservation inside the droplet is given by (taking bubble temperature T_b = surface temperature $T_s = T_d$)

$$\dot{Q}_l - \dot{m}_b L(T_d) = (mc_v)_d \frac{\mathrm{d}T_d}{\mathrm{d}t}. \qquad (4.53)$$

The five equations in this subsection are solved for the five unknowns: $\dot{m}_l$, $\dot{m}_b$, V_b, V_l, and T_d. The preceding formulation neglects any resistance to the flow of fuel vapor through the pores. The mean free path of n-octane vapors at 10-atm pressure is estimated to be ~ 0.04 μm, which is much less than the mean pore size of ~ 1 μm, thus justifying the neglect of pore resistance.

Bhatia (1993) made calculations with an expanding-bubble–constant-outer-liquid-radius model, in which the bubble is allowed to grow into the inside pores and the outer surface of the droplet is kept wet with liquid. This results in an $\sim 10\%$ reduction in the liquid-fuel-vaporization time. However, the expanding-bubble–regressing-liquid formulation is preferred over the expanding-bubble–constant-liquid model.

GAS-PHASE ANALYSIS. The gas-phase-conservation equations determine the fuel-vaporization and the bubble-inflow rates and the heat penetrating into the droplet interior. In this study, we modify the droplet-vaporization model of Abramzon and Sirignano (1989) to account for the gas inflow in the bubble corresponding to the vaporization of liquid fuel during Stages II and III. The model assumes quasi-steady gas-phase heat and mass transfer and no pressure drop, and the thermal properties (evaluated at reference conditions) are treated as constant. Film theory is used to include the effect of the convective transport. The formulation is subsequently briefly explained.

The molar and mass fuel-vapor fractions at the droplet surface are given by

$$X_{Fs} = p_{Fs}/p, \tag{4.54}$$

$$Y_{Fs} = X_{Fs} W_F \bigg/ \sum_i X_i W_i, \tag{4.55}$$

where p_{Fs} is the fuel-vapor saturated pressure, obtained from the Clausius–Clapeyron relation.

The Nusselt and the Sherwood numbers for the vaporizing droplet are calculated following the work of Abramzon and Sirignano [see Eqs. (3.24)–(3.29)].

Consider a spherical control volume of radius $r(r_l < r < r_f)$. Let n_1 and n_2 denote the mass fluxes (in units of grams per square centimeter times seconds) of fuel and air flowing radially outward. The fuel-vaporization rate $\dot{m}_l$ is related to n_1 and n_2 by

$$n_{1_s} = \frac{\dot{m}_l - Y_{Fs} \dot{m}_e}{4\pi R_l^2}, \qquad n_{2_s} = \frac{-(1 - Y_{Fs})\dot{m}_e}{4\pi R_l^2}. \tag{4.56}$$

Considering quasi-steady mass transfer, we have

$$r^2 n_1 = R_l^2 n_{1_s}, \qquad r^2 n_2 = R_l^2 n_{2_s}. \tag{4.57}$$

Note that, for a nonslurry droplet and also during Stage I vaporization from a slurry droplet, $n_2 = 0$. Using Fick's law, we get

$$-r^2 \rho_g D_g \frac{dY_F}{dr} + Y_F R_l^2 (n_{1_s} + n_{2_s}) = R_l^2 n_{1_s}. \tag{4.58}$$

Making the substitutions $\eta = R_l/r$ and $\omega = (b_M Y_F - 1)/(b_M Y_{Fs} - 1)$ and defining mass and heat correction factors to the transfer numbers as

$$b_M = 1 + \frac{n_{2_s}}{n_{1_s}}, \qquad b_T = 1 + \frac{n_{2_s} c_{pg}}{n_{1_s} c_{pF}}, \tag{4.59}$$

we obtain

$$\frac{\rho_g D_g}{n_{1_s} R_l b_M} \frac{d\omega}{d\eta} + \omega = 0. \tag{4.60}$$

Integrating Eq. (4.60) from $r = R_l$, $\omega = 1$ to $r = r_f$, $\omega = (b_M Y_{F\infty} - 1)/(b_M Y_{Fs} - 1)$, and by using Eqs. (4.57), we now have the corrected equation:

$$\dot{m}_l - \dot{m}_e = 2\pi \bar{\rho}_g \bar{D}_g R_l \text{Sh}^* \ln(1 + B_M). \tag{4.61}$$

The mass and heat transfer numbers are calculated as

$$B_m = \frac{Y_{Fs} - Y_{F\infty}}{1/b_m - Y_{Fs}}, \tag{4.62}$$

$$B_T = \frac{\bar{c}_{pf} b_T (T_\infty - T_s)}{L(T_s) + \dot{Q}_l/\dot{m}_l}. \tag{4.63}$$

The procedure is iterative until convergence is reached by b_T. The parameter φ is given by

$$\varphi = \frac{\bar{c}_{pF}}{\bar{c}_{pg}} \frac{\text{Sh}^*}{\text{Nu}^*} \frac{1}{\text{Le}} \frac{b_T}{b_M}, \tag{4.64}$$

$$\text{Le} = \bar{\lambda}_g / (\bar{\rho}_g \bar{D}_g \bar{c}_{pg}). \tag{4.65}$$

Finally, the heat penetrating into the liquid droplet is given by

$$\dot{Q}_l = 4\pi R_l^2 n_{1_s} \left[\frac{\bar{c}_{pF} b_T (T_\infty - T_s)}{B_T} - L(T_s) \right]. \tag{4.66}$$

DROPLET DYNAMICS. Consider a slurry droplet of radius R consisting of N metal spheres of radius r_m and density ρ_m. Then the metal volume and mass fractions are given by

$$\phi_m = \frac{\frac{4}{3}\pi r_m^3 N}{\frac{4}{3}\pi R^3} = \frac{N}{(R/r_m)^3}, \tag{4.67}$$

$$Y_m = \frac{1}{1 + \left(\dfrac{1}{\phi_m} - 1\right)\dfrac{\rho_l}{\rho_m}}. \tag{4.68}$$

Following Abramzon and Sirignano (1989), we consider the gas flow to be one-dimensional and the initial droplet velocity parallel to the gas-flow direction. For the surface-regression stage, the drop motion and radius reduction are governed by the following equations:

$$\frac{\mathrm{d}X}{\mathrm{d}t} = U_l, \tag{4.69}$$

$$\frac{\mathrm{d}U_l}{\mathrm{d}t} = \frac{3}{16} \frac{\bar{\mu}_\infty}{\rho_l} \frac{(U_\infty - U_l)}{R^2} \text{Re} C_D \frac{1 - Y_m}{1 - \phi_m}, \tag{4.70}$$

$$\text{Re} = \frac{2\rho_\infty |U_\infty - U_l| R}{\bar{\mu}_\infty}, \tag{4.71}$$

$$C_D = \frac{24}{\text{Re}} \left\{ 1 + \frac{\text{Re}^{2/3}}{6} \right\}. \tag{4.72}$$

It has been shown (Yuen and Chen, 1976) that the drag coefficient for evaporating droplets may be approximated by the standard drag curve, provided the gas viscosity is evaluated at some reference temperature and fuel concentration:

$$T_{\text{ref}} = T_s + A_r(T_\infty - T_s), \tag{4.73}$$

$$Y_{F,\text{ref}} = Y_{Fs} + A_r(Y_{F\infty} - Y_{Fs}), \tag{4.74}$$

where $A_r = 1/3$ for the one-third averaging rule.

Results have been obtained for two sets of parameters. In the first set, the ratio of particle size to droplet size was varied with constant initial volume of the solid

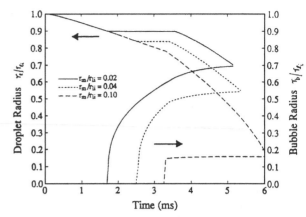

Figure 4.17. Droplet radius and bubble radius for different metal-particle sizes and fixed initial metal volume fraction (0.25) (Bhatia and Sirignano, 1993b).

particles at 25%. In the second set, the particle size was fixed and the initial volume fraction varied. Figure 4.17 shows the results for liquid vaporization from the slurry droplets initially containing 25% by volume (~51% by weight) of aluminum particles. The droplet radius and the bubble radius are shown, both normalized by the initial droplet radius. The relative durations of the three stages of vaporization are dependent on the initial solid loading.

Figure 4.18 shows the normalized shell thickness and the bubble pressure as functions of time for various metal-particle radii. The shell thickness increases as vaporization of the liquid occurs. The imposed criterion is that when the shell thickness becomes six particle radii, the shell becomes rigid, with the outer-droplet radius becoming fixed and an interior bubble beginning to form. However, shell growth continues until all of the liquid is vaporized and dry-pore heating begins. The bubble pressure increases because of the decrease in the pore size as the shell forms. When the liquid is totally vaporized, the effect of surface tension disappears and the pressure inside the shell equals the ambient pressure. During the surface-regression period, and after a short time, the $R_l^{1.5}$ versus time curve is found to be linear because the heat and mass exchange between the gas and the slurry droplet is convection dominated [$Re \sim O(100)$].

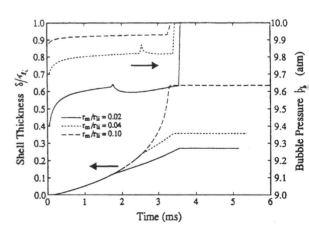

Figure 4.18. Shell thickness and bubble pressure for different metal-particle sizes and fixed initial metal volume fraction (0.25) (Bhatia and Sirignano, 1993b).

Ray and Sirignano (1992) analyzed the effect of droplet internal circulation on the shell formation and report that the circulation may promote shell growth rate. In conclusion, the phenomena of transient droplet heating, droplet surface regression, shell and bubble formation, and pore drying of the shell are shown to be salient features of fine-metal-slurry-droplet vaporization, and model formulations are presented. Relative fractions of vaporization are dependent on the initial solid loading and the metal-particle size. The droplet behavior is transient in nature, and the classical d^2-law-type analysis cannot be used to describe vaporization of liquid fuel from slurry droplets.

4.3.3 Metal-Particle Combustion with Oxide Condensation

Metal-particle combustion has relevance across other applications in addition to metal-slurry combustion. It is pertinent to solid-propellant combustion and accidental metal-dust deflagrations. Pioneering efforts in the study of metal-particle combustion were performed by Glassman and co-workers (Glassman, 1987). See also the recent review by Williams (1997).

After the melting of this oxide, aluminum burns mostly in the gas phase. The burning mechanism of aluminum is complicated by product condensation because of the nature of the volatility differences between the fuel and its combustion products. A critical aspect in the earlier studies (e.g., Law, 1973) has been the avoidance of the thermodynamic phase-equilibrium relation between the condensed and the vapor phases; the condensation is assumed to occur only at fixed radial positions: at the flame and at the droplet surface. In addition, arbitrary specification of the droplet surface temperature and the flame temperature are required, rather than being determined in the model.

A more recent and more thorough analysis reveals some interesting features. Although the gas-phase behavior is quasi-steady, the radial mass flux for the gas flow will vary with radial position because of condensation of the oxide product. The variation is so large that the flow outside of the condensation region is directed radially inward whereas the flow inside the condensation region is directed radially outward. The condensed particles are dragged by the gaseous flow so that they remain within the condensation region. Thermophoresis does not significantly modify the behavior. The governing isobaric, spherically symmetric, and quasi-steady gas-phase mass and energy equations are simplified by use of the Shvab–Zel'dovich transformation and analytically integrated. Here the condensed particles do not diffuse so that the Shvab–Zel'dovich formulation does not reduce the equations to identical form. However, a canonical description of the system is achieved. The advantage of this model over earlier theoretical work is that a knowledge of the droplet, flame, and condensation temperatures is not needed a priori. For all the cases analyzed, in a zone bounded by the droplet radius and a radius that can be greater than or equal to flame radius, the gas temperatures are higher than the saturation temperatures of the oxide vapor. Hence this zone is condensate free. Details can be found in Bhatia (1993).

Results from Bhatia and Sirignano (1993b) for droplet-surface and flame temperatures, flame radius, and residual particle radius are compared with available

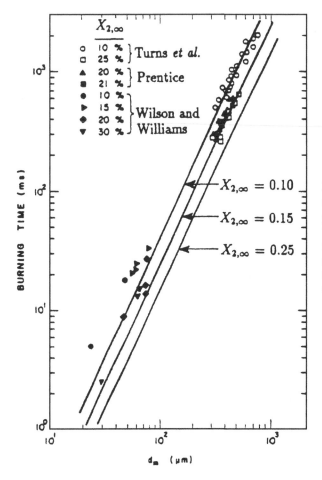

Figure 4.19. Comparison of predicted (Bhatia and Sirignano, 1993b) and experimental (Wong and Turns, 1987) values of the slurry-droplet-burning time: solid lines are predicted results for three ambient oxygen mole fractions (0.10, 0.15, and 0.25). $T_\infty = 1660$ K, $p_\infty = 1$ atm (Bhatia, 1993b).

experimental data. Good agreement with published experimental results for droplet temperature, final droplet size, flame temperatures, and burning times is obtained. The agreement on droplet-burning times between the experimental and the theoretical observations is shown in Fig. 4.19. A limitation of the model is that the effects of heat of dissociation are not considered.

Some interesting experiments by Molodetsky et al. (1996, 1998) for titanium and zirconium metal-droplet combustion raise new perspectives. For example, they show that the combustion products (e.g., oxides and nitrides) can dissolve into the droplet interior, causing a microexplosion. The burning of aluminum (and aluminum alloy) droplets in CO_2 atmospheres was studied experimentally by Rossi et al. (2001). The problem has potential relevance to power generation in the Martian environment. The droplet lifetime obeyed a $d^{2.5}$ law rather than the classical d^2 law. The droplets burned faster in CO_2 than in air.

4.4 Emulsified-Fuel-Droplet Vaporization and Burning

Emulsified liquids are heterogeneous mixture of immiscible liquids. In the binary case, one component forms the continuous phase and the other component forms the discrete phase. The discrete phase consists of globules of the one fluid that are suspended in the continuous-phase fluid. Water-in-oil emulsions are commonly considered for combustion applications because emulsions can result in microexplosion, thereby reducing the average droplet diameter and enhancing the vaporization rate. Also, water addition to the combustion process can result in reduced soot formation. The water is the discrete phase in the water-in-oil emulsion; note that oil-in-water emulsions can also be created. An excellent review of emulsified-fuel combustion was given by Dryer (1976).

Limited studies on the modelling of emulsified-liquid-droplet vaporization have been published. The studies of Jackson and Avedisian (1996,1998), Law et al. (1980b), and Law (1977) are noteworthy. All of the studies consider spherically symmetric vaporization; forced convection is not addressed. The gas phase is considered to be quasi-steady. The droplet temperature is considered to be uniform at the wet-bulb value. Law (1977) and Law et al. (1980b) used constant properties and one-step chemical kinetics for hydrocarbon oxidation. Results were presented for octane, dodecane, and hexadecane as the oil in the water-in-oil emulsion. Jackson and Avedisian considered variable properties and detailed oxidation kinetics for water/*n*-heptane emulsions. The kinetics were taken from Warnatz (1984).

Law (1977) and Law et al. (1980b) treat the gas phase essentially in the quasi-steady manner described in Section 4.1. Water and a hydrocarbon are the two vaporizing components; the water vapor is chemically inert whereas the hydrocarbon vapor oxidizes in a thin flame. Law (1977) considers that vaporization occurs as the surface regresses in a "frozen" manner, that is, the vaporization rate of each component is proportional to the amount of each component in the liquid phase:

$$\varepsilon_m = Y_{l,m} = Y_{l,m}(t = 0). \tag{4.75}$$

This implies that no motion of the discrete and the continuous phases occurs within the droplet. Note that, for an emulsified fuel, the local mass or molar fraction is determined as the average over a domain that is larger than the size of the discrete-phase particles and larger than the distance between neighboring discrete particles.

Law assumes that the gas-phase molar (or mass) fraction of each vaporizing species is given at the droplet surface by a Clasius–Clapeyron relation. This is incorrect because the sum of the two vaporizing molar (or mass) species could be greater than unity. Recognizing a difficulty, Law used only one of the two Clasius–Clapeyron relations for the component that is arbitrarily declared to be the dominant vaporizing species.

This author finds that ad hoc correction still unsatisfactory; it would be better to use a surface-area weighting for each component. Multiplication of the Arrhenius factor in the phase-equilibrium relation by the fraction of surface area identified with the liquid component would be more appealing and would avoid a summation greater than unity. This use of the two phase-equilibrium relationships would require a relaxation of Eq. (4.75). Some relative motion of the two liquid phases is then required. In general, $\varepsilon_m \neq Y_{l,m}(t = 0)$.

Law discusses the potential for microexplosion and the two limits of superheat: the kinetic limit and the thermodynamic limit. At the kinetic temperature limit (pressure dependent), spontaneous boiling can be initiated. The thermodynamic temperature exceeds the kinetic limit and is a transition point that separates the metastable and unstable liquid states. Law uses the thermodynamic limit of super- heat to provide a criterion for microexplosion. The microexplosion can occur if the water superheat limit is less than the oil boiling point. Law shows that the thermody- namic water superheat limit is weakly dependent on ambient pressure whereas the oil boiling point increases with pressure. Therefore microexplosion becomes more likely as pressure increases. For a given pressure, microexplosion is more probable with the heavier, less volatile hydrocarbon oils.

Law predicted that the water addition dilutes the combustion process, reducing the peak temperatures. Reductions in pollutant formation are then suggested. Law clarifies that the fractional amount of water in the liquid is the only emulsion char- acteristic identified. For example, the model does not require identification of the discrete-phase component or of the continuous-phase component.

Law et al. (1980b) extends the study of Law (1977) by examining three differ- ent variations or modes of the liquid phase and interface modelling. One variation described as the "frozen, steady-depletion mode" simply relaxes the uniform and constant liquid temperature. The heat-diffusion equation governs the liquid tem- perature so that spatial and temporal variations occur.

Another mode is described as the "frozen, accumulation mode." Here, droplet temperature is uniform but time varying. Law suggests that this mode can occur only when water is the less volatile species. It is assumed that the surface tension would cause caps or globules of water to form at the surface, leaving some open area for the continued vaporization of the oil. In the analysis of this mode, the phase-equilibrium relationship is applied to each of the two species independently.

In the third mode, identified as the "distillation mode," the liquid-phase mass fraction is uniform but time varying. The time variation of each liquid mass fraction is related to the vaporization rate of that component through a mass-balance rela- tionship. Otherwise, the modelling of temperature behavior and phase equilibrium is identical to the modelling for the frozen accumulation mode. The uniform mass fractions and temperature are sustained through a rapid internal liquid convection according to Law et al.

The frozen, steady-depletion mode suffers from the same shortcoming as the Law (1977) model. An arbitrary choice must be made about which liquid species is dominant, and thereby one of the phase-equilibrium relationships must be sup- pressed. The other two modes have a theoretical flaw because the independent ap- plication of the two phase-equilibrium laws allows for the possibility of the gas- phase mass fractions summing to a value greater than unity. Again, it is suggested by this author that a weighting based on the fraction of interface area occupied by the species should be applied to the phase-equilibrium relations. The rapid-mixing concept for the distillation mode cannot be justified. As discussed in Chapter 3 and Section 4.2, internal convection does not eliminate stratification of temperature or concentration. Furthermore, internal convection should not occur in a spherically symmetric case in which there is no gas-phase forced convection to drive the inter- nal flow.

It is seen that a need exists for a better model of liquid behavior and phase equilibrium for the vaporizing emulsified droplet. A phase-equilibrium relationship must be applied that satisfies two simple rational criteria: (1) the vapor mass fractions of the vaporizing species cannot, under any realized or potential conditions, sum to greater than unity; (2) there must be no need for an a priori determination of the dominant or preferred vaporizing species. This author suggests a relationship analogous to the Raoult Law given by Eq. (4.3a). The weighting factor for each vaporizing species should be the fraction of surface area exposed to each species. An assumption could be made that the surface-area fraction equals the volume fraction at the surface. Given the mass fraction and the density of each species, the volume fraction can be readily determined.

The model should not force the relative values of the species-vaporization rates through the frozen model or through other assumptions. Instead, the modelling used for the slurry discussed in Section 4.3 could be followed with regard to certain (but not all) aspects. As the surface regresses, the less volatile substance is left in higher concentration near the surface. Some maximum packing in terms of a maximum volume fraction must be specified for that less volatile phase. Experiments could help in determining this maximum volume fraction; it should depend not only on the particular properties for the two liquids but also on whether the less volatile substance is in the discrete or the continuous phase. This proposed model differs from the slurry model in two ways: (a) no formation of a hard shell with subsequent bubble formation is proposed; (b) the discrete phase and the continuous phase can simultaneously vaporize.

5 *Droplet Behavior under Near-Critical, Transcritical, and Supercritical Conditions*

High pressures and supercritical conditions in liquid-fueled diesel engines, jet engines, and liquid rocket engines present a challenge to the modelling and the fundamental understanding of the mechanisms controlling the mixing and combustion behavior of these devices. Accordingly, there has been a reemergence of investigations to provide a detailed description of the fundamental phenomena inherent in these conditions. Unresolved and controversial topics of interest include prediction of phase equilibria at high and supercritical pressures (Curtis and Farrell, 1988; Litchford and Jeng, 1990; Hsieh et al., 1991; Delplanque and Sirignano, 1993; Poplow, 1994; Yang and Lin, 1994; Delplanque and Potier, 1995; Haldenwang et al., 1996), including the choice of a proper equation of state, definition of the critical interface, importance of liquid diffusion, significance of transport-property singularities in the neighborhood of the critical mixing conditions, and influence of convection (including secondary atomization); d^2-law behavior at supercritical conditions (Daou et al., 1995); droplet-lifetime predictions (Yang et al., 1992; Delplanque and Sirignano, 1993, 1994; Yang and Lin, 1994; Delplanque and Potier, 1995; Haldenwang et al., 1996); dense spray behavior (Delplanque and Sirignano, 1995; Jiang and Chiang, 1994a, 1994b, 1996); combustion-product condensation (Litchford and Jeng, 1990; Litchford et al., 1992; Delplanque and Sirignano, 1994; Daou et al., 1995); and flame structures at high and supercritical pressures (Daou et al., 1995). The actual combustion process is characterized by the supercritical combustion of relatively dense sprays in a highly convective environment. However, most studies considered decoupled problems in order to isolate a limited set of issues. Consequently, most results were derived in the case of an isolated droplet gasifying (no reaction) in a quiescent environment. Convective effects, influence of neighboring droplets, detailed chemical kinetics, or product condensation have received less and more recent attention.

There are key challenges associated with operation at near-critical and supercritical conditions in order to increase efficiency and combustion-rate processes. The distinction between liquids and gases disappears at high pressures above the thermodynamic critical point, which has a strong nonlinear dependence on the composition. This introduces some crucial phenomena that were neglected decades ago

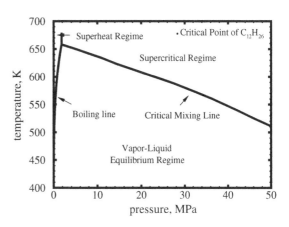

Figure 5.1. Pressure–temperature diagram for the phase behavior of $O_2/C_{12}H_{26}$ system in equilibrium ($P_r = p/p_{cr}, C_{12}H_{26}$) (Yang, 2008).

when the compositional distinction between the original liquid and its surrounding gases in the combustor was neglected. Also, the reduced surface tension can cause a new mechanism to be the rate-controlling factor for energy conversion.

Many practical applications involve the introduction of the spray into an environment where the ambient pressure is supercritical for the liquid but the liquid starts with a subcritical temperature below the ambient-gas temperature. In this case, there is a distinction between the phases in spite of the supercritical pressure; discontinuities in density and composition occur across the interface between the droplets in the spray and the surrounding gas. Typically the ambient gas is a different species from the liquid. The case in which the liquid is introduced into an environment that consists solely of its own vapor is of very limited interest. Generally, heat is supplied from the higher-temperature gas to vaporize the liquid. There is no instantaneous thermodynamic instability of the liquid. Heat transfer takes a finite, albeit very short, time. While the liquid is heating, some vaporization occurs. The liquid temperature at the interface can eventually rise to the critical temperature for the given pressure. Once that surface temperature reaches the critical value, there is a continuous variation of density and other properties across the "interface." Quotation marks are used because the lack of any discontinuity there removes its right to be called a true interface. Anyway, this "interface" becomes a surface along which the critical temperature exists with liquid on one side (for subcritical temperature) and the supercritical fluid on the other side. With time and continued heating, this surface propagates into the liquid until all of the liquid reaches critical temperature.

Figure 5.1 comes from the analysis by Yang (2008). It shows regimes with distinct phases for the oxygen gas and dodecane liquid. The behavior is qualitatively typical for many common liquids in common gases such as air, oxygen, or nitrogen. The same type of behavior will be seen for liquid oxygen in a hydrogen environment. We primarily address here the behavior at high pressures; the critical mixing line is shown on the right-hand side of the figure. If the initial pressure and temperature are above the critical line, no distinctions between phases exist. We focus here on the more interesting case in which the liquid temperature is initially subcritical but the pressure is supercritical, initially placing the liquid below the critical line in the graph.

A situation that is not addressed in detail in this chapter is portrayed by the left-hand side of the graph in Fig. 5.1; the possibility for an instantaneous thermodynamic instability of the liquid exists in a certain configuration. It is commonly identified in the field as flash vaporization. If the liquid were suddenly expanded to low pressure, bringing it leftward almost instantaneously in Fig. 5.1 and crossing the boiling-point curve, it would suddenly become a liquid in the superheat region. That is unstable and no heat transfer was needed to bring it above the boiling point; rather, that change was accomplished by instantaneously lowering the boiling point through the pressure drop. It is obvious from Fig. 5.1 that, if the high pressure is maintained, flash vaporization will not occur.

5.1 High-Pressure Droplet Behavior in a Quiescent Environment

High-pressure and supercritical ambient conditions have considerable influence on the mechanisms controlling engine behavior and performance. Most of these effects are related to droplet behavior. When liquid is injected into a combustion chamber that is filled with a gas at supercritical thermodynamic conditions, all aspects of the combustion process, from atomization to chemical reaction, can be expected to depart significantly from the better-known subcritical patterns. Over the past two decades, numerous studies have investigated how and to what extent supercritical conditions may affect these various aspects of the combustion of an isolated droplet in a quiescent environment. A detailed review of these investigations was recently contributed by Givler and Abraham (1996).

Consider a spherically symmetric, constant-pressure situation in which Fick's Law governs mass diffusion. The gas solubility in the liquid becomes important near the critical point so that, even if the liquid phase is initially monocomponent, it is necessary to consider multicomponent behavior in the liquid phase. The liquid density must also be considered as variable rather than constant. Therefore the same unsteady forms of the continuity, species, and energy equations are used in both the liquid and the gas phases. See Eqs. (A.20), (A.22), and (A.23). The form of the interface mass-balance conditions are given by Eqs. (4.1a), (4.1b), and (4.5b), and the interface energy-balance conditions are given by Eq. (2.6) with an appropriate modification of the energy of vaporization. Note, however, that the energy of vaporization will decrease as the critical point is approached; now, it must be considered to be strongly dependent on the thermodynamic state.

The thermal, mechanical, and chemical equilibria at the interface are expressed by the continuity across the interface of the temperature, pressure, and chemical potential μ, respectively.

The chemical potential is expressed in terms of the fugacity f for each species. Specifically,

$$\mu_i - \mu_i^0 = RT \ln \frac{f_i}{f_i^0}, \quad i = 1 \cdots N. \tag{5.1}$$

In the limit of pressure going to zero, the fugacity goes to the partial pressure. The fugacity can be related to temperature and volume (or to any two other independent thermodynamic variables) by basic thermodynamic principles; we have

$$RT \ln \frac{f_i}{X_i p} = \int_v^\infty \left[\left(\frac{\partial p}{\partial n_i} \right)_{T,V,n_i} - \frac{RT}{V} \right] dV - RT \ln \frac{pV}{nRT}. \quad (5.2)$$

The enthalpy of vaporization to be used in the energy-balance condition at the interface is the energy per unit mass (or per mole if preferred) required for vaporizing at the given temperature and pressure and into the particular surrounding gaseous mixture. On the other hand, the latent heat used at conditions well below initial is the energy per unit mass required for vaporizing the liquid into an environment composed of its own vapor. The enthalpy of vaporization for each species can be determined as a function of interface temperature, mole fractions on both sides of the interface, and the molecular weight (see Delplanque and Sirignano, 1993).

Typical models (for example, Delplanque and Sirignano, 1993) include a detailed computation of the high-pressure phase equilibria based on a cubic equation of state. A prevalent cubic equation of state used in this range of pressures and temperatures by the spray combustion community has been the Redlich–Kwong (1949) equation of state:

$$p = \frac{RT}{(v - b)} - \frac{a}{T^{0.5}v(v + b)}, \quad (5.3)$$

where v is the specific volume. This empirical cubic equation has only two parameters, a and b. This equation was then modified by Chueh and Prausnitz (1967) to include the dependence of a and b on composition and temperature. Although derived for pure components, these equations are often used for mixtures by means of mixing rules. Note that when hydrogen is involved [LOX/H_2 system (where LOX is liquid oxygen) for example], its quantum nature must be taken into account in the definition of the mixture rules. Phase equilibrium is expressed in terms of mechanical, thermal, and chemical equilibria. The equation of state is used in the computation of the fugacity of each component in each phase required for expressing the chemical equilibrium. The resulting system of equations is solved iteratively to yield the equilibrium composition in each phase at the given pressure and temperature. Note that the numerical solution of the system defining phase equilibrium requires a carefully designed iteration scheme (Delplanque and Sirignano, 1992). The equilibrium compositions thus computed are generally in good agreement with the scarce experimental data; see Fig. 5.2.

More recently, variations of the Redlich–Kwong equation such as the Soave–Redlich–Kwong equation allow a to be a function of temperature (Poplow, 1994; Yang and Lin, 1994) and are more accurate in predicting phase equilibria for particular mixtures. There is no evidence, however, that these improvements significantly affect the predicted overall droplet behavior. Another challenging issue inherent in the simulation of transcritical-phase processes is the evaluation of transport properties. Some transport properties (e.g., thermal conductivity) are expected to diverge at the critical transition. To quantify this singular behavior, a given transport property is considered to be the sum of a low-density value, an excess value that is due to high-pressure effects, and a critical enhancement, including the

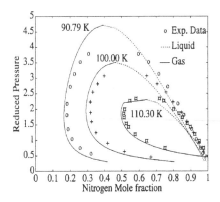

Figure 5.2. High-pressure phase equilibrium for the nitrogen/hydrogen system. Comparison of the predicted equilibrium mole fraction with the experimental data of Street and Calado (1978). (Delplanque and Sirignano, 1993, with permission of the *International Journal of Heat and Mass Transfer.*)

singular effects at the critical transition. Many authors duly note this issue, but, in general, actual models rarely include critical enhancement corrections because the data are still scarce and our understanding in this area limited. More details about this issue can be found in the works of Sengers et al. (1981), Sengers and Sengers (1977), Cheng et al. (1997), Luettmerstrathmann and Sengers (1996), and Abdulagatov and Rasulov (1997). Although these transport-property singularities are important to our understanding of critical phenomena, their macroscopic effects on droplet behavior in conditions relevant to actual processes (e.g., convective droplet heating and vaporization) might not be so pertinent.

The predictions of current models (Delplanque and Sirignano, 1993; Poplow, 1994; Yang and Lin, 1994; Haldenwang et al., 1996) are qualitatively consistent. Consider a liquid droplet just after introduction into a hot, supercritical, quiescent environment. The droplet is heated by conduction, and its diameter increases because the liquid density decreases as the temperature rises. Yang et al. (1993) noted that, because the density inside the droplet is nonuniform, liquid convection inside the droplet should be considered. The droplet-surface temperature rises until it reaches the computed critical mixture value. During this phase, the mixture composition on either side of the liquid–gas interface is imposed by the chemical equilibrium and mass diffusion that occur in the droplet; see Fig. 5.3. Note that the number of degrees of freedom of the corresponding thermodynamic system, as given by the Gibbs phase rule, is 2 (two components minus two phases plus two). The case considered here is that of a mixture for which the pressure p is prescribed. Hence only 1 degree of freedom is left. In the subcritical case, the interface temperature is the last degree of freedom. Once it is known, the phase-equilibrium conditions yield the mixture's composition.

Alternatively, requiring that the mixture be critical adds one condition to the system that is then completely defined, i.e., the corresponding values of temperature and composition, Y_c and T_c, are fully determined. See Fig. 5.4, which is taken from the analysis by Delplanque and Sirignano (1993) that relates to the vaporization in LOX/H$_2$ liquid-propellant rocket motors. It shows regimes with distinct phases for the oxygen liquid and hydrogen gas. P_r is the pressure divided by the critical pressure for pure oxygen. At pressures above the critical pressure (e.g., $P_r = 2$ through 7 in the figure) but temperatures below the critical temperature (about 154 K), it is

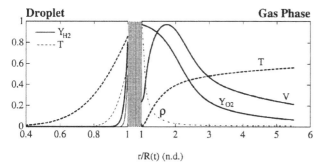

Figure 5.3. LOX droplet (100 K) vaporizing in hydrogen (1500 K, 100 atm). Left-hand side, temperature $[(T - T_0)/(T_{\text{ref}} - T_0), T_{\text{ref}} = 150$ K] and hydrogen mass fraction $(Y_{H2}/Y_{H2,\text{ref}}, /Y_{H2,\text{ref}} = 0.02)$ profiles at $t = 0.1$ ms (thin curves) and 0.9 ms (thicker curves). Right-hand side, temperature $[(T - T_s)/(T_\infty - T_s)]$, mass fraction, velocity $(u/u_{\text{ref}}, u_{\text{ref}} = 5$ cm/s), and density $(\rho/\rho_{\text{max}}, \rho_{\text{max}} = 323$ kg/m^3) profiles. (Delplanque and Sirignano, 1993, with permission of the *International Journal of Heat and Mass Transfer.*)

still possible to obtain a phase equilibrium with, of course, distinction between the phases. When the liquid exists with the ambient gas differing from the pure vapor of the liquid, some mass exchange occurs; the vapor of the liquid enters the gas phase and some of the gas molecules enter the liquid. This exchange becomes more important as the critical point is approached. The figure shows that, as P_r increases, the equilibrium has a decreasing fraction of hydrocarbon and an increasing fraction of oxygen in the liquid at the interface. That is, more oxygen has dissolved. The temperature value above which the phases are no longer distinct will decrease with increasing pressure. The energy of vaporization is nonzero in this domain but decreases with increasing temperature until it reaches zero at the temperature value where phase distinction disappears. The figure shows that, for $P_r = 1$ or less, the

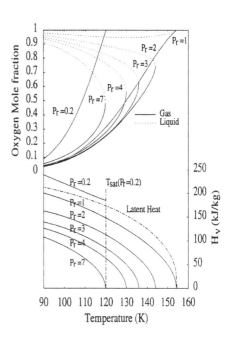

Figure 5.4. Computed phase equilibrium of the oxygen/hydrogen system. (Delplanque and Sirignano, 1993, with permission of the *International Journal of Heat and Mass Transfer.*)

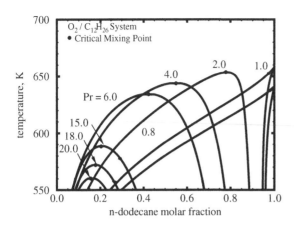

Figure 5.5. Vapor–liquid phase-equilibrium compositions for $O_2/C_{12}H_{26}$ system at various pressures. $P_r = p/p_{cr}$, $C_{12}H_{26}$ (Yang, 2008).

liquid is nearly pure oxygen whereas the gas at the liquid interface might have a significant fraction of the vapor from the liquid. The vapor fraction goes to unity as the liquid temperature goes to the saturation value. This saturation temperature increases with pressure, reaching the critical temperature as the critical pressure is reached. Above that critical temperature value, the distinction between phases is lost.

Yang and co-workers (2008) made similar calculations for liquid dodecane $C_{12}H_{26}$ surrounded by oxygen gas O_2 and for liquid methane CH_4 with surrounding oxygen gas. Figures 5.5, 5.6, and 5.7 show results from their computations for dodecane. The compositional behavior given by Fig. 5.5 and the energy of vaporization data of Fig. 5.6 are qualitatively similar to the O_2/H_2 results of Fig. 5.4. In addition, Fig. 5.7 shows that, under near-critical conditions, the dependence of the liquid density on pressure and temperature is profound.

These thermodynamic results are well known and have been used for analysis of spray vaporization. See Delplanque and Sirignano (1993), Hsieh et al. (1991), and Yang (2008). Properties (e.g., viscosity and diffusivity) for these gas and liquid mixtures at high pressures can be found in the literature; see, for example, Poling et al. (2001).

Y_c and T_c do not coexist at the same point in space except when the critical mixing condition is reached at the interface. Once this happens, the concept of an

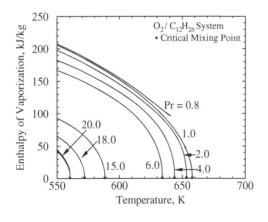

Figure 5.6. Effect of pressure on enthalpy of vaporization of $C_{12}H_{26}$ in an equilibrium mixture of O_2 and $C_{12}H_{26}$. $P_r = p/p_{cr}$, $C_{12}H_{26}$ (Yang, 2008).

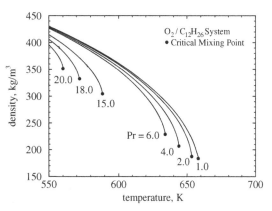

Figure 5.7. Liquid density as function of temperature for an equilibrium mixture of O_2 and $C_{12}H_{26}$. $P_r = p/p_{cr,C_{12}H_{26}}$ (Yang, 2008).

interface ceases to be relevant. If a droplet lifetime is to be determined, the part of the computation domain constituting the supercritical droplet must be defined. Two definitions have been chosen in the literature, and neither really by itself corresponds to the critical mixing conditions. Yang et al. (1993) chose the isotherm at the critical mixing temperature as the supercritical droplet boundary, whereas Haldenwang et al. (1996) chose the composition isopleth at the critical mass fraction.

Consider a droplet of liquid component A in gaseous component B at pressure p_a. The liquid initial temperature is T_l and the initial ambient temperature is $T_a > T_l$. For conditions well below the critical conditions, temperature varies continuously throughout the surrounding gas and liquid interior and is continuous across the liquid–gas interface. A negligible amount of the ambient gas dissolves in the liquid; composition and densities are discontinuous across the interface but piecewise continuous in the gas and in the liquid.

If $p_a > p_c$ and $T_a > T_c$, the temperature is at the critical value T_c at some point, but not necessarily at the interface. However, the gaseous mass fraction of A is initially smaller than the critical value Y_c whereas the liquid mass fraction of A is initially larger than the critical value. Hence, although an isotherm with the critical value will immediately appear, a mass fraction isopleth with the critical value will appear only when the interface discontinuities disappear. See Figs. 5.8 and 5.9.

As the critical conditions are approached, ambient gas dissolves in the liquid and the magnitudes of the interface discontinuities in density and in composition decrease. Still, the critical temperature and the critical composition do not occur at the same point in space and time. At the point when and where the discontinuity first disappears, the temperature, composition, and density assume their critical values simultaneously. After that time, the critical isotherm and the critical mass fraction isopleth must coincide. This critical surface that has replaced the interface will regress as further heating and mixture occur, until it reaches the droplet center. See Figs. 5.8 and 5.9. The phenomena of mixing and vaporization can involve situations therefore in which both a subcritical spatial domain and a supercritical spatial domain can exist simultaneously. This situation is described as transcritical (Delplanque and Sirignano, 1993, 1994, 1995).

The scarce experimental data (Haldenwang et al., 1996) show that the droplet lifetime exhibits a "transcritical minimum." The droplet-lifetime decrease with the

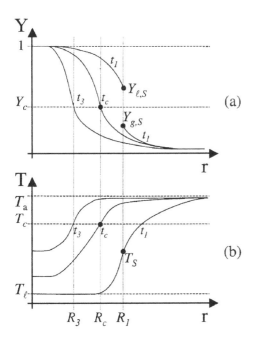

Figure 5.8. Sketch of the transcritical behavior of the (a) mass fraction isopleths, (b) isotherms; t_c is the time at which the droplet surface reaches the critical mixing conditions and $t_1 < t_c < t_3$, R_1 and R_c are the droplet radii at t_1 and t_c, respectively, and R_3 denotes the location of the critical interface at $t = t_3$.

subcritical pressure increase is attributed to the liquid–gas density ratio's decreasing to 1. For supercritical pressure, the supercritical droplet boundary regression rate is controlled by mass diffusion and thus increases with pressure. Yang et al. (1993) have shown that, in vaporization-only cases (no combustion), the quasi-steady state is never reached. However, the work of Daou et al. (1995) on supercritical combustion of a LOX droplet, using a detailed reaction mechanism, indicates that, after an initial transient corresponding to the ignition phase characterized by the fast propagation of a premixed flame, a diffusion flame is established and combustion proceeds in a quasi-steady manner. Daou et al. found that during this latter stage the d^2 law is approximately valid. Note that these conclusions cannot be directly compared with the classical model for particle combustion at high pressure proposed by Spalding (1959) because that early model really dealt with a different problem: that of the combustion of a gas puff. This model did not include a high-pressure phase-equilibrium model and considered infinite reaction rate. A complete discussion of the validity of the d^2 law for supercritical droplet combustion is provided by Givler and Abraham (1996).

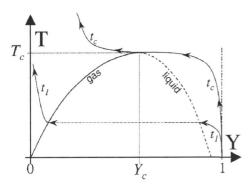

Figure 5.9. Sketch of the trajectories that a fluid element in the droplet that is initially liquid can follow through a T/Y plot of the binary phase equilibrium. t_1 corresponds to the vaporization behavior of a fluid element that is initially closer to the interface and thus vaporizes before the interface reaches the critical mixing conditions. t_c corresponds to the vaporization behavior of a fluid element that is initially closer to the droplet center such that it crosses the interfaces just when it is at the critical mixing conditions.

Harstad and Bellan (1998a, 1998b, 1999, 2000, 2001a, 2001b) examined an oxygen droplet vaporizing in a hydrogen environment and a heptane droplet vaporizing in a nitrogen environment at high pressure, with special interest in the behavior and effect of Lewis number and the Soret and Dufour effects. They attempted to approximate drop interaction in a group but maintained spherical symmetry by using a sphere of influence for each droplet. In a series of papers, Zhu and Aggarwal (2000), Zhu et al. (2001), Aggarwal et al. (2002), and Yan and Aggarwal (2006) considered the transient supercritical vaporization of n-heptane/ nitrogen and n-hexane/nitrogen systems. They made comparisons between calculations with quasi-steady gas-phase and fully transient calculations, confirming previous findings that the quasi-steady assumption breaks down at high pressures.

In summary, most theoretical and experimental investigations of spray combustion at supercritical combustion have focused on the case of one or a few droplets in a quiescent gas. The more advanced models now provide consistent predictions. The droplet-gasification process at these conditions is essentially unsteady. The interface regression does not follow the d^2 law at these conditions, but the d^2 law does provide an order-of-magnitude approximation. See the next section for droplets with convection and Section 9.7 for sprays under critical and near-critical conditions.

5.2 Convective Effects and Secondary Atomization

Practical environments such as those found in engines are characterized by significant relative velocity between the gas and the droplet with subsequent strong convection. The first effect is that heat and mass transfer are enhanced by the convection. With near-critical and supercritical environments, the surface-tension coefficient of a droplet decreases to zero as the interface temperature approaches the critical conditions. Hence the second effect is that droplet deformation and secondary atomization can be initiated by smaller values of the droplet–gas relative velocity.

Different modes of droplet breakup or secondary atomization have been shown to exist, depending on the particular balance of the forces; see Ferrenberg et al. (1985) and Fornes (1968). The important nondimensional groupings are the Weber number We and the Bond number Bo, where

$$\text{We} = \frac{2R\rho(\Delta U)^2}{\sigma},$$

$$\text{Bo} = \frac{\rho_l a R^2}{\sigma}.$$

The acceleration of the droplet is denoted by a, and the relative gas–liquid velocity is ΔU. The Weber number gives the ratio of the aerodynamic force (which is due to pressure variation over the droplet) to the surface-tension forces, and the Bond number gives the ratio of a body force (or reversed D'Alambert force for an accelerating droplet) to the surface-tension force.

The aerodynamic consequence of droplet motion is that the fore and the aft pressures on the surface become greater than the pressures on the side. There is a tendency therefore for the droplet to deform in an axisymmetric manner to the shape of a lens.

The surface-tension force will resist this deformation because it increases the surface area. For small values of We and Bo, there is some vibration but no significant deformation or breakup. Above We = 5, aerodynamic forces have some effect on the droplet shape. A critical value of We occurs in the range 10–20, above which continuous deformation of the droplet occurs; the droplet has a convex side and a concave side and takes the shape of a bag or an umbrella. Viscosity does not play a significant role in this deformation. When We is above the critical value, the shear on the droplet surface will cause stripping of liquid from the surface. The critical value for the Bond number is 11.22 according to Harper et al. (1972); above that value, surface waves grow exponentially. The first unstable mode appears at and above Bo = 11.22. Other unstable modes will appear in sequence as Bo is continually increased. However, they remain small enough in magnitude until Bo $= O(10^4)$ so that the aerodynamic forces dominate. Experimental data for shock-wave interaction with a water droplet (Ranger and Nicholls, 1972) indicate that stripping occurs above Bo $= 10^2$, filling the near wake with a mist.

An approximate criterion for stripping can be determined by balancing the surface-tension force on the element of liquid in the surface viscous boundary layer with the centrifugal force. This balance gives a critical value of $We/\sqrt{Re}$, which is of the order of unity. The Reynolds number appears because it determines the amount of fluid moving in the boundary layer and thereby experiencing the centrifugal effect. Often the critical value is taken to be $We/\sqrt{Re} = 0.5$ with stripping occurring above this value.

Convective effects on the behavior of a droplet in a supercritical environment have often been limited to heat and mass transfer effect or droplet deformation without breakup (Lee et al., 1990). Litchford and Jeng (1990) included the effect of convection on heat and mass transfer rates in their model for LOX droplet vaporization at supercritical conditions by using the film model for droplet vaporization. However, the validity of their results is somewhat undermined by their neglecting the difference between the liquid propellant and the surrounding gas specific heats, which reaches 1 order of magnitude in the LOX/H$_2$ case (Delplanque and Sirignano, 1993). Litchford and Jeng (1990) did note that stripping was likely to occur in environments typical of cryogenic rocket engines. They performed an order-of-magnitude analysis to evaluate the stripping rate and concluded that stripping would be important only after the droplet interface reached the critical mixing conditions.

Delplanque and Sirignano (1993) followed a similar path. Their model for droplet vaporization in a supercritical convective environment (with neglect of secondary atomization), based on the Abramzon and Sirignano (1989) film model, showed that a LOX droplet injected into a rocket engine is likely to reach the critical state before it disappears, much sooner than in a quiescent atmosphere. However, they showed that, because of the behavior of the surface-tension coefficient under near-critical conditions, a LOX droplet in a rocket engine is likely during its lifetime to undergo secondary atomization in the stripping regime (mass removal from its surface by aerodynamic shearing) before the droplet interface reaches the critical mixing conditions; see Fig. 5.10.

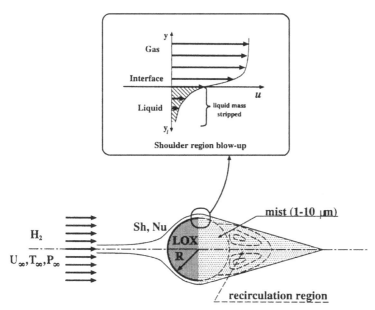

Figure 5.10. Sketch of a droplet undergoing boundary-layer stripping. (Delplanque and Sirignano, 1994, with permission of *Atomization and Sprays.*)

Prior analyses by Ranger and Nicholls (1969, 1972) and Litchford and Jeng (1990) of the magnitude of the stripping rate indicated that the stripping rate might be much larger than the convective vaporization rate with an obvious consequence on the droplet lifetime. Delplanque and Sirignano (1994) expanded on the work of Ranger and Nicholls to develop an integral analysis of the coupled gas–liquid boundary layers at the droplet interface and obtain an expression for the mass removal rate, including blowing effects:

$$\dot{m}_{\mathrm{BLS}} = \pi D \int_0^\infty \rho_l u_l \, \mathrm{d}y = 3\pi R \rho_l U_\infty A \alpha_l \sqrt{\frac{\pi R}{2}}, \tag{5.4}$$

where A (the nondimensional interfacial velocity) and α_l (a liquid boundary-layer velocity-profile parameter) are functions of the droplet size, its velocity relative to the gas, and gas and liquid properties. Their modified film model uses a corrected heat transfer number to include the effect of stripping on the heat transfer into the droplet:

$$B_T = \frac{c_{pg,O_2}(T_\infty - T_s) + \varepsilon_{\mathrm{mist}}/\dot{m}}{\Delta H_v(T_s) + Q_l/\dot{m}}, \tag{5.5}$$

where $\varepsilon_{\mathrm{mist}}$ is a correction to the driving potential that is due to the presence of the mist:

$$\varepsilon_{\mathrm{mist}} = \dot{m}_{\mathrm{BLS}} c_{pl,O_2}(T_\infty - T_s) + \dot{m}_2 \left[\Delta H_v(T_l) + c_{pg,O_2}(T_\infty - T_l)\right], \tag{5.6}$$

and $\dot{m}_2$ is the portion of $\dot{m}_{\mathrm{BLS}}$ that vaporized from the mist. The first term in Eq. (5.6) accounts for heating the liquid in the mist, and the second term accounts for the vaporization of the mist and the heating of the vapors.

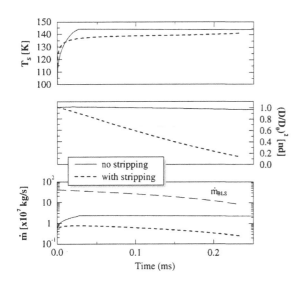

Figure 5.11. Influence of boundary-layer stripping on predicted droplet behavior under supercritical conditions. Surface temperature, diameter squared, primary vaporization rate, and stripping-rate history for a 100-μm LOX droplet injected at 100 K in gaseous hydrogen at 1000 K, 100 atm, with a relative velocity of 20 m/s. (Delplanque and Sirignano, 1994, with permission of *Atomization and Sprays*.)

Results obtained with this model confirmed that the stripping rate is much larger than the gasification rate. The predicted droplet lifetime was found to be reduced by at least 1 order of magnitude (Fig. 5.11) when stripping occurred. An important consequence is that, in most cases, the droplet disappears before the interface can reach the critical mixing conditions.

In summary, the effect of forced convection on the behavior of a droplet under supercritical conditions is considerable because it is coupled with the effects of a significantly reduced surface-tension coefficient. When this phenomenon is taken into account, the droplet lifetime is found to be controlled by secondary atomization in the stripping regime, yielding a droplet lifetime 1 order of magnitude smaller than those predicted in the absence of stripping. Furthermore, the occurrence of stripping minimizes the importance of transcritical phenomena such as the expected singularity of some transport properties so that good predictions can be obtained even with a rather crude modelling of these singularities.

A direct numerical method for the calculation of problems with distorting free surfaces or interfaces is a major computational challenge. There are several basic approaches. One method is a Lagrangian approach that places markers on the interface or equivalently uses a grid that deforms with the heavier fluid. See, for example, Nichols and Hirt (1971). These methods tend to avoid problems with numerical diffusion across the interface but fail under severe distortion such as occurs in secondary atomization problems. The second approach utilizes an Eulerian grid but tracks the intersecting interface. An additional field variable, typically the fractional volume of fluid, is introduced to indicate the extent of the presence or absence of one of the fluids. These volume-of-fluid (VOF) methods are designed to minimize numerical diffusion. See Hirt and Nichols (1981). Liang (1990) extended the VOF method so that a continuous curve could be uniquely defined to represent the interface. This is important in determining the surface curvature and the effect of surface tension. Liang (1990) used his methodology to predict the behaviors of (i) a falling water droplet striking a surface, spreading in a film over the surface, and then contracting back because of the effect of surface tension, (ii) a kerosene droplet undergoing normal-mode oscillations, and (iii) a kerosene droplet in a moving air stream undergoing a flattening type of distortion and the shedding of liquid ligaments because of shearing. A third, more modern method is the level-set method, which creates a new function that maintains a constant value at a certain interface. By calculating contours of this level-set function at different times, one can determine the location of the interface.

5.3 Molecular-Dynamics Simulation of Transcritical Droplet Vaporization

Clearly some of the crucial problems associated with the critical point, such as the singularity of some transport coefficients, would benefit from a most fundamental approach. As such an approach, molecular dynamics (MD) has been the subject of increasing interest over the past decade. This interest has also been increased by the availability of ever-increasing computing power. Here, the MD approach is briefly described; note the significant progress that has recently been made to elucidate some of the issues relevant to the prediction of droplet vaporization under supercritical conditions. An excellent introduction to this method is provided by Rapaport (1995). The MD approach simulates the macroscopic behavior of matter by computing the behavior of the molecules. This requires extremely large resources of computational power. The simplest models are based on spherical particles, but more complex configurations can be considered. Typically, the particle interaction potential used is the Lennard–Jones 6–12 potential. MD simulations have been attempted since the early 1950s. However, only small numbers of particles could be considered at that time (fewer than 100).

Two types of MD studies are relevant to the topic of droplet vaporization at supercritical conditions: MD simulation of supercritical fluids (Luo et al., 1995; Nwobi et al., 1998) and MD simulation of droplet behavior (Rusanov and Brodskaya, 1977; Thompson and Gubbins, 1984; Long et al., 1996). It is only very recently that MD has been successfully applied to the problem of droplet behavior under supercritical conditions (Kaltz et al., 1998).

There is a growing interest in using MD simulations to investigate supercritical fluids from both the chemistry (e.g., physical chemistry of supercritical fluids, structure and dynamics of clusters; Yoshii and Okazaki, 1997) and chemical engineering (e.g., hazardous-waste conversion by use of oxidation in supercritical water; Seminario et al., 1994) perspectives. Although most of the literature on this topic focuses on simulating the behavior of supercritical fluids, such simulations can also be a powerful tool in the study of singular transport behavior near the critical point. Luo et al. (1995) successfully used two-dimensional MD simulations to predict the unusual behavior of highly compressible fluids in the near-critical region, where the hydrodynamic theory is still valid. Their analysis of the temperature–temperature and density–density time-correlation functions showed that, in this region, the temperature relaxation is acoustically driven and the density behavior is mostly diffusive. In their investigation, they limited themselves to a two-dimensional fluid system in order to be able to consider larger wavelengths than a three-dimensional system would allow for a given number of particles. Closer to the critical point, the

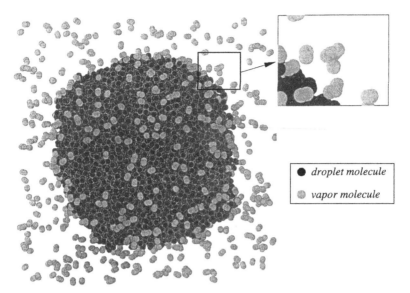

Figure 5.12. MD simulation of a saturated LOX droplet at 100 K. Droplet molecules are colored dark gray, and vapor molecules are light gray. This droplet contains roughly 9300 molecules. (Kaltz et al., 1998, with permission of *Combustion Science and Technology.*)

correlation length becomes comparable to the system size and the density fluctuations are so large that the computational requirements become unrealistic.

Rusanov and Brodskaya (1977) used a method developed by Alder and Wainwright (1959) to investigate the dependence of the surface-tension coefficient on the surface-tension radius. They found that the surface-tension coefficient increases when the droplet size decreases. Furthermore, they noted that the omission of nonadditive interactions, which are important in real liquids, may lead to significant errors in the prediction of macroscopic quantities such as the phase-transition temperature. Rusanov and Brodskaya's simulations were limited to 5000 molecules by the computational power available in the mid-1970s. In the mid-1980s, Thompson and Gubbins (1984) were able to perform MD simulations with up to 2004 molecules. Their results, although more detailed, are qualitatively similar to those of Rusanov and Brodskaya.

More recently, Long et al. (1996) used MD to simulate the vaporization of submicrometer droplets at subcritical conditions. This approach recently was extended to supercritical (Kaltz et al., 1998) conditions. In the subcritical regime, their results agree with the Knudsen theory of aerosols (e.g., vaporization rate). These simulations were extended to consider the vaporization of three-dimensional submicrometer LOX drops at pressures from 2 to 20 MPa and temperatures from 200 to 300 K. Furthermore, these simulations confirmed that the assumptions typically used in the subcritical regime (e.g., quasi-steady approximation) are valid at much higher pressures. See Fig. 5.12, taken from the work of Kaltz et al. (1998). Consilini et al. (2003) used the MD approach to study submicrometer xenon droplets vaporizing in nitrogen, for both subcritical and supercritical conditions. Whereas subcritical droplets that were initially nonspherical quickly deformed toward a spherical shape, the supercritical droplets did not.

Kaltz et al. also noted that, because the cases considered typically have high Knudsen numbers, the results of MD simulations are not suitable for comparison with the classical results of the d^2 law. Furthermore, other important phenomena such as gas solubility and convection have yet to be included in MD simulations. Finally, and more important, the simulation of the behavior of three-dimensional fluids in the near-critical region where the hydrodynamic theory is not valid presents theoretical and computational challenges that have yet to be tackled.

6 *Droplet Arrays and Groups*

To this point in this book, we have discussed only isolated droplets. In a practical situation, of course, many droplets are present in a spray, and the average distance between droplets can become as low as a few droplet diameters. A typical droplet therefore will not behave as an isolated droplet; rather, it will be strongly influenced by immediately neighboring droplets and, to some extent, by all droplets in the spray.

There are three levels of interaction among neighboring droplets in a spray. If droplets are sufficiently far apart, the only impact is that neighboring droplets (through their exchanges of mass, momentum, and energy with the surrounding gas) will affect the ambient conditions of the gas field surrounding a given droplet. As the distance between droplets becomes larger, the influence of neighboring droplets becomes smaller and tends toward zero ultimately. At this first level of interaction, the geometrical configuration of the (mass, momentum, and energy) exchanges between a droplet and its surrounding gas is not affected by the neighboring droplets. In particular, the Nusselt number, Sherwood number, and lift and drag coefficients are identical in values to those for an isolated droplet. This type of interaction will be fully discussed in Chapter 9.

At the next level of interaction, droplets are closer to each other, on average, and the geometrical configurations of the exchanges with the surrounding gas are modified. In addition to modification of the ambient conditions, the Nusselt number, Sherwood number, and the lift and drag coefficients are modified. Here, a droplet cannot be treated as if it were an isolated droplet; the neighboring droplets are within the gas film or wake of the droplet. In a convective situation, a droplet can influence a second droplet at substantial distances of many tens of droplet radii if the latter is in its wake. If the droplets are placed side by side in a convective situation, significant influence occurs over only short distances of a few droplet radii. This type of dense-spray interaction is the major subject of discussion for this chapter.

The third level of interaction among droplets involves collisions whereby the liquids of the different droplets actually make contact with each other. Here, the droplets might coalesce into one droplet or emerge from the collision as two or more droplets. Some discussion of this subject is made in Section 6.4.

From another perspective, Sirignano (1983) classified interactive droplet studies into three categories: droplet arrays, droplet groups, and sprays. Arrays involved a few interacting droplets with ambient gaseous conditions specified. There are many droplets in a group but gaseous conditions far from the cloud are specified and are not coupled with the droplet calculations. The spray differs from the group in that the total gas-field calculation in the domain is strongly coupled to the droplet calculation. In the spray, either the droplets penetrate to the boundaries of the gas or, while the droplets may not penetrate, the impact of the exchanges of mass, momentum, and energy extends throughout the gas. A useful review on droplet interactive processes can be found in Annamalai and Ryan (1992). Note that unpublished information from Annamalai and Ryan indicates that radiative transfer between droplets can be significant even if the separation distance is 10 diameters or greater.

To understand the distinctions among arrays, groups, and sprays, it must be visualized that a cloud or collection of droplets occupies a certain volume. These primary ambient-gas conditions are defined as those conditions in the gas surrounding the cloud of droplets. Each droplet in the cloud has a gas film surrounding it. The local ambient conditions are defined as the gas properties at the edge of the gas film but within the volume of the cloud. This definition becomes imprecise when the gas films of neighboring droplets overlap; one can cite that fact only as evidence of conceptual weakness in the theory. In such a case, the local ambient conditions would be replaced with some average over the gas in the droplet neighborhood.

Three different phenomena may be identified (following Sirignano, 1983) in the vaporization and burning of fuel-droplet clouds or sprays: (i) The primary ambient-gas environment is strongly affected by the presence of droplets. Because typical combustors are finite in size, conditions in the gas surrounding a cloud of droplets will be coupled strongly to the gas and liquid properties within the cloud. The properties in the liquid phase and in the gas film surrounding the droplet do not affect the ambient conditions in the isolated-droplet studies but, in the realistic situation, the primary ambient-gas temperature, momentum, and composition are strongly influenced by the presence of the droplets. (ii) The local ambient environment is also affected by the droplets. This will directly affect the heat and mass transfer rates between individual droplets and the local gas environment. Vaporization rates will also be influenced. (iii) The geometry and the scale of the diffusion field surrounding each individual droplet will be affected. For example, we expect that the Nusselt number and the functional form of the relationship between vaporization rates and local ambient conditions would be affected by the droplet spacing. This functional relationship is called the vaporization law. Its limiting form is that given for isolated droplets; see, for example, Eqs. (2.9). This effect of droplet spacing will also modify heat and mass transfer and vaporization rates. It is seen that the second and the third phenomena each affect transfer rates. We would conjecture that a decrease in droplet spacing leads to an increase in local ambient fuel-vapor concentration and a decrease in local ambient temperature. The implication is that this, by itself, would tend to decrease the heat and mass transfer rates. The third phenomenon should result in a decrease in the gas diffusion length scale as droplet spacing decreases. If r_c were the characteristic diffusion length in the gas film, the surface area through which diffusion occurs would be proportional to the square of r_c whereas the diffusion rate per unit surface area would be inversely proportional to r_c. The net result is that the heat and mass transfer rates and (under diffusion control) the vaporization rate would be proportional to r_c. Here again, the effect of a decrease in droplet spacing is to decrease the transfer and vaporization rates. The second and the third phenomena each would have the same qualitative effect as droplet spacing (or, equivalently, number density) varies. A number of analyses have together addressed the first and the second phenomena (Polymeropoulas, 1974; Seth et al., 1978; Aggarwal and Sirignano, 1980; Dukowicz, 1980; Gosman and Johns, 1980; Aggarwal et al., 1981, 1983). Some considered only the second phenomenon; see Suzuki and Chiu (1971), Chiu and Liu (1977), Chiu et al. (1983), Correa and Sichel (1983), Labowsky and Rosner (1978), and Samson et al. (1978). Others examined the second and the third phenomena together; see Twardus and Brzustowski (1977), Labowsky (1978), Umemura et al. (1981a, 1981b), Tal and Sirignano (1982, 1984), and Tal et al. (1983, 1984a, 1984b). This last group of investigators examined a few droplets or spherical particles in a well-defined geometry or a large number of droplets in a periodic configuration. Let us define these arrangements as droplet arrays. These arrays are artificial and contrived but can be useful in obtaining information about the third phenomenon and, to some extent, about the second phenomenon. Because the number of droplets in an array is typically small, the impact on the primary ambient conditions is not significant, and arrays are not useful for studying the first phenomenon. Another interesting approach taken by several investigators involves group vaporization and burning of droplets. The group combustion and vaporization theory takes the primary ambient conditions as given and proceeds to determine the local ambient conditions, droplet properties, and vaporization rates. Droplet group theory is distinct (from array theory) in that a statistical description of droplet spacing (rather than a precise geometrical description) is used. In particular, a number density of the droplets is considered. As a practical matter, group theory can deal with many more droplets than array theory. (An exception occurs when the array is geometrically periodic because then array theory can be used to analyze one cycle.) Group theory is useful in studying the second phenomenon but not useful in studying the first phenomenon because the primary ambient-gas conditions are uncoupled and prescribed. Array theory is conducive to the analysis of the third phenomenon but group theory is not; in group theory, the functional relationship between the local ambient conditions and the vaporization rate must be prescribed whereas in array theory it can be determined by analysis. This is a consequence of the detailed field analysis of array theory versus the averaging or statistical approach of group theory. Array theory can be used to determine the Nusselt number for a droplet in the array or to obtain the mathematical relationship between vaporization rate and local conditions; then the result of that analysis can be postulated as an input for a group-theory analysis. The current analyses have not included the effect described here as the third phenomenon but rather considered only vaporization laws obtained for isolated droplets. No inherent limitation exists that prevents a better representation of the vaporization law for use in group theory.

From the work of Sirignano (1983), spray vaporization theory will be understood to be distinct from the theory of droplet-array vaporization or the theory of

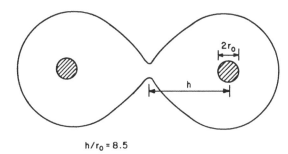

$h/r_0 = 8.5$

Figure 6.1. Effect of droplet spacing on the burning of two droplets (Sirignano, 1983, redrawn from Twardus and Brzustowski, 1977).

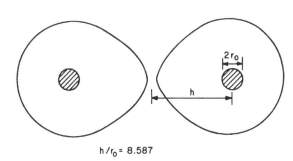

$h/r_0 = 8.587$

droplet group vaporization in that the full coupling among the ambient-gas properties, the local-gas properties, and the droplet properties is considered. Again, a statistical or average representation of properties is made on account of the large number of droplets considered. The first and the second phenomena are treated and analyzed. The effects of droplet spacing on Nusselt number and vaporization law can be included in spray theory so that the third phenomenon may be represented. However, no investigators have included that phenomenon in their model. Obviously this is a serious neglect for dense sprays, i.e., sprays for which the average distance between droplets is comparable with or less than the size of the gas film surrounding an average droplet.

6.1 Heating and Vaporization of Droplet Arrays

Two burning droplets of equal size were analyzed by Twardus and Brzustowski (1977), who used a bispherical coordinate system to facilitate their analysis. Stefan convection and forced convection were not considered in the analysis. As shown in Fig. 6.1, there is a critical value of the ratio of the distance between droplet centers to the droplet radius, above which the droplets burn with two individual and separate flames and below which the two droplets burn with one envelope flame. The critical value depends on the particular stoichiometry; for normal heptane, the value was found by Twardus and Brzustowski to be 17.067. The vaporization rate remains diffusion controlled, and the rate of diffusion (and therefore the vaporization rate) decreases as the droplet spacing decreases. In the limit as the droplets come into contact, the vaporization becomes a factor of $\ln 2 (= 0.693)$ of the value for two distant and isolated droplets.

Two important conclusions can immediately be seen. First, spacing between droplets in typical fuel sprays will be smaller than this critical value. It is noteworthy that Chigier and McCreath (1974) cited experimental evidence that droplets in combustors rarely burn in an isolated fashion; rather, a flame envelops many droplets. Also, the interaction between (or effectively the merging of) the gas films surrounding droplets leads to a reduction in heat transfer, mass transfer, and vaporization rates. In the quasi-steady, nonconvective situation considered, a d^2 law for vaporization still exists but the coefficient is less than the value for an isolated droplet.

Labowsky (1978) was able to show that Stefan convection can be easily included by means of a transformation that reduces the mathematical form to that for slow vaporization. Laplace's equation for an array of droplets is then solved by means of a method of images. In the numerical example, three droplets of equal size and temperature are arranged with their centers at the apices of an equilateral triangle. The theory is more general, however, and allows for an arbitrary number of particles of varying initial size and initial temperature. The quasi-steady assumption is made, and thermodynamic equilibrium is assumed for the fuel vapor at the surfaces. The general conclusions are that (i) the vaporization rate is proportional to the vaporization rate for an equivalent isolated droplet times a corrective factor, (ii) the corrective factor decreases as droplet spacing decreases (obviously it is unity for infinitely large spacing), and (iii) the droplet temperatures are independent of droplet spacing. (The last conclusion may be correct only under the assumption of unitary Lewis number because heat transfer and mass transfer have opposite effects on gas temperature.) Labowsky's approach has been generalized by Sirignano and co-workers to address vaporization with burning, multicomponent liquids, nonunitary Lewis numbers, and a far greater number of droplets in the array. Details of the extended theory are discussed in Section 6.3.

Umemura et al. (1981a) extended the analysis of two equisized burning droplets by Twardus and Brzustowski (1977) by accounting for the Stefan convection. They also needed bispherical coordinates to analyze the field. Their results were qualitatively identical to the Twardus and Brzustowski solutions shown in Fig. 6.1. Umemura et al. (1981b) considered two interacting burning droplets of unequal size. The vaporization for an interacting droplet is always reduced from the isolated-droplet value given by Eq. (2.10) or (2.15b). The isolated-droplet values are approached by each droplet as the spacing between the two droplets becomes infinite. The fractional decrease from its isolated-droplet vaporization-rate value is greater for the smaller droplet. Xiong et al. (1985) found experimentally that theory overpredicted the corrections for interaction; their results indicate that, above 5-diameter spacing, the droplets behave as if they were isolated. Their experiments had droplet diameters greater than 1 mm; the authors acknowledge that buoyancy played a major role. This effect should decrease the diffusion-layer thickness on the sides and bottoms of the horizontally aligned droplets, thereby reducing the distance over which interactions can occur. These experimental findings should not be inferred to indicate a lack of validity in the application of the theories of Twardus and Brzustowski, Labowsky, and Umemura et al. to submillimeter droplets.

Labowsky and Rosner (1978) and Samson et al. (1978) studied arrays in which the geometry is well defined but each droplet is considered to be a monopole source.

Labowsky and Rosner showed that the difference in results between consideration of a continuous distribution of monopole sources of fuel vapor and consideration of a discrete distribution of monopole sources is not significant whenever the number of droplets in the cloud is large. They referred to the continuous distribution of sources as the continuum-theory approach but, in the terminology of this book, it is referred to as a group-combustion approach. The use of discrete sources is a first approximation within the realm of array theory. The fields around these monopole point sources are spherically symmetric. On account of this neglect of higher-order terms that represent the deviation from spherical symmetry, the third phenomenon cannot be treated by this approximation to array theory. In other words, the use of monopole sources in a well-defined array is somewhat pointless; once monopole sources are used in large arrays, the effect of averaging the discrete sources locally to obtain a continuous source distribution does not have a significant effect on a scale larger than the average distance between droplets. It is recommended therefore that one should use the group-combustion (continuous or averaged) approach rather than monopole arrays. On a large scale, the resolutions are identical, as shown by Labowsky and Rosner. The monopole arrays can achieve a finer resolution on the small scale but could be quite inaccurate because of the neglect of the nonsymmetric effects. If arrays are to be used, a much better representation of the vaporizing droplet than a simple monopole should be used wherever possible and practical to be consistent (that is, to make the finely resolved features more believable). As already mentioned, array analysis can provide insight into the small-scale phenomena, which should provide useful input to group analysis. For example, a Nusselt number relationship can be determined from array theory and then used in a group theory.

The discussion on theories in which arrays with discrete monopole sources are used is deferred until Section 6.2 on group vaporization and combustion of fuel droplets. For the reasons just mentioned, it is rational to separate array theories with monopole sources from the more exact representations of the local conditions around droplets. Here, we speak only of analyses in which one monopole represents each droplet. The use of the method of images to solve exactly for a potential flow in the gas surrounding the droplets would yield many monopoles of varying location and strength for each droplet; such a technique has been used for droplet arrays by Labowsky (1976, 1978, 1980) and Imaoka and Sirignano (2005a).

Some work on heat and mass transfer to an array of nonvaporizing droplets with forced convection was performed by Tal and Sirignano (1982) and Tal et al. (1983, 1984a, 1984b). These works are a precursor to work on vaporizing droplets; however, some interesting effects have already been observed that should be relevant to vaporizing droplets. In these studies, a cylindrical-cell model is used to take advantage of the periodic structure of the array. Reynolds numbers are of the order of 100, based on the diameter of the nonvaporizing sphere in the array of uniformly sized and spaced spheres. The finite-difference calculations give excellent agreement with experimental results for drag coefficients (defined by use of local ambient conditions), indicating that drag increases as longitudinal (streamwise) and lateral (transverse) spacings decrease in identical fashion. These phenomena are explained later in this section. It has been shown that, when the structure of the array

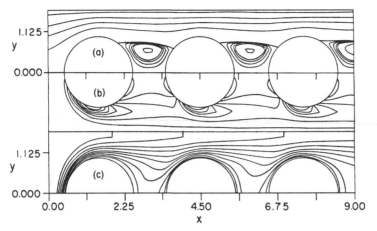

Figure 6.2. (a) Stream function, (b) vorticity, (c) isotherm contours for three spheres in a cell with 1.5-diameter spacing, Re = 100, Pr = 1. (Tal et al., 1983, with permission of the *International Journal of Heat and Mass Transfer.*)

is periodic, the problem for the flow-field properties may be cast into a periodic form (although all properties are not exactly periodic, e.g., vorticity and temperature). Figures 6.2 and 6.3, taken from the work of Tal et al. (1983), show (a) stream function, (b) vorticity contours, and (c) isotherms for spheres in an array at two different spacings. Strong interaction between the flow fields surrounding neighboring droplets occurs when the droplet centers are less than 2 diameters apart. For example, at Re = 100 with droplets in tandem in the flow stream, the recirculation zone aft of the first sphere would make contact with the forward portion of the next sphere.

Table 6.1 shows typical results for the friction drag coefficient, pressure drag coefficient, and total drag coefficient. A calculation with three spheres in tandem confined laterally within a cylindrical cell (contrived to represent the periodic

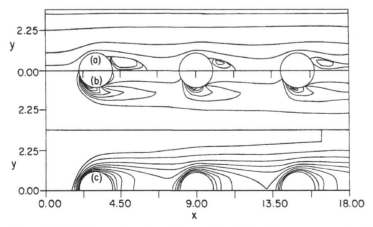

Figure 6.3. (a) Stream function, (b) vorticity, (c) isotherm contours for three spheres in a cell with 3.0-diameter spacing, Re = 100, Pr = 1. (Tal et al., 1983, with permission of the *International Journal of Heat and Mass Transfer.*)

Table 6.1. *Drag coefficients of spheres in cylindrical cells (Re = 100, b/a = 1.5)*

Type of solution	C_{DF}			C_{DP}			C_D		
Three spheres in cell	1st	2nd	3rd	1st	2nd	3rd	1st	2nd	3rd
	0.966	0.692	0.677	0.977	0.693	0.671	1.964	1.385	1.349
Periodic	0.673			0.657			1.330		
Unconfined	0.675			0.639			1.314		

structure in the lateral direction) is compared with that of a periodic cell with one sphere (contrived to represent periodicity of the array structure in both lateral and streamwise directions). Results for a single unconfined sphere are also indicated.

Table 6.2 indicates that heat transfer to each sphere decreases as the position of the sphere becomes farther downstream of the first sphere. This effect is due to the decrease in gas temperature as it flows downstream. This can be seen when we define the Nusselt number based on the primary ambient conditions (the inlet conditions to the cell of the first sphere). If we define the Nusselt number based on the local ambient conditions (inlet conditions at each cell), the Nusselt number becomes approximately constant after the first sphere; its value here is greater than in the first definition but still less than the value for an unconfined sphere.

The same studies produced Table 6.3, which shows that the Nusselt number depends on lateral (or transverse) spacing in one way and on longitudinal (or streamwise) spacing in an opposite way. As lateral spacing decreases in the array, the flow field accelerates and the Nusselt number increases; as longitudinal spacing decreases, droplets are protected by the wakes of other droplets and the Nusselt number decreases. Note that the friction drag coefficient is expected to increase as the lateral spacing decreases and the pressure drag should increase as longitudinal spacing increases because of accelerated flow over the individual sphere (a Venturi effect). With lateral and longitudinal spacing identical in the array, the drag increases as spacing decreases and exceeds the drag for an unconfined sphere. However, the Nusselt number decreases as spacing decreases and is below the value for an unconfined sphere.

Two spheres in tandem were studied for an intermediate Reynolds numbers by use of bispherical coordinates (Tal et al., 1984b). The general qualitative conclusions are similar to those obtained by Twardus and Brzustowski for the zero Reynolds number case, namely, transport rates decrease as spacing decreases. However, with

Table 6.2. *Nusselt number in cylindrical cells (Re = 100, b/a = 1.5)*

Type of solution	Nu			Remarks
Three spheres in a cell	1st	2nd	3rd	Nu based on inlet
	7.644	4.765	4.061	to first sphere
Three spheres in a cell	1st	2nd	3rd	Nu based on inlet
	7.644	6.151	6.270	each cell unit
Periodic	5.935			
Unconfined	7.910			

Table 6.3. *Average Nusselt number for the second
of three spheres in a cell (Re = 100, Pr = 1)*

Lateral spacing	Longitudinal spacing	Nu
1.5	1.5	4.765
1.06	1.5	5.708
1.5	1.35	4.594
1.5	1.2	4.372

a mean flow, the two spheres are not identical in their transport characteristics; the upstream sphere has a higher Nusselt number than the downstream sphere but both values are reduced from that for an isolated or unconfined sphere.

Tal and Sirignano (1984) and Patnaik et al. (1986) have shown by numerical calculations that internal circulation for a liquid droplet in the vicinity of other droplets can still be well approximated by a Hill spherical vortex. The implication is that the approach used to study transient heating and multicomponent vaporization for isolated droplets may still be followed for droplets in sprays. The effect of the droplet interactions can be properly represented through the drag coefficient and Nusselt number for the gas film. Although the coupling between gas and liquid phases will change the quantitative behavior of the liquid, there seems to be no qualitative change in liquid behavior that is due to interaction among droplets.

The preliminary evidence indicates that transport rates and vaporization rates will be reduced from the values for isolated droplets for two reasons. First, the local ambient environment is affected by the presence of droplets in the neighborhood. The droplets cool the environment, thereby decreasing heat transfer to the droplet and supply fuel vapor to the environment, so that mass transport also decreases. These two effects may be expected to increase as droplet number density increases or, equivalently (for a given droplet size), spacing decreases. Second, the Nusselt number based on local ambient conditions will be less than the corresponding value for an isolated droplet and will decrease as spacing decreases.

Patnaik and Sirignano (1986) extended the work of Patnaik et al. (1986) to consider two droplets moving in tandem at constant spacing. The solution of the first droplet was calculated as if that droplet were isolated. Then that solution for the wake was used as a free-stream input to the downstream droplet. The weaknesses of this model are the omission of variation in spacing and upstream influence of the second droplet. Tong and Chen (1988) extended the cylindrical-cell model of Tal and Sirignano (1982) to include vaporization. A three-droplet linear (tandem) array was studied to obtain correlations for the Nusselt number. Kleinstreuer et al. (1989) used a finite-element analysis of the linear array to yield the drag coefficients of interacting spheres. Also, they used a boundary-layer analysis for vaporizing droplets to simulate coupled transfer processes for three droplets in tandem. The analysis involved individual computations for the three spheres with coupling only through the determination of an effective temperature for the gas flow approaching the downstream droplets. Tsai and Sterling (1990) determined the Nusselt number and drag coefficients for steady flow past a linear array of nonvaporizing spheres. They obtained results that were in qualitative agreement with those of Tal et al. (1983)

and Tong and Chen (1988). All of the preceding analyses take advantage of axisymmetric flow.

Raju and Sirignano (1990b) developed a transient axisymmetric finite-difference analysis with a grid-generation scheme for two vaporizing droplets moving in tandem. They considered variable density but otherwise constant thermophysical properties in a fashion similar to that of Patnaik et al. (1986). The extension beyond other studies for tandem droplets involves transient behavior (including unequal regression rates of the two droplet surfaces and temporal variation in droplet spacing that is due to differences in droplet drag and mass), a fully coupled Navier–Stokes solution (allowing for complete coupling of internal liquid flow and gas flow and for complete elliptic behavior with upstream influence), and different initial sizes for the upstream and the downstream droplets. They studied a limited range of initial values for Reynolds number, droplet spacing, and droplet-radii ratio. The most interesting result was that there is a critical ratio of the two initial droplet diameters below which droplet collision does not occur. If the ratio of the downstream-droplet initial diameter to the upstream-droplet initial diameter is larger than the critical ratio, the reduced drag coefficient of the downstream droplet causes less deceleration and greater relative velocity with the gas for the downstream droplet. Therefore collision is likely because the spacing decreases with time. Below the initial ratio, the reduced inertia of the downstream droplet causes the spacing to increase. The critical ratio is found to be less than unity and weakly dependent on the initial Reynolds number. Recall, however, that Chiang et al. (1992) showed that constant thermophysical properties could lead to errors in the detailed prediction of the drag coefficients. These errors, of course, make the precise determination of the critical ratio questionable.

Later, Chiang and Sirignano (1993a) and Chiang (1990) extended the two-tandem-droplet calculation of Raju and Sirignano (1990b) to account for variable thermophysical properties. A wider range of initial values for Reynolds number, droplet spacing, and droplet-radii ratio was considered. Also, correlations of the numerical results for drag coefficient, Nusselt number, Sherwood number with Reynolds number, transfer number, spacing, and radii ratio were obtained. Furthermore, Chiang and Sirignano (1993b) extended the analysis to three droplets moving in tandem.

The two-droplet results of Chiang and Sirignano are in qualitative agreement with the findings of Raju and Sirignano. Figure 6.4 shows gas-phase velocity vector, liquid-phase stream function, and vorticity contours for both phases. The downstream sphere is seen to be within the wake of the upstream sphere. Hence the effective Reynolds number for the downstream sphere is less than the Reynolds number for the upstream sphere. The strength of the liquid-phase vortex and the transport rates on the forward side are less for the downstream droplet than for the upstream droplet. Transport rates on the aft side of the droplet are greater for the downstream droplet than for the upstream droplet, as shown by Fig. 6.5.

The drag coefficient is displayed as a function of instantaneous Reynolds number in Fig. 6.6. The reduction of the drag coefficient for the downstream droplet and the general overprediction of the constant-property solution are shown. The similarity between the drag coefficient of an isolated droplet and that of the lead droplet is shown also. Center-to-center spacing as a function of time is shown in Fig. 6.7 for

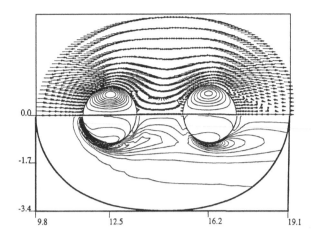

Figure 6.4. Gas-phase velocity vector, liquid-phase stream function, and vorticity contours for both phases in two-tandem-droplet case. Nondimensional time $= 3.00$, $\mathrm{Re}_1 = 80.31$, $\mathrm{Re}_2 = 85.84$, $R_1 = 1.00$, nondimensional spacing $= 3.71$. (Chiang and Sirignano, 1993a, with permission of the *International Journal of Heat and Mass Transfer.*)

GAS AND LIQUID-PHASE VORTICITY
Contour Interval: 2.36E-01 Min: -4.49E+00 Max: 2.23E-01

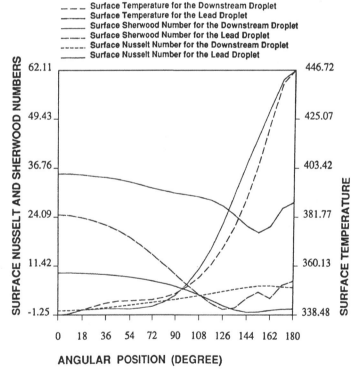

Figure 6.5. Surface temperatures, Nusselt number, and Sherwood number versus position on droplet surface for the two-tandem-droplet case. *N*-decane liquid 1000 K ambient temperature, 300 K initial droplet temperature, 100 initial Reynolds number, 1.00 initial R_1 and R_2, 8.0 initial spacing. (Chiang and Sirignano, 1993a, with permission of the *International Journal of Heat and Mass Transfer.*)

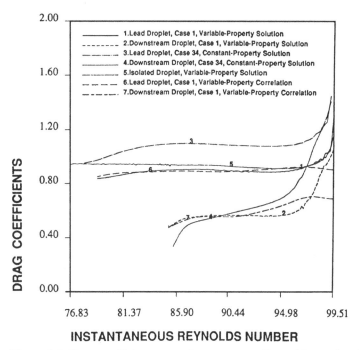

Figure 6.6. Drag coefficient versus instantaneous Reynolds number for two-tandem-droplet case. (Chiang and Sirignano, 1993a, with permission of the *International Journal of Heat and Mass Transfer.*)

various initial values of the spacing D and various initial values of the droplet-radii ratio R_2. The prediction of strong influence on the downstream droplet at a spacing of 16-droplet radii is consistent with the experimental findings of Temkin and Ecker (1989), who found influences up to 30-radii spacing. It is seen that a critical initial value of R_2 can exist that distinguishes between increasing spacing and decreasing spacing with time.

Chiang and Sirignano (1993a) obtained correlations for instantaneous drag coefficients with other instantaneous parameter values from their computational results, as shown in Fig. 6.6; also, correlations for Nusselt number and Sherwood number were obtained. A linear regression model was used to fit over 3000 data points. The correlations are normalized by the isolated droplet correlations given by Eq. (3.55) of Chapter 3. The correlations for each of the droplets follow. The Reynolds number Re_m is based on instantaneous droplet–gas relative velocity, droplet diameter, free-stream gas density, and average gas-film viscosity.

For the lead droplet,

$$\frac{C_{D1}}{C_{D\,\mathrm{iso}}} = 0.877\,\mathrm{Re}_m^{0.003}(1 + B_H)^{-0.040}D^{0.048}(R_2)^{-0.098},$$

$$\frac{\mathrm{Nu}_1}{\mathrm{Nu}_{\mathrm{iso}}} = 1.245\,\mathrm{Re}_m^{-0.073}\mathrm{Pr}_m^{0.150}(1 + B_H)^{-0.122}D^{0.013}(R_2)^{-0.056},$$

$$\frac{\mathrm{Sh}_1}{\mathrm{Sh}_{\mathrm{iso}}} = 0.367\,\mathrm{Re}_m^{0.048}\mathrm{Sc}_m^{0.730}(1 + B_M)^{0.709}D^{0.057}(R_2)^{-0.018}, \qquad (6.1)$$

where $0 \leq B_H \leq 1.06, 0 \leq B_M \leq 1.29, \quad 11 \leq \text{Re}_m \leq 160, 0.68 \leq \text{Pr}_m \leq 0.91, \ 1.47 \leq \text{Sc}_m \leq 2.50, \ 2.5 \leq D \leq 32,$ and $0.17 \leq R_2 \leq 2.0.$

For the downstream droplet,

$$\frac{C_{D2}}{C_{D\,\text{iso}}} = 0.549 \, \text{Re}_m^{-0.098} (1 + B_H)^{0.132} D^{0.275} (R_2)^{0.521},$$

$$\frac{\text{Nu}_2}{\text{Nu}_{\text{iso}}} = 0.528 \text{Re}_m^{-0.146} \text{Pr}_m^{-0.768} (1 + B_H)^{0.356} D^{0.262} (R_2)^{0.147},$$

$$\frac{\text{Sh}_2}{\text{Sh}_{\text{iso}}} = 0.974 \text{Re}_m^{0.127} \text{Sc}_m^{-0.318} (1 + B_M)^{-0.363} D^{-0.064} (R_2)^{0.857}, \qquad (6.2)$$

where $0 \leq B_H \leq 2.52, \ 0 \leq B_M < 1.27, \ 11 \leq \text{Re}_m \leq 254, \ 0.68 \leq \text{Pr}_m < 0.91,$ and $1.48 \leq \text{Sc}_m \leq 2.44.$

Chiang and Sirignano (1993b) considered three-tandem-vaporizing droplets of equal initial diameters. The qualitative conclusions of the two-droplet study generally apply to the behavior of the first two droplets. The drag coefficients and the transport rates for the second and the third droplets are significantly reduced below the lead-droplet values, which remain close to the isolated-droplet values. The transport rates of the second and the third droplets differ by a very small amount in general compared with the difference between the rates of the firstand second

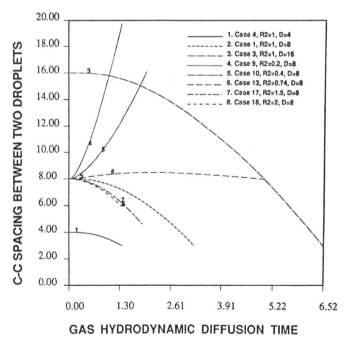

Figure 6.7. Center-to-center (C-C) droplet spacing versus nondimensional time for various initial values of spacing D and radii ratio R_2. (Chiang and Sirignano, 1993a, with permission of the *International Journal of Heat and Mass Transfer.*)

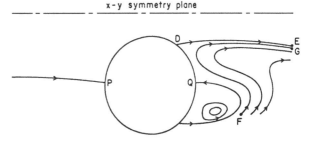

Figure 6.8. Streamlines in symmetry plane for flow past two side-by-side spheres (Kim et al., 1993).

droplets. However, the second droplet has a slightly lower drag coefficient than the third droplet. Correlations are reported in their paper.

The two-tandem- and three-tandem-vaporizing-droplet computational results are consistent with the nonvaporizing-sphere results of Tal et al., who found little difference in drag coefficient for downstream particles in a tandem stream. Also, many experimental investigators found that, in a long stream of equisized droplets moving in tandem with uniform initial spacing among the droplets, the spacing remains constant with time, implying equal drag coefficients (Sangiovanni and Kesten, 1976; Nguyen et al. 1991; and Nguyen and Dunn-Rankin, 1992).

Asano et al. (1988) analyzed two tandem spheres in a steady flow with constant properties. The interactions are described in a simple manner by use of geometric factors. Nguyen et al. (1991) performed a computational and experimental study of droplets moving in tandem. Vaporization was not significant in their experiment because of low ambient temperatures and saturated vapors in the droplet-stream vicinity. They found that a lead droplet and a trailing droplet of equal size will collide for small initial spacings, which agrees with theoretical predictions, including their own prediction. Further experimental studies on drag reduction and collisions for a small number of droplets moving in tandem were reported by Nguyen and Dunn-Rankin (1992).

The interaction of two spheres moving side by side in parallel, or approximately in parallel, is a three-dimensional phenomenon. Only a limited number of three-dimensional flow calculations for individual spheres have been made. Dandy and Dwyer (1990) considered steady, uniform shear flow past a heated sphere. Tomboulides et al. (1991) considered flow past a sphere with unsteady wakes.

More recently, Kim et al. (1993) studied three-dimensional flow over two identical spheres moving in parallel at constant velocity. The separation distance between the two sphere centers was held constant, and the line connecting the centers was normal to the free-stream velocity vector. Both liquid and solid spheres were considered at Reynolds numbers between 50 and 150. The transient finite-difference calculations were made in one quadrant of the flow field, taking advantage of the two perpendicular planes of symmetry. The asymptotic results yielded the steady-flow results.

Figure 6.8 shows streamlines in the plane of symmetry containing the two sphere centers. In this case, the separation distance is small. It is seen that the flow through the gap between the neighboring spheres has a jetlike character and entrains the outer flow. On the foreside of the sphere in Fig. 6.8, the stagnation point is

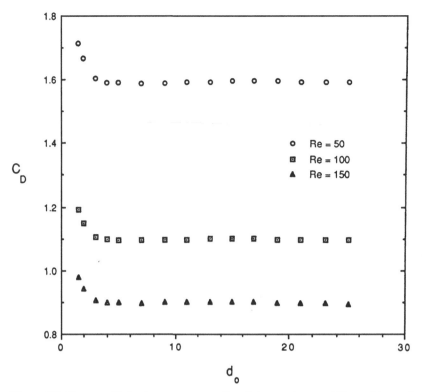

Figure 6.9. Drag coefficients versus separation distance at several Reynolds numbers for flow past two side-by-side spheres (Kim et al., 1993).

perturbed from its position in an axisymmetric flow. As a result, the flow on the side closer to the neighboring sphere has a lower velocity and higher pressure than the axisymmetric case; on the other side, the velocity is higher and the pressure is lower than the axisymmetric reference. On account of asymmetric deviations of pressure and surface friction, the spheres experience lift and torque as well as a modification of the drag that is due to the mutual interaction. The lift forces on the spheres result in a weak attraction at large separation distances and a strong repulsion at small separation distances. The drag coefficient increases as the separation decreases. Figures 6.9 and 6.10 display drag and lift coefficients for liquid spheres. Moment coefficients are also reported.

In another recent study, Kim et al. (1992) considered the effects of drag and lift on the trajectories of two droplets of equal size and equal and parallel initial velocity. At a large initial separation, the two droplets moved slightly closer whereas, at a small initial separation, the deflection that was due to repulsion was more significant. In both cases, the translation of the two droplets was nearly parallel.

Stapf et al. (1998) considered a three-dimensional array of droplets vaporizing and burning. Navier–Stokes numerical solutions were obtained. The unsteady computation allowed for relative motion among the droplets. The interaction among the droplets was found to be important. Jeffrey and Onishi (1984) considered two

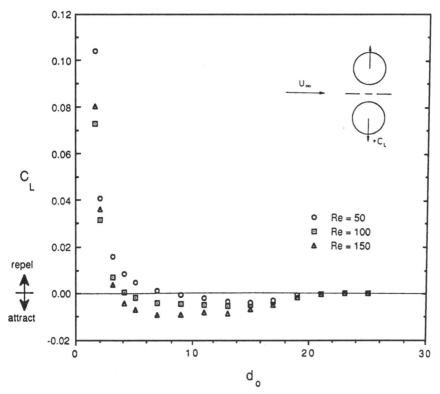

Figure 6.10. Lift coefficients versus separation distance at several Reynolds numbers for flow past two side-by-side spheres (Kim et al., 1993).

unequal spheres interacting at low Reynolds number. Batchelor and Green (1972) considered a linear flow for two freely moving, interacting spheres.

6.2 Group Vaporization and Combustion

In this section, we discuss representations of vaporization of clouds for which the precise position of each droplet is not important and only such average quantities as droplet number density (or, equivalently, the average distance between droplets) are important. Theories (Suzuki and Chiu, 1971; Chiu and Liu, 1977; Labowsky and Rosner, 1978; Chiu et al., 1983; Correa and Sichel, 1983) that use continuum representations of droplets as distributed monopole sources of fuel vapor are included herein. Also included are representations (Labowsky and Rosner, 1978; Samson et al., 1978) of droplets as discrete monopole sources with such a high number density of droplets that the distinction between discrete and continuum representations is not significant.

Two major shortcomings can be cited in the existing theories. First, the theories do not account for the fact that the Nusselt number and the vaporization law for each droplet will depend on the spacing between droplets (with one exception, subsequently mentioned). Second, the theories are quasi-steady and do not consider transient droplet heating or unsteady gas-phase conduction across the clouds. Because the time scale for droplet heating and the time and length scales for conduction across the cloud can be large, these types of unsteadiness can be profound. With this in mind, the present theories should be viewed as a base for further studies rather than as complete theories.

The most interesting result of these theories is that certain distinct regimes of combustion have been discovered: isolated-droplet combustion, internal group combustion, external group combustion, and sheath combustion. The particular regime of operation is determined by droplet number density, the primary ambient conditions, droplet radius, and fuel volatility. Chemical kinetics is, of course, an important factor as well, but it is typically considered to be quite fast.

Chiu and co-workers (Suzuki and Chiu, 1971; Chiu and Liu, 1977; Chiu et al., 1983) provided an extremely interesting approach to group combustion. They consider a quasi-steady vaporization and diffusion process with infinite kinetics and show the importance of a group-combustion number given by

$$G = 3(1 + 0.276 \mathrm{Re}^{1/2} \mathrm{Sc}^{1/3}) \mathrm{Le}\, N^{2/3} (R/s), \qquad (6.3)$$

where Re, Sc, Le, N, R, and s are the Reynolds numbers based on diameter, Schmidt number, Lewis number, total number of droplets in the cloud, instantaneous average droplet radius, and average spacing between the centers of the droplets, respectively. This Chiu number G increases with Reynolds number, Prandtl number, and the size of the cloud (measured in droplet numbers) whereas it decreases with increasing Schmidt number and spacing between droplets.

The value of the Chiu number G has been shown to have a profound effect on the flame location and distributions of temperature, fuel vapor, and oxygen. Four types of behavior are found: for large G numbers, external sheath combustion occurs; then, as G is progressively decreased, external group combustion, internal group combustion, and isolated droplet combustion occur. These regimes are identified in Fig. 6.11. Isolated-droplet combustion involves a separate flame enveloping each droplet. Typically a group number of less than 10^{-2} is required. Internal group combustion involves a core within the cloud where vaporization is occurring with the core totally surrounded by a flame. Outside of the core, each droplet is enveloped by individual flames. This occurs for G values above 10^{-2} and somewhere below unity. As G increases, the size of the core increases. When the single flame envelops all droplets, we have external group combustion. This phenomenon begins with G values close to the order of unity so that many industrial burners and most gas-turbine combustors will be in this range. With external group combustion, the vaporization rate of individual droplets increases with distance from the center of the core. At very high G values (above 10^{-2}), only the droplets in a thin layer at the edge of the cloud are vaporizing. This regime has been called external sheath combustion by Chiu and co-workers.

Labowsky and Rosner (1978) found similar results. They use a different terminology; their "incipient group combustion" is internal group combustion and their "total group combustion" is external group combustion. They use a quasi-steady continuum approach similar to that of Chiu and co-workers. As discussed in Section 6.1, they also use a superposition method with discrete monopole sources.

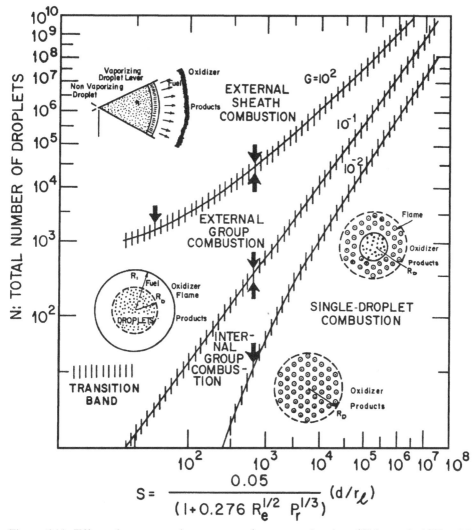

Figure 6.11. Effect of group number on type of spray combustion. (Chiu et al., 1983, with permission of The Combustion Institute.)

This has the advantage of giving the details of the flame shape or the field around each droplet; however, because the monopole approximations produce errors on that fine scale, this is a questionable advantage. The authors did not advocate the noncontinuum approach but rather were interested in gaining insight through comparison.

Labowsky and Rosner (1978) also showed that the Thiele modulus $(3\phi)^{1/2} R_c / R$ is an important parameter in the determination of the onset of internal group combustion. Here, ϕ is the volume fraction of droplets and R_c is the radius of a spherical cloud of droplets. The square of that factor, $(3\phi) R_c^2 / R^2$, is a Damkohler number, as stated by the authors. Although the authors do not indicate it, this Damkohler number can be readily shown to be equal to $3N^{2/3} R / S$ so that it equals the group-combustion number of Chiu and co-workers (Suzuki and Chiu, 1971;

Chiu and Liu, 1977; Chiu et al., 1983) for the case of zero Reynolds number and unitary Lewis number.

Correa and Sichel (1983) performed an asymptotic analysis for large G values and obtained external sheath-combustion results that agreed well with those of Chiu and co-workers. A d^2 law based on cloud size is obtained in this quasi-steady sheath limit. They do raise concerns about whether the quasi-steady formulation will apply for lower values of G.

Kerstein and Law (1983) presented an interesting model of group combustion based on percolation theory. The presentation contains no quantitative results for profiles, vaporization rates, and flame locations, so that it is quite difficult to evaluate their contribution. However, they do seem to be attempting to represent the third phenomenon mentioned at the beginning of this chapter. Also, as we discuss in Section 7.3, there does seem to be a need for new approaches based on stochastic methods.

All the approaches to group-combustion theory were based on infinite chemical-kinetic rates. Obviously, finite rates should produce significant quantitative differences. Only Chiu introduced a convective effect into his model; he used an ad hoc extension of the empirical correction for single droplets. This procedure could be improved in the future.

6.3 Generalized Theory for Droplet-Array Vaporization and Burning

Imaoka and Sirignano (2005a, 2005b, 2005c), Sirignano (2007), and Sirignano and Wu (2008) developed a theory for droplet-array vaporization and burning based on the approach of Labowsky (1978). A mass-flux potential function is defined and is the solution to Laplace's equation in the gas field. This approach is powerful because the scalar properties in the gas surrounding the droplets in the array become uniform valued across each surface of constant mass-flux potential. So the three-dimensional scalar field becomes one dimensional in terms of the potential. The normalized potential function is solved in the three-dimensional space; it depends on only the geometrical configuration of the droplets in the spray and not on other parameters of the problem, giving the three-dimensional solution for the potential a more universal character.

6.3.1 Basic Formulation

The generalized formulation relies on many of the assumptions used in the cited works. A quasi-steady gas phase and one-step chemical reaction are required. Fourier heat conduction and Fickian mass diffusion apply. All liquids have identical single-component compositions. Radiation is neglected. Phase equilibrium exists at the liquid surfaces, and the gas is negligibly soluble in the liquid. Kinetic energy, viscous dissipation, and other terms of the order of the square of the Mach number will be neglected. The momentum equation subject to these assumptions yields that the pressure is of the order of the square of the Mach number and, in the calculation of the scalar properties, it can be considered as constant. The steady-state continuity equation, energy-conservation equation, and species-conservation equation apply:

$$\nabla \cdot \left(\rho \vec{V} \right) = 0, \tag{6.4}$$

$$\nabla \cdot \left(\rho \vec{V} h - \lambda \nabla T - \sum_{m=1}^{N} \rho D_m h_m \nabla Y_m \right) = -\dot{\omega}_F Q, \tag{6.5}$$

$$\nabla \cdot \left(\rho \vec{V} Y_m - \rho D_m \nabla Y_m \right) = \dot{\omega}_m, \qquad m = 1, \ldots, N. \tag{6.6}$$

Sensible mixture enthalpy h, species enthalpy h_m, specific heat c_p, and mixture thermal conductivity λ are computed with the following equations:

$$h = \int_{T_{\text{ref}}}^{T} c_p(T') \mathrm{d}T'; \quad h_m = \int_{T_{\text{ref}}}^{T} c_{p,m}(T') \mathrm{d}T'; \quad c_p = \sum_{m} c_{p,m}(T) Y_m;$$

$$\lambda = \sum_{m} \lambda_m(T) Y_m. \tag{6.7}$$

It will be shown in the next subsection that, if there is no tangential velocity at the liquid surface, if scalar properties are uniform on the liquid surface, and if the gas flow is irrotational, the gas-phase mass flux is governed by a potential function ϕ, such that

$$\rho \vec{V} = \nabla \phi. \tag{6.8}$$

From Eq. (6.4), ϕ satisfies Laplace's equation and the following boundary conditions:

$$\nabla^2 \phi = 0 \quad \begin{cases} \phi = 0 & \text{at liquid surfaces} \\ \phi = \phi_\infty & \text{far from the liquid} \end{cases}. \tag{6.9}$$

With a potential function governing mass flux, it then follows that

$$\nabla \times \rho \vec{V} = \rho \left(\nabla \times \vec{V} \right) - \vec{V} \times \nabla \rho = 0. \tag{6.10}$$

Therefore the existence of a potential function requires an irrotational velocity field and the alignment or counteralignment of the density gradient and velocity vectors throughout the gas field. As a consequence, velocities at the liquid–gas interface will not have a tangential component because scalar properties are uniform over the liquid surface. Species and energy balances at the liquid surfaces indicate that the quantity $L_{\text{eff}}/(1 - Y_{Fs})$ is spatially uniform over all liquid surfaces. Therefore the instantaneous liquid surface temperature T_s and the potential function ϕ will also be spatially uniform at all liquid surfaces. As mentioned, verification of the existence of a potential function will be given in Subsection 6.3.2. The equations shown here apply to vaporization with and without combustion. Simplifications and additional assumptions will be made as needed.

In the remainder of this section, the analysis of the gas phase is emphasized. The liquid-phase heating can be represented by any of the models discussed in Subsection 2.2.2. The phase-equilibrium interface condition given by Eq. (2.4) can be used for single-component fuels, whereas, for multicomponent liquids, Raoult's Law given by Eq. (4.3) can be used. For multicomponent liquids, the liquid-phase mass diffusion equations of the type given by Eq. (4.5) must also be used.

6.3.2 Analysis of Vaporization Without Combustion

In the absence of combustion, no additional assumptions are necessary. Equations (6.5) and (6.6) are reduced to

$$\nabla \cdot \left(\rho \vec{V} h - \lambda \nabla T - \sum_{m=1}^{N} \rho D_m h_m \nabla Y_m \right) = 0, \tag{6.11}$$

$$\nabla \cdot \left(\rho \vec{V} Y_m - \rho D_m \nabla Y_m \right) = 0, \qquad m = 1, \ldots, N. \tag{6.12}$$

From Equations (6.4) and (6.12), it follows that

$$\nabla \cdot \left(\rho \vec{V} Y_m - \rho D_m \nabla Y_m - C_m \rho \vec{V} \right) = 0, \qquad m = 1, \ldots, N, \tag{6.13}$$

where C_m is a constant. Assume Y_m is uniform over the liquid surface and $\vec{V}$ is normal to the liquid surface. Choose C_m so that

$$\rho \vec{V} Y_m \Big)_s - \rho D_m \nabla Y_m)_s - C_m \rho \vec{V} \Big)_s = 0, \qquad m = 1, \ldots, N, \tag{6.14}$$

is satisfied at the liquid boundary. Then, define

$$\rho \vec{V} Y_m - \rho D_m \nabla Y_m - C_m \rho \vec{V} = \vec{\Psi}_m, \qquad m = 1, \ldots, N, \tag{6.15}$$

where $\vec{\Psi}_m = 0$ at the liquid surface from (6.14), and from (6.15) $\nabla \cdot \vec{\Psi}_m = 0$ throughout the gas field. Note further that $\rho \vec{V}$ and ∇Y_m go to zero at infinity; so $\vec{\Psi}_m = 0$ there. The only solution is $\vec{\Psi}_m = 0$. Therefore,

$$(Y_m - C_m) \rho \vec{V} - \rho D_m \nabla Y_m = 0, \qquad m = 1, \ldots, N, \tag{6.16}$$

throughout the field. Applying the same analysis to the energy equation with a uniform-temperature value over the liquid surface, and noting that ∇h goes to zero at infinity, we obtain

$$(h - C) \rho \vec{V} - \lambda \nabla T - \sum_{m=1}^{N} \rho D_m h_m \nabla Y_m = 0. \tag{6.17}$$

Consequently, the heat and mass diffusion flux vectors are aligned (or counter-aligned) with the streamlines. The same analysis can be applied in the next subsection to the equation governing the scalar coupling functions. Because the mass fractions and enthalpy have gradients aligned with the velocity vector and the pressure variation is insignificant, the density gradient will be locally parallel to the streamlines. Note that, for a quasi-steady inviscid flow, the fractional pressure variations are of the order of the square of the Mach number. As shown by Joseph and Liao (1994) and Joseph (2003), constant-density flows with normal viscous stress can be irrotational. In that case, fractional variations in the thermodynamic pressure can be of the order of the Mach number. So viscosity could be considered in our potential-flow analysis if the constant-density assumption were invoked. With an irrotational flow, Eq. (6.10) is satisfied and the mass-flux potential may be used:

$$(Y_m - C_m) \nabla \phi - \rho D_m \nabla Y_m = 0, \qquad m = 1, \ldots, N, \qquad (6.18)$$

$$(h - C) \nabla \phi - \lambda \nabla T - \sum_{m=1}^{N} \rho D_m h_m \nabla Y_m = 0. \qquad (6.19)$$

Species mass balance at the liquid–gas interface indicates that $C_m = 0$ for $m \neq F$, and $C_F = 1$. Equation (6.18) is multiplied by h_m and summed for all species. The result is subtracted from (6.19). Energy balance at the interface yields the constant C, and noting that $\nabla T = (\nabla h_F)/c_{p,F}$,

$$\nabla \ln (h_F - h_{Fs} + L_{\text{eff}}) = \frac{\nabla \phi}{\lambda/c_{p,F}}, \qquad (6.20)$$

$$L_{\text{eff}} = L + \frac{\dot{q}_l}{|\rho \vec{V}|_s}. \qquad (6.21)$$

The term $\dot{q}_l$ is the magnitude of the conductive heat flux into the liquid when droplets are not at wet-bulb temperatures. With the arbitrariness of a constant in the determination of ϕ, taking $\phi_s = 0$ incurs no loss of generality. We obtain the implicit relation between h_F and ϕ by integrating Eq. (6.20) along a pathline to yield

$$1 + \frac{h_F - h_{Fs}}{L_{\text{eff}}} = (1 + B_H) e^{-\int_\phi^{\phi_\infty} \frac{d\phi'}{\lambda/c_{p,F}}}, \qquad (6.22)$$

where

$$B_H = \frac{h_{F\infty} - h_{Fs}}{L_{\text{eff}}}. \qquad (6.23)$$

Evaluating Eq. (6.22) at infinity, we obtain

$$\phi_\infty = \overline{\lambda/c_{p,F}} \ln (1 + B_H), \qquad (6.24)$$

where the definition is made that

$$\overline{\lambda/c_{p,F}} = \frac{\phi_\infty}{\int_0^{\phi_\infty} \frac{d\phi'}{\lambda/c_{p,F}}}. \qquad (6.25)$$

A normalized potential function Φ is defined such that

$$\Phi = \frac{\phi}{\phi_\infty}. \qquad (6.26)$$

Equations (6.22) and (6.25) can now be written as functions of Φ:

$$1 + \frac{h_F - h_{Fs}}{L_{\text{eff}}} = (1 + B_H)^{\overline{\lambda/c_{p,F}} \int_0^{\Phi} \frac{d\Phi'}{\lambda/c_{p,F}}} \qquad (6.27)$$

$$\overline{\lambda/c_{p,F}} = \left(\int_0^1 \frac{d\Phi'}{\lambda/c_{p,F}} \right)^{-1}. \qquad (6.28)$$

From Eqs. (6.4), (6.9), and (6.26), the normalized potential function in any geometry satisfies Laplace's equation with the following boundary conditions:

$$\nabla^2 \Phi = 0 \quad \begin{cases} \Phi = 0 & \text{at liquid surfaces} \\ \Phi = 1 & \text{far from the liquid} \end{cases}. \tag{6.29}$$

For a given geometrical description of the liquid surfaces, Φ can be determined. There is no explicit dependence on liquid-fuel choice, transport properties, and scalar boundary values; these parameters appear only through the normalization factor.

The species equations are treated in a process similar to the energy equation. Integrating (6.18) for each of the N species, and using the relations given in Eqs. (6.24) and (6.26), we obtain

$$\frac{Y_F - 1}{Y_{Fs} - 1} = (1 + B_H)^{\overline{\lambda/c_{p,F}} \int_0^{\Phi} \frac{d\Phi'}{\rho D_F}} \qquad m = F, \tag{6.30}$$

$$\frac{Y_m}{Y_{m\infty}} = (1 + B_H)^{\overline{\lambda/c_{p,F}} \int_0^{\Phi} \frac{d\Phi'}{\rho D_m} - \text{Le}_m}, \qquad m \neq F, \tag{6.31}$$

The following relations are defined:

$$\text{Le}_m = \frac{\overline{\lambda/c_{p,F}}}{\overline{\rho D_m}}, \tag{6.32}$$

$$\overline{\rho D_m} = \left(\int_0^1 \frac{d\Phi'}{\rho D_m} \right)^{-1}, \tag{6.33}$$

$$B_M = \frac{Y_{Fs} - Y_{F\infty}}{1 - Y_{Fs}}. \tag{6.34}$$

The relationship between B_M and B_H is given by

$$1 + B_M = (1 + B_H)^{\text{Le}_F}. \tag{6.35}$$

Solutions to Eqs. (6.27), (6.30), and (6.31) yield the relationship between the gas-field properties and the normalized potential function Φ with a variable Lewis number and variable ρD, where Φ is governed by Eq. (6.29).

Note that, throughout the preceding theory, the relevant specific heat is the specific heat of the transported species, namely the fuel vapor.

Y_{Fs} is obtained as a function of surface temperature at atmospheric pressure by use of a phase-equilibrium relation such as the Clausius–Clapeyron relation, which is independent of configuration and ambient conditions other than pressure. The mass fraction of fuel vapor at the interface increases substantially with time as the droplet heats. This will have a dramatic effect on the temporal behavior of the vaporization rate. For any configuration, the vaporization rate can be shown to be directly proportional to $\ln(1 + B_M)$, which by itself is independent of configuration (i.e., droplet radius and spacing) or transport properties. Of course, the vaporization rate will depend on both transport properties and configuration through other parameters, as shown later in Eqs. (6.50) and (6.93). Realize from Equation (6.35) that the vaporization rate will be directly proportional to $\ln(1 + B_H)$ only when the Lewis number has unity value, making $B_H = B_M$.

The quantity L_{eff} is a measure of the energy conducted from the gas to the liquid surface per unit mass of vaporized fuel. The product of L_{eff}/L and $\ln(1 + B_M)$, which we define as w, is a critical factor in the evaluation of the normalized energy per unit time conducted to the liquid surface; w, which is independent of configuration but dependent on the Lewis number, would be multiplied by factors dependent on transport properties and configuration to obtain the actual rate.

6.3.3 Combustion Analysis

The formulation presented in the previous subsection applies here with several additional assumptions. Fast chemical kinetics prevents oxygen from diffusing to the liquid surface, and a unitary Lewis number, $\rho D = \lambda/c_p$, is required. The Shvab–Zel'dovich form of the species- and energy-conservation equations apply:

$$\nabla \cdot \left(\rho \vec{V} \alpha_i - \rho D \nabla \alpha_i \right) = 0, \qquad i = 1, 2. \tag{6.36}$$

The coupling functions are defined as

$$\alpha_1 = h + \nu Q Y_O, \qquad \alpha_2 = Y_F - \nu Y_O. \tag{6.37}$$

Consistent with the analysis of the previous sections, the advection and diffusion of the scalar variables (6.37) are aligned with the flow:

$$\rho \vec{V} \alpha_i - \rho D \nabla \alpha_i = A_i \rho \vec{V}, \qquad i = 1, 2. \tag{6.38}$$

Substituting (6.8) into (6.38) and rearranging, we then find it follows that

$$\nabla \ln (\alpha_i - A_i) = \frac{\nabla \phi}{\rho D}, \qquad i = 1, 2. \tag{6.39}$$

Integrating along any path, and setting $\phi = 0$ at the liquid surfaces, we obtain

$$\frac{\alpha_i - A_i}{\alpha_{i,s} - A_i} = e^{\int_0^\phi \frac{d\phi'}{\rho D}}, \qquad i = 1, 2. \tag{6.40}$$

Note that this solution for the coupling function α_i is consistent with the literature when $\rho D = \text{constant}$. Evaluation of Eq. (6.40) in the farfield yields the relation for ϕ_∞:

$$\phi_\infty = \overline{\rho D} \ln \left(\frac{\alpha_{i,\infty} - A_i}{\alpha_{i,s} - A_i} \right) = \overline{\rho D} \ln (1 + B), \qquad i = 1, 2. \tag{6.41}$$

The definitions are made that

$$\overline{\rho D} = \frac{\phi_\infty}{\int_0^{\phi_\infty} \frac{d\phi'}{\rho D}}, \tag{6.42}$$

$$B = \frac{h_\infty - h_s + \nu Q Y_{O\infty}}{L_{\text{eff}}} = \frac{\nu Y_{O\infty} + Y_{Fs}}{1 - Y_{Fs}}, \tag{6.43}$$

where L_{eff} is given by Eq. (6.21). Because Le=1 implies that $\rho D = \lambda/c_p$, these values will be used interchangeably wherever necessary. Equations (6.26), (6.40), and (6.41) are combined to yield

$$\frac{\alpha_i - A_i}{\alpha_{i,s} - A_i} = (1 + B)^{\overline{\rho D} \int_0^\Phi \frac{d\Phi'}{\lambda/c_p}}. \tag{6.44}$$

Application of boundary conditions provides the values of A_i. Then, from Eqs. (6.37) and (6.44),

$$(1 + B)^{\overline{\rho D} \int_0^\Phi \frac{d\Phi'}{\lambda/c_p}} = 1 + \frac{h - h_s + \nu Q Y_O}{L_{\text{eff}}} = 1 + \frac{Y_F - Y_{Fs} - \nu Y_O}{Y_{Fs} - 1}. \tag{6.45}$$

Noting that Le = 1, the combination of Eqs. (6.26), (6.41), and (6.42) yields the following relation:

$$\overline{\rho D} = \overline{\lambda/c_p} = \left(\int_0^1 \frac{d\Phi'}{\lambda/c_p} \right)^{-1}. \tag{6.46}$$

In the limit of infinite-rate chemical kinetics, the flame surface will lie on the constant Φ surface denoted by Φ_F. This value is determined implicitly by the following expression:

$$\frac{\ln(1 - Y_{Fs})}{\ln(1 + B)} = -\overline{\rho D} \int_0^{\Phi_F} \frac{d\Phi'}{\lambda/c_p}. \tag{6.47}$$

The system of Eqs. (6.45)–(6.47) yields the four relations necessary to determine the values of $\overline{\rho D}$, the flame contour Φ_F, mixture enthalpy h, and fuel-vapor mass fraction Y_F. Fuel type, T_∞, $Y_{O\infty}$, and T_s are treated as parameters in the calculations. Equations (6.45) and (6.46) are coupled because transport properties and specific heat depend on temperature and composition. When the liquid surface temperature is less than the wet-bulb value, phase equilibrium dictates that B be calculated with the last term in Eq. (6.43). The first term then provides the liquid-heating rate through L_{eff}. Note that ρD was not assumed constant in the analysis. With the assumption $\rho D = \lambda/c_p = \overline{\rho D} = $ constant, Eqs. (6.45)–(6.47) are simplified and uncoupled. Then,

$$(1 + B)^\Phi = 1 + \frac{h - h_s + \nu Q Y_O}{L_{\text{eff}}} = 1 + \frac{Y_F - Y_{Fs} - \nu Y_O}{Y_{Fs} - 1}, \tag{6.48}$$

$$\Phi_F = \frac{-\ln(1 - Y_{Fs})}{\ln(1 + B)}. \tag{6.49}$$

Equations (6.48) and (6.49) would still result if the constant value of $\rho D = \lambda/c_p$ were defined arbitrarily, without accordance with Eq. (6.46), for example, if the value were taken at infinity or at the liquid surface.

In either the variable ρD or constant ρD situation, the vaporization rate of the jth droplet (or any portion of the liquid surface designated as the jth segment) is obtained by integrating the mass flux over the droplet (segment) surface:

$$\dot{m}_j = \int \int \nabla \phi \cdot d\vec{A}_j = \overline{\rho D} \ln(1 + B) \int \int \nabla \Phi \cdot d\vec{A}_j \tag{6.50}$$

It is not uncommon for droplet-vaporization rates to be normalized by the vaporization rate of an isolated droplet at the wet-bulb temperature. In the literature, this has often been referred to as a burning-rate correction factor or an interaction coefficient, η. Then, for the jth droplet,

$$\eta_j = \frac{\dot{m}_j}{\dot{m}_{\text{iso}}} = \frac{1}{4\pi a_j} \int \int \nabla \Phi \cdot d\vec{A}_j. \tag{6.51}$$

The vaporization rate of an isolated droplet at wet-bulb temperature with variable $\rho D = \lambda/c_p$ is given by

$$\dot{m}_{\mathrm{iso}} = 4\pi \left(\int_a^\infty \frac{\mathrm{d}r}{(\lambda/c_p)\,r^2} \right)^{-1} \ln(1+B). \tag{6.52}$$

For a single, isolated droplet, the solution to Eq. (6.29) yields $\Phi_{\mathrm{iso}} = 1 - a/r$. Upon substitution with (6.46), Eq. (6.52) can be expressed as

$$\dot{m}_{\mathrm{iso}} = 4\pi a \overline{\rho D}\, \ln(1+B). \tag{6.53}$$

Therefore, within this generalized analysis, a single, isolated droplet is a special case for which the solution Φ to Eq. (6.29) can be obtained analytically. The preceding formulas for vaporization rate are the same with and without an envelope flame. The value of the transfer number B would differ in those two cases.

The use of normalized vaporization rates as in Eq. (6.51) has been the standard for authors studying multiple-droplet arrays. Although this practice does provide an assessment of the effect of droplet interactions, several key aspects are obscured. To obtain an actual (dimensional) vaporization rate, one would refer to Eq. (6.51) with η known for a specific geometry. However, the vaporization rate of an isolated droplet, and more specifically the value of $\rho D = \lambda/c_p$, is not obvious. In all of the cited work, ρD is assumed constant, yet in practice it will vary spatially in the gas phase. No mention of the appropriate value, or an appropriate average value, has ever been presented. This is possibly due to the absence of liquid heating, in which the numerical value of $\rho D = \lambda/c_p$ would have been required. Law (1976b) and Law and Sirignano (1977) include liquid-phase heating for an isolated drop but use a constant value for ρD. Another shortcoming of a normalized vaporization rate involves its applicability to array geometry optimization. For example, would a larger number of smaller droplets result in higher burning rates? This issue, along with a transient liquid-phase heating analysis, is discussed in Imaoka and Sirignano (2005c).

Normalizations of lengths appearing in Eq. (6.51) will show that the results for η_j are independent of the choice of the reference length. This means that η_j will depend on only length ratios, e.g., droplet-diameter-to-droplet-spacing ratio, and not on actual size. So, if all lengths are scaled upward or downward in proportion, η_j will not change in value.

The average nondimensional vaporization rate for a droplet array of N droplets can be found as $\eta_A = (\sum_{j=1}^N \eta_j)/N$. In these nondimensional forms, the vaporization rates η_j and η_A are independent of liquid-fuel choice, transport properties, and scalar boundary conditions. They depend on only geometrical configuration so that previous computational results can be used. In particular, the computational correlation $\eta(\xi)$ of Imaoka and Sirignano (2005a, 2005c) can be used; that is,

$$\eta_A = 1 - \frac{1}{1 + 0.725671 \xi^{0.971716}},$$

$$\xi = \frac{\left[\frac{4\pi V_A N}{3 V_l} \right]^{1/3}}{[N^{1/3} - 1] N^{0.72}}, \tag{6.54}$$

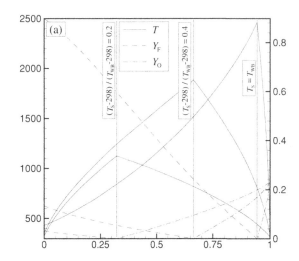

Figure 6.12. Temperature, fuel, and oxidizer mass fractions versus Φ with $T_{\infty} = 298$ and $Y_{O\infty} = 0.231$ for various surface temperatures for decane, $T_{WB} = 429.73$ K. (Imaoka and Sirignano, 2005b, with permission of the *International Journal of Heat and Mass Transfer*.)

where V_A, V_l, and N are the array volume, total liquid volume, and droplet number, respectively. So if only vaporization rates are desired, it is not necessary to obtain the field solution by solving Laplace's equation. Note that, after the volume ratio in Eq. (6.54) is related to the ratio of droplet spacing to droplet radius, the reciprocal of ξ can be shown to be a constant times the Chiu number in the limit of very large N and zero Reynolds number. However, although Chiu's theory produces a similar dimensionless group, it does not account for changes in Nusselt number or Sherwood number that are due to the proximity of other droplets.

Note that Eq. (6.54) and much of the previous formulation apply only if every droplet has the same instantaneous surface temperature. That assumption will not hold during transient heating of a large, closely packed array in which the inner droplets are given heat protection by the outer droplets.

The analysis applies universally to all droplet-array sizes and geometries. As previously mentioned, other liquid–gas interface problems are also included. Geometrical effects are calculated separately through the potential function Φ and are independent of gas-phase transport properties and boundary conditions. Consequently, the problem for the scalar properties is one dimensional for any configuration whereas the three-dimensional analysis is necessary only in solving Eq. (6.29). Coupled equations (6.45)–(6.47) are solved numerically with variable properties computed with Eqs. (6.7). The iterative solution takes $\rho D = \lambda/c_p = \overline{\rho D}$ as a first approximation. Mixture composition and enthalpy are then determined as functions of Φ from Eq. (6.45) and used in (6.46) and (6.47) to determine $\overline{\rho D}$ and Φ_F. The same mixture composition is used in (6.45), but with the newly computed $\overline{\rho D}$ and Φ_F to yield updated mixture compositions and enthalpies. The process is repeated, with good convergence after approximately 20 iterations.

The scalar quantities T, Y_F, and Y_O are shown versus Φ in Fig. 6.12 for decane fuel, with $T_{\infty} = 298$ K and $Y_{O\infty} = 0.231$. The ambient pressure is 1 atm in all of the calculations. A large variation in flame location for decane compared with lower-molecular-weight fuels is caused by the larger wet-bulb temperature and lower volatility resulting in lower values of Y_{Fs} at 298 K than for the other fuels. An

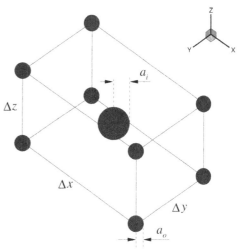

Figure 6.13. The droplet arrangement and spacing for a 9-drop rectangular array with variable radii and nonuniform spacing. (Imaoka and Sirignano, 2005c, with permission of the *International Journal of Heat and Mass Transfer.*)

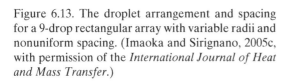

increase in ambient oxidizer mass fraction leads to higher flame temperatures and brings the flame closer to the liquid surface, as expected. The proximity of the flame to the liquid surface for a low-volatility fuel introduces the possibility of individual droplet flames for low droplet-surface temperatures. However, the occurrence of individual flames will depend strongly on the array geometry.

Gas-phase scalar variables, transport properties, specific heats, and flame locations for a specified fuel type and boundary conditions depend on only the potential function and liquid-surface temperature. Therefore the problem for the scalar properties becomes one dimensional for any configuration, whereas the three-dimensional analysis is limited to the solution of Laplace's equation. Flame standoff distances are found to decrease by more than a factor of 2 when the quantity ρD is not assumed constant.

To find the effects of droplet spacing, an N-droplet array with constant droplet radius was stretched initially in one coordinate direction, then simultaneously in two coordinate directions. Calculations were performed for more than 50 different droplet arrays of 9, 27, and 64 droplets, as well as droplets arranged in a 30-drop pyramid to study the effects of asymmetry. 27- and 64-drop arrays were cubic or rectangular arrays with 3 and 4 droplets along an edge of the rectangular volume. A 9-drop array consisted of an 8-drop array with 1 additional droplet at the center of the array. The arrays were elongated in various increments so that the largest droplet spacing exceeded the smallest droplet spacing by at least a factor of 3; however, the droplet spacing was uniform in each of the three directions within the array. Droplet spacings varied between 3 and 50 radii. The configuration of a rectangular 9-drop array is shown in Fig. 6.13.

Other results from Imaoka and Sirignano (2005a) have shown that, in a 125-drop monosized cubic array with $d/a = 3$, the outermost droplet will vaporize more than 5000 times faster than the central droplet of the array. However, although not previously mentioned, the induced Stefan velocity leaving the array might further enhance the burning rates of the outer droplets, making the factor of 5000 a conservative estimate. At that same spacing but with 1000 droplets, the difference is more

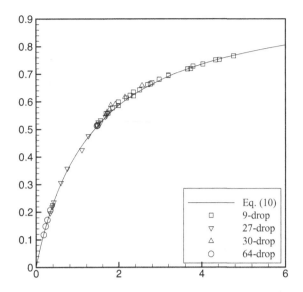

Figure 6.14. Normalized array vaporization rates for various arrays with nonuniform spacing versus the similarity parameter ξ. (Imaoka and Sirignano, 2005c, with permission of the *International Journal of Heat and Mass Transfer*.)

than a factor of 10^7. Because the current problem formulation permits only normal velocities at the droplet surfaces, the effects of very strong blowing velocities from the inner droplets creating boundary layers over the outer droplets are not included in the analysis.

Let us now consider an array with nonuniform droplet spacing. The average spacing is calculated on a volumetric basis: $d = V_A^{1/3}/(N^{1/3} - 1)$, where V_A is the array volume of a polyhedron surface drawn using the centers of the outermost droplets. The level of agreement between the vaporization data yielded through solution of Laplace's equation and the curve fit for uniform droplet spacing given by Eq. (6.54) and shown in Fig. 6.14 depends on the value of the parameter d_{max}/d_{min}, where the two spacings are the extreme values in the array. When d_{max}/d_{min} is larger than ~ 3, vaporization rates may not be accurately predicted by (6.54). This disagreement arises when a three-dimensional array is stretched substantially in one or two directions and affects both uniform and nonuniform arrays. Therefore vaporization rates for arrays consisting of droplets in a line or in a plane are not predicted well. If only a few droplets within the array have exceptional spacing but the overall external geometry of the array is still somewhat three dimensional, (6.54) might still be valid because a local extremum of spacing will have a small effect on average. Furthermore, the validity of Eq. (6.54) does not depend strongly on the positions of individual droplets within the array and therefore does not consider the details of any nonuniformity in droplet spacing within the array.

For a finite number of droplets with infinite droplet spacing, the total vaporization rate of the array will be $4\pi \overline{\rho D} \ln(1 + B)$ multiplied by the sum of the droplet radii. Therefore it is not surprising that a good correlation is achieved with a linear average of the nonuniform radii: $a_{avg} = (1/N) \sum_{j=1}^{N} a_j$. By varying a_i/a_o with different array geometries and taking the linear average of the droplet radii, the vaporization results compared well with previous results for monosized arrays. These results and Eq. (6.54) are shown versus ξ in Fig. 6.15. A linear average of the droplet

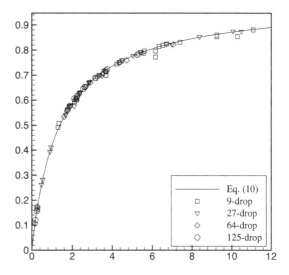

Figure 6.15. Normalized array vaporization rates versus the similarity parameter ξ for variations in droplet radii. (Imaoka and Sirignano, 2005c, with permission of the *International Journal of Heat and Mass Transfer*.)

radii gives good agreement with Eq. (6.54) except for 9-drop arrays with large values of a_i/a_o. However, more emphasis is placed on larger, more realistic arrays, for which a better correlation was achieved. The error between computed data and Eq. (6.54) for most cases was $<1\%$ and did not exceed 5%.

6.3.4 Array Combustion with Nonunitary Lewis Number

When the scalar gradients and velocity are parallel locally, we can define the normalized mass-flux fraction ϵ_n so that

$$\rho \vec{V} \varepsilon_m = \rho \vec{V} Y_m - \rho D_m \nabla Y_m = \rho (\vec{V} + \vec{V}_m) Y_m. \tag{6.55}$$

Then the species-continuity and energy equations (6.6) and (6.5) become

$$\nabla \cdot (\rho \vec{V} \varepsilon_m) = \rho \, \omega_m; \quad n = 1, \ldots, N; \tag{6.56}$$

$$\nabla \cdot (\rho \vec{V} \sum_{m=1}^{N} \varepsilon_n (h_m + h_{m,f}) - \lambda \nabla T) = 0. \tag{6.57}$$

Equations (6.4) and (6.56) together imply that, outside of the reaction zone(s), ε_m will be constant along a streamline with allowable discontinuities across the reaction zone. With piecewise-constant values of ε_m, it is convenient to define a mass-flux-weighted specific sensible enthalpy h_ε and a mass-flux-weighted specific heat $c_{p,\varepsilon}$ such that

$$h_\varepsilon = \sum_{m=1}^{N} \varepsilon_m h_m = \sum_{m=1}^{N} \varepsilon_m \int c_{p,m} dT$$

$$= \int \left(\sum_{m=1}^{N} \varepsilon_m c_{p,m} \right) dT = \int c_{p,\varepsilon} dT. \tag{6.58}$$

With known ε_m values, h_ε is a function only of temperature, making it more convenient than the mass-weighted specific sensible enthalpy $h = \sum_{m=1}^{N} Y_m h_m =$ which varies with both mass fractions and temperature.

Mass-flux-weighted (and mole-flux-weighted) enthalpy has been used for more than half a century in the transport, combustion, and vaporizing-droplet literature; see, for example, Hirschfelder et al. (1954) and El-Wakil et al. (1956). It was also used by Law and Law (1976, 1977) for their nonunitary Lewis number, single-droplet work. However, new importance for it is identified here. Some new features and insights that this enthalpy function allows: e.g., formation of conserved scalars, discontinuities in this flux-weighted enthalpy across the flame (except in special cases such as when all species have identical specific heats), the proper (natural) definitions of Lewis number and averaged transport properties across the field, and the formation of a constraint on the surface temperature and surface heat flux.

The analysis of Sirignano (2007) will be followed. We can now simplify the energy equation outside of the reaction zones where the mass-flux fractions are constant:

$$\nabla \cdot (\rho \vec{V} h_\varepsilon - \lambda \nabla T) = 0. \tag{6.59}$$

The momentum equation will be replaced with a uniform-pressure assumption that is consistent with neglect of terms of the order of Mach number squared and of the order of Mach number squared divided by Reynolds number when compared with unity.

Consider that vaporization occurs for only a single-component F from the liquid phase. The boundary conditions at the liquid–gas interface for mass fluxes become

$$\rho \vec{V} \cdot \vec{e}_n = (\rho \vec{V} Y_F - \rho D_F \nabla Y_F) \cdot \vec{e}_n$$
$$= \rho(\vec{V} + \vec{V}_F) \cdot \vec{e}_n Y_F, \tag{6.60}$$
$$(\rho \vec{V} Y_m - \rho D_m \nabla Y_m) \cdot \vec{e}_n = \rho(\vec{V} + \vec{V}_m) \cdot \vec{e}_n Y_m$$
$$= 0; \ m \neq F, \tag{6.61}$$

where $\vec{e}_n$ is the unit normal vector at the liquid–gas interface. With the assumption that the velocity and the scalar gradients at that interface are normal to the surface, we can compare Eqs. (6.60) and (6.61) with Eq. (6.103) to conclude that between the liquid–gas interface and the reaction zone, $\epsilon_F = 1$ and $\epsilon_m = 0$ for $m \neq F$. Without chemical reaction, these values apply throughout the complete gaseous domain. With flames surrounding the liquid surfaces, jumps in some of the ε_m values occur at the flame. The values outside the flames (i.e., on the oxidizer side) can be determined by simple mass balances at the flame.

As an example here, we consider five species: hydrocarbon $C_x H_y$ fuel vapor, oxygen O_2, nitrogen N_2, carbon dioxide CO_2, and water vapor H_2O. The molar balance for the one-step reaction is $C_x H_y + (x + y/4)O_2 \longrightarrow xCO_2 + (y/2)H_2O$. Nitrogen does not react in this simple model. The molecular weights of the four reacting species are $12x + y$, 32, 44, and 18, respectively. The stoichiometric mass ratios of oxygen, carbon dioxide, and water vapor with fuel are $\nu_{O_2} = 32(x + y/4)/$

$(12x + y)$, $v_{CO_2} = 44x/(12x + y)$, and $v_{H_2O} = 9y/(12x + y)$, respectively. On the oxidizer side, the mass-flux fraction for oxygen must be $\varepsilon_{O_2} = -v_{O_2}$, with the minus sign indicating that the net flux is opposed to the Stefan flow. The other fluxes are $\varepsilon_{CO_2} = v_{CO_2}$ and $\varepsilon_{H_2O} = v_{H_2O}$. These three mass-flux fractions sum to unity, indicating a continuity of total mass flux across the flame.

Define the quantity H such that

$$\rho \vec{V} H = \rho \vec{V} h_\varepsilon - \lambda \nabla T. \tag{6.62}$$

Comparison with Eq. (6.59) indicates that H will not vary along a streamline, except for a jump at the flame. With neglect of radiation, the energy-balance boundary condition at the liquid–gas interface is

$$\lambda(\nabla T) \cdot \vec{e}_n = \rho \vec{V} \cdot \vec{e}_n L_{\text{eff}}, \tag{6.63}$$

where $L_{\text{eff}} = L + \dot{q}_l/(\rho \vec{V} \cdot \vec{e}_n)$ contains the effect of both the latent heat L and the heat flux to the liquid interior $\dot{q}_l$. With the velocity and the scalar gradients at that interface normal to the surface, we can compare Eqs. (6.62) and (6.63) to obtain for the fuel side of the flame that $H_- = h_{\varepsilon,s} - L_{\text{eff}} = h_{F,s} - L_{\text{eff}}$, where the subscript s denotes the gas property at the liquid–gas surface and the minus subscript refers to the fuel side of the flame.

An energy balance at the flame gives that $H_+ - H_- = Q$, where H_+ refers to the constant H on the oxygen side of the flame and the definition is made that $Q = \varepsilon_{F,-} h_{F,f} - \varepsilon_{O_2,+} h_{O_2,f} - \varepsilon_{CO_2,+} h_{CO_2,f} - \varepsilon_{H_2O,+} h_{H_2O,f} = v_F h_{F,f} + v_{O_2} h_{O_2,f} - v_{CO_2} h_{CO_2,f} - v_{H_2O} h_{H_2O,f}$.

We can conclude from Eqs. (6.103), (6.62), and (6.63) that, outside of the reaction zones,

$$\rho \vec{V} = \nabla \phi = \frac{\lambda \nabla T}{h_\varepsilon - H} = \frac{(\lambda/c_{p,\varepsilon}) \nabla h_\varepsilon}{h_\varepsilon - H}$$

$$= \frac{\rho D_m \nabla Y_m}{Y_m - \varepsilon_m}; \quad m = 1, \dots, N. \tag{6.64}$$

So logarithmic forms naturally appear for h_ε and Y_m. Also, the scalar variables are one dimensional, i.e., functions only of ϕ, consistent with the previous results of Imaoka and Sirignano for Le $= 1$. Integrals can readily be obtained for these scalars:

$$\frac{h_\varepsilon - H}{h_{\varepsilon,\text{ref}} - H} = e^{\int_{\phi_{\text{ref}}}^{\phi} d\phi'/(\lambda/c_{p,\varepsilon})}, \tag{6.65}$$

$$\frac{Y_m - \varepsilon_m}{Y_{m,\text{ref}} - \varepsilon_m} = e^{\int_{\phi_{\text{ref}}}^{\phi} d\phi'/(\rho D_m)}. \tag{6.66}$$

References values are taken at the liquid–gas interface, the flame, or infinity. T_∞ and $Y_{m,\infty}$ are given at infinity. ϕ_∞ remains to be determined. ϕ at the liquid–gas interface will be conveniently chosen to be zero. It is useful to form the

Shvab–Zel'dovich variable $\alpha = Y_F - Y_{O_2}/\nu_{O_2} = Y_F - \nu Y_{O_2}$, where $\nu = 1/\nu_{O_2}$ is the fuel-to-oxygen mass stoichiometric ratio. This variable will be continuous through the reaction zone. Combining two of the species-continuity equations, we can write

$$\nabla \cdot [\rho \vec{V}(\varepsilon_F - \nu \varepsilon_{O_2})] = \nabla \cdot (\rho \vec{V}\alpha - \rho D_* \nabla \alpha) = 0. \tag{6.67}$$

Here, $D_* = D_F$ on the fuel side of the reaction zone and $D_* = D_{O_2}$ on the oxidizer side. The equation applies outside of the reaction zones. Even if $D_F \neq D_{O_2}$ through the flame, the quantities in parentheses in Equation (6.67) are continuous through the flame. So, contrary to common belief, a coupling function α built on the mass fractions of the two reactants can be constructed and serve a purpose even if the Lewis number does not possess the value of unity. Using boundary conditions at the liquid–gas interface and the previous arguments about the scalar gradients being parallel to the velocity, we can readily derive that

$$\rho \vec{V}(\varepsilon_F - \nu \varepsilon_{O_2}) = \rho \vec{V} = \rho \vec{V}\alpha - \rho D_* \nabla \alpha. \tag{6.68}$$

For the limiting case of infinite chemical-kinetic rate or for the limiting case of finite-rate kinetics with the same value of diffusivity $D_n = D$ for all species, the equation applies continuously through the flame. In either of these limiting cases, we can integrate through the reaction zone. Using the mass-flux potential, we can write

$$\nabla \phi = \frac{\rho D_* \nabla \alpha}{\alpha - 1}, \tag{6.69}$$

which, after integration, yields

$$\frac{\alpha - 1}{\alpha_{\text{ref}} - 1} = \frac{1 - Y_F + \nu Y_{O_2}}{1 - Y_{Fs}} = e^{\int_0^\phi d\phi'/(\rho D_*)}, \tag{6.70}$$

where the reference value has been taken at the liquid surface.

Equation (6.70) can be used to determine ϕ_∞ and allow normalization of the potential function. We use Eqs. (6.26) and (6.42) with

$$\phi_\infty = \overline{\rho D_*} \ln \frac{1 + \nu Y_{O_2,\infty}}{1 - Y_{Fs}} = \overline{\rho D_*} \ln[1 + B_M], \tag{6.71}$$

where the Spalding mass transfer number is given by

$$B_M = \frac{Y_{Fs} + \nu Y_{O_2,\infty}}{1 - Y_{Fs}}. \tag{6.72}$$

A phase-equilibrium constraint at the liquid surface will prescribe Y_{Fs} as a function of the surface temperature $T_S s$. We take for our calculations an approximate form of the Clausius–Clapeyron relation, Eq. (2.4), so that $Y_{Fs} = e^{L/RT_b} e^{-L/RT_s}$, where T_b is the boiling temperature (a function of pressure), L is the latent heat of vaporization, and R is the gas constant.

Equations (6.26), (6.70), and (6.95) can be combined to yield

$$\frac{1 - Y_F + \nu Y_{O_2}}{1 - Y_{Fs}} = [1 + B_M]^{\overline{\rho D_*} \int_0^\Phi d\Phi'/(\rho D_*)}. \tag{6.73}$$

This equation gives the fuel and oxygen mass fractions directly at any position for the infinite kinetics (with infinitesimal flame thickness) because only one of them can be nonzero at any position.

The solution for Φ will be identical to solutions obtained in the previous unitary Lewis number studies.

The thin-flame position is given by the Φ value in which both mass fractions become zero:

$$\overline{\rho D_*} \int_0^{\Phi_{\text{flame}}} d\Phi'/(\rho D_*) = -\frac{\ln[1 - Y_{Fs}]}{\ln[1 + B_M]}. \tag{6.74}$$

Properties for $\Phi > \Phi_{\text{flame}}$ will affect the flame position implicitly through the values of $\overline{\rho D_*}$ and Y_{Fs}.

Given that $Y_{CO_2,\infty} = Y_{H_2O,\infty} = 0$ and using Eqs. (6.66), (6.26), and (6.95), we can write for $\Phi > \Phi_{\text{flame}}$ that

$$\frac{Y_{CO_2}}{\nu_{CO_2}} = 1 - e^{-\overline{\rho D_*} \ln[1+B_M] \int_\Phi^1 d\Phi'/(\rho D_{CO_2})}$$

$$= 1 - [1 + B_M]^{-\overline{\rho D_*} \int_\Phi^1 d\Phi'/(\rho D_{CO_2})}, \tag{6.75}$$

$$\frac{Y_{H_2O}}{\nu_{H_2O}} = 1 - e^{-\overline{\rho D_*} \ln[1+B_M] \int_\Phi^1 d\Phi'/(\rho D_{H_2O})}$$

$$= 1 - [1 + B_M]^{-\overline{\rho D_*} \int_\Phi^1 d\Phi'/(\rho D_{H_2O})}. \tag{6.76}$$

From these two relations, we can calculate $Y_{CO_2,\text{flame}}$ and $Y_{H_2O,\text{flame}}$ at Φ_{flame}. Then we can use Eq. (6.66) with reference values at Φ_{flame} to obtain values for $\Phi < \Phi_{\text{flame}}$:

$$\frac{Y_{CO_2}}{Y_{CO_2,\text{flame}}} = e^{-\overline{\rho D_*} \ln[1+B_M] \int_\Phi^{\Phi_{\text{flame}}} d\Phi'/(\rho D_{CO_2})}$$

$$= [1 + B_M]^{-\overline{\rho D_*} \int_\Phi^{\Phi_{\text{flame}}} d\Phi'/(\rho D_{CO_2})}, \tag{6.77}$$

$$\frac{Y_{H_2O}}{Y_{H_2O,\text{flame}}} = e^{-\overline{\rho D_*} \ln[1+B_M] \int_\Phi^{\Phi_{\text{flame}}} d\Phi'/(\rho D_{H_2O})}$$

$$= [1 + B_M]^{-\overline{\rho D_*} \int_\Phi^{\Phi_{\text{flame}}} d\Phi'/(\rho D_{H_2O})}. \tag{6.78}$$

As previously mentioned, the nitrogen mass fraction is calculated with global continuity, i.e., $Y_{N_2} = 1 - Y_F - Y_{O_2} - Y_{CO_2} - Y_{H_2O}$.

Now we can examine the temperature field by using Eq. (6.65) and realizing that for $\Phi < \Phi_{\text{flame}}$ only fuel vapor is transported so that $h_\varepsilon = h_F$ and $c_{p,\varepsilon} = c_{p,F}$:

$$\frac{h_F - h_{Fs} + L_{\text{eff}}}{L_{\text{eff}}} = e^{\overline{\rho D_*} \ln[1+B_M] \int_0^\Phi d\Phi'/(\lambda/c_{p,F})}$$

$$= [1 + B_M]^{\overline{\rho D_*} \int_0^\Phi d\Phi'/(\lambda/c_{p,F})}. \tag{6.79}$$

If there is no flame, this solution gives the flux-weighted enthalpy and thereby the temperature over the infinite gas domain. We can use the equation together with Eq. (6.73) to obtain a constraint on the enthalpy h_ϵ at the flame:

$$h_{F,\text{flame}} = H_- + L_{\text{eff}}[1 + B_M]^{[\overline{\rho D_*} \int_0^{\Phi_{\text{flame}}} d\Phi'/(\rho D_*)]/\text{Le}_{\varepsilon,-}}$$

$$= H_- + \frac{L_{\text{eff}}}{[1 - Y_{Fs}]^{1/\text{Le}_{\varepsilon,-}}}, \tag{6.80}$$

where we define $\text{Le}_{\varepsilon,-} = (\lambda/c_{p,F})/\rho D_F$ and assume that it is constant on the fuel side of the flame.

Now let us use Eq. (6.65) to obtain the flux-weighted enthalpy on the oxygen side of the flame. Here

$$h_\varepsilon = \nu_{CO_2} h_{CO_2} + \nu_{H_2O} h_{H_2O} - \nu_{O_2} h_{O_2}, \tag{6.81}$$

so that

$$\frac{h_\varepsilon - H_+}{h_{\varepsilon,\text{flame}+} - H_+} = e^{\overline{\rho D_*} \ln[1+B_M] \int_{\Phi_{\text{flame}}}^{\Phi} \text{d}\Phi'/(\lambda/c_{p,\varepsilon})}$$

$$= [1 + B_M]^{\overline{\rho D_*} \int_{\Phi_{\text{flame}}}^{\Phi} \text{d}\Phi'/(\lambda/c_{p,\varepsilon})}. \tag{6.82}$$

At infinity, the mass-flux-weighted enthalpy is $h_{\varepsilon,\infty} = \nu_{CO_2} h_{CO_2,\infty} + \nu_{H_2O} h_{H_2O,\infty} - h_{O_2,\infty}/\nu$. So we substitute into Eq. (6.82) at $\Phi = 1.0$ and use Eq. (6.73) to obtain

$$h_{\varepsilon,\text{flame}+} = H_+ + [h_{\varepsilon,\infty} - H_+][1 + B_M]^{-[\overline{\rho D_*} \int_{\Phi_{\text{flame}}}^{1} \text{d}\Phi'/(\rho D_*)]/\text{Le}_{\varepsilon,+}}$$

$$= H_+ + \frac{[h_{\varepsilon,\infty} - H_+]}{[1 + B_M]^{1/\text{Le}_{\varepsilon,+}}[1 - Y_{Fs}]^{1/\text{Le}_{\varepsilon,+}}}$$

$$= H_+ + \frac{[h_{\varepsilon,\infty} - H_+]}{[1 + \nu Y_{O_2,\infty}]^{1/\text{Le}_{\varepsilon,+}}}, \tag{6.83}$$

where we define $\text{Le}_{\varepsilon,+} = (\lambda/c_{p,\varepsilon})/\rho D_{O_2}$ and assume that it is constant on the oxidizer side of the flame. Realize that, because of the use of a mass-flux-weighted specific heat rather than a mass-weighted specific heat, the $\text{Le}_{\varepsilon,+}$ and $\text{Le}_{\varepsilon,-}$ here are different from the Lewis numbers presented in other literature, except for the case in which species specific heats have the same value. So a unitary Lewis number situation with mass-weighted mixture properties in the other literature need not be a unitary Lewis number situation with the mass-flux-weighted specific heat used here. For example, on the fuel side of the flame for a hydrocarbon fuel with large molecules, the Lewis number with the mass-flux-weighted specific heat will be lower than the Lewis number that uses a mass-weighted specific heat.

For the purpose of calculations, specific enthalpies for each species are expressed as functions of temperature using the polynomials presented in the Appendix of Turns (2000). Note that, in this analysis, there is no need for the assumption made by Law and Law (1977) about the separation of the transport coefficients into the product of a function of spatial position times a function of temperature.

For piecewise-constant values of the modified Lewis number and for infinite chemical kinetics (thin flame), Eqs. (6.65), (6.66), and (6.69) can yield a relationship that extends the superscalar concept of Sirignano (2002) originally developed for $\text{Le} = 1$. Namely, in each region where the modified Lewis number has a constant value,

$$\frac{h_\varepsilon - H}{h_{\varepsilon,\text{ref}} - H} = \left(\frac{\alpha - 1}{\alpha_{\text{ref}} - 1}\right)^{1/\text{Le}_\varepsilon} = \left(\frac{Y_m - \varepsilon_m}{Y_{m,\text{ref}} - \varepsilon_m}\right)^{1/\text{Le}_\varepsilon}. \tag{6.84}$$

The first equation in Eq. (6.84) applies everywhere, although Le_ε, H, and the reference values will vary from one side of the flame to the other. In the second equation, the modified Lewis number is taken to be the local piecewise-constant value based on the diffusion of the nth species in nitrogen.

On the fuel side of the flame, the superscalar is a nonlinear combination of h_F and Y_F:

$$h_F - h_s + L_{\text{eff}} - \frac{L_{\text{eff}}}{(1 - Y_{Fs})^{1/\text{Le}_{\varepsilon-}}}(1 - Y_F)^{1/\text{Le}_{\varepsilon-}} = 0. \qquad (6.85)$$

On the oxygen side, the superscalar is a a nonlinear combination of h_ε and Y_{O_2}:

$$
\begin{aligned}
& h_\varepsilon - h_s + L_{\text{eff}} - Q \\
& \quad - \frac{(h_{\varepsilon,\infty} - h_s + L_{\text{eff}} - Q)}{(1 + \nu Y_{O_2,\infty})^{1/\text{Le}_{\varepsilon+}}}(1 + \nu Y_{O_2})^{1/\text{Le}_{\varepsilon+}} = 0.
\end{aligned}
\qquad (6.86)
$$

In this nonunitary Lewis number case, the superscalar is developed under the assumptions of quasi-steadiness and Stefan flow only, which were not required for the unitary Lewis number case discussed in previous subsections.

If T_∞, $Y_{O_2,\infty}$, L_{eff}, and fuel properties are given and the phase-equilibrium relation gives Y_{Fs} as a function of T_s, Eqs. (6.80) and (6.83), after substitution of a phase-equilibrium relation for Y_{Fs}, determine independently h_ε on each side of the flame as a function of the interface temperature T_s and of L_{eff}. Note that, if it is preferred to avoid the piecewise-constant Lewis number assumption, Eqs. (6.98) with $\Phi = \Phi_{\text{flame}}$ and (6.82) with $\Phi = 1$ can be used instead. Given equation-of-state relationships between temperature and enthalpies for the various species, Eqs. (6.80) and (6.83) [or (6.98) and (6.82)] lead to two independent relations, each presenting flame temperature T_{flame} as a function of T_s and L_{eff}. Each value of h_ε implies a flame temperature value. Whereas h_ε can be discontinuous across the thin flame, temperature must be continuous. Continuity of temperature at the flame allows the combination of these two relations to yield a direct relation between T_s and L_{eff}. That is, the continuity of flame temperature becomes a constraint on the surface temperature T_s. So, if L_{eff} is known, then T_s is prescribed. If L_{eff} is not known before a solution of the liquid-transient-heating problem is made, the constraint provides a relation between T_s and L_{eff} that is a satisfactory boundary condition on the liquid-transient-heating problem.

A merged form of the integrated energy relationship can be obtained from the combination of Eqs. (6.80) and (6.83). Defining $\tilde{Q} = Q + h_{F,\text{flame}} - h_{\varepsilon,\text{flame}+}$, we obtain by subtraction of (6.80) from (6.83)

$$\frac{L_{\text{eff}}}{[1 - Y_{Fs}]^{1/\text{Le}_{\varepsilon,-}}} = \tilde{Q} + \frac{[h_{\varepsilon,\infty} - H_+]}{[1 + \nu Y_{O_2,\infty}]^{1/\text{Le}_{\varepsilon,+}}}. \qquad (6.87)$$

We can define $B_{H\varepsilon}$, so that Eq. (6.87) becomes

$$
\begin{aligned}
\frac{[1 + \nu Y_{O_2,\infty}]^{1/\text{Le}_{\varepsilon,+}}}{[1 - Y_{Fs}]^{1/\text{Le}_{\varepsilon,-}}} &= 1 + B_{H\varepsilon} \\
&= 1 + \frac{h_{\varepsilon,\infty} - h_{Fs} + \tilde{Q}[1 + \nu Y_{O_2,\infty}]^{1/\text{Le}_{\varepsilon,+}} - Q}{L_{\text{eff}}}.
\end{aligned}
\qquad (6.88)
$$

The phase-equilibrium relation together with Eq. (6.88) is insufficient to prescribe the surface temperature T_s because the flame enthalpy is required through the $\tilde{Q}$ term. So Eqs. (6.80) and (6.83) must still be used except for the case of equal specific heats, which will be subsequently discussed.

We can solve Eq. (6.88) for L_{eff} to obtain a different form

$$
\begin{aligned}
L_{\text{eff}} = & \frac{Q([1 + \nu Y_{O_2,\infty}]^{1/\text{Le}_{\varepsilon,+}} - 1)[1 - Y_{Fs}]^{1/\text{Le}_{\varepsilon,-}}}{[1 + \nu Y_{O_2,\infty}]^{1/\text{Le}_{\varepsilon,+}} - [1 - Y_{Fs}]^{1/\text{Le}_{\varepsilon,-}}} \\
& + \frac{(h_{\varepsilon,\infty} - h_{Fs})[1 - Y_{Fs}]^{1/\text{Le}_{\varepsilon,-}}}{[1 + \nu Y_{O_2,\infty}]^{1/\text{Le}_{\varepsilon,+}} - [1 - Y_{Fs}]^{1/\text{Le}_{\varepsilon,-}}} \\
& + \frac{\Delta h([1 - Y_{Fs}]^{1/\text{Le}_{\varepsilon,-}})[1 + \nu Y_{O_2,\infty}]^{1/\text{Le}_{\varepsilon,+}}}{[1 + \nu Y_{O_2,\infty}]^{1/\text{Le}_{\varepsilon,+}} - [1 - Y_{Fs}]^{1/\text{Le}_{\varepsilon,-}}}
\end{aligned}
\tag{6.89}
$$

where $\Delta h = h_{F,\text{flame}} - h_{\varepsilon,\text{flame}+}$.

Because Δh does depend on L_{eff}, Eq. (6.89) is not an explicit solution for L_{eff}; however, it can be a useful form for an iterative process to calculate L_{eff} as a function of T_s. This relationship, when combined with the phase-equilibrium relationship, relates the temperature gradient on the liquid side of the liquid surface to the surface temperature. It thereby presents a boundary condition for the heat diffusion equation governing transient heating in the liquid.

In the special case in which $\text{Le}_{\varepsilon,+} = \text{Le}_{\varepsilon,-} = \text{Le}_{\varepsilon}$, we find that

$$
[1 + B_M]^{1/\text{Le}_{\varepsilon}} = 1 + B_{H\varepsilon}.
\tag{6.90}
$$

In the case in which specific heats are identical for all species, including nitrogen, we have $h_{\varepsilon} = h$, which is continuous across the flame, $h_{\varepsilon,\infty} = h_{\infty}$, $h_{Fs} = h_s$, and $\tilde{Q} = Q$. Then, $\text{Le}_{\varepsilon} = \text{Le}$ (the classical Lewis number). Furthermore, if D_* is also continuous, Le can be assumed to be constant (rather than piecewise constant) and the result is

$$
B_{H\varepsilon} = \frac{h_{\infty} - h_s + Q([1 + \nu Y_{O_2,\infty}]^{1/\text{Le}} - 1)}{L_{\text{eff}}}.
\tag{6.91}
$$

If $\text{Le}_{\varepsilon} = 1$, then $B_{H\varepsilon} = B_H$, the classical heat transfer number proposed by Spalding. In that case, we obtain an explicit relationship for L_{eff}:

$$
L_{\text{eff}} = \frac{(h_{\infty} - h_s + Q\nu Y_{O_2,\infty})[1 - Y_{Fs}]}{\nu Y_{O_2,\infty} + Y_{Fs}}.
\tag{6.92}
$$

It is remarkable that the liquid-surface boundary condition given by Eq. (6.89), (6.99), or (6.92) is independent of the geometrical configuration (e.g., shape of liquid surfaces or number of droplets). The nonlinear algebraic equations were solved by iteration using MATLAB. For the heavy hydrocarbon molecules, we expect slow diffusion and a large Lewis number on the fuel side of the flame. So we consider $\text{Le}_{\varepsilon-} > 1.0$.

Figure 6.16 shows the result for L_{eff}/L as a function of the surface-temperature value T_s for various values of $\text{Le}_{\varepsilon-}$, $\text{Le}_{\varepsilon+}$, T_{∞}, and $Y_{O_2,\infty}$ for heptane fuel burning in air. The parameter choices for each of the calculated cases is indicated in Table 6.4. Lewis number and other values were chosen to be in the practical range for hydrocarbon/air combustion. The value of L_{eff} and thereby the magnitude of the heat transfer (per unit mass) of vaporizing fuel from the gas to the liquid will decrease as the surface temperature increases. At the wet-bulb temperature, $L_{\text{eff}}/L = 1$.

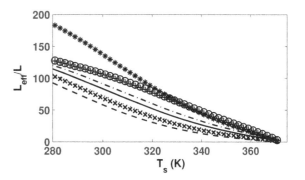

Figure 6.16. Normalized effective latent heat of vaporization L_{eff}/L versus surface temperature T_s for heptane fuel burning in air. (Sirignano, 2007, with permission of *Combustion and Flame*.)

Calculations for values of $\text{Le}_{\varepsilon-} = 1.0, 1.5, 2.0,$ and 2.5 indicate that, at a given value of T_s, L_{eff} increases with increasing fuel-side Lewis number $\text{Le}_{\varepsilon-}$ because increasing heat is conducted to the liquid. L_{eff} also increases with decreasing oxygen-side Lewis number because decreasing heat is lost to the surroundings. Increase of the ambient values of oxygen mass fraction or temperature will also increase the heat feedback to the droplet. The fractional differences of L_{eff} at low surface temperatures in this range of $\text{Le}_{\varepsilon-}$ are modest. Close to the wet-bulb temperature, the fractional differences are substantial although the magnitudes of L_{eff} are lower there. The figure indicates how the heat flux to the droplet surface will decrease with time as the droplet temperature rises; the particular temporal dependence of the T_s can be determined by the solution of the heat diffusion equation in the droplet. In this section, we are not addressing that portion of the problem, which was well addressed in Chapter 2.

Once $L_{\text{eff}}(T_s)$ is known, Eq. (6.80) or (6.83) determines the flame temperature using the polynomial expressions for specific enthalpies from Turns (2000). Figure 6.17 shows the flame temperature T_{flame} as a function of the liquid-surface temperature for the same parameters as given for Fig. 6.16. As the T_s increases with time, T_{flame} also increases. Equations (6.89) plus the phase-equilibrium relationship yield the droplet wet-bulb temperature. For the parameter cases previously considered, that temperature T_{wb} is given in the table, which shows that neither the wet-bulb temperature nor the flame temperature at the wet-bulb condition is very sensitive to the Lewis number on the fuel side of the flame. The oxygen-side Lewis number has a very large effect on flame temperature but no significant effect on the

Table 6.4. *Parameter matrix for droplet-array calculations*

Graph lines Fig. 6.16 and 6.17	$\text{Le}_{\varepsilon-}$	$\text{Le}_{\varepsilon+}$	T_∞ (K)	$Y_{O_2,\infty}$	T_{wb} (K)	$T_{\text{flame}}(T_{\text{wb}})$ (K)
Dashed	1.0	1.0	298	0.232	367.8	1944.3
Dotted	1.5	1.0	298	0.232	370.4	1944.5
Solid	2.0	1.0	298	0.232	371.2	1944.6
Dot-dash	2.5	1.0	298	0.232	371.5	1944.6
Plus	2.0	0.5	298	0.232	371.5	3222.8
Cross	2.0	1.5	298	0.232	370.8	1497.0
Asterisk	2.0	1.0	1000	0.232	371.5	2824.9
Circle	2.0	1.0	298	0.500	371.5	3325.1

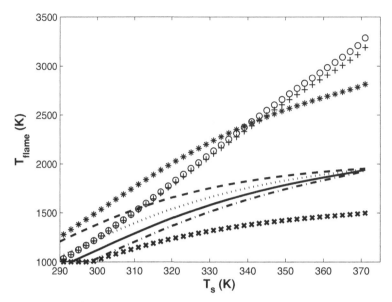

Figure 6.17. Flame temperature T_{flame} versus surface temperature T_s for heptane fuel burning in air. (Sirignano, 2007, with permission of *Combustion and Flame*.)

wet-bulb temperature. However, for the liquid-surface temperature between room temperature and the wet-bulb temperature, substantial influences of either Lewis number value occur on the droplet heating and the flame temperature. For any particular value of the surface temperature T_s, the heat conducted to the liquid per unit mass of fuel vaporized L_{eff} increases with increasing $\text{Le}_{\varepsilon-}$ (and with decreasing $\text{Le}_{\varepsilon+}$) while the flame temperature T_{flame} decreases. A decrease in $Le_{\varepsilon+}$ can be seen as a relative decrease in the heat conducted to the ambience, causing a higher flame temperature and allowing more energy to be conducted from the flame to the fuel droplet.

An increase in ambient values of oxygen mass fraction or temperature causes an increase in flame temperature. It is noteworthy that flame temperature and wet-bulb temperature will not depend on the geometrical configuration.

The global vaporization rate with nonunitary Lewis number is given by

$$\dot{m} = \int_S \int \rho \vec{V} \cdot \vec{e}_n \mathrm{d}S$$

$$= \int_S \int \nabla\phi \cdot \vec{e}_n \mathrm{d}S = \overline{\rho D_*} \ln[1 + B_M] \int_S \int \nabla\Phi \cdot \vec{e}_n \mathrm{d}S. \qquad (6.93)$$

Equations (6.51) and (6.54) still apply. However, $\dot{m}_{j,\text{iso}}$ is now the instantaneous vaporization rate for an isolated droplet of the same radius but with the same gas-phase properties, i.e., nonunitary Lewis number. So Eq. (6.53) should be replaced with

$$\dot{m} = 4\pi \overline{\rho D_*} a \ln[1 + B_M]. \qquad (6.94)$$

6.3.5 Array Vaporization with Multicomponent Liquids

Consider now an array of droplets formed from a liquid that is a blend of several miscible components. Following Sirignano and Wu (2008), we now have

$$\phi_\infty = \overline{\rho D_m} \ln\left[\frac{\varepsilon_m}{\varepsilon_m - Y_{m,s}}\right] = \overline{\rho D_m} \ln[1 + B_{M,m}], \tag{6.95}$$

where the Spalding mass transfer number for each vaporizing species is given by

$$B_{M,m} = \frac{Y_{m,s}}{\varepsilon_m - Y_{m,s}}, \tag{6.96}$$

$$\frac{Y_m - \varepsilon_m}{Y_{m,s} - \varepsilon_m} = [1 + B_{M,m}]^{\overline{\rho D_m} \int_0^\Phi d\Phi'/(\rho D_m)}. \tag{6.97}$$

For each vaporizing species, a phase-equilibrium constraint at the liquid surface will prescribe $Y_{m,s}$ as a function of the surface temperature T_s. We will use Raoult's Law for our calculations, given by Eq. (4.3):

$$\frac{h_\varepsilon - h_{\varepsilon,s} + L_{\text{eff}}}{L_{\text{eff}}} = [1 + B_{H,\varepsilon}]^{\overline{\lambda/c_{p,\varepsilon}} \int_0^\Phi d\Phi'/(\lambda/c_{p,\varepsilon})}$$

$$= [1 + B_{M,m}]^{\overline{\rho D_m} \int_0^\Phi d\Phi'/(\lambda/c_{p,\varepsilon})}. \tag{6.98}$$

Note that any value of n may be used be in Eq. (6.98) in accordance with Eq. (6.95).

We define a new average Lewis number here for each species diffusing through the nitrogen: $\text{Le}_{\varepsilon,m} = (\overline{\lambda/c_{p,\varepsilon}})/(\overline{\rho D_m})$. Realize that, because of the use of a mass-flux-weighted specific heat rather than a mass-weighted specific heat, the $\text{Le}_{\varepsilon,m}$ here is different from the Lewis number presented in other literature, except for the case in which species specific heats have the same value. So a unitary Lewis number situation with mass-weighted mixture properties in the other literature need not be a unitary Lewis number situation with the mass-flux-weighted specific heat used here. For the purpose of calculations, specific enthalpies for each species can be expressed as functions of temperature using the NASA polynomials given in the appendix of the book by Gardiner (1984).

From Eq. (6.98) and the definition of the average Lewis number, we find that

$$[1 + B_{M,m}]^{1/\text{Le}_{\varepsilon,m}} = 1 + B_{H,\varepsilon}. \tag{6.99}$$

Equation (6.99) holds for any value of the index m (i.e., any vaporizing species) whereas the right-hand side does not vary with m. So,

$$[1 + B_{M,j}]^{1/\text{Le}_{\varepsilon,j}} = [1 + B_{M,k}]^{1/\text{Le}_{\varepsilon,k}}. \tag{6.100}$$

This means that the particular function of mass transfer number and Lewis number is the same for any two values j, k of the parameter n. Also, we can say that

$$[1 + B_{M,j}] = [1 + B_{M,k}]^{\text{Le}_{\varepsilon,j}/\text{Le}_{\varepsilon,k}} = [1 + B_{M,k}]^{\overline{\rho D_k/\rho D_j}}. \tag{6.101}$$

Obviously, if the mass diffusivity is the same for any two species, their $B_{M,m}$ values are identical. For the case in which all gas-phase diffusivities for the vaporizing

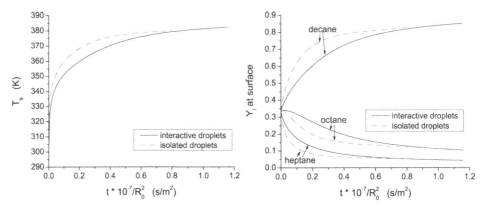

Figure 6.18. Surface temperature and species mass fraction at the liquid surface in transient behavior for both interactive and isolated droplets, at the ambient temperature 2000 K, and initial mass fraction for heptane, octane and decane: 1/3, 1/3, 1/3. (Sirignano and Wu, 2008, with permission of the *International Journal of Heat and Mass Transfer.*)

species are equal, then all $B_{M,m}$ values are identical and Eqs. (6.96) and (6.101) yield the $N-3$ relations:

$$\varepsilon_1/Y_{1,s} = \varepsilon_2/Y_{2,s} = \cdots = \varepsilon_{N-2}/Y_{N-2,s}. \tag{6.102}$$

Given that $\sum_{k=1}^{N-2} \varepsilon_k = 1$ and $\sum_{k=1}^{N} Y_{k,S} = 1$, Eq. (6.102) yields

$$\varepsilon_m = \frac{Y_{m,s}}{\sum_{k=1}^{N-2} Y_{k,s}}; \quad m = 1, 2, \ldots, N-2. \tag{6.103}$$

The theory has been developed for a very general configuration and for an arbitrary number of species in the liquid phase. Here, for a sample calculation, we choose an 8-droplet array with triple symmetry; in particular, the droplets have identical radii and are at the corners of an abstract cube so center-to-center spacing between droplets is uniform at 10 initial droplet radii. The droplets also have identical initial temperatures of 300 K and initial composition, a blend of equal amounts of heptane, octane, and decane. Ambient pressure is 1 atm. Ambient temperatures will be varied. The transient behavior is considered.

Figure 6.18 compares the change of surface temperature and mass fractions of heptane, octane and decane at the liquid surface with time for our interactive case with the isolated-droplet (infinite-droplet-spacing) case. The surface temperature rises sharply at the beginning and then increases more and more gently before the droplet is totally vaporized. At the liquid surface, the mass fraction of the most volatile component, heptane, drops steadily for a long time and then remains nearly constant because of the achieved balance between heptane's volatility and its large gradient of concentration in the liquid near the surface. The isolated droplets vaporize faster and have greater rates of change of surface scalars than interactive droplets. The values for surface scalars for interactive and isolated droplets come very close to each other near the end of the lifetime when the droplet spacing is rather large compared with the droplet radius.

During the transient behavior, there is heat conduction and species diffusion inside the droplets. Figures 6.19–6.21 show the profiles of interior temperature

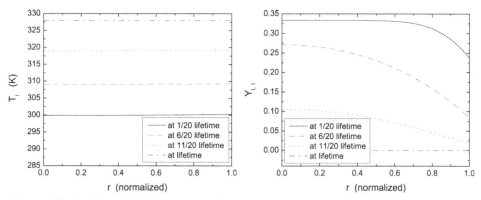

Figure 6.19. Profiles of temperature and heptane mass fraction in the liquid phase during the transient behavior, with identical initial mass fraction for heptane, octane, and decane. $T_\infty = 350$ K. (Sirignano and Wu, 2008, with permission of the *International Journal of Heat and Mass Transfer*.)

and heptane mass fraction at different times in the droplet lifetime. These profiles change with time whereas the changes of surface temperature and surface composition become slower with time. As liquid-phase heat conduction is much faster than liquid-phase mass diffusion, the temperature becomes nearly uniform and constant after some time whereas the species mass fraction still vary over the droplet interior. As seen from Fig. 6.19, lower ambient temperature always leads to more uniform profiles because it results in a longer lifetime and allows more time for heat and species diffusion in the droplets. For 350 K ambient temperature, the temperature profiles are nearly uniform throughout the lifetime but the mass fraction profiles are not; so the slow vaporization limit is not strictly satisfied, even at a low ambient temperature of 350 K. For 2000 K and 3000 K ambient temperatures, as shown in Figs. 6.20 and 6.21, the temperature and mass fraction profiles become steeper but still do not produce a sufficiently thin diffusion layer to satisfy strictly the fast vaporization limiting conditions.

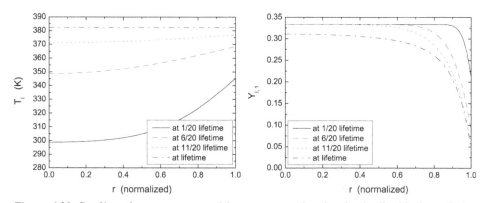

Figure 6.20. Profiles of temperature and heptane mass fraction in the liquid phase during the transient behavior, with identical initial mass fraction for heptane, octane, and decane. $T_\infty = 2000$ K. (Sirignano and Wu, 2008, with permission of the *International Journal of Heat and Mass Transfer*.)

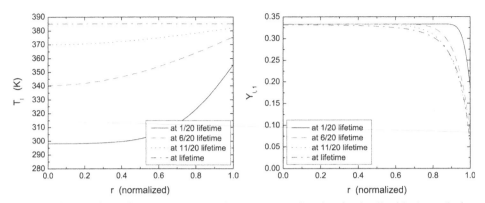

Figure 6.21. Profiles of temperature and heptane mass fraction in the liquid phase during the transient behavior, with identical initial mass fraction for heptane, octane, and decane. $T_\infty = 3000$ K. (Sirignano and Wu, 2008, with permission of the *International Journal of Heat and Mass Transfer*.)

In the multicomponent case, the mass-flux fraction of each component equals its initial mass fraction in the fast vaporization limit. Figure 6.22 shows the change of mass-flux fraction with time during the transient behavior compared with the theoretical constant value in the fast vaporization limit. The curves at the ambient temperature 3000 K are closer to the constant value of 1/3 than at 2000 K, but still have an apparent difference, because the fast vaporization limit is still not well approached, as shown in Figure 6.21.

6.4 Droplet Collisions

The collision between two droplets and the collision of a droplet with a wall or substrate are of wide practical interest. In dense-spray situations for many applications, the possibility of droplet collisions exists. In applications of sprays in small volumes, droplet collision with walls can be an interestingfactor. In droplet-based

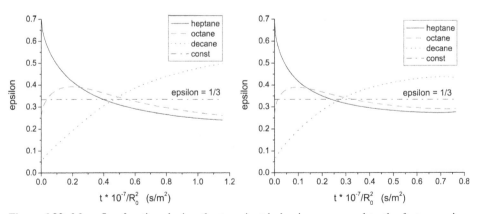

Figure 6.22. Mass-flux fraction during the transient behavior compared to the fast vaporization limit, with identical initial mass fraction for heptane, octane, and decane. (Sirignano and Wu, 2008, with permission of the *International Journal of Heat and Mass Transfer*.)

manufacturing and material processing, the collision of droplets with a substrate is designed. We discuss droplet–droplet collisions and droplet–wall collisions in the next two subsections.

The collision of two droplets or the collision of a droplet with a wall involves conservation of energy, conservation of momentum, and conservation of angular momentum. In a collision, the droplet loses kinetic energy as the droplet strains and deforms. The strain leads to viscous dissipation, accounting for some conversion of mechanical energy to heat. Often more important, the droplet surface increases and surface energy (because of surface tension) increases as well. The surface energy can be viewed as a potential energy, and the conversion of kinetic energy to surface energy can be viewed as a conservative process. The increase of surface energy during the early part of a collision can result in recoiling or rebounding later through the conversion of the surface energy back to kinetic energy. The momentum balance occurs through a force imposed on the droplet by the other droplet or by the wall in a collision as the droplet loses velocity and possibly rebounds in the other direction. For collisions between droplets that are not head-on or for collisions with a wall at angles other than 90°, we can expect that conservation of angular momentum is displayed through a torque imposed during collision. This torque will result in a rotation of the colliding droplet.

The collision problem is still very much a virgin area where much more is unknown than is known. Some limited studies exist and are discussed here, but the full range of parameters and the intermolecular scale phenomena have not been adequately studied.

6.4.1 Droplet–Droplet Collisions

A general conclusion from the several studies is that collisions among droplets seem to have a low probability in a spray in which the droplets are moving in a parallel direction or along divergent paths. The stability of droplets in a linear stream and the repulsive lift forces of nearby droplets moving in parallel tend to support this conclusion.

Arguments are given by O'Rourke and Bracco (1980) that coalescence is an important factor in the dense-spray region near an injector orifice. They developed an elaborate spray model that accounted for collision and coalescence of droplets. Bracco (1985) and Reitz (1987) also discuss and use this model. The model predicts increasing average droplet size with downstream distance from the injector; some experimental evidence of growing droplet size with distance is also cited. However, no fundamental experimental study resolving the phenomena on the scale of the droplets within a spray context has been made. Several other mechanisms could cause increasing average droplet size with downstream distance: (i) smaller droplets vaporize faster, leaving the larger droplets; (ii) condensation occurs in the cold, vapor-rich, dense spray region near the injector; and (iii) the longer-wavelength disturbances on the jet will take a longer Lagrangian time to grow and to yield the larger droplets.

Collisions between two droplets are possible in other spray environments, where droplet streams are oriented in interesting directions. Weak collisions such as

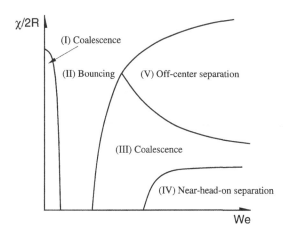

Figure 6.23. Schematic of various collision regimes of hydrocarbon droplets in 1-atm air. (Redrawn from Qian and Law, 1997.)

grazing collisions can occur. We can have stronger head-on collisions, as determined by relative velocity (i.e., both droplets can be moving in any absolute direction). The stronger collisions can lead to one of several phenomena: permanent coalescence, coalescence with a vibration that leads to separation into two other droplets, coalescence and vibration with separation into two droplets plus smaller satellite droplets, shattering into many smaller droplets, or bouncing. The 3-droplet collision is less interesting on account of its lower probability of occurrence.

A number of interesting fundamental experimental studies of two droplets in a collision configuration have been made. See Ashgriz and Poo (1990), Jiang et al. (1992), and Qian and Law (1997). Qian and Law identify five distinct regimes for collisions in a plot of droplet Weber number $2R\rho U^2/\sigma$ and impact parameter $\chi/2R$, where χ is the shortest distance at the time of impact between the tangent lines to the trajectories of two equal-sized droplets of radius R, relative velocity U, and surface-tension coefficient σ. Clearly, only $\chi \leq 2R$ is interesting. Figure 6.23 schematically indicates the results of Qian and Law.

Regime I involves coalescence of the two droplets after minor deformation. During the final instances of the approach, the thin gas film between the two droplets is squeezed and discharged in sufficient time for the two droplets to merge under the force of the surface tension. In Regime II, the Weber number is higher and the allowed time for discharge of the gas film is smaller. There, insufficient time for discharge results and the surface tension cannot act to coalesce the droplets. Rather, a rebound or bounce occurs in this regime.

In Regime III, the higher-approach velocity apparently forces the discharge to occur in sufficient time, resulting in coalescence. Here, in contrast to Regime I, substantial deformation occurs before coalescence. The pressure in the closing gap mounts to high levels, flattening each droplet. The intermolecular liquid forces of attraction work at a distance of 100 Å, so that the gap must become that small in order to make coalescence probable. Gas and liquid properties such as density and viscosity and the nature of the intermolecular forces are expected to affect the boundaries between different regimes. In fact, although experiments with hydrocarbon droplets in air at 1 atm yielded all five regimes, the comparable experiments for water droplets yielded only three regimes, omitting Regimes II and III.

Regime IV occurs for small values of the impact parameter (head-on and near-head-on collisions) and is characterized by a temporary coalescence followed by separation. The Weber number is higher than in the other regimes (at the same impact parameter value), and enough kinetic energy remains in the internal liquid flow after coalescence to overcome the surface tension and result in eventual separation into two larger droplets plus a smaller satellite droplet.

Regime V occurs for off-center collisions and involves temporary coalescence followed by eventual separation. Rotation of the coalesced mass results because of conservation of angular momentum. Multiple satellite droplets plus the two larger droplets result from the separation.

Zhang and Law (2007) developed a theory that describes the behavior found in the experiments of Jiang et al. and Qian and Law. The theory includes rarefied gas effects in the thin gas film between colliding droplets. The linearized Boltzmann equation together with the Krook model for the molecular collision integral is used. The internal liquid motion and the intermolecular forces at the colliding surfaces are considered.

Numerical modelling of two colliding droplets was performed by Nobari and Tryggvason (1996) and Nobari et al. (1996). The authors used a modified form of Navier–Stokes equation (A.2) with the following surface-tension term subtracted from the right-hand side:

$$\sigma \int_s \kappa n_i \delta(x_1 - x_{1f}) \delta(x_2 - x_{2f}) \delta(x_3 - x_{3f}) \, \mathrm{d}a,$$

where σ is the surface-tension coefficient; κ is twice the mean curvature of the droplet interface at the position x_{1f}, x_{2f}, x_{3f}; n_i is the unit normal vector component pointing outward from the droplet; δ is the Dirac delta function; and $\mathrm{d}a$ is the element of the droplet surface. The incompressible-continuity equation yielding a divergence-free velocity field is also used. In this manner, the same equation is used for both phases. A finite-difference technique with front tracking (for the droplet surface) is utilized.

The equations imply that density and viscosity will remain constant along a particle path in either fluid because heat and mass transfer are not considered. The Reynolds number, the Weber number, and the impact parameter are the three major nondimensional parameters.

The numerical technique spreads the theoretical discontinuity in density at the droplet surface over several grid points. The grid points are reconstructed at every time step, and numerical diffusion is eliminated. The thin film of gas between two colliding droplets cannot be accurately resolved in these calculations. The gas film can be completely discharged here only in an ad hoc imposed manner.

Nobari et al. (1996) considered head-on collisions with axisymmetric collisions, and Nobari and Tryggvason (1996) examined three-dimensional collisions. Without artificial removal of the thin gas film, the droplets always rebound in both analyses; see, for example, Fig. 6.24. With the removal of the thin film, coalescence occurs, as indicated by Fig. 6.25. Because the thin-film removal is artificially imposed, a direct comparison with the experiment cannot be made. The modelling of the thin-film

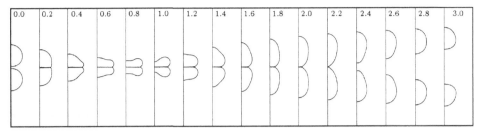

Figure 6.24. Collision of two drops. Here We = 32, Re = 98, $r = 15$, and $\lambda = 350$. The nondimensional time (scaled by the initial velocity and the drop diameter) is noted in each frame. The grid used here is 64×256 grid points. (Nobari et al., 1996, with permission of *Physics of Fluids*.)

behavior is not only an issue of resolution. On this scale of a few hundred angstroms, the intermolecular forces must be modelled as attempted by Zhang and Law (2007).

Wang et al. (2003) examined two-component fuel droplets formed by droplet–droplet collision. They ignited the newly formed droplets. An air bubble was trapped in the interfacial region when two droplets collided, with or without merging. This air bubble would serve as a heterogeneous nucleation site for the onset of microexplosion during the burning process. Wang et al. (2004) extended this work to the nonmiscible case in which a fuel droplet and a water droplet collided and merged, with microexplosions occurring in some cases.

A recommended review of droplet collisions and coalescence is given by Orme (1997).

6.4.2 Droplet–Wall Collisions

Collisions between droplets in a spray (or in a droplet stream) and a wall or substrate have been studied in recent years. The motivating application often involves

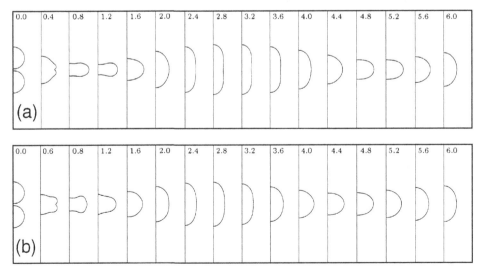

Figure 6.25. The evolution following rupture of the interface separating the drops for the simulation in Fig. 6.22. In both cases the drops coalesce permanently. (a) Rupture at $t = 0.4$, (b) rupture at $t = 0.6$. (Nobari et al., 1996, with permission of *Physics of Fluids*.)

materials processing for which solidification of the molten material in the droplet is a vital factor. Noteworthy papers are those of Fukai et al. (1993), Liu et al. (1994), Fukai et al. (1995), Delplanque et al. (1996), Zhao et al. (1996), and Delplanque and Rangel (1998).

Material scientists are most interested in the microstructure of the solidified material on the substrate following droplet impact. Those issues are not emphasized here. For an overview from the materials perspective, the reader is referred to the work of Lavernia and Wu (1996).

The existing studies have focused on axisymmetric cases in which the droplet approaches the wall at a right angle. The approach at other angles has not been modelled. Also, the models never predict a rebound from the wall. One might expect that a domain of parameters exists whereby the colliding droplet will rebound away from the wall. It is not clear whether the existing models are deficient or the parameter range studied is too limited. Some of the former is suspected because the researchers never discuss the issue of how the gas layer is squeezed from between the droplet and the wall just before collision. The same type of issue was dominant for the droplet–droplet collision. The droplet–wall collision researchers typically start their calculations with the droplet in contact with the wall whereas the droplet–droplet collision researchers begin their calculations before contact is made. In the droplet–droplet computations, some distortion was observed before contact.

Fukai et al. (1995) performed experimental and computational studies of an initially spherical droplet hitting a flat surface at a right angle. They used water droplets and surfaces of varying wettability. Their computations involved the use of finite elements to solve a Lagrangian formulation of the Navier–Stokes equations. Results were sensitive to the wetting model. A contact angle was prescribed at the circular line representing the meeting of the three phases: liquid, surrounding gas, and solid wall. The contact angle is the angle between the tangent to the liquid–gas interface (at the solid-wall surface) and the solid wall. This contact line moved in time during the process; its radius (the splat radius) increased during the early portion of the impact and later decreased, possibly with subsequent oscillation. Clearly, in the early collision, kinetic energy is converted to surface energy and, through viscous dissipation, to heat. Later, the surface energy will cause some recoil and conversion back to kinetic energy. Fukai et al. (1995) used different values of the dynamic contact angle for the spreading and recoiling stages, thereby extending the work of Fukai et al. (1993). Fukai et al. (1995) reported that the maximum splat radius decreased as the contact angle during the spreading stage increased. The rate of change of splat height during the early collision was directly proportional to the impact velocity.

Liu et al. (1994) considered a droplet hitting a wall with solidification occurring during the impact. They solved a two-phase (liquid and solid) system of Navier–Stokes-based equations by means of a finite-difference technique. A volume-of-fluid method was used to track the gas–liquid interface. The energy equation and equations governing phase change were not solved directly. Rather, an integrated solution of the one-dimensional Stefan problem of phase change was used to describe the movement of the liquid–solid interface in a direction from the substrate toward the top of the splat liquid. Emphasis was placed on the microporosity of the solidified material. They showed that, during the solidification stage, the liquid flow could separate temporarily from the wall surface, creating gas bubbles that result in pores for the solidified material. Note that various unexplained differences exist between the two-phase flow equations of Liu et al. and those presented in Chapter 7. The two-phase equations of Liu et al. were constructed for two adjacent continuous phases whereas the equations of Chapter 7 apply to a situation with a discrete phase and a continuous phase.

Zhao et al. (1996) considered heat transfer by means of convection and conduction in the splat droplet; solidification was not considered. They extended the work of Fukai et al. (1993, 1995) by adding the axisymmetric energy equation to the system. They found that the temperature profiles were truly two dimensional and questioned the use of a one-dimensional approximation. At later times, the radial dependence of the temperature was the more prominent dependence. Mass tended to accumulate at the periphery of the splat.

Delplanque et al. (1996) extended the dynamic-splat-with-solidification model of Liu et al. (1994) to consider multidirectional solidification. They constrained the solid–liquid interface to move locally in a direction normal to the interface.

Delplanque and Rangel (1998) made comparisons among experiments, Navier–Stokes-based computations, and simpler models. In both cases with and without solidification, the agreement between experiments and numerical simulations was viewed as satisfactory. A simplified model assumed that the splat immediately formed and maintained the shape of a cylindrical disk with time-varying radius and height. The energy equation with neglect of radial conduction was solved in this liquid disk. Although this model had various shortcomings, it did predict reasonably well the time to reach maximum splat radius and the solidification time. See also Deplanque and Rangel (1999), Delplanque et al. (2000), and Chung and Rangel (2000, 2001).

Rein (1993,1995,1996) provided some interesting discussion and analysis of a droplet striking a liquid surface. He considered the transition between the regime where coalescence of the drop with the larger liquid body occurs and the regime where splashing occurs. Various intermediate phenomena such as the rise of jets, crater formation, the entrainment of gas bubbles, and the creation of splash droplets are discussed by Rein (1993,1996). High-speed compressible effects are addressed by Rein (1995).

Pan and Law (2007) consider droplets impacting a surface that is already wet by previous droplets. The merging and bouncing responses are nonlinear, depending on the film thickness. Small droplet Weber number We favors bouncing whereas large Weber number favors absorption of the impacting droplet. Interesting results are found with the nondimensional parameter H_f, which is film-thickness-to-droplet-radius ratio. Absorption is favored in two ranges: $H_f << 1$ and in the vicinity of $H_f = 1$. As a result, a range of We exists where increasing of H_f results in multiple bifurcations: starting with adsorption, going to bouncing, back to absorption, and then back to bouncing.

7 Spray Equations

We focus here on the formulation of the equations for describing the dynamics of a spray, considering the discrete phase to be strictly a liquid and the continuous phase to be strictly a gas. Of course, it is simple to generalize the formulation to include dusty flows and bubbly flows.

The spray equations are presented in three different but related constructions: the two-continua or multicontinua formulation, the discrete-particle formulation, and the probabilistic formulation. Both Lagrangian and Eulerian methods are examined. The relationships among these various formulations and methods are emphasized and are a major contribution here. Previous research on the multicontinua formulation is reported by Sirignano (1972, 1986, 1993a, 1993b), Crowe et al. (1998), Crowe et al. (1977), Crowe (1978, 1982), Dukowicz (1980), and Druzhinin and Elghobashi (1998). The discrete-particle method discussed by Sirignano (1986, 1993a, 1993b), and the probabilistic formulation is discussed by Williams (1985) and Sirignano (1986, 1993a, 1993b). The extensions of the current book beyond those previous formulations will be identified as we proceed.

In developing the spray equations, it is convenient to define various density functions. Let ρ be the material density of the gas that is the mass per unit volume in a volume that includes only gas; $\bar{\rho}$ is the gas bulk density that is the mass of the gas in a volume that includes both gas and liquid. Similarly, ρ and $\bar{\rho}_l$ are the material density and the bulk density of the liquid. Furthermore, θ is defined as the void fraction of the gas that is the ratio of the equivalent volume of gas to a given volume of a gas and liquid mixture. Obviously, $1 - \theta$ is the void fraction of liquid. In our terminology, void fraction simply means θ. It follows that $\bar{\rho} = \theta\rho$ and $\bar{\rho}_l = (1 - \theta)\rho_l$. For very dilute sprays, $\theta \to 1$ and $\bar{\rho} \approx \rho$. However, because $\rho_l \gg \rho$ in typical cases, it could be poor in that limit to declare that $\bar{\rho}_l$ is negligible compared with ρ. Note that ρ_l will be considered to be constant although $\bar{\rho}_l$ can vary spatially and temporally.

There are various types of spray calculations that can be of interest. A most important issue involves the length scale of resolution. First, there is the question of whether droplets can be viewed as point sources with respect to gas-phase consideration. This can be a valid approximation in a domain if the collective volume of all liquid droplets and of all of their immediate gas films in that domain is small compared with the total gas volume in that same domain. The approximation can fail locally in regions of dense spray even if the preceding criteria are met globally. Only in the limiting case of a spherically symmetric transport field around the droplet does the point-source approximation give the exact influence on the far gas field. We almost always make the point-source approximation.

Another issue concerns the desired resolution compared with the average spacing between droplets. If the resolution is smaller than that spacing, we must account for each droplet physically present in the flow. This clearly limits the total number of droplets that can be considered. Because droplets are considered here as discrete particles, a numerical method that uses an Eulerian scheme for the gas phase and a Lagrangian scheme for the liquid phase is utilized.

If resolution on a scale larger than the droplet spacing is sufficient, only average droplets in each neighborhood need to be considered. Then, we can obviously consider domains that include a much larger number of droplets. Eulerian calculations are used for the gas phase, and either an Eulerian or Lagrangian scheme can be used for the dispersed liquid phase. The Lagrangian scheme is preferred because it reduces numerical error that is due to artificial diffusion, as discussed in Chapter 8.

With most practical problems involving sprays, the largest scale and the total number of droplets to be considered are typically so large that computational resources are insufficient to resolve the dynamical and thermal behavior of the liquid internal to the droplets, the gas films surrounding the droplets, and the gas regions between neighboring droplets. That is, these phenomena can be described only by use of subgrid scales. So modelling of those subgrid phenomena is required. These subgrid models are essentially the models discussed in Chapters 2–6. The subgrid modelling avoids the costly solution (for a large volume with many droplets) of a large system of equations on a grid size smaller than the droplet diameter. In the case of a dense spray, the droplet diameter becomes comparable with the spacing, and therefore resolution of the gas phase and resolution of the liquid phase are just as costly. If the gas-phase grid size and the average droplet diameter become comparable, it becomes inappropriate to use the discrete-particle method with its inherent subgrid modelling for these dense sprays. If fine resolution is required, solution of the Navier–Stokes equations with account for phase boundaries is suggested. However, computational resources will limit the total volume of mixture that can be treated. If average properties were sufficient, the two-continua method could be used for these dense sprays with modified representation of mass, momentum, and energy exchanges between the phases. We will discuss that averaging method. Note that the equations formulated in this chapter do not account for collisions that might occur in very dense sprays.

7.1 Averaging Process for Two-Continua Formulations

In this section and Section 7.2, we develop governing equations for the two-continua and multicontinua formulations. As a consequence of these formulations, each dependent variable at any spatial point will be an instantaneous average value over a

neighborhood of that point. The neighborhood includes both liquid and gas. Therefore both liquid properties and gas properties will be defined at any point in a spray
flow, regardless of whether that point is actually in the gas or in the liquid at that
instant. So a continuum of gas properties and a continuum of liquid properties are
defined.

In both laminar and turbulent spray flow computations that relate to practical
situations, it is usual to have flow regions with many more droplets than the achievable number of computational cells. Therefore we normally define an "average"
droplet for each locality that has properties that are representative of the droplets
in its neighborhood. Although the volume defining the neighborhood for droplet-
property averaging is arbitrary, it is rational to make the averaging volume of the
order of the computational-cell volume; resolution is lost by both the averaging process and the computational discretization so resolution is not gained by making one
scale smaller than the other. The length scale for the averaging volume (and therefore for the computational mesh) should be orders of magnitude larger than the
droplet diameter or the average spacing between droplets. If each of these droplet
length scales is at least an order of magnitude larger than the mesh size, the fields
within and around the droplets can be resolved; so there is no point in using average
quantities. If the droplet scales and mesh size are comparable, very few droplets are
involved in the averaging process and very large fluctuations in the average quantities can be expected.

After this averaging, we have two continua overlaying each other. The two-
phase-flow equations have been in use for decades, especially for sprays; see the
overviews by Spalding (1980) and Sirignano (1986, 1993a, 1993b). They were first
used by Crocco and Cheng (1956) and Crocco et al. (1962) in the study of combustion instability in liquid-propellant rocket motors. Williams (1962, 1985) presented
a derivation of these equations by using droplet distribution functions. Both gas and
liquid properties are averaged over a neighborhood so that, regardless of whether
liquid or gas exists at a particular point in space at a particular instant, both gas-
averaged properties and liquid-averaged properties will have continuous values at
that point and instant. If we categorize our droplets into many classes (determined
for example by initial size, velocity, or composition or by point of injection) and
average over each class, there will be a separate continuum for each class plus a
continuum for the gas. We classify this approach as a multicontinua approach. The
gas-phase continuum is normally solved with an Eulerian formulation for the governing equations whereas either Lagrangian or Eulerian formulations are used for
the droplet equations.

Spray flow science can be considered as part of a broader science named two-
phase flows, which include additional suspensions, bubbly flows, dusty flows, and
porous flows. Overviews of this field can be found in Soo (1967), Wallis (1969),
Marble (1970), Bear (1972), and Drew (1983). Various portions of our analysis will
resemble or reproduce elements in the literature; however, significant distinctions
occur because of the attempt here to unify the averaging method for two-phase flows
with averages constructed for turbulent flows and for computational purposes and
because of the special distinctions of rapidly vaporizing, high-liquid-mass sprays.
Our interest in sprays here will include liquid-fuel combustion, which has some

characteristics that demand special analytical treatment: The mass and momentum of the liquid (discrete phase) can be within an order of magnitude of the values for the gas (continuous phase) although the volume fraction of liquid is orders of magnitude smaller; the vaporization time scale in the high-temperature environment can be as small as or smaller than the times for the liquid droplets to reach kinematic or thermal equilibrium with the gas; the Reynolds number (and Peclet number) based on relative gas–droplet velocity and droplet dimension can be $O(10)$ to $O(100)$, causing substantial stratification and large gradients within the gaseous microstructure surrounding droplets and thereby requiring more information than mere averages of the microstructure field; droplet vaporization and heating rates can depend on internal spatial and temporal variations of liquid temperature, composition, and velocity, thereby requiring more resolution than simply average properties; and the smallest turbulent length scales can be as small as the droplet size. In the applications of interest, the exchanges of mass, momentum, and energy between the phases will all have highly significant impacts.

Important contributions to the theory of averaging for two-phase flows can be found in the papers of Gray and Lee (1977), Prosperetti and Jones (1984), Zhang and Prosperetti (1994a, 1994b, 1997), Bulthuis et al. (1995), and Prosperetti and Zhang (1995).

Drew and Passman (1999) give a very broad and detailed overview of averaging methods for multicomponent flows; they discuss ensemble averaging by various methods: constructing a large number of realizations, volume averaging, and time averaging. Vaporizing sprays are not emphasized in those papers; so a more general approach is needed. For our purposes, volume averaging has some critical advantages and will be used. Because spatial averaging or filtering methods are widely used by the turbulence community in their large-eddy simulations (LESs), the treatment of turbulent spray flows integrates better with those studies if spatial averaging is used. Furthermore, we also are interested in unsteady flows for which time averaging will not be useful. Finally, we ultimately solve difference equations rather than differential equations. Finite-volume methods involve averaging over the computational-cell volume and its surfaces. In other numerical differencing methods, there is an implied spatial averaging; so, if we use another averaging method for the two-phase character, we are effectively applying two averages to the equations with no particular gain.

Important developments in LESs and so-called "direct numerical simulations" (DNSs) for particle-laden turbulent flows have been made in recent years. See Squires and Eaton (1991), Elghobashi and Truesdell (1992, 1993), Wang and Squires (1996), Boivin et al. (1998), Druzhinin and Elghobashi (1998, 2001), and Ferrante and Elghobashi (2003). Useful overviews of LES methods for single-phase flows are provided by Piomelli (1999) and Givi (2003). Some recent applications of LES methodology to spray combustion problems are addressed by Sankaran and Menon (2002a, 2002b). See Sections 10.3 and 10.4 for more discussion and citations of other works. The turbulent-flow community is well aware of the important relationship between the volume of the computational cell and the length scales for the phenomena that are filtered or averaged.

Generally, the literature on the development of the averaged equations for two-phase flows, e. g., Gray and Lee (1977) and Prosperetti and Jones (1984), has not been specific about the volume size or shape to define the neighborhood over which averaging occurs. In particular, the relationship between the averaging process for multifluid flows and the high-wave-number filtering process for turbulent flows, as well as the relationship between the computational-cell shape and size and the shape and size of the averaging volume, have not been well treated yet. Nor has the error been fully evaluated for the common gas-phase approximation that the volume average of a product equals the product of the volume averages; the impact of this on highly stratified-flow microstructures will be shown to have potential importance. A substantial amount of research by Zhang and Prosperetti (1994a, 1994b, 1997), Bulthuis et al. (1995), and Prosperetti and Zhang (1995) is based on multirealization ensemble averaging rather than volume averaging. The literature on LES for single-phase flows (Piomelli, 1999; Givi, 2003) also discusses the process of averaging over a neighborhood to create a new continuum and modified governing equations. They tend to refer to the averaging process as a "filtering" process because they filter out the portion of the spectrum associated with smaller length scales. In principle, of course, averaging and filtering are the same here. The spray researchers can learn from the LES researchers who do address the specification of the averaging-volume size and shape and the modelling and evaluation of the difference between the average of a product and the product of the averages. Here, we attempt to make that advance first for laminar spray flows and then for turbulent spray flows.

Researchers have begun to treat turbulent spray and particle-laden flows by using either LES or DNS. Particle-, droplet-, or bubble-laden flows in cases of homogeneous turbulence (Elghobashi and Truesdell, 1992, 1993; Boivin et al., 1998; Druzhinin and Elghobashi, 1998; Ferrante and Elghobashi, 2003), temporal mixing layers (Ling et al., 1998), and spatially developing mixing layers (Druzhinin and Elghobashi, 2001) have been treated by DNS. The LES approach has been used for turbulent, particle-laden channel flow (Wang and Squires, 1996) and turbulent combustion (Sankaran and Menon, 2002a, 2002b). Both DNS and LES analyses typically begin with equations that treat the droplet and gas properties as continuous variables. So an implicit averaging has been made on the droplet scale. Then, for LES, another averaging (or filtering) is performed explicitly to avoid resolution of the smallest scales of turbulence. The modelling of the droplet behavior has neglected any direct interaction of small eddies with the droplets. Rather, the resulting equations are based implicitly on the assumption that the smallest eddies of the turbulence are larger than the largest droplet scales. In this chapter, we attempt to unify the droplet averaging process and the LES length-scale filtering process into one process. We discuss the implications of the situation in which the smallest eddy scales are comparable to the droplet scales.

The so-called DNS calculation of spray or particle-laden flows is also built on modified equations for the flow. (For that reason, this author finds the "DNS" label inappropriate for its current use in two-phase-flow calculations.) Whether the droplets are averaged together as a group and represented by an average droplet or they are represented individually as point (monopole) sources, there is a loss of resolution on the scale of the droplet and its surrounding film. So the DNS calculations require modelling of the behavior at the droplet scales and implicitly are restricted to situations in which the smallest eddy scales are at least an order magnitude larger than the droplet scales. Many interesting situations, particularly in practical combustors, involve comparable sizes for droplet scales and the Kolmogorov scale of turbulence. For these situations, current DNS and LES methods would both be inadequate.

Appendix A presents the basic equations governing the gas and liquid phases of the flow. In this section, the averaging process is defined and the averaged equations are developed for each phase, following an approach advocated by Sirignano (2005a, 2005b). The liquid-phase equations are presented in both Lagrangian and Eulerian forms. Computational implications are discussed. Certain newly identified (at least within the spray literature) flux terms in the equations, each related to the difference between an average of a product and a product of the averages, are analyzed and quantified by use of simplistic models of the flow on the scale smaller than that of the averaging volume (i.e., subgrid scale or microstructural scale). The microstructural behavior and analytical challenge are discussed for laminar flows, for flows in which the smallest turbulent eddy is larger than the microstructure, and for flows in which the smallest turbulent eddies appear in the microstructure.

7.1.1 Averaging of Dependent Variables

Let us consider the variable $\theta(\vec{x}, t)$ to have a value of unity in the gas phase and zero in the liquid phase. θ has been named a characteristic function or a component indicator function (Drew and Passman, 1999). The function was introduced as a void-volume distribution function in a series of papers on porous flows (Whitaker, 1966, 1967) that used volume averaging over the microstructure that included both the solid porous material and the fluid in the pores. A spatial derivative of θ is therefore zero everywhere except at a liquid–gas interface where it is a Dirac delta function. Note that the time derivative will be zero except at a moving interface. We now define a weighting factor $G(\vec{x} - \vec{\xi})$ for the averaging of quantities over the neighborhood of any particular point $\vec{x}$ at any particular instant t; $\vec{\xi}$ varies to identify the points in the neighborhood of the point $\vec{x}$. $\vec{x}$ and $\vec{\xi}$ have the same reference origin so that $\vec{x} - \vec{\xi}$ is a relative position. G depends on only the relative position; it does not separately depend on $\vec{x}$ or $\vec{\xi}$. Nor does it have a temporal dependence. We also require that

$$\int_{-\infty}^{\infty} \int_{-\infty}^{\infty} \int_{-\infty}^{\infty} G(\vec{x} - \vec{\xi}) \mathrm{d}\vec{\xi} = 1, \qquad (7.1)$$

where $\mathrm{d}\vec{\xi} \overset{\text{def}}{=} \mathrm{d}\xi_1 \mathrm{d}\xi_2 \mathrm{d}\xi_3$. Three examples of symmetric choices for G are

$$G \overset{\text{def}}{=} \begin{cases} \frac{3}{4\pi a^3} & \text{if } 0 \leq |\vec{x} - \vec{\xi}| \leq a \\ 0 & \text{if } |\vec{x} - \vec{\xi}| > a \end{cases}; \qquad (7.2)$$

$$G \overset{\text{def}}{=} \begin{cases} \frac{1}{8abc} & \text{if} \quad -a \leq x_1 \leq a \,, \; -b \leq x_2 \leq b \,, \; -c \leq x_3 \leq c \\ 0 & \text{otherwise} \end{cases} \,; \qquad (7.3)$$

$$G \overset{\text{def}}{=} \frac{1}{2} \left(\frac{b}{\pi} \right)^{3/2} e^{-b|\vec{x} - \vec{\xi}|^2}. \qquad (7.4)$$

In the choice presented by Eq. (7.2) or (7.3), the spherical or rectangular volume is named the averaging volume. In the choice given by Eq. (7.4), the integration should be performed over the infinite domain. If the exponential function is "clipped" to become zero outside of some finite symmetrical domain (such as a sphere or rectangular box) centered at the point $\vec{x}$, the constants in the G function must be adjusted to satisfy Eq. (7.1). The domain where the clipped G is positive is named the averaging volume for this case. We could proceed with an infinite averaging volume but, as noted, there is disadvantage in making the averaging volume larger than the computational cell volume. The choice of Eq. (7.3) is commonly used for LES computations and fits well with a rectangular gridding scheme.

Note that, in the preceding examples, two types of symmetry are displayed for the function G: for spherical symmetry, $G(|\vec{x} - \vec{\xi}|) = G(|\vec{\xi} - \vec{x}|)$, and, for symmetry about each of three orthogonal planes, $G(x_1 - \xi_1, x_2 - \xi_2, x_3 - \xi_3) = G(\xi_1 - x_1, x_2 - \xi_2, x_3 - \xi_3)$, $G(x_1 - \xi_1, x_2 - \xi_2, x_3 - \xi_3) = G(x_1 - \xi_1, \xi_2 - x_2, x_3 - \xi_3)$, and $G(x_1 - \xi_1, x_2 - \xi_2, x_3 - \xi_3) = G(x_1 - \xi_1, x_2 - \xi_2, \xi_3 - x_3)$. In the following analyses, the results still apply for either symmetry type. Furthermore, the condition of symmetry of G is not required. The results in this chapter would still apply, for example, if the point $\vec{x}$ were not in the center of the averaging volume or if the averaging volume had a nonsymmetric shape.

$G(\vec{x} - \vec{\xi})$ has three important properties that will be useful in this analysis. They apply for either symmetric or asymmetric G functions. One is the condition that $\partial G / \partial x_i = -\partial G / \partial \xi_i$. This condition will be valuable in determining the average value of a spatial gradient. Another property is that $G \to 0$ as $|\vec{x} - \vec{\xi}| \to \infty$. The third property is that it has units of reciprocal volume (reciprocal length cubed) and is normalized so that, when used as a weighting factor in an integration over the volume, it maintains the proper dimensional units and implicitly divides the integral by the weighted volume. The importance of the constraints on the choice of G has been well recognized by researchers in the LES field but not as well recognized by two-phase-flow researchers. Define the void volume (fraction of volume occupied by gas) as

$$\overline{\theta(\vec{x}, t)} \overset{\text{def}}{=} \int_{-\infty}^{\infty} \int_{-\infty}^{\infty} \int_{-\infty}^{\infty} G(\vec{x} - \vec{\xi}) \theta(\vec{\xi}, t) \mathrm{d}\vec{\xi}. \qquad (7.5)$$

The overbar implies an average over the volume. Clearly, the volume fraction occupied by liquid will be

$$\overline{1 - \theta(\vec{x}, t)} = 1 - \overline{\theta(\vec{x}, t)} = \int_{-\infty}^{\infty} \int_{-\infty}^{\infty} \int_{-\infty}^{\infty} G(\vec{x} - \vec{\xi})(1 - \theta(\vec{\xi}, t)) \mathrm{d}\vec{\xi}. \qquad (7.6)$$

In averaging the gas-phase flow variables, some variables will be averaged with normalized density in the weighting factor whereas others will be averaged without the use of density. The choice of whether or not to use density weighting for any particular variable here follows conventional practice, is intended to produce the simpler form of the averaged equations, and is based on how the specific term appears in the governing equations. When density weighting is used, the averaging is over the mass rather than simply over the volume; this is known as Favre averaging. Note that the two types of averaging (volume and Favre) produce identical results for the situation in which density is uniform such as we shall see later for the averaged liquid-phase equations.

The average gas density will be

$$\overline{\rho(\vec{x}, t)} \stackrel{\text{def}}{=} \int_{-\infty}^{\infty} \int_{-\infty}^{\infty} \int_{-\infty}^{\infty} G(\vec{x} - \vec{\xi}) \theta(\vec{\xi}, t) \rho(\vec{\xi}, t) \mathrm{d}\vec{\xi}. \tag{7.7}$$

The scalars Y_n, h, and ω_n and the vector u_i are averaged with a mass (Favre) weighting. If ψ is a generic scalar or vector, we obtain

$$\overline{\rho(\vec{x}, t) \psi(\vec{x}, t)} \stackrel{\text{def}}{=} \overline{\rho(\vec{x}, t)} \ \langle \psi(\vec{x}, t) \rangle$$

$$\stackrel{\text{def}}{=} \int_{-\infty}^{\infty} \int_{-\infty}^{\infty} \int_{-\infty}^{\infty} G(\vec{x} - \vec{\xi}) \theta(\vec{\xi}, t) \rho(\vec{\xi}, t) \psi(\vec{\xi}, t) \mathrm{d}\vec{\xi}. \tag{7.8}$$

The angle brackets $\langle \ \rangle$ are subsequently used to denote a mass-weighted average of a property or of the product of properties. Now, consider $\rho Y_m u_i$, $\rho Y_m V_{nm,i}$, and $\rho h u_i$. If we take the product of a scalar ψ and a vector w_i, we obtain

$$\overline{\rho(\vec{x}, t) \psi(\vec{x}, t) w_i(\vec{x}, t)} \stackrel{\text{def}}{=} \overline{\rho(\vec{x}, t)} \ \langle \psi(\vec{x}, t) w_i(\vec{x}, t) \rangle$$

$$\stackrel{\text{def}}{=} \int_{-\infty}^{\infty} \int_{-\infty}^{\infty} \int_{-\infty}^{\infty} G(\vec{x} - \vec{\xi}) \theta(\vec{\xi}, t) \rho(\vec{\xi}, t) \psi(\vec{\xi}, t) w_i(\vec{\xi}, t) \mathrm{d}\vec{\xi}. \tag{7.9}$$

The mass-weighted average of the product of two velocity vectors produces the tensor

$$\overline{\rho(\vec{x}, t) u_i(\vec{x}, t) u_j(\vec{x}, t)} \stackrel{\text{def}}{=} \overline{\rho(\vec{x}, t)} \ \langle u_i(\vec{x}, t) u_j(\vec{x}, t) \rangle$$

$$\stackrel{\text{def}}{=} \int_{-\infty}^{\infty} \int_{-\infty}^{\infty} \int_{-\infty}^{\infty} G(\vec{x} - \vec{\xi}) \theta(\vec{\xi}, t) u_i(\vec{\xi}, t) u_j(\vec{\xi}, t) \mathrm{d}\vec{\xi}. \tag{7.10}$$

For variables such as Φ, q_i, p, and the viscous stress tensor τ_{ij}, the averaging is performed with a volume weighting to obtain

$$\overline{\Phi(\vec{x}, t)} \stackrel{\text{def}}{=} \int_{-\infty}^{\infty} \int_{-\infty}^{\infty} \int_{-\infty}^{\infty} G(\vec{x} - \vec{\xi}) \theta(\vec{\xi}, t) \Phi(\vec{\xi}, t) \mathrm{d}\vec{\xi}, \tag{7.11}$$

$$\overline{q_i(\vec{x}, t)} \stackrel{\text{def}}{=} \int_{-\infty}^{\infty} \int_{-\infty}^{\infty} \int_{-\infty}^{\infty} G(\vec{x} - \vec{\xi}) \theta(\vec{\xi}, t) q_i(\vec{\xi}, t) \mathrm{d}\vec{\xi}, \tag{7.12}$$

$$\overline{\theta(\vec{x}, t)} \ \widehat{p(\vec{x}, t)} \stackrel{\text{def}}{=} \overline{p(\vec{x}, t)} \stackrel{\text{def}}{=} \int_{-\infty}^{\infty} \int_{-\infty}^{\infty} \int_{-\infty}^{\infty} G(\vec{x} - \vec{\xi}) \theta(\vec{\xi}, t) p(\vec{\xi}, t) \mathrm{d}\vec{\xi}, \tag{7.13}$$

$$\overline{\theta(\vec{x}, t)} \ \widehat{\tau_{ij}(\vec{x}, t)} \stackrel{\text{def}}{=} \overline{\tau_{ij}(\vec{x}, t)} \stackrel{\text{def}}{=} \int_{-\infty}^{\infty} \int_{-\infty}^{\infty} \int_{-\infty}^{\infty} G(\vec{x} - \vec{\xi}) \theta(\vec{\xi}, t) \tau_{ij}(\vec{\xi}, t) \mathrm{d}\vec{\xi}. \tag{7.14}$$

The averages for pressure and viscous stress indicated by the "hat" symbol differ from the others with the explicit appearance of $\overline{\theta}$ in order to facilitate the collection of terms that contribute to the aerodynamic forces on the droplets. In the case of Fickian diffusion, it follows that the integral in Eq. (7.9) becomes the average of the triple product of density ρ, mass diffusivity D, and the gradient of mass fraction $\partial Y_n / \partial x_i$. Furthermore, if the product ρD were constant, the integral would become the product of ρD and the average gradient.

We obtain the average values for the liquid properties by using the liquid-properties fields and replacing $\theta(\vec{\xi}, t)$ with $[1 - \theta(\vec{\xi}, t)]$ in the integrands of Eqs. (7.7)–(7.14). Also, $\overline{\theta(\vec{x}, t)}$ is replaced with $[1 - \overline{\theta(\vec{x}, t)}]$ in Eqs. (7.13) and (7.14). For the constant-density liquid considered here, there is no distinction in the results between volume-weighted averages and mass-weighted (Favre) averages. So only the overbar is used to indicate liquid-property averages. We do not write those liquid-phase relations here because they are obvious.

The application of boundary conditions will present some problems. The specified value of a flow variable at a boundary is not exactly equal to the average value of that variable in a neighborhood adjacent to the boundary. So some higher-order error is made when that condition of equality is applied. It will not be a problem if the weighting factor G is nonzero outside of the boundary, provided that the flow variables are prescribed to be zero outside of the boundaries.

7.1.2 Averaging of Derivatives

The relationships between the volume average of the derivative of a dependent variable and the derivative of a volume-averaged dependent variable was shown for porous flow problems by Whitaker (1966, 1967) and Slattery (1967). Later, Whitaker (1973) extended the treatment to multiphase systems with mass exchange. In those early works, G was not identified but implicitly was set equal to unity over the averaging volume. The possibility of volume changes with $\vec{x}$ was not considered. The time derivative of any averaged (scalar, vector, or tensor) gas-phase quantity ϕ can be related to the average of the time derivative as

$$\overline{\frac{\partial \phi(\vec{x}, t)}{\partial t}} = \int_{-\infty}^{\infty} \int_{-\infty}^{\infty} \int_{-\infty}^{\infty} G(\vec{x} - \vec{\xi}) \theta(\vec{\xi}, t) \frac{\partial \phi(\vec{\xi}, t)}{\partial t} d\vec{\xi}$$

$$= \frac{\partial}{\partial t} \int_{-\infty}^{\infty} \int_{-\infty}^{\infty} \int_{-\infty}^{\infty} G(\vec{x} - \vec{\xi}) \theta(\vec{\xi}, t) \phi(\vec{\xi}, t) d\vec{\xi}$$

$$- \int_{-\infty}^{\infty} \int_{-\infty}^{\infty} \int_{-\infty}^{\infty} G(\vec{x} - \vec{\xi}) \frac{\partial \theta(\vec{\xi}, t)}{\partial t} \phi(\vec{\xi}, t) d\vec{\xi}. \qquad (7.15)$$

So, recognizing that θ will change with time only because droplet motion, distortion, vaporization, or condensation causes the gas–liquid interface to move, we can relate the time derivative of θ to the velocity of the interface as

$$\frac{\partial \theta(\vec{\xi}, t)}{\partial t} = -u_{\theta, j} \frac{\partial \theta(\vec{\xi}, t)}{\partial \xi_j}, \qquad (7.16)$$

where $u_{\theta, i}$ is the ith component of the normal interface velocity. With n taken as the local normal coordinate at the surface, $\partial \theta(\vec{\xi}, t) / \partial n$ is a Dirac delta function; so the three-dimensional integral can immediately be integrated to give a two-dimensional surface integral over all interfaces in the averaging volume. The result is that

$$\overline{\frac{\partial \phi(\vec{x}, t)}{\partial t}} = \frac{\partial \overline{\phi(\vec{x}, t)}}{\partial t} + \int_S \int G(\vec{x} - \vec{\zeta}) \phi(\vec{\zeta}, t) u_{\theta, j}(\vec{\zeta}, t) \mathrm{d}A_j. \qquad (7.17)$$

Here, dA_i is the ith component of the normal interfacial area vector, pointing into the continuous phase (gas), and the $\vec{\zeta}$ vectors are the subset of $\vec{\xi}$ vectors that locate points on the interfaces. In Eqs. (7.15), (7.16), and (7.17), ϕ can be a scalar, vector, or tensor.

Let us now examine the average of a spatial derivative. Again, ϕ can be a scalar, vector, or tensor:

$$\overline{\frac{\partial \phi(\vec{x}, t)}{\partial x_i}} = \int_{-\infty}^{\infty} \int_{-\infty}^{\infty} \int_{-\infty}^{\infty} G(\vec{x} - \vec{\xi}) \theta(\vec{\xi}, t) \frac{\partial \phi(\vec{\xi}, t)}{\partial \xi_i} \mathrm{d}\vec{\xi}$$

$$= \int_{-\infty}^{\infty} \int_{-\infty}^{\infty} \int_{-\infty}^{\infty} \frac{\partial [G(\vec{x} - \vec{\xi}) \theta(\vec{\xi}, t) \phi(\vec{\xi}, t)]}{\partial \xi_i} \mathrm{d}\vec{\xi}$$

$$- \int_{-\infty}^{\infty} \int_{-\infty}^{\infty} \int_{-\infty}^{\infty} \frac{\partial G(\vec{x} - \vec{\xi})}{\partial \xi_i} \theta(\vec{\xi}, t) \phi(\vec{\xi}, t) \mathrm{d}\vec{\xi}$$

$$- \int_{-\infty}^{\infty} \int_{-\infty}^{\infty} \int_{-\infty}^{\infty} G(\vec{x} - \vec{\xi}) \frac{\partial \theta(\vec{\xi}, t)}{\partial \xi_i} \phi(\vec{\xi}, t) \mathrm{d}\vec{\xi}$$

$$= 0 + \frac{\partial}{\partial x_i} \int_{-\infty}^{\infty} \int_{-\infty}^{\infty} \int_{-\infty}^{\infty} G(\vec{x} - \vec{\xi}) \theta(\vec{\xi}, t) \phi(\vec{\xi}, t) \mathrm{d}\vec{\xi}$$

$$- \int_S \int G(\vec{x} - \vec{\zeta}) \phi(\vec{\zeta}, t) \mathrm{d}A_i. \qquad (7.18)$$

The first integral on the right-hand side is zero because $G \to 0$ as $|\vec{x} - \vec{\xi}| \to \infty$. So we have

$$\overline{\frac{\partial \phi(\vec{x}, t)}{\partial x_i}} = \frac{\partial \overline{\phi(\vec{x}, t)}}{\partial x_i} - \int_S \int G(\vec{x} - \vec{\zeta}) \phi(\vec{\zeta}, t) \mathrm{d}A_i. \qquad (7.19)$$

The computational cell in a calculation need not be identical in shape to the averaging volume but should have the same approximate magnitude for volume because the largest volume determines the resolution for the problem. The averaging volume is distinct from the computational volume in principle because there are an infinite number of averaging volumes (before discretization of the differential equations into difference equations), one corresponding to each point in space, whereas a finite number of computational cell volumes exist. In other words, averaging volumes can overlap each other but cell volumes can not. Of course, in practice, there is no advantage to considering more than one point for each computational cell because resolution cannot be improved by consideration of more points. So, after discretization, the number of averaging volumes engaged in the calculation equals the number of cells.

If, for convenience, averaging volumes are made identical to computational cell volumes in uniform-sized Cartesian grids, the averaging-volume shapes can be symmetric types such as the rectangular boxes considered with Eq. (7.3); all volume sizes would be uniform. However, with nonuniform Cartesian grids, cylindrical or spherical grids, and unstructured grids, including finite elements, the size and shape of the averaging volume would change if it were set equal to the computational-cell volume locally. In some situations, the averaging volume might be varied spatially because the computationally resolved, physical-length scales vary. The spatial variation means that $G = G(\vec{x}, \vec{x} - \vec{\xi})$; namely, a separate dependence on $\vec{x}$ appears that modifies the gradient formulation from the form given by Eqs. (7.18) and (7.19). An option is to correct Eqs. (7.18) and (7.19) and the resulting conservation equations to account for a variation in the averaging volume. For example, suppose we have a nonuniform or unstructured grid and use $G = 1/V$, where V is the averaging volume that is now a function of $\vec{x}$. Then Eq. (7.19) is modified to be

$$\frac{\overline{\partial \phi(\vec{x}, t)}}{\partial x_i} = \frac{\partial \overline{\phi(\vec{x}, t)}}{\partial x_i} + \frac{\partial (\ln V)}{\partial x_i} \overline{\phi(\vec{x}, t)}$$

$$- \int_{A^*} \int \phi(\vec{\eta}, t) \theta(\vec{\eta}, t) \frac{1}{V} \frac{\partial n(\vec{\eta})}{\partial x_i} dA^* - \int_S \int \frac{1}{V} \phi(\vec{\zeta}, t) dA_i, \quad (7.20)$$

where A^* is the external surface area of the averaging volume, $\vec{\eta}$ is a vector locating a point on that surface, and n is the positive-outward length extension of that surface locally as V changes with $\vec{x}$. The second and third terms on the right-hand side are the result of the variation in the volume. The second term is caused by the change in the magnitude of G over the volume as Eq. (7.1) is obeyed. The other term results from the change in the domain of the integration. They can be combined easily into one term when the rate of local volume change is uniform over the bounding surface of the averaging volume. That is, $\partial n/\partial x_i$ is uniform over the surface. Then, we obtain

$$\frac{\overline{\partial \phi(\vec{x}, t)}}{\partial x_i} = \frac{\partial \overline{\phi(\vec{x}, t)}}{\partial x_i} + \frac{\partial (\ln V)}{\partial x_i} \left\{ \overline{\phi(\vec{x}, t)} - \overline{\phi_S(\vec{x}, t)} \right\} - \int_S \int \frac{1}{V} \phi(\vec{\zeta}, t) dA_i, \quad (7.21)$$

where the average of ϕ over the bounding surface is given by

$$\overline{\phi}_S = \frac{1}{A^*} \int_{A^*} \int \phi(\vec{\eta}, t) \theta(\vec{\eta}, t) dA^*. \quad (7.22)$$

The new term can be zero in special cases, e.g., when ϕ is uniform over the boundary and equal to $\overline{\phi}$, but generally it is not zero. The problem with this situation is that a model for the value of ϕ on the volume boundary is needed. If the averaging volume were a sphere centered at $\vec{x}$, the use of the mean-value theorem (see page 276 of Courant and Hilbert, 1962) leads to the conclusion that $\overline{\phi} - \overline{\phi}_S = O(R^2 \nabla^2 \phi)$, where R is the sphere radius. This implies that the ratio of the second term on the right-hand side of Eq. (7.21) to the first term on that side is $O(R^2/(L_1 L_2))$. Here, L_1 is the physical length scale for change in ϕ and L_2 is the designed numerical length scale for change in the averaging volume. We can expect this order of magnitude to remain the same if the averaging volume is not spherical. So, clearly, if L_2 is chosen to be too small (i.e., change in volume size is too abrupt), significant errors can occur if the second term on the right-hand side of Eq. (7.21) is not considered. Note that this need for correction would apply to any calculation with averaging, including single-phase LES calculations, as recognized by Ghosal and Moin (1995). There also has been recognition of the potential for error by the computational fluid dynamics community. See, for example, pages 60–63 of the second volume of the book by Fletcher (1991), where nonuniform grids are discussed. It is known that if a computational mesh size varies too abruptly with position, errors can be introduced that are of first order in an otherwise second-order-accurate differencing scheme.

Independence of grid size is a necessary condition. For calculations with filtering or averaging, this condition should be interpreted as the requirement that, for two averaging-volume choices of the same order of magnitude, the two computations should agree for the portion of the spectrum with length scales larger than the larger of the two filtering lengths for the two choices. That is, only the mutually unfiltered portion can be expected to agree. Clearly, independence of grid size depends not only on the correct choice of grid size but also on the correct choice of subgrid models.

7.1.3 Averaged Gas-Phase Equations

We will now multiply each one of Eqs. (A.1), (A.2), (A.3), (A.4), and (A.6) by the product $G\theta$ and integrate term-by-term over the volume. The relations given by Eqs. (7.5)–(7.14) and (7.17) and (7.19) will be used to substitute for the various integrals. Then, some rearrangement of terms will be made to yield the averaged equations. Finally, we obtain the averaged gas-phase continuity equation,

$$\frac{\partial \overline{\rho}}{\partial t} + \frac{\partial(\overline{\rho}\,\langle u_j\rangle)}{\partial x_j} = \dot{M}, \qquad (7.23)$$

and the averaged gas-phase species-continuity equation

$$\frac{\partial(\overline{\rho}\,\langle Y_m\rangle)}{\partial t} + \frac{\partial(\overline{\rho}\,\langle Y_m\rangle\,\langle u_j\rangle)}{\partial x_j} + \frac{\partial(\overline{\rho}\,\langle Y_m\rangle\,\langle V_{m,j}\rangle)}{\partial x_j}$$

$$= \overline{\rho}\,\langle \omega_m\rangle + \dot{M}\epsilon_m + \frac{\partial(\overline{\rho}\,\alpha_{m,j})}{\partial x_j} + \frac{\partial(\overline{\rho}\,\beta_{m,j})}{\partial x_j}, \quad m = 1, \dots, N. \qquad (7.24)$$

In these equations, we have used the definitions

$$\dot{M} \stackrel{\text{def}}{=} \int_S \int G(\vec{x} - \vec{\zeta})\rho(\vec{\zeta}, t)[u_j(\vec{\zeta}, t) - u_{\theta,j}(\vec{\zeta}, t)]\mathrm{d}A_j, \qquad (7.25)$$

$$\dot{M}\epsilon_m \stackrel{\text{def}}{=} \int_S \int G(\vec{x} - \vec{\zeta})\rho(\vec{\zeta}, t)Y_m(\vec{\zeta}, t)[u_j(\vec{\zeta}, t)$$

$$-u_{\theta,j}(\vec{\zeta}, t) + V_{m,j}(\vec{\zeta}, t)]\mathrm{d}A_j, \qquad (7.26)$$

$$\alpha_{m,i} \stackrel{\text{def}}{=} \langle Y_m\rangle\,\langle u_i\rangle - \langle Y_m u_i\rangle, \qquad (7.27)$$

$$\beta_{m,i} \stackrel{\text{def}}{=} \langle Y_m\rangle\,\langle V_{m,i}\rangle - \langle Y_m V_{m,i}\rangle. \qquad (7.28)$$

Note that $[u_j(\vec{\zeta}, t) - u_{\theta,j}(\vec{\zeta}, t)]$ in Eqs. (7.25) and (7.26) is the relative velocity of the gas at the liquid surface; so the normal component of this velocity is the Stefan velocity at the surface caused by the vaporization (or condensation). Therefore $\dot{M}$ is the total vaporization rate per unit volume that is due to all of the droplets in the neighborhood. ϵ_m is the fractional vaporization rate of the mth species. Models for

$\dot{M}$ and ϵ_m (for situations in which the smallest scales of turbulence are larger than the droplet scales) can be found in the earlier chapters. The quantities $\alpha_{m,i}$ and $\beta_{m,i}$ require modelling and are discussed later.

A Reynolds stress term will appear in the momentum equation. It is given as

$$\Gamma_{ij} \overset{\text{def}}{=} \langle u_i \rangle \langle u_j \rangle - \langle u_i u_j \rangle. \tag{7.29}$$

With the use of Eq. (7.5), it can be shown that

$$\frac{\partial \overline{\theta}}{\partial x_i} = -\int_{-\infty}^{\infty} \int_{-\infty}^{\infty} \int_{-\infty}^{\infty} \frac{\partial G(\vec{x} - \vec{\zeta})}{\partial \zeta_i} \theta(\vec{\zeta}, t) \mathrm{d}\vec{\zeta}$$

$$= \int_{-\infty}^{\infty} \int_{-\infty}^{\infty} \int_{-\infty}^{\infty} G(\vec{x} - \vec{\zeta}) \frac{\partial \theta(\vec{\zeta}, t)}{\partial \zeta_i} \mathrm{d}\vec{\zeta}$$

$$= \int_S \int G(\vec{x} - \vec{\zeta}) \mathrm{d} A_i. \tag{7.30}$$

Now, following Prosperetti and Jones (1984), we can combine a portion of the terms related to the averaged viscous stress and pressure to yield F_i, the aerodynamic force (per unit volume) on the droplets,

$$F_i \overset{\text{def}}{=} \int_S \int G(\vec{x} - \vec{\zeta}) \left\{ [\tau_{ij}(\vec{\zeta}, t) - \delta_{ij} p(\vec{\zeta}, t)] - [\widehat{\tau_{ij}(\vec{x}, t)} - \delta_{ij} \widehat{p(\vec{x}, t)}] \right\} \mathrm{d} A_j, \tag{7.31}$$

where δ_{ij} is the Kronecker delta symbol. Models for F_i, including models for unsteady effects, can be found in Chapter 3.

The term representing the momentum exchange between the phases associated with vaporization (condensation) is an integral over the liquid surface of the product of the local vaporization rate per unit area and the velocity of the gas at the surface. If the difference between the liquid velocity and the gas velocity at the surface is small compared with the droplet velocity, this term can be modelled as the product of the global vaporization rate and the average droplet velocity. Even if the velocity difference at the surface is not small compared with the droplet velocity, there should be substantial cancellation from the integration of the relative velocity vector over the closed surfaces of the droplets. Therefore we have for the vaporization case

$$\int_S \int G(\vec{x} - \vec{\zeta}) \rho(\vec{\zeta}, t) u_i(\vec{\zeta}, t) [u_j(\vec{\zeta}, t) - u_{\theta,j}(\vec{\zeta}, t)] \mathrm{d} A_j = \dot{M} \overline{u}_{l,j}, \tag{7.32}$$

where $\overline{u}_{l,i}$ is the averaged liquid velocity. In the condensation case, $\overline{u}_{l,i}$ should be replaced with $\langle u_i \rangle$. So the gas-phase momentum equation can be cast as

$$\frac{\partial(\overline{\rho} \langle u_i \rangle)}{\partial t} + \frac{\partial(\overline{\rho} \langle u_i \rangle \langle u_j \rangle)}{\partial x_j} + \overline{\theta} \frac{\partial \widehat{p}}{\partial x_i} = \overline{\theta} \frac{\partial \widehat{\tau}_{ij}}{\partial x_j} + \overline{\rho} g_i - F_i + \dot{M} \overline{u}_{l,j} + \frac{\partial(\overline{\rho} \Gamma_{ij})}{\partial x_j}. \tag{7.33}$$

In his porous-flow analysis, Whitaker (1967) found the $\alpha_{m,i}$ term, which he named a dispersion vector. $\beta_{m,i}$ did not appear because Fickian diffusion was assumed. Inertial effects were neglected so the Γ_{ij} term was not developed in the porous-flow analyses. Whitaker (1973) also neglected the inertial effects in the more general multiphase systems transport study. Most of the terms in these equations have been recognized before for spray flows, bubbly flows, and particle-laden flows. However, the last two terms in Eq. (7.24) and the last term in Eq. (7.33) are new terms for spray flow theory. They have been recognized in turbulent-flow theory but we shall discuss in Subsection 7.1.7 how stratification in the microstructure for laminar two-phase flows can make these terms significant.

For the development of the averaged energy equation, we note that the term that gives the rate of energy exchange (per unit volume) with the droplets due to heat conduction, mass transfer, and radiation can be modelled as

$$\int_S \int G(\vec{x} - \vec{\zeta}) \left\{ \rho(\vec{\zeta}, t) h(\vec{\zeta}, t)[u_j(\vec{\zeta}, t) - u_{\theta, j}(\vec{\zeta}, t)] + q_j(\vec{\zeta}, t) \right\} \mathrm{d} A_j$$
$$= \dot{M} [\langle h_{g,s} \rangle - L_{\text{eff}}], \tag{7.34}$$

where

$$q_i = -\lambda \frac{\partial T}{\partial x_i} + q_{\text{rad},i} + \sum_{n=1}^{N} \rho V_{n,i} h_n Y_n, \tag{7.35}$$

and $\langle h_{g,s} \rangle$ and L_{eff} are the averaged specific gas enthalpy at the liquid surface and the heat per unit mass for vaporization and interior heating of the droplet, respectively. L_{eff} is the sum of the latent heat of vaporization L and $(\dot{Q}_l/\dot{M})$, where $\dot{Q}_l$ is the heating rate of the droplet interior.

Now we define

$$S_* \stackrel{\text{def}}{=} \int_S \int G(\vec{x} - \vec{\zeta}) \left\{ p(\vec{\zeta}, t) - p(\widehat{\vec{x}}, t) \right\} u_{\theta, j}(\vec{\zeta}, t) \mathrm{d} A_j. \tag{7.36}$$

Here, $u_{\theta, j}(\vec{\zeta}, t) \mathrm{d} A_j$ is the rate of change of infinitesimal volume that is due to liquid–gas interface motion. After weighting with G and integrating over all liquid–gas interfaces in the volume, the negative of the time derivative of the void volume is obtained. So if a droplet is locally in a spherically symmetric situation relative to the surrounding gas, this droplet's contribution to the S_* integral is proportional to the product of the pressure difference and the time derivative of the ratio of the droplet volume to the averaging volume. If there is only a Stefan velocity in the region and no imposed velocity, the gas velocity decreases with distance from the droplet and gas pressure increases with that distance. So the differential surface pressure in the integral for S_* is negative. In this case, the S_* term represents differential pressure work done on the gas by the liquid which is changing volume. For vaporization (condensation), it is positive (negative) and acts as an energy source (sink) for the gas. The difference in pressure appears in the S_* term because of the definition of the average pressure given by Eq. (7.13).

For a translating particle with fore-and-aft symmetry in the surface-pressure distribution and without volume change or shape distortion, $S_* = 0$. If the flow separates over the translating particle, the integral for S_* is not equal to zero and work is done on the gas by the particle or on the particle by the gas. The fore–aft orientation and therefore the orientation of the net pressure drag force on the particle are determined by the vector $\langle u_i \rangle - \overline{u}_{l,i}$ and not by $\overline{u}_{l,i}$ alone. So, if the dot product $\overline{u}_{l,i}(\langle u_i \rangle - \overline{u}_{l,i})$ is negative, the particle is moving in a direction opposed to the net force and a positive value appears for S_*. That means that the particle is doing work on the continuous fluid (gas). If the dot product is positive, the integral is negative and work is done on the particle. The order of magnitude of the S_* term here is given by the combined pressure drag force of all particles in the averaging volume times the average particle velocity divided by the averaging volume. Interestingly, it is not the total drag that contributes to this work term; the friction acts parallel to the surface; so, unlike the pressure, it does no work as the infinitesimal element of surface area sweeps over an infinitesimal volume.

These simple understandings about the pressure work can guide the creation of a model for the evaluation of S_*. For example, we can add the two effects just discussed to yield

$$S_* = \frac{1}{4\,\overline{\rho}_s}\left\{\frac{\dot{m}}{4\pi R}\right\}^2 \frac{D(\overline{\theta})^2}{Dt} - \frac{3(1-\overline{\theta})}{8\overline{\theta}R}\overline{\rho}C_{Dp}|\langle u\rangle - \overline{u}_l|\overline{u}_{l,j}(\langle u_j\rangle - \overline{u}_{l,j}), \quad (7.37)$$

where $\dot{m}$ is the the vaporization rate of an average droplet, R is the instantaneous droplet radius, the time derivative following the gas is $D(\ )/Dt = \partial(\ )/\partial t + \tilde{u}_j\partial(\ )/\partial x_j$, and C_{Dp} is the coefficient of pressure drag for the droplet. It is only suggested here that this model is worthy of testing; it is not endorsed yet. The S_* term is the flux of a velocity-squared term, so it should be of the order of Mach number squared times a thermal flux term such as the divergence of E_i; so, for low-speed subsonic flows, it can be negligible.

Now the averaged gas-phase energy equation becomes

$$\frac{\partial(\overline{\rho}\langle h\rangle)}{\partial t} + \frac{\partial(\overline{\rho}\langle u_j\rangle \langle h\rangle)}{\partial x_j} + \frac{\partial\,\overline{q}_j}{\partial x_j} - \overline{\theta}\left\{\frac{\partial\,\widehat{p}}{\partial t} + \langle u_j\rangle\frac{\partial\,\widehat{p}}{\partial x_j}\right\}$$

$$= \overline{\Phi} + \sum_{n=1}^{N}\overline{\rho}\,\langle\omega_n\rangle\,Q_n + \dot{M}\left[\langle h_{g,s}\rangle - L_{\text{eff}}\right] + S_* - \Delta_g + \frac{\partial(\overline{\rho}\,E_j)}{\partial x_j}, \quad (7.38)$$

where

$$\Delta_g \stackrel{\text{def}}{=} \langle u_j\rangle\,\overline{\theta}\,\frac{\partial(\widehat{p})}{\partial x_j} - \overline{u_j\frac{\partial p}{\partial x_j}}, \quad (7.39)$$

$$E_i \stackrel{\text{def}}{=} \langle u_i\rangle\,\langle h\rangle - \langle u_i h\rangle. \quad (7.40)$$

The last four terms in Eq. (7.38) have not been recognized for spray flows before. The last term has been used in the turbulence community. These flux terms are discussed further in Subsection 7.1.7.

The averaged gas-phase equation of state may be expressed as

$$\overline{\theta}\,\widehat{p} = \overline{\rho}\,\langle RT\rangle = \overline{\rho}\,[\,\langle h\rangle - \langle e\rangle\,], \quad (7.41)$$

where e is the specific internal energy. Note that for a multicomponent mixture, $\tilde{h}$ and $\tilde{e}$ must eventually be related to the average temperature $\tilde{T}$ and the average mass fractions $\tilde{Y}_m$, presenting another closure challenge. If the energy equation is reformulated by use of Equation (A.6) so that $\langle e\rangle$ becomes the dependent variable instead of $\langle h \rangle$, it will be seen that the total pressure at the droplet surface is critical in determining pressure work associated with volume change of the phase. This, of course, is consistent with the First Law of Thermodynamics.

The gas-phase species-continuity equations, momentum equation, and energy equation can individually be placed in nonconservative form by a combination of Eq. (7.24), (7.33), or (7.38) with Equation (7.23).

The same results for Eqs. (7.24), (7.33), and (7.38) would not be obtained in a two-step averaging process. If the first step averaged over droplet scales only, the same form as those equations would result with the same formal definition of the new flux terms. Of course, there should be differences in the modelling of the new terms. The second averaging over the turbulent scales would produce still more flux terms. The results of the two-step averaging would also be different if the turbulent-scale averaging were performed first.

7.1.4 Averaged Vorticity and Entropy

We can divide the nonconservative form of the momentum equation by the average density and then take the curl to yield an evolution equation for the average vorticity Ω_i, which is here the curl of the mass-weighted average velocity and not the volume-weighted average of ω_i, the curl of velocity. The relationship between the two averages is given by

$$\overline{\omega}_i = \epsilon_{ijk} \overline{\frac{\partial u_j}{\partial x_k}} = \epsilon_{ijk} \frac{\partial \overline{u}_j}{\partial x_k} - \epsilon_{ijk} \int_S \int G(\vec{x} - \vec{\zeta}) u_j(\vec{\zeta}, t) \mathrm{d} A_k$$

$$= \Omega_i + \epsilon_{ijk} \frac{\partial (\overline{u}_j - \langle u_j \rangle)}{\partial x_k} - \epsilon_{ijk} \int_S \int G(\vec{x} - \vec{\zeta}) u_j(\vec{\zeta}, t) \mathrm{d} A_k, \qquad (7.42)$$

where the permutation symbol $\epsilon_{ijk} = 0$ if two or three indices are identical, $+1$ if the indices reflect an even permutation of 123, and -1 if the indices reflect an odd permutation of 123. Part of the distinction (between the curl of the average velocity and the average of the curl) disappears if the integral of the cross product of the normal surface vector and the tangential velocity vector over the droplet surface is zero. For a solid particle with no slip at the surface and no rotation, the surface velocity is the velocity of the particle's mass center and its tangential component varies from zero to the particle velocity value. The contribution of that particle to the surface integral in Eq. (7.42) becomes zero. If the particle is rotating in addition to translating, the surface integration, for that particle alone, of the cross product of that tangential velocity and the normal surface vector will have a nonzero contribution whose value is proportional to the angular velocity and to the ratio of the particle's volume to the averaging volume. For a liquid droplet with internal circulation and/or a distortion that is symmetric about an axis, the surface integration will produce a zero value.

As noted by Sirignano (1972), the droplet aerodynamic force term and the droplet-momentum source term will produce or modify the average vorticity even in the case of barotropic flow, negligible shear forces (away from the droplet surfaces), and zero initial vorticity. Our modification here indicates that Γ_{ij} will also affect average vorticity Ω_i. The resulting vorticity evolution equation will be

$$\frac{\partial \Omega_i}{\partial t} + \langle u_j \rangle \frac{\partial \Omega_i}{\partial x_j} = \Omega_j \frac{\partial \langle u_i \rangle}{\partial x_j} - \Omega_i \frac{\partial \langle u_j \rangle}{\partial x_j} - \epsilon_{ijk} \frac{\partial (\overline{\theta}/\overline{\rho})}{\partial x_k} \frac{\partial \widehat{p}}{\partial x_j}$$

$$+ \epsilon_{ijk} \left\{ \frac{\overline{\theta}}{\overline{\rho}} \frac{\partial^2 \widehat{\tau}_{jr}}{\partial x_k \partial x_r} + \frac{\partial (\overline{\theta}/\overline{\rho})}{\partial x_k} \frac{\partial \widehat{\tau}_{jm}}{\partial x_m} \right\}$$

$$- \epsilon_{ijk} \left\{ \frac{\partial F_j}{\partial x_k} + \dot{M} \left\{ \frac{\partial \langle u_j \rangle}{\partial x_k} - \frac{\partial \overline{u}_{l,j}}{\partial x_k} \right\} + \frac{\partial \dot{M}}{\partial x_k} \left\{ \langle u_j \rangle - \overline{u}_{l,j} \right\} \right\}$$

$$+ \epsilon_{ijk} \frac{\partial^2 (\overline{\rho} \, \Gamma_{jr})}{\partial x_k \partial x_r}. \tag{7.43}$$

An alternative choice of a method that obtains $\overline{\omega}_i$ directly is to combine primitive equations (A.1) and (A.2) to obtain a nonconservative form, divide the result by density, then take the curl term-by-term to yield an evolution equation for ω_i, and finally perform the volume averaging. Thereby, it would be necessary to evaluate more product terms; so this author prefers the other route just outlined.

Sirignano (1972) also obtained an evolution equation for the average entropy by combining the momentum and energy equations, neglecting some terms of the order of the Mach number squared (as appropriate for the combustion instability application of that publication) and assuming the same value of specific heat for each species. It was shown that the droplet interactions (exchange of mass and energy between the phases) produce entropy even in the absence of chemical reaction, heat and mass diffusion, and viscosity. Now we improve on the accuracy of that relationship by averaging Equation (A.11), still neglecting any variation in the specific heats across the species. The inclusions of the new terms and the higher-order (in Mach number) effects now cause the averaged entropy-conservation equation to be

$$\frac{\partial (\overline{\rho} \langle s \rangle)}{\partial t} + \frac{\partial (\overline{\rho} \langle u_j \rangle \langle s \rangle)}{\partial x_j} = \frac{\partial (\overline{\rho} H_j)}{\partial x_j} + \frac{\overline{R}_1}{\widetilde{T}} + J$$

$$+ \int_S \int G(\vec{x} - \vec{\zeta})(u_j - u_{\theta,j}) \rho s \, dA_j, \tag{7.44}$$

where, from the analysis of the averaged energy equation, we have

$$\overline{R}_1 = -\frac{\partial \overline{q}_j}{\partial x_j} + \overline{\Phi} + \sum_{n=1}^{N} \overline{\rho} \langle \omega_n \rangle Q_n - \dot{M} L_{\text{eff}} \stackrel{\text{def}}{=} R_2 - \dot{M} L_{\text{eff}}. \tag{7.45}$$

Furthermore, we define

$$H_i \stackrel{\text{def}}{=} \langle u_i \rangle \langle s \rangle - \langle u_i s \rangle, \tag{7.46}$$

$$J \stackrel{\text{def}}{=} \left\{ \frac{R_1}{T} \right\} - \frac{\overline{R}_1}{\langle T \rangle}. \tag{7.47}$$

The integral in Eq. (7.44) reflects an entropy flux associated with mass transfer from the discrete phase. We can model that integral as $\dot{M} \langle s_{g,s} \rangle$, where the subscripts imply that the quantity is evaluated in the gas phase at the interface between the phases.

The averaged entropy-conservation equation can be combined with the averaged continuity equation to obtain an averaged entropy-evolution equation:

$$\overline{\rho}\frac{\partial \langle s \rangle}{\partial t} + \overline{\rho} \langle u_j \rangle \frac{\partial \langle s \rangle}{\partial x_j} = \dot{M}\left(\langle s_{g,s} \rangle - \langle s \rangle - \frac{L_{\text{eff}}}{\langle T \rangle} \right) + \frac{\partial (\overline{\rho} H_j)}{\partial x_j} + \frac{R_2}{\langle T \rangle} + J. \quad (7.48)$$

A reasonable model for the averaged entropy difference is

$$\langle s_{g,s} \rangle - \langle s \rangle - \frac{L_{\text{eff}}}{\langle T \rangle} = \frac{\langle h_{g,s} \rangle - \langle h \rangle - L_{\text{eff}}}{\langle T \rangle}. \quad (7.49)$$

Clearly, now the effect of mass and energy exchanges between the phases, as indicated by $\dot{M}[\langle h_{g,s} \rangle - \langle h \rangle - L_{\text{eff}}]$, is seen to affect entropy production.

7.1.5 Averaged Liquid-Phase Partial Differential Equations

The averaged equations for the liquid phase can be developed by multiplying every term in Eqs. (A.1)–(A.4) by the product of G and $(1 - \theta)$, followed by integration over the neighborhood. For the constant-density liquid, mass weighting and volume weighting produce the same average. First, we obtain the averaged liquid-phase continuity equation:

$$\frac{\partial \overline{\rho}_l}{\partial t} + \frac{\partial (\overline{\rho}_l \overline{u}_{l,j})}{\partial x_j} = -\dot{M}. \quad (7.50)$$

Next, the averaged liquid-phase species-continuity equation can be written. The Gauss divergence theorem will show that there will be a contribution to the divergence of $\overline{\rho_l Y_{l,m} V_{l,m,i}}$ coming only from the portions of the boundary of the averaging volume that intersect the volumes of droplets located on the boundary. A large Peclet number can be considered; liquid-phase mass diffusion is slow compared with advection of the liquid-phase species across the averaging-volume boundaries. Therefore we can neglect this diffusion contribution compared with other terms for spray applications. (Liquid-phase mass diffusion might be important, however, for some other multiphase application.) The average diffusion flux goes to zero. So it can be shown that

$$\overline{\rho_l Y_{l,m} V_{l,m,i}} = \overline{\rho}_l \overline{Y_{l,m} V_{l,m,i}} = \overline{\rho}_l \overline{Y}_{l,m} \overline{V}_{l,m,i} - \overline{\rho}_l \beta_{l,m,i} = 0. \quad (7.51)$$

It follows that

$$\frac{\partial (\overline{\rho}_l \overline{Y}_{l,m})}{\partial t} + \frac{\partial (\overline{\rho}_l \overline{Y}_{l,m} \overline{u}_{l,j})}{\partial x_j} = -\dot{M}\epsilon_{l,m} + \frac{\partial (\overline{\rho}_l \alpha_{l,m,j})}{\partial x_j} \quad m = 1, \ldots, N. \quad (7.52)$$

The averaged liquid-phase momentum equation for the vaporizing case is

$$\frac{\partial (\overline{\rho}_l \overline{u}_{l,i})}{\partial t} + \frac{\partial (\overline{\rho}_l \overline{u}_{l,i} \overline{u}_{l,j})}{\partial x_j} + (1 - \overline{\theta})\frac{\partial \widehat{p}_l}{\partial x_i}$$

$$= (1 - \overline{\theta})\frac{\partial \widehat{\tau}_{l,ij}}{\partial x_j} + \overline{\rho}_l g_i + F_i - \dot{M} \overline{u}_{l,i} + \frac{\partial (\overline{\rho}_l \Gamma_{l,ij})}{\partial x_j}. \quad (7.53)$$

Note that the next-to-last term in Eq. (7.53) must be modified to read $\dot{M} \langle u_i \rangle$ for the case of condensation. This momentum equation can be developed further by making a standard assumption that the gradient of the averaged pressure and the gradient of the averaged viscous stress tensor at any point $\vec{x}$ are identical for the two phases. Note that capillary effects can prevent the matching of pressure and viscous stress across the gas–liquid interface. Capillary pressure causes a jump in pressure across the interface; however, for spherical droplets, that jump quantity is uniform along the interface; so the pressure gradients on the two sides match. If surface temperature, surface composition, or both, vary along the surface, viscous stress can jump across the interface and the pressure jump can become nonuniform along the surface. However, if the droplet interior field is axisymmetric, these effects average to zero effect on the gradients. Solving for these gradients from Eq. (7.33) and substituting into Eq. (7.53), we find

$$
\frac{\partial(\overline{\rho}_l\,\overline{u}_{l,i})}{\partial t} + \frac{\partial(\overline{\rho}_l\,\overline{u}_{l,i}\,\overline{u}_{l,j})}{\partial x_j} = \frac{1-\overline{\theta}}{\overline{\theta}}\overline{\rho}\frac{D\,\langle u_i\rangle}{Dt} + \left\{\overline{\rho}_l - \overline{\rho}\frac{1-\overline{\theta}}{\overline{\theta}}\right\}g_i
$$

$$
+ \frac{F_i}{\overline{\theta}} - \dot{M}\left\{\frac{\overline{u}_{l,i}}{\overline{\theta}} - \frac{1-\overline{\theta}}{\overline{\theta}}\langle u_i\rangle\right\}
$$

$$
+ \frac{\partial(\overline{\rho}_l\,\Gamma_{l,ij})}{\partial x_j} - \frac{1-\overline{\theta}}{\overline{\theta}}\frac{\partial(\overline{\rho}\,\Gamma_{ij})}{\partial x_j}. \tag{7.54}
$$

For the liquid-phase energy equation, we can again use the Gauss divergence theorem to show that there will be a contribution to the divergence of $\overline{q}_{l,i}$ coming from only the portions of the boundary of the averaging volume that intersect the volumes of droplets located on the boundary. We can consider this contribution to be negligible compared with other terms for spray applications. This could be important, however, for some other multiphase application. The averaged liquid-phase energy equation can now be written as

$$
\frac{\partial(\overline{\rho}_l\,\overline{h}_l)}{\partial t} + \frac{\partial(\overline{\rho}_l\,\overline{h}_l\,\overline{u}_{l,j})}{\partial x_j} = (1-\overline{\theta})\left\{\frac{\partial\,\overline{p}_l}{\partial t} + \overline{u}_{l,j}\frac{\partial\,\overline{p}_l}{\partial x_j}\right\} + \overline{\Phi}_l - S_{l,*} - \Delta_l
$$

$$
- \dot{M}\left[\langle h_{g,s}\rangle - L_{\text{eff}}\right] + \frac{\partial(\overline{\rho}_l\,E_{l,j})}{\partial x_j}. \tag{7.55}
$$

Again, the last term in Eq. (7.52), the last term in Eq. (7.53), and the last three terms in Eq. (7.55) have not been recognized for two-phase flows before. The equations are constructed so that energy flux into the droplet surface from the gas (or work done by the gas on the droplet surface) does not necessarily equal in magnitude the energy flux into the liquid (or work done on the liquid) because capillary action causes a pressure jump across the interface and surface energy can change as the droplet size, shape, temperature, and/or composition change. $S_{l,*}$ is determined by a modified form of Eq. (7.36) with the surface pressure defined on the liquid side of the interface and the average pressure is the average for the liquid phase. Because of the capillary actions, these liquid-pressure values will be higher than the corresponding gas-pressure values indicated in that equation. So, although a positive S_* represents work (per unit volume per unit time) done on the gas by the differential pressure at the gas side of the interface as the liquid volume changes, it is not generally equal to the value of $S_{l,*}$, which represents work done by the differential pressure force on the liquid side of the interface. The difference is caused by the integral of the product of interface velocity $u_{\theta,j}$ and the difference between the local capillary pressure at the surface and the average capillary pressure. In the special case of spherical droplets, $u_{\theta,j}$ and the capillary pressures are uniform in magnitudes over the interface with the average and local capillary pressures equal; so the difference between S_* and $S_{l,*}$ disappears.

7.1.6 Averaged Liquid-Phase Lagrangian Equations

In developing the Lagrangian form of the equations, we define the time derivative following the liquid $d(\ )/dt = \partial(\ )/\partial t + \bar{u}_{l,j}\partial(\ )/\partial x_j$. The liquid density is considered to be constant so that $\bar{\rho}_l = [1 - \bar{\theta}\,]\rho_l$.

Recognizing that, for constant-liquid density, $\bar{\rho}_l = \rho_l[1 - \bar{\theta}\,] = n\rho_l V_{\mathrm{drop}}$, where $n(\vec{x}, t)$ is the droplet number density and $V_{\mathrm{drop}}(\vec{x}, t)$ is the average droplet volume, we may construct two other equivalent forms of the liquid-phase continuity equation:

$$\frac{d\bar{\theta}}{dt} = \frac{\partial\bar{\theta}}{\partial t} + \bar{u}_{l,j}\frac{\partial\bar{\theta}}{\partial x_j} = \frac{\dot{M}}{\rho_l} + (1-\bar{\theta})\frac{\partial\bar{u}_{l,j}}{\partial x_j} \tag{7.56}$$

or

$$\frac{dV_{\mathrm{drop}}}{dt} = \frac{\partial V_{\mathrm{drop}}}{\partial t} + \bar{u}_{l,j}\frac{\partial V_{\mathrm{drop}}}{\partial x_j} = -\frac{\dot{M}}{\rho_l n} = -\frac{\dot{m}}{\rho_l}, \tag{7.57}$$

where $\dot{m}(\vec{x}, t)$ is the mass vaporization rate of an average droplet. In the last equation, it has been assumed that the total number of droplets is conserved; i.e., there is no coalescence or shattering and $\partial n/\partial t + \partial(n\,\bar{u}_{l,j})/\partial x_j = 0$.

Equations (7.24), (7.53), and (7.55) can be reorganized into nonconservative and Lagrangian forms. The species-continuity equation becomes

$$\bar{\rho}_l\frac{d\bar{Y}_{l,m}}{dt} = \bar{\rho}_l\frac{\partial\bar{Y}_{l,m}}{\partial t} + \bar{\rho}_l\bar{u}_{l,j}\frac{\partial\bar{Y}_{l,m}}{\partial x_j} = -\dot{M}[\epsilon_{l,m} - \bar{Y}_{l,m}] + \frac{\partial(\bar{\rho}_l\,\alpha_{l,m,j})}{\partial x_j},$$

$$m = 1, \ldots, N. \tag{7.58}$$

The averaged liquid-phase momentum equation in nonconservative and Lagrangian forms becomes

$$\bar{\rho}_l\frac{d\bar{u}_{l,i}}{dt} = \bar{\rho}_l\frac{\partial\bar{u}_{l,i}}{\partial t} + \bar{\rho}_l\bar{u}_{l,j}\frac{\partial\bar{u}_{l,i})}{\partial x_j} = \frac{1-\bar{\theta}}{\bar{\theta}}\bar{\rho}\frac{D\,\langle u_i\rangle}{Dt} + \left\{\bar{\rho}_l - \bar{\rho}\frac{1-\bar{\theta}}{\bar{\theta}}\right\}g_i$$

$$+ \frac{F_i}{\bar{\theta}} - \dot{M}[\,\langle u_i\rangle - \bar{u}_{l,i}\,]\frac{1-\bar{\theta}}{\bar{\theta}} + \frac{\partial(\bar{\rho}_l\,\Gamma_{l,ij})}{\partial x_j} - \frac{1-\bar{\theta}}{\bar{\theta}}\frac{\partial(\bar{\rho}\,\Gamma_{ij})}{\partial x_j}. \tag{7.59}$$

The averaged liquid-phase energy equation in nonconservative or Lagrangian form is

$$\bar{\rho}_l\frac{d\bar{h}_l}{dt} - (1-\bar{\theta})\frac{d\widehat{p}_l}{dt} = \bar{\rho}_l\frac{\partial\bar{h}_l}{\partial t} + \bar{\rho}_l\bar{u}_{l,j}\frac{\partial\bar{h}_l}{\partial x_j} - (1-\bar{\theta})\left\{\frac{\partial\widehat{p}_l}{\partial t} + \bar{u}_{l,j}\frac{\partial\widehat{p}_l}{\partial x_j}\right\}$$

$$= \bar{\Phi}_l + \dot{Q}_l - S_{l,*} - \Delta_l + \frac{\partial(\bar{\rho}_l\,E_{l,j})}{\partial x_j}. \tag{7.60}$$

The following relationships have been used: $L_{\text{eff}} = L + \dot{Q}_l/\dot{M} = \langle h_{g,s} \rangle - \overline{h}_l + \dot{Q}_l/\dot{M}$.

It is somewhat a misnomer to describe this system as a Lagrangian system of equations. Normally in a Lagrangian tracking, we follow a fixed element of mass. (This in contrast to an Eulerian calculation, in which we deal with a fixed volume.) The Lagrangian element normally can change its volume and shape, changes of phase and chemistry can occur, and exchanges of equal amounts of mass can occur with its environment can occur; but it remains the amount of same mass. However, in the system of equations here, the averaging volume at each instant of Lagrangian time will have the same magnitude and shape but the amount of mass in this volume can change. It is better therefore in mathematical terms to describe this as a method of characteristics. Note that we could create a truly Lagrangian system for the liquid-phase equations by making the vectors $\vec{x}$, $\vec{\xi}$, and $\vec{\zeta}$ in the averaging integration reflect initial positions (rather than instantaneous positions) and thereby always identify the same points of mass. The problem would come then with the match with the gas-phase equations; namely, the averagings would not occur over the same field. So, in a true Lagrangian formulation, it would be very difficult to sensibly describe exchanges of mass, momentum, and energy between the phases. So we remain with the pseudo-Lagrangian scheme that locally and instantaneously uses a fixed volume for the averaging integration.

The number of droplets in the averaging volume can change in accordance with the droplet number conservation equation. Written in Lagrangian form, that equation becomes

$$\frac{\mathrm{d}n}{\mathrm{d}t} = -n\frac{\partial \overline{u}_{l,j}}{\partial x_j}. \tag{7.61}$$

So one approach to the calculations is to determine the varying droplet number density by using Eq. (7.61) and to relate $\dot{M}$ to $\dot{m}$ through the use of n as indicated by Eq. (7.57). Another approach that will also conserve droplet numbers is to fix the number of droplets associated with each characteristic path at the value given by an initial condition or inflow boundary condition.

An alternative analysis of the liquid properties could be developed following an approach in which the droplets are separated into N^* different classes based on their initial properties, e.g., diameter, velocity, or point of injection. A set of conservation equations (or evolution equations) could be developed for each droplet class. This approach is discussed later in this chapter.

It must be understood that the problem of a spray with rapidly vaporizing droplets in a high-temperature gaseous environment requires more knowledge of the microstructure than is provided by the preceding averaged liquid-phase equations. For example, the temporal and spatial variations of temperature and composition in the liquid droplet must be resolved by modelling for accurate prediction of heat and mass exchange between the phases. The internal fluid motion of the droplet can affect the transport and must also be resolved by modelling. These models can be used to predict liquid temperature and composition in place of Eqs. (7.52) and (7.55) or, equivalently, (7.58) and (7.60).

7.1.7 The Microstructure

Let us now analyze the contribution of the gas-phase flux terms $\alpha_{m,i}$, $\beta_{m,i}$, Γ_{ij}, Δ_g, and E_i. The results will depend on the microstructure of the flow within the averaging volume. The flux terms will also depend on the magnitude of the averaging volume and, through these terms and a few other source terms in the equations, the size of the averaging volume will affect the averaged quantities governed by the equations: e.g., ρ, u_i, Y_m, and h. In spectral terms, the larger the averaging volume is, the larger is the minimum wavelength left unfiltered by the averaging. These terms are known to be important in LES calculations, in which the the averaging volume is larger than the smallest turbulent eddies. We focus at first here on cases in which the smallest vortical eddies are larger than the averaging volume; so the microstructure is laminar. In the next subsection, we discuss the case in which some eddies are smaller than the volume.

Some of these flux terms have been reported in the literature on two-phase flows and porous-media flows. Zhang and Prosperetti (1994a, 1994b) used multi-realization averages to treat disperse flows without mass exchange between phases, viscosity, heat transfer, or compressibility. They considered potential flow in the microstructure and used an expansion with $1 - \bar{\theta}$ as the perturbation parameter. The present analysis is different in that we consider the continuous fluid and the disperse fluid to be multicomponent; we account for viscosity, heat transfer, and chemical reaction; and we allow the continuous-flow density to vary because of compression (or expansion) and/or heating (or cooling). Also, the perturbation expansion is not useful here because, although the gas volume is orders of magnitude larger than the liquid volume, the mass, momentum, and energy of the discrete liquid phase are of the same order of magnitude or not more than 1 order less than the counterpart continuous gas properties. In a later paper, Zhang and Prosperetti (1997) added the the effects of heat conduction and viscosity but still did not allow compressibility, mass exchange, chemical reaction, or multicomponent character. They also advocated the use of particle equations for the discrete phase that track global characteristics of the particles without requiring resolution of the microstructure of the discrete-fluid field. As noted earlier, these particle equations might "wash away" interesting physics for certain problems, e.g., rapidly vaporizing droplets.

Early papers by Whitaker (1966, 1967) and Slattery (1967) on flows through porous media used volume averaging to develop equations governing the hydrodynamics (e.g., Darcy's Law). Whitaker (1967) discussed the dispersion vector α_i for a single, nonadsorbing, nonreacting species in an incompressible porous flow. Later, Whitaker (1973) extended the transport equations to consider multicomponent diffusion, chemical reaction, and change of phase; the paper focused on mass transfer and did not consider the coupling with the equations for momentum and energy conservation. So, in addition to unifying the averaging methods for two-phase flows, LES analyses, and practical computation, the current work does extend and strengthen the theoretical foundations for two-phase flows.

Sirignano (2005a, 2005b) also examined some of the flux terms in the averaged equations to find where they might have quantitative importance. A few very simple

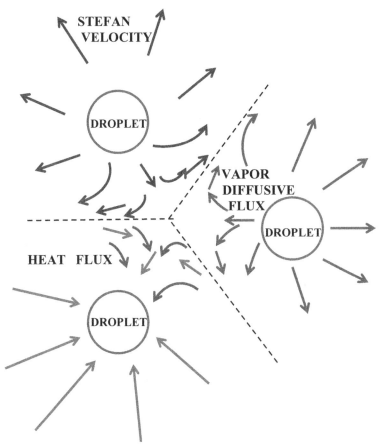

Figure 7.1. Vaporizing-droplet array with Stefan flow. (Sirignano, 2005a, with permission of the *International Journal of Multiphase Flow*.)

descriptions of droplet arrays undergoing heat, momentum, and mass exchanges with the gas were created as model problems intended to reflect the most essential physics and to determine roughly the magnitudes of these flux terms. In particular, combustion applications in which a liquid mass of fuel is injected at roughly stoichiometric proportions into air at a pressure $O(10 \text{ atm})$ were examined in two general cases: one case with a cloud of vaporizing droplets without forced or natural convection but only Stefan convection, and another case with relative motion between the droplets in a cloud and the ambient gas.

In the first case with only Stefan convection, the simple asymmetrical situation is sketched in Fig. 7.1. It is seen that, with significant asymmetry of the droplet distribution, the scalar fluxes could become of the order of the product of the average scalar and the average velocity. In that situation, the flux term can have importance. Calculations from Imaoka and Sirignano (2005a, 2005b, 2005c) indicate that, for two droplets (with only Stefan convection) placed asymmetrically within a rectangular control volume, the $\alpha_{m,i}$ and E_i flux terms will be less than 10% of the largest flux terms in the equation. The term Γ_{ij} can be significant for symmetric or asymmetric microstructural configurations. The quantity $u_i u_j$ is everywhere positive for $i = j$

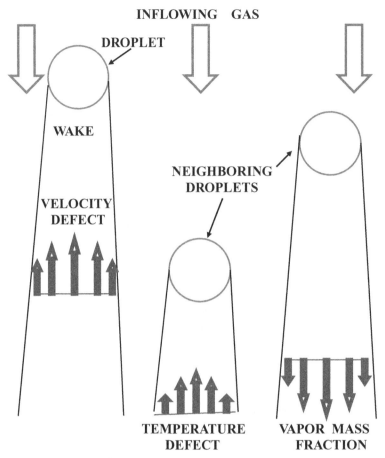

Figure 7.2. Vaporizing-droplet-array wake flows. (Sirignano, 2005a, with permission of the *International Journal of Multiphase Flow.*)

and thereby $\overline{u_i u_j} > 0$. For a symmetric case, $\overline{u}_i = 0$; clearly, Γ_{ij} can be significant in the momentum equation, i.e., Eq. (7.33).

In the second problem, as portrayed in Fig. 7.2, the wakes of an array of vaporizing droplets were examined. The linearized approach for slender axisymmetric far wakes described in many texts (see, for example, White, 1991) were followed. The magnitude of Γ_{11} was found to be roughly a factor $O(10^{-2})$ to $O(10^{-1})$ times the square of the average velocity component in the range $10 < \mathrm{Re} < 100$. Therefore it might be sufficiently important to consider as a correction for laminar two-phase flows. For typical fuels and the conditions considered, $E_1/(\overline{u}\,\overline{h})$ are possibly an $O(10^{-1})$ quantity and $\alpha_{F,1}/(\overline{u}\,\overline{h})$ is an $O(10^{-1})$ or higher quantity. So they can be important corrections. More attention must be paid to the microstructure behavior in order to have good predictions on a larger scale.

The Δ_g term depends on the pressure gradient. So if that pressure gradient is balanced primarily by the inertial term (spatial acceleration), it will differ from the thermal energy flux term such as the the divergence of E_i by a factor of the order of the square of the flow Mach number. If the pressure gradient is balanced primarily by the local acceleration term in unsteady cases or by the viscous stress term, we can expect the factor to be of the order of the Mach number. So, for the wide range of cases in which the flow in the microstructure is at low subsonic speeds, the Δ_g term should be insignificant compared with other terms.

Each of these same flux terms discussed here for the gas phase has a counterpart in the liquid phase. The discrete phase for sprays has the advantage in that it is less difficult to model than the gas phase. Models for heat and mass transport and internal fluid motion within droplets do exist in the literature. So models for the five averaged flux terms could be developed with some reasonable effort and would be an interesting future exercise. However, because decent microscale models exist for the internal liquid behavior, another route for calculations can be been followed. The internal velocity, temperature, and composition have been spatially and temporally resolved by means that are not excessively computationally intensive, as shown in Chapters 2 and 3. As a result, Eqs. (7.52) and (7.55) or, equivalently, (7.58) and (7.60) can be bypassed. Instead, the models provide the spatially resolved solutions for an average droplet, including the properties on the droplet surface. Still, to determine the trajectories of these average droplets, Eq. (7.54) or (7.59) must be solved simultaneously with the equations of the model.

7.2 Two-Continua and Multicontinua Formulations

This section presents the two-continua system of equations in the forms that are commonly used. In particular, the higher-order quantities that relate to differences between products (and a ratio) of averages and averages of products (and a ratio) are neglected: $\alpha_{m,i}$, $\beta_{m,i}$, Γ_{ij}, Δ_g, E_i, H_i, J, and S_*. If future research demonstrates the importance of any of these neglected terms, they will have to be reinserted into analyses. Furthermore, for the sake of simplicity, brackets, hat symbols, and bar symbols indicating averages are removed in this section and subsequent chapters. The only exception is for the bar symbol over the density in order to distinguish between bulk density and the mass per unit volume of the mixture. We understand generally now that averaged quantities are being considered.

In the following subsections, we examine the various conservation principles for the gas and liquid phases. Separation of the liquid phase into various classes dependent on certain distinguishing features is discussed, leading to a multicontinua formulation. The hyperbolic nature of the liquid-phase equations and the consequences are examined. Finally, the relations with subgrid models are discussed.

7.2.1 Continuity Equations

In this subsection, we consider the various forms of mass continuity or conservation equations for the gas and liquid phases. Mass conservation of individual chemical species or of individual classes of liquid droplets is also considered. The conservation of droplet numbers is explored. From Eqs. (7.23) and (7.50), the gas-phase mass-conservation equation is

$$\frac{\partial \bar{\rho}}{\partial t} + \frac{\partial}{\partial x_j}(\bar{\rho} u_j) = \dot{M}, \tag{7.62}$$

and the liquid-phase mass-conservation equation is

$$\frac{\partial \bar{\rho}_l}{\partial t} + \frac{\partial}{\partial x_j}(\bar{\rho}_l u_{l,j}) = -\dot{M}, \tag{7.63}$$

where $\dot{M}$ is the mass vaporization rate per unit volume. Models for evaluating the vaporization rate are discussed in other sections.

The mixture mass-conservation equation is given by

$$\frac{\partial(\bar{\rho} + \bar{\rho}_l)}{\partial t} + \frac{\partial}{\partial x_j}(\bar{\rho} u_j + \bar{\rho}_l u_{l,j}) = 0. \tag{7.64}$$

Clearly, only two of the three preceding equations are independent. They indicate that, because of vaporization or condensation, mass is not conserved for each phase but the mass of the mixture is conserved.

For a constant bulk liquid density ρ_l, Eqs. (7.5) and (7.7) yield that $\overline{\rho_l} = \theta \rho_l$, where the bar above θ is no longer explicit. Then Eq. (7.63) can be recast as an equation governing θ instead of $\bar{\rho}_l$. It becomes

$$\frac{\partial \theta}{\partial t} + \frac{\partial}{\partial x_j}(\theta u_{l,j}) = \frac{\dot{M}}{\rho_l} + \frac{\partial u_{l,j}}{\partial x_j}. \tag{7.65}$$

The nonzero right-hand side indicates that θ is not conserved for two reasons: (i) vaporization or condensation changes the liquid volume, and (ii) droplet trajectories diverge (or converge), thereby increasing (or decreasing) the distances between droplets.

Often it is convenient to divide the droplets into many classes according to initial values of velocity, position, diameter, and/or composition. Then our two-continua approach is expanded to a multicontinua approach. In such a case, it is not usually convenient to solve Eq. (7.63) or Eq. (7.65) for $\bar{\rho}_l$ or θ. Rather, we distinguish each class of droplets according to an integer value k and set

$$n = \sum_k n^{(k)}; \qquad \dot{M} = \sum_k \dot{M}^{(k)} = \sum_k \dot{m}^{(k)} n^{(k)}, \tag{7.66}$$

where $\dot{m}^{(k)}$ is the vaporization rate of an average droplet in the kth class, $n^{(k)}$ is the number density (number per unit volume) of droplets in that kth class, and n is the global droplet number density. The methods discussed in Chapters 2 and 3 can be used to determine $\dot{m}^{(k)}$. See, for example, Eq. (2.10), Eq. (3.30), or Eq. (3.34c). The determination of $n^{(k)}$ is discussed later in this section.

In the limit of a very large number of different droplet classes, the summation of Eqs. (7.66) can be replaced with an integral. This extension to a multicontinua approach was first suggested by Dukowicz (1980) and pursued by Sirignano (1986); it allows for better resolution of the variations among the many droplets that can be present in a spray calculation. Note that, although Dukowicz (1980) refers to his method as a discrete-particle method, it does not meet the discrete-particle definition used here. It is a multicontinua method.

We can define $\psi^{(k)}$ as the liquid volume fraction of the kth class of droplets and $\bar{\rho}_l^{(k)}$ as the bulk liquid density of that class. Then,

$$1 - \theta = \sum_k \psi^{(k)}, \qquad \bar{\rho}_l^{(k)} = \psi^{(k)} \rho_l^{(k)} = n^{(k)} \frac{4\pi}{3} \left[R^{(k)} \right]^3 \rho_l^{(k)}. \qquad (7.67)$$

Note that we are allowing for the possibility that different droplet classes can possess different material densities. The effects of different liquid-material densities and the effect of void volume were considered by Dukowicz (1980) but not contained in the original presentations by Sirignano (1972, 1986, 1993a, 1993b). Crowe et al. (1977) did consider the effect of void volume but did not consider separate classes of droplets.

It follows that

$$\frac{\partial \bar{\rho}_l^{(k)}}{\partial t} + \frac{\partial}{\partial x_j} \left[\bar{\rho}_l^{(k)} u_{l,j}^{(k)} \right] = -n^{(k)} \dot{m}^{(k)} = -\dot{M}^{(k)}, \qquad (7.68)$$

where

$$u_{lj} = \sum_k \frac{\bar{\rho}_l^{(k)}}{\bar{\rho}_l} u_{l,j}^{(k)}$$

is a mass-weighted average liquid-droplet velocity. When the material density is the same for all droplet classes, it also becomes a volume-weighted average velocity. Furthermore, we have

$$\frac{\partial \psi^{(k)}}{\partial t} + \frac{\partial}{\partial x_j} \left[\psi^{(k)} u_{l,j}^{(k)} \right] = -\frac{\dot{M}^{(k)}}{\rho_l^{(k)}}. \qquad (7.69)$$

We can solve either Eq. (7.68) or Eq. (7.69) (simultaneously with other equations discussed later in this section) and then use Eqs. (7.67) to get both $\psi^{(k)}$ and $\bar{\rho}_l^{(k)}$. However, we still require either $n^{(k)}$ or $R^{(k)}$ to determine the right-hand side of Eq. (7.68) or Eq. (7.69). Realizing that

$$\dot{m}^{(k)} = -\frac{4\pi}{3} \rho_l^{(k)} \frac{d}{dt} \left[R^{(k)} \right]^3,$$

where the time derivative is taken in the Lagrangian sense following the droplet, we can manipulate Eqs. (7.67) and (7.68) to obtain

$$\frac{\partial n^{(k)}}{\partial t} + \frac{\partial}{\partial x_j} \left[n^{(k)} u_{l,j}^{(k)} \right] = 0, \qquad (7.70)$$

which is a conservation equation for the droplet number for each class. A global conservation equation is obtained by summation of Eq. (7.70) over all classes. Then,

$$\frac{\partial n}{\partial t} + \frac{\partial}{\partial x_j} (n \tilde{u}_{l,j}) = 0, \quad \text{where} \quad \tilde{u}_{lj} = \sum_k \frac{n^{(k)} u_{l,j}^{(k)}}{n} \qquad (7.71)$$

is a number-weighted average velocity.

It is now seen that Eqs. (7.67), (7.68) [or (7.69)], and (7.70) provide four (two differential plus two algebraic) equations that will yield $\bar{\rho}_l^{(k)}$, $\psi^{(k)}$, $n^{(k)}$, and $R^{(k)}$.

Necessary inputs are $u_{l,i}^{(k)}$, which will come from the solution of the momentum equation and a mathematical model for the vaporization rate $\dot{m}^{(k)}$.

Consider now the simplifying restatement of Eq. (7.24) for the gas-phase species-mass conservation. The integer index m represents the particular species. The mass fraction Y_m is described by

$$\frac{\partial}{\partial t}(\bar{\rho} Y_m) + \frac{\partial}{\partial x_j}(\bar{\rho} u_j Y_m) - \frac{\partial}{\partial x_j}\left(\bar{\rho} D \frac{\partial Y_m}{\partial x_j}\right) = \dot{M}_m + \bar{\rho}\dot{w}_m = \sum_k n^{(k)} \dot{m}_m^{(k)} + \bar{\rho}\dot{w}_m,$$

(7.72)

where

$$\dot{M} = \sum_m \dot{M}_m = \sum_m \varepsilon_m \dot{M}, \qquad \dot{m}^{(k)} = \sum_k \dot{m}_m^{(k)}, \qquad \sum_m \dot{w}_m = 0.$$

The mass diffusivity is assumed to be the same for all species. Other options have been discussed in Appendix A and in previous chapters. ε_m is the fractional vaporization rate for species m and, for a quasi-steady gas phase, becomes a species mass-flux fraction. Obviously, $\sum_m \varepsilon_m = 1$. These mass-flux fractions can be determined locally with Eq. (3.4). Summation over all components in Eq. (7.72) yields continuity equation (7.62). Therefore, if we have N different species, only $N-1$ species-conservation equations need be solved together with Eq. (7.62). Note that Eqs. (7.62) and (7.72) can be combined to yield a nonconservative form:

$$\bar{\rho}\left[\frac{\partial Y_m}{\partial t} + u_j \frac{\partial Y_m}{\partial x_j}\right] - \frac{\partial}{\partial x_j}\left(\bar{\rho} D \frac{\partial Y_m}{\partial x_j}\right) = (\varepsilon_m - Y_m)\dot{M} + \bar{\rho}\dot{w}_m.$$

(7.73)

Generally, the conservative form of the equations is preferred for computation. Note that here conservative equation (7.72) still has source terms.

7.2.2 Momentum Conservation

Now, let us consider the gas-phase momentum equation obtained from Eq. (7.33) and constructed in a simpler form:

$$\frac{\partial}{\partial t}(\bar{\rho} u_i) + \frac{\partial}{\partial x_j}(\bar{\rho} u_j u_i) + \theta \frac{\partial p}{\partial x_i} - \theta \frac{\partial \tau_{ij}}{\partial x_j j} = \sum_k n^{(k)} \dot{m}^{(k)} u_{l,i}^{(k)} - F_{Di} + \bar{\rho} g_i,$$

(7.74)

where

$$F_{Di} = \sum_k n^{(k)} F_{Di}^{(k)}, \qquad \tau_{ij} = \mu\left(\frac{\partial u_i}{\partial x_j} + \frac{\partial u_j}{\partial x_i}\right) - \frac{2}{3}\delta_{ij}\frac{\partial u_i}{\partial x_j}.$$

Equation (7.74) includes momentum sources and sinks that are due to droplet-vapor mass sources, reaction to droplet drag, and body forces on the gas. The drag and lift forces in Eq. (7.74) can be related to the relative droplet–gas velocity and the droplet radius through the use of drag and lift coefficients. See the discussions in Subsections 3.2.5 and 3.2.7. A combination of Eqs. (7.62) and (7.74) yields another nonconservative form of the momentum equation:

$$\bar{\rho}\left[\frac{\partial u_i}{\partial t} + u_j\frac{\partial u_i}{\partial x_j}\right] + \theta\frac{\partial p}{\partial x_i} - \theta\frac{\partial \tau_{ij}}{\partial x_j} = \sum_k n^{(k)}\dot{m}^{(k)}\left[u_{l,i}^{(k)} - u_i\right] - F_{Di} + \bar{\rho}g_i. \quad (7.75)$$

The simplified liquid-phase momentum equation can be written from Eq. (7.53) as

$$\frac{\partial}{\partial t}(\bar{\rho}_l u_{l,i}) + \frac{\partial}{\partial x_j}(\bar{\rho}_l u_{l,j}u_{l,i})$$

$$= -\sum_k n^{(k)}\dot{m}^{(k)}u_{l,i}^{(k)} + F_{Di} + \bar{\rho}_l g_i - (1-\theta)\frac{\partial p}{\partial x_i} + (1-\theta)\frac{\partial \tau_{ij}}{\partial x_j}. \quad (7.76a)$$

Typically for dilute sprays, the last two terms in Eq. (7.76a) are replaced with the result that follows from Eq. (7.75) if we neglect the effect of the droplets on the gas flow and replace θ with $1 - \theta$. Note that the replacement of θ with $1 - \theta$ is strictly correct only if θ varies much more slowly in space and time than the stress terms do. Then,

$$-(1-\theta)\frac{\partial p}{\partial x_i} + (1-\theta)\frac{\partial \tau_{ij}}{\partial x_j} = \rho(1-\theta)\left[\frac{\partial u_i}{\partial t} + u_j\frac{\partial u_i}{\partial x_j} - g_i\right].$$

Now Eq. (7.76a) becomes

$$\frac{\partial}{\partial t}(\bar{\rho}_l u_{l,i}) + \frac{\partial}{\partial x_j}(\bar{\rho}_l u_{l,j}u_{l,i}) = -\sum_k n^{(k)}\dot{m}^{(k)}u_{l,i}^{(k)} + F_{Di} + [\bar{\rho}_l - \rho(1-\theta)]g_i$$

$$+ \rho(1-\theta)\left[\frac{\partial u_i}{\partial t} + u_j\frac{\partial u_i}{\partial x_j}\right]. \quad (7.76b)$$

The last term in Eq. (7.76b) implies that the acceleration that would have been given to the gas (if it were to exist in the fractional volume occupied by liquid), because of the pressure and the viscous stresses transferred from the neighboring gas, is transmitted as a force on the droplets. The buoyancy effect and this last term are each negligible when $\rho \ll \rho_l$. The neglected additions, because of droplet interactions to the last term in Eq. (7.76b), can be shown to be of the order of $(1-\theta)^2$. Note that the reaction to this force is already implied in gas momentum equation (7.75) and need not be explicitly represented.

An exact substitution without neglect of the effect of the droplets on the gas flow yields

$$\frac{\partial}{\partial t}(\bar{\rho}_l u_{l,i}) + \frac{\partial}{\partial x_j}(\bar{\rho}_l u_{l,j}u_{l,i})$$

$$= -\frac{1}{\theta}\sum_k n^{(k)}\dot{m}^{(k)}u_{l,i}^{(k)} + \frac{1-\theta}{\theta}u_i\sum_k n^{(k)}\dot{m}^{(k)} + \frac{1}{\theta}F_{Di} + [\bar{\rho}_l - \rho(1-\theta)]g_i$$

$$+ \rho(1-\theta)\left[\frac{\partial u_i}{\partial t} + u_j\frac{\partial u_i}{\partial x_j}\right]. \quad (7.77)$$

Actually, it is preferable to avoid Eqs. (7.76) and to use a separate momentum equation for each class of droplets. In particular, we have

$$\frac{\partial}{\partial t}\left[\bar{\rho}_l^{(k)}u_{l,i}^{(k)}\right] + \frac{\partial}{\partial x_j}\left[\bar{\rho}_l^{(k)}u_{l,j}^{(k)}u_{l,i}^{(k)}\right]$$

$$= -n^{(k)}\dot{m}^{(k)}u_{l,i}^{(k)} + n^{(k)}F_{Di}^{(k)} + \left[\bar{\rho}_l^{(k)} - \rho\psi^{(k)}\right]g_i + \rho\psi^{(k)}\left[\frac{\partial u_i}{\partial t} + u_j\frac{\partial u_i}{\partial x_j}\right], \quad (7.77)$$

or, combining Eq. (7.77) with Eq. (7.68), we obtain

$$\bar{\rho}_l^{(k)}\left[\frac{\partial u_{l,i}^{(k)}}{\partial t} + u_{l,j}^{(k)}\frac{\partial u_{l,i}^{(k)}}{\partial x_j}\right] = n^{(k)}F_{Di}^{(k)} + \left[\bar{\rho}_l^{(k)} - \rho\psi^{(k)}\right]g_i + \rho\psi^{(k)}\left[\frac{\partial u_i}{\partial t} + u_j\frac{\partial u_i}{\partial x_j}\right].$$
$$(7.78)$$

Note that Eqs. (7.74) and (7.76a) could be added to yield a mixture momentum equation that eliminates all right-hand-side source terms except for the body forces. For more details, see the discussion about the equation of particle motion (7.78) and the aerodynamic forces F_{Di} given in Section 3.5. Also, see that section for a brief discussion of electrostatic forces on charged droplets.

7.2.3 Energy Conservation

The perfect-gas law is taken from Eq. (7.41) to become

$$\theta p = \bar{\rho}RT, \qquad \bar{\rho}e = \bar{\rho}h - \theta p. \tag{7.79}$$

Then energy equation (7.38) can be rewritten as

$$\frac{\partial}{\partial t}(\bar{\rho}h) + \frac{\partial}{\partial x_j}(\bar{\rho}u_jh) - \frac{\partial}{\partial x_j}\left(\lambda\frac{\partial T}{\partial x_j}\right) - \frac{\partial}{\partial x_j}\left(\bar{\rho}D\sum_m h_m\frac{\partial Y_m}{\partial x_j}\right)$$

$$= \theta\frac{dp}{dt} + \Phi + \bar{\rho}\sum_m \dot{w}_m Q_m - \sum_k n^{(k)}\dot{m}^{(k)}L_{\text{eff}}^{(k)} + \sum_k n^{(k)}\dot{m}^{(k)}h_s^{(k)}. \quad (7.80)$$

See Eq. (2.6) for the definition of L_{eff}. It can be noted that

$$h = \int_{T_{\text{ref}}}^{T} c_p\, dT' = \int_{T_{\text{ref}}}^{T}\left[\sum_m Y_m c_{pm}(T')\right]dT'.$$

Furthermore, we can combine continuity equation (7.62) with Eq. (7.80) to obtain a nonconservative form:

$$\bar{\rho}\left[\frac{\partial h}{\partial t} + u_j\frac{\partial h}{\partial x_j}\right] - \frac{\partial}{\partial x_j}\left(\frac{\lambda}{C_p}\frac{\partial h}{\partial x_j}\right) + \frac{\partial}{\partial x_j}\left[\bar{\rho}D(\text{Le}-1)\sum_m h_m\frac{\partial Y_m}{\partial x_j}\right]$$

$$= \theta\frac{dp}{dt} + \Phi + \bar{\rho}\sum_m \dot{w}_m Q_m - \sum_k n^{(k)}\dot{m}^{(k)}\left[h - h_s^{(k)} + L_{\text{eff}}^{(k)}\right]. \quad (7.81)$$

Note that the chemical source terms would be eliminated from Eqs. (7.80) and (7.81) if the enthalpy were redefined to include the summation of the mass fraction times the heat of formation over all species. The effective latent heat of vaporization in

Eq. (7.81) can be determined with the help of Eq. (2.6) plus Eqs. (2.16b), (3.48b), or (3.52a). For low Mach number, the viscous dissipation Φ can be neglected and the Lagrangian time derivative on the right-hand side of Eq. (7.80) or (7.81) can be replaced with a local time derivative.

The liquid-phase temperature will generally vary spatially and temporally within the liquid droplet. A Navier–Stokes solver or some approximate algorithms, as described in Chapters 2 and 3 by Eqs. (2.16), (3.48), or (3.51), can be used to determine the temperature field in the droplet, including the surface temperature. In the special case of a uniform but time-varying liquid temperature in the droplet, an equation for the thermal energy contained in the droplet can be useful. If e_l is the liquid internal energy per unit mass, then $\bar{\rho}e_l$ is the liquid internal energy per unit volume of mixture. In the case in which a spacial variation of temperature occurs in the droplet, e_l could be considered as the average over the droplet. However, an equation for e_l would not be so useful here because the difference between the average value and the surface value is not specified but yet the results are most sensitive to the surface temperature. Viscous dissipation can be neglected in the liquid-phase energy equations. The liquid energy equation can be rewritten for each class of droplets:

$$\frac{\partial}{\partial t}\left[\bar{\rho}_l^{(k)}e_l^{(k)}\right] + \frac{\partial}{\partial x_j}\left[\bar{\rho}_l^{(k)}u_{l,j}^{(k)}e_l^{(k)}\right]$$

$$= n^{(k)}\dot{q}_l^{(k)} - n^{(k)}\dot{m}^{(k)}e_{ls}^{(k)} + \frac{\partial}{\partial x_j}\left[\psi^{(k)}\lambda\frac{\partial T}{\partial x_j}\right] + \frac{\partial}{\partial x_i}\left[r\psi^{(k)}D\sum_m h_m\frac{\partial Y_m}{\partial x_j}\right]. \quad (7.82)$$

The last two terms in Eq. (7.82) are developed in analogous fashion to the last two terms in Eq. (7.77). Equation (7.82) can be combined with Eq. (7.68) to yield an alternative form:

$$\bar{\rho}_l^{(k)}\left[\frac{\partial e_l^{(k)}}{\partial t} + u_{l,j}^{(k)}\frac{\partial e_l^{(k)}}{\partial x_j}\right] = n^{(k)}\dot{m}^{(k)}\left[e_l^{(k)} - e_{ls}^{(k)} + \frac{\dot{q}_l^{(k)}}{\dot{m}^{(k)}}\right] + B^{(k)}, \quad (7.83)$$

where the definitions have been made that $\dot{q}_l^{(k)}$ is the conductive heat flux from the gas interface of the droplet toward its interior, as given by Eq. (2.6) and

$$B^{(k)} \equiv \frac{\partial}{\partial x_j}\left[\psi^{(k)}\lambda\frac{\partial T}{\partial x_j}\right] + \frac{\partial}{\partial x_i}\left[r\psi^{(k)}D\sum_m h_m\frac{\partial Y_m}{\partial x_j}\right].$$

The term $B^{(k)}$ represents the energy flux to the volume occupied by the liquid if gas filled that volume. This is an effect of the nonuniform surroundings that is added to the energy flux to the droplet determined for a uniform ambiance. This nonuniform-field effect has not previously been recognized for a vaporizing droplet. Michaelides and Feng (1994) developed the term for a nonvaporizing sphere without mass transport. They mistakenly claim, however, that the term is analogous to the added (or apparent) mass term in the particle-motion equation. [See Eq. (3.63) and its following discussion, for example.] This is not correct because it appears only because of variations in the free stream or undisturbed conditions. Any effect analogous to the added-mass term would appear through the $\dot{q}_l$ term; it

was missed by Michaelides and Feng because they considered a spherically symmetric heat transport in the gas phase. They do capture an unsteady gas-phase effect
analogous to the history integral of Eq. (3.63) and an effect that is due to the curvature of the nonuniform-temperature field, analogous to the Faxen effect discussed in
Section 3.5. Nonuniform-temperature-field effects are neglected in the discussions
in other chapters. We can expect them to be negligible when $\rho \ll \rho_l$ and when the
temperature-field curvature is small. The unsteady gas-phase effects are discussed
in Chapter 5 but they are not placed in an integral form. Note that, during droplet
heating, we expect that $e_l^{(k)} - e_{ls}^{(k)}$ is negative (maximum liquid temperature is at the
interface) or zero so that some of the conductive heat flux must be used to heat the
liquid above the average droplet temperature before it is vaporized.

7.2.4 Hyperbolic Character of Liquid-Phase Equations

The liquid-phase equations form a hyperbolic subsystem of partial differential equations. In particular, we define the operation

$$\frac{d^{(k)}}{dt} = \frac{\partial}{\partial t} + u_{l,j}^{(k)} \frac{\partial}{\partial x_j}.$$

This is a Lagrangian time derivative following an average droplet in the kth droplet
class. Then mass conservation states that the droplet radius is described by

$$\frac{d^{(k)} R^{(k)}}{dt} = -\frac{\dot{m}^{(k)}}{4\pi \rho_l^{(k)} \left[R^{(k)} \right]^2}. \tag{7.84}$$

The droplet position is governed by

$$\frac{d^{(k)} x_{l,i}^{(k)}}{dt} = u_{l,i}^{(k)}, \tag{7.85}$$

and the droplet velocity is given by

$$\bar{\rho}_l^{(k)} \frac{d^{(k)} u_{l,i}^{(k)}}{dt} = n^{(k)} F_{Di}^{(k)} + \left[\bar{\rho}_l^{(k)} - \rho \psi^{(k)} \right] g_i + \rho \psi^{(k)} \frac{du_i}{dt}. \tag{7.86}$$

Energy equation (7.83) can be written as

$$\bar{\rho}_l^{(k)} \frac{de_l^{(k)}}{dt} = n^{(k)} \dot{m}^{(k)} \left[e_l^{(k)} - e_{ls}^{(k)} + \frac{\dot{q}_l^{(k)}}{\dot{m}^{(k)}} \right] + B^{(k)}. \tag{7.87}$$

Equations (7.86) and (7.87) are the characteristic equations for hyperbolic partial differential equations (7.78) and (7.81). Equation (7.85) describes the characteristic lines through space for those hyperbolic equations. The Lagrangian method for
the averaged liquid properties is therefore fully equivalent to the method of characteristics, as explained by Sirignano (1986, 1993a). The Dukowicz (1980) analysis
is actually a method of characteristic approach to the multicontinua formulation
rather than the type of discrete-particle formulation discussed in Section 7.3. This
method of characteristics allows the partial differential equations to be converted to
ordinary differential equations that are solved more readily with a reduction in the

numerical error that is due to integration. Specifically, the solution of the partial differential equations can produce artificial diffusion through the difference equation
approximation; this type of error does not occur with the differencing of the ordinary differential equations.

These equations are solved simultaneously with gas-phase equations (7.62),
(7.72) or (7.73), (7.74), or (7.75), and (7.80) or (7.81) together with equations of state
(7.79). These gas-phase equations are typically solved by finite-difference methods
on an Eulerian mesh that is either fixed or defined through some adaptive grid
scheme. The terms representing exchanges of mass, momentum, and energy between the two phases appear as source and sink terms in both the gas-phase and
the liquid-phase equations. They provide the mathematical coupling between the
two subsystems of equations representing the two phases. The liquid Lagrangian
equations are effectively solved on a different grid from that of the gas-phase equations so that interpolation is continually used in the evaluation of the source and sink
terms. Care must be taken in avoiding other numerical errors of the same order as
that of the artificial diffusion that has been eliminated; these other errors can be created by a low-order interpolation scheme. Details about the method and analyses of
its performance can be found in the works of Aggarwal et al. (1981, 1983, 1985).

The Lagrangian or characteristic path is the trajectory of an average droplet that
represents a group of droplets that (at least initially) is in the same neighborhood.
The droplet number density can vary along this path but the number of droplets
represented remains fixed (in the absence of droplet shattering or coalescence).

In this two-continua method, two distinct characteristic lines (from the same or
different droplet classes) can intersect without implication of a collision. The characteristic lines or trajectories represent the paths of average droplets; each average
droplet represents a beam or moving cloud of droplets. If average spacing between
neighboring droplets is sufficiently large, two beams can pass through each other
without any collisions. An important implication here is that multivalued liquid
properties can occur; at a given point in space and time, more than one value of
a liquid property for a given class of droplets can exist. Note that the direct solution of hyperbolic partial differential equations for the liquid properties will smear
any multivalued solutions on numerical integration. In this regard, the Lagrangian
method is substantially more powerful and reliable.

In the Lagrangian method, we bypass the use of Eq. (7.70) for the droplet number conservation. We must still evaluate the droplet number density that appears
in source and sink terms in Eqs. (7.62), (7.63), (7.65), (7.68), (7.69), (7.72)–(7.78),
(7.80), and (7.81)–(7.83). Typically, we determine number density locally in a two-
continua finite-difference computation by (i) determining the number of average
droplets for each droplet class in each computational cell at each temporal point
in the calculation, (ii) determining, based on the initial condition for each average
droplet, the total number of actual droplets of each class present in the computational cell, and (iii) dividing this total number by the volume of the computational
cell to obtain the number density.

The gas-phase and the liquid-phase equations described in this section can be
applied to turbulent flows, but first, Reynolds averaging or Favre averaging of the
equations should be performed. Also, some approximations are required for closing

the equations. As a result of this process, averaged forms of the previous equations plus additional equations for the turbulent kinetic energy and other second-order quantities result. See the discussion in Chapter 10 for details on these turbulent spray analyses.

Initial conditions are required for the solution of the unsteady form of the equations. Often, the unsteady form is used even when we wish to obtain the steady-state solution. For example, if the system of differential equations (7.62), (7.72), (7.74), (7.80), and (7.84)–(7.87) were to be solved, initial conditions would be required for the gas-phase velocity, density, mass fractions, and enthalpy and for the velocities, positions, radii, and thermal energies of the average droplets in each class.

Boundary conditions are also required. Equations (7.72), (7.74), and (7.80) have elliptic operators so that, for velocity, mass fractions, and enthalpy, conditions on the quantities or their gradients are required for every boundary point. Equation (7.62) is a first-order hyperbolic equation so that density needs to be specified at only the inflow boundaries. Boundary conditions on Eqs. (7.84)–(7.87) for the average droplet properties are required only for inflow. When the droplet passes through an open boundary, we discontinue the calculation for that particular droplet. When a droplet strikes a solid wall, there are several options: (i) the wall can be assumed to be so cold that the droplet sticks to the wall and no further vaporization occurs (effectively removing the droplet from the field of computation), (ii) the wall is so hot that the droplet immediately vaporizes, providing a local gaseous mass source, and (iii) the droplet rebounds, perhaps shattering into a number of small droplets. Clearly, the last condition is the most difficult to implement in a calculation. Naber and Reitz (1988) claimed that better agreement with experimental data can be obtained for some calculations if the droplet is assumed to flow along the wall with the local-gas velocity near the wall.

One valid and interesting computational approach often found in the literature is the sectional approach; see Tambour (1984) and Greenberg et al. (1993). It uses an Eulerian representation for the droplet properties. This method classifies droplets according to their instantaneous size. As an individual droplet vaporizes or condenses, it will move from one section to another. A system of hyperbolic partial differential equations applies for each section. These equations have sources and sinks representing the passage of droplets from one section to another. The sectional approach can be seen as an approximation to a distribution function formulation for the droplets.

This author is uncomfortable with the sectional approach for two reasons. First, it introduces many source and sink terms that are physically artificial and numerically cumbersome. Second, transient droplet heating and vaporization requires the analysis of an average droplet throughout its history. The movement of this droplet many times from one section to another makes the transient analysis unnecessarily difficult.

7.2.5 Subgrid Models for Heat, Mass, and Momentum Exchange

The use of the two-phase equations requires specific models for exchange of heat, mass, and momentum between the two phases. These models are presented in

Chapters 2, 3, and 4. When the integrations of the two-phase equations of this chapter are being performed, an additional integration of the subgrid model for droplet heating and vaporization is required. In principle, an integration across the gas film surrounding the droplet and an integration across the droplet-liquid interior are required. Equations (2.15), (3.29) and (3.30), (3.34), or (3.62) offer the advantage of an integrated form for the gas-film representation. Equations (2.6) and (2.14) allow us to relate $\dot{q}_l$ to Nu and $\dot{m}$. Additional integration is required for the liquid-phase properties when spatial variations through the droplet interior are to be resolved. Typically, droplet drag coefficients are used to represent the momentum exchange between phases. See Chapters 3 and 6 for further details on drag coefficients.

Often, we bypass the global droplet energy equation as expressed by Eq. (7.82), Eq. (7.83), or Eq. (7.87) when we resolve the temperature profile (spatially and temporally) within the droplet. Then an energy equation of the form of Eq. (2.16), Eq. (2.70), or Eq. (2.73) is used, depending on the selected internal liquid-heating models. In the limit of infinite conductivity, each of those models is identical to Eq. (7.82) and its derivative equations with $e_l = e_{ls}$. When the models of Chapter 2 with partial differential equations are used, the matching condition at the droplet–gas interface is given by Eq. (2.16b), Eq. (3.48b), or Eq. (3.52a). Of course, the temperature is assumed continuous across the interface. When $e_l \neq e_{ls}$, further information would be required about internal droplet heat transfer in order to close the system of equations. Typically, either infinite liquid conductivity is assumed or a partial differential energy equation for the liquid is solved.

When multicomponent liquids are present, we must account for varying composition within the droplet. Then additional equations to those previously provided are required. Additional subgrid models are required. A liquid-phase mass diffusion equation of the type presented by Eq.(4.5a), Eq. (4.6a), or Eq. (4.8a) is required. Droplet–gas interface conditions of the type of Eqs. (4.1) are needed for each component present in the gas phase. (For those components not also present in the liquid, $\varepsilon_m = 0$.) For the multicomponent case, a gas-film subgrid model can be obtained by use of Eq. (4.4), in place of Eq. (3.33), together with Eqs. (3.34).

7.3 Discrete-Particle Formulation

The two-continua or multicontinua approach allows resolution only on a scale larger than the average spacing between neighboring droplets. Often, however, we must resolve a spray behavior on a finer scale. In combustion applications, ignition and flame structure are examples for which that level of resolution is required. An alternative to the two-continua approach is the discrete-particle approach that serves the purpose of higher resolution. This method follows each individual droplet and resolves the liquid field within each droplet and the gas field surrounding each droplet. Obviously, it is limited to a smaller number of droplets and to a smaller volume of mixture than the two-continua approach. Because the discrete phase is being resolved here, only gas properties or only liquid properties exist at a given point in space or time. The hyperbolic partial differential equations or their characteristic differential equations have no meaning in this approach. Although the droplet trajectories are not characteristic lines in this approach, certain ordinary differential equations are still written along these trajectory lines so that we have a Lagrangian method here, albeit of another type than that previously discussed. Here we do not distinguish droplets by class because, obviously, each droplet forms its own class.

The governing equations for each droplet are

$$\frac{\mathrm{d}R}{\mathrm{d}t} = -\frac{\dot{m}}{4\pi\rho_l R^2}, \tag{7.88}$$

$$\frac{\mathrm{d}x_i}{\mathrm{d}t} = u_{l,i}, \tag{7.89}$$

$$\frac{\mathrm{d}u_{l,i}}{\mathrm{d}t} = \frac{3\tilde{F}_{Di}}{4\pi\rho_l R^3} + g_i\left(1 - \frac{\rho}{\rho_l}\right) + \frac{\rho}{\rho_l}\frac{\mathrm{d}u_i}{\mathrm{d}t}, \tag{7.90}$$

$$\frac{\mathrm{d}e_l}{\mathrm{d}t} = \frac{3\dot{m}}{4\pi\rho_l R^3}\left(e_l - e_{ls} + \frac{\dot{q}_l}{\dot{m}}\right) + \frac{\sum_k B^{(k)}}{\rho_l(1-\theta)}. \tag{7.91}$$

Equation (7.91) is often replaced with the energy equation for the liquid that governs diffusion and convection in the droplet. The comments following Eq. (7.87) about the use of models for droplet heating and for multicomponent liquid vaporization also apply here. The functions for mass vaporization rate $\dot{m}$, drag force $\tilde{F}_{Di}$, and droplet-heating flux $\dot{q}_l$ can be found in Chapters 2 and 3.

The gas-phase equations must be solved simultaneously with the preceding droplet equations. The discrete-particle approach is sensible only when the resolution is finer than the spacing between neighboring droplets. Therefore the computational mesh size must be significantly smaller (by at least an order of magnitude) than the spacing. In the dilute (or nondense) sprays, this interdroplet spacing is also much greater than the average droplet diameter, so therefore we expect the computational mesh size to be comparable to or greater than the droplet diameter. Here we expect that most computational cells will not contain a droplet but only gas; only a small fraction of the cells will contain a droplet, and typically only one droplet each, as average spacing is so large. Equations (7.62), (7.72) or (7.73), (7.74) or (7.75), (7.79), and (7.80) or (7.81) describe the gas phase and can be placed into finite-difference form. Here, $\theta = 1$ and $\bar{\rho} = \rho$ are taken. Also, the source terms representing exchanges of mass, momentum, and energy between phases go to zero in those equations for these cells. This is definitely accurate for the great majority of cells that do not contain any droplet. In the cells containing droplets, we can allow θ to deviate from unity to account for the droplet volume. Whenever the droplet diameter is small compared with the mesh size, θ can be assumed to be unity, neglecting droplet volume from the perspective of the coupling with the gas phase.

7.4 Probabilistic Formulation

It is sometimes convenient to use a probabilistic formulation for the spray analysis. Whenever we have spatial resolution on a scale comparable with or smaller than the average spacing between droplets, a given computation cell has a significant probability of not containing any droplet at any particular instant of time. Furthermore, in dealing with a typically large number of droplets in a spray, there is no practical way to know exactly where each droplet is located at each instant; a probabilistic formulation is practical therefore in a high-resolution analysis.

A probability density function can be defined as $f(t, x_i, R, u_{li}, e_l)$, and an infinitesimal hypervolume in eight-dimensional space is given by

$$dV = dx_1 \, dx_2 \, dx_3 \, du_{l1} \, du_{l2} \, du_{l3} \, dR \, de_l.$$

Then $f \, dV$ is the probability of finding a droplet in the hypervolume dV at a particular instant of time. Note that, for a very fine resolution, $f \, dV$ is less than unity. However, for a very coarse resolution, that product can become much larger than unity because many droplets can be in the hypervolume; in that case, f is more commonly called a distribution function. Williams (1985) discusses the distribution function. Sirignano (1986, 1993a) extends the independent variable space, relates the probability density function to the distribution function, and separates droplets by class according to initial values. Recent work by Archambault et al. (2003a, 2003b) extends the probability density function approach of Williams (1962, 1985). It relies on multirealization ensembles, thereby not unifying LES and two-phase-flow averaging processes. Also, it neglects the effects of heat and mass exchanges between the phases.

We can separate the droplets into distinct classes, depending on their initial size, velocity, or composition. Then a probability density function or a distribution function can be defined for each class of the droplets. Conservation of droplet numbers (neglecting shattering or coalescence) leads to the following equation, which governs the probability density function $f^{(k)}$ for the kth class of droplets:

$$\frac{\partial f^{(k)}}{\partial t} + \frac{\partial}{\partial x_j}\left[u_{l,j} f^{(k)}\right] + \frac{\partial}{\partial u_{l,j}}\left[a_{l,j} f^{(k)}\right] + \frac{\partial}{\partial R}\left[\dot{R} f^{(k)}\right] + \frac{\partial}{\partial e_l}\left[\dot{e}_l f^{(k)}\right] = 0,$$

(7.92)

where

$$a_{l,i} \equiv \frac{\tilde{F}_{Di}}{\frac{4}{3}\pi \rho_l R^3} + g_i\left(1 - \frac{\rho}{\rho_l}\right) + \frac{\rho}{\rho_l}\frac{du_i}{dt}$$

is the droplet acceleration. Note that Eq. (7.92) is equivalent to the statement that $f \, dV$ is a constant.

The characteristics of hyperbolic partial differential equation (7.92) are given by

$$\frac{d^{(k)} x_i}{dt} = u_{l,i}, \qquad \frac{d^{(k)} u_{l,i}}{dt} = a_{l,i}, \qquad \frac{d^{(k)} R}{dt} = \dot{R} = -\frac{\dot{m}}{4\pi \rho_l R^2},$$

$$\frac{d^{(k)} e_l}{dt} = \dot{e}_l = \frac{\dot{m}}{\frac{4\pi}{3}\rho_l R^3}\left[e_l - e_{ls} + \frac{\dot{q}}{\dot{m}}\right] + \frac{B^{(k)}}{\rho_l \psi^{(k)}}.$$

(7.93)

These characteristics define particle paths or probable particle paths for each class of droplet.

Integration of the distribution function gives the averaged values used in the multicontinua approach. The droplet number density $n^{(k)}$ is given by the five-dimensional integral

$$n^{(k)} = \int f^{(k)} \, du_{l,i} \, dR \, de_l. \tag{7.94}$$

Any of the variables $\varphi = u_{l,i}, a_{l,i}, R, \dot{R}, e_l$, or $\dot{e}_l$ can be integrated to give the average quantity

$$\varphi^{(k)} = \left[\int f^{(k)} \varphi \, du_{l,i} \, dR \, de_l \right] \Big/ n^{(k)}, \tag{7.95}$$

where $\varphi^{(k)}$ represents $u_{l,i}^{(k)}, a_{l,i}^{(k)}, R^{(k)}, \dot{R}^{(k)}, e_l^{(k)}$, or $\dot{e}_l^{(k)}$.

Integration of the divergence form of equation governing $f^{(k)}$, namely Eq. (7.92), leads to Eq. (7.70) for conservation of the droplet number that governs droplet number density. Note that, because both $u_{l,i}$ and x_i are independent variables in Eq. (7.92), the second term in that equation can be simplified. In particular, within that term, u_{li} can be brought forward of the derivative sign.

Equation (7.92) can be multiplied by φ (term by term), rearranged, and integrated over the five-dimensional volume. Here it is assumed that $f^{(k)}$ goes to zero as $u_{l,i}$ or R becomes infinite. Also, $\dot{R}$ becomes zero when $R = 0$. Finally, with respect to the averaged quantities defined by Eq. (7.95), it is assumed that the average of a product of two quantities equals the product of the averages of the two quantities. The error in this assumption reduces as the band of the independent variables $(u_{l,i}, R, e_l)$ for a nonzero probability density function in a given class k is reduced. When $\varphi = 1$, Eq. (7.70) is reproduced by this process, and when $\varphi = u_{l,i}$, Eq. (7.78) is reproduced. With modest effort, use of $\varphi = R$ and $\varphi = e_l$ can yield Eqs. (7.63) and (7.83), respectively.

The probabilistic formulation or, equivalently, the distribution function formulation is of primary value because it strengthens the theoretical foundations of the spray equations. In the case in which coarse resolution is sought, both gas-phase and liquid-phase properties are averaged over a neighborhood containing many droplets. Then any uncertainty associated with the precise locations of individual droplets is removed from the final analytical form by the averaging process. In the fine-resolution case, uncertainty in droplet position remains a factor; the probability density function describes this uncertainty. There is also an uncertainty in the gas-phase values here on account of the coupling between the phases. For very dilute sprays with few or no droplets in a given neighborhood, the uncertainty in the gas-phase properties can be neglected. In denser sprays, these uncertainties in the gas-phase properties appear in high-resolution practical spray problems; theory has not yet fully addressed this interesting problem. As explained in Section 9.3, Aggarwal and Sirignano (1985a) discuss the uncertainty in ignition delay associated with the uncertain distance of the droplet nearest to the ignition source.

8 Computational Issues

There are some important computational issues associated with spray computations. One issue concerns improved computation for droplet-heating and -vaporization subgrid models that are time consuming in spray calculations because of numerical stiffness introduced through the source terms. Another issue relates to the optimization of the finite-difference schemes used in the computation of the spray equation. Third, numerical errors associated with the grid interpolation necessary to account for mass, momentum, and energy exchanges between the phases can be important. Discussion of these three general issues are provided in the three sections of this chapter.

8.1 Efficient Algorithms for Droplet Computations

Some attention has been given to computational improvement of the droplet-heating and -vaporization models discussed in Chapters 2 and 3. Tong and Sirignano (1986a) have shown that Eq. (3.48a), the liquid-phase heat diffusion equation, is approximately

$$\frac{\partial T_l}{\partial \tau} = \phi \frac{\partial^2 T_l}{\partial \phi^2} + (1 + C\phi)\frac{\partial T_l}{\partial \phi}, \tag{8.1}$$

where ϕ is the normalized stream function and T_l is the liquid temperature. Note that

$$d\tau = \frac{b_1}{b_0}\frac{\alpha_l}{R^2}\,dt, \qquad C = C(\tau) = 2\left(\frac{R_0}{R}\right)^{3/2}\frac{d}{d\tau}\left(\frac{R}{R_0}\right)^{3/2}. \tag{8.2}$$

To obtain the form of Eq. (8.1), the definition of the nondimensional time has been modified from that of Eq. (3.48). Furthermore, an approximate behavior has been used for the functions $g_1(\phi)$ and $g_2(\phi)$ that appear in Eq. (3.48a). The boundary conditions are given as

$$\text{(i) at } \tau = 0, \quad T_l = 0,$$

$$\text{(ii) at } \phi = 0, \quad \frac{\partial T_l}{\partial \phi} = \frac{\partial T_l}{\partial \tau},$$

$$\text{(iii) at } \phi = 1, \quad \frac{\partial T_l}{\partial \phi} = q_l. \tag{8.3}$$

In the preceding equations, t is the time variable, α_l is the thermal diffusivity, R is the droplet radius, R_0 is the initial radius, q_l is the normalized heat flux at the surface, and b_1 and b_0 are constants.

On account of C, Eq. (8.1) is nonlinear. However, C will be considered as slowly varying compared with T_l and will be taken as a piecewise constant in our integration calculation. The approximately linear differential operator allows us to develop a Green's function and reduce the solution to an integral form whereby, by means of quadrature, the temperature at any point can be determined as a function of the heat flux at the liquid surface. In particular, an integral equation can result that relates the surface temperature to the surface flux. The Green's function or kernel function is given by

$$K(\tau, \phi) = \frac{\tau + F_s(0, C_0, \phi)}{G_s(0, C_0, 1)} - \frac{1}{2} \frac{G_{ss}(0, C_0, 1)}{G_s^2(0, C_0, 1)} + \sum_{n=1}^{\infty} \frac{F(\lambda_n, C_0, \phi)}{\lambda_n G_s(\lambda_n, C_0, 1)} e^{\lambda_n \tau}, \tag{8.4}$$

where F and G are analytic functions of ϕ:

$$F(S, C, \phi) = \sum_{k=1}^{\infty} \frac{S(S-C)[S-(k-1)C]\phi^k}{(k!)^2},$$

$$G(S, C, \phi) = \sum_{k=1}^{\infty} \frac{S(S-C)[S-(k-1)C]\phi^{k-1}}{k[(k-1)!]^2},$$

$$F_s(0, C, \phi) = \sum_{k=0}^{\infty} \frac{(-1)^k C^k \phi^{(k+1)}}{(k+1)(k+1)!},$$

$$G_s(0, C, \phi) = \sum_{k=0}^{\infty} \frac{(-1)^k C^k \phi^k}{(k+1)!},$$

$$G_{ss}(0, C, \phi) = \sum_{k=1}^{\infty} \frac{2(-1)^{(k-1)} a_k C^{(k-1)} \phi^k}{(k+1)(k!)^2}.$$

The eigenvalues λ_n are determined by

$$G(\lambda_n, C_0, 1) = 0. \tag{8.5}$$

Now the temperature anywhere in the liquid may be determined as a function of the surface heat flux:

$$T_l(\tau, \phi) = \int_0^{\tau} f(x) \frac{\partial K(\tau - x, \phi)}{\partial \tau} \, dx, \tag{8.6}$$

where

$$f(\tau) = \frac{\partial T_l}{\partial \phi} \quad \text{at } \phi = 1.$$

In particular, at the surface, $\phi = 1$, we find that

$$T_s = \alpha \int_0^\tau f(x)\, dx + \sum_{n=1}^\infty \beta_n \int_0^\tau f(x) e^{\lambda_n(\tau - x)}\, dx, \tag{8.7}$$

where

$$\alpha = \frac{1}{G_s(0, C_0, 1)},$$

$$\beta_n = \frac{F(\lambda_n, C_0, 1)}{G_s(\lambda_n, C_0, 1)}.$$

From the gas-phase analysis, we also have a relationship between the surface temperature and B', where B' does depend on both surface temperature and surface heat flux. This allows for complete determination of the surface temperature and heat flux on coupling with the liquid-phase solution. The solutions are numerical and time dependent.

The integral equation at the surface can be replaced by the following system of ordinary differential equations (ODEs). First, we define

$$T_s = \sum_{n=0}^\infty g_n(\tau),$$

where

$$g_n(\tau) = \beta_n \int_0^\tau f(x) e^{\lambda_n(\tau - x)} dx, \quad n \geq 1.$$

Then it follows that

$$\frac{dg_0}{d\tau} = \alpha f(\tau), \tag{8.8}$$

$$\frac{dg_n}{d\tau} = \lambda_n g_n(\tau) + \beta_n f(\tau); \quad n \geq 1. \tag{8.9}$$

Note that, by definition,

$$g_n(0) = 0 \quad \text{for all } n \geq 0.$$

The independent time variable in Eqs. (8.8) and (8.9) is a Lagrangian time so that an advantage of these equations is that they can be solved as we follow the vaporizing droplets through the spray combustor. Less computer time is required for the ODE system solution than for the direct numerical solution (DNS) of Eq. (8.1). The improved efficiency measured in CPU (computer central processing unit) time occurs at any desired accuracy. See Tong and Sirignano (1986b).

It is necessary to solve mass diffusion equations for $n - 1$ of the n components in the liquid phase. Species-conservation equations for gas-phase fuel-vapor components must also be solved. The same type of algorithms can be used here as was used in the solution for the gas and the liquid temperatures. Although interface conditions and diffusivity coefficients differ, the same types of differential operators are present.

The simplified analysis developed for the single-component droplet has been extended to the multicomponent droplet (Tong and Sirignano, 1986a, 1986b). As for the single-component droplet, the resulting coupled formulation for the droplet vaporization can be expressed as a parabolic partial differential equation (PDE) or a Volterra-type integral equation. The integral equation formulation can be transformed into a system of coupled first-order ODEs.

The fractional vaporization rate ε_m and the latent heat of vaporization for a multicomponent liquid are given by Eqs. (4.1b) and (4.2). The phase-equilibrium condition is given by Eq. (4.3a). In the gas-phase analysis, the modified Spalding transfer number B_m is given by Eq. (4.4).

For the interface conditions, the liquid-phase heat flux at the droplet surface is given by

$$\lambda_l \frac{\partial T_l}{\partial y}\Bigg)_s = \rho v (L_{\text{eff}} - L). \tag{8.10}$$

Similarly, the liquid-phase concentration gradient for species m at the droplet surface is given by

$$\frac{\partial Y_{m,l}}{\partial y}\Bigg)_s = \mathrm{Le}\frac{c_{p,l}}{\lambda_l}(Y_{m,ls} - \varepsilon_m)\rho v. \tag{8.11}$$

Note that mass diffusivity is taken as identical for each species. For the liquid-phase analysis, the unsteady heat and mass diffusion equations are, respectively, Eq. (8.1) and

$$\frac{\partial Y_{m,l}}{\partial \tau} = \frac{\phi}{\mathrm{Le}}\frac{\partial^2 Y_{m,l}}{\partial \phi^2} + \left(\frac{1}{\mathrm{Le}} + C\phi\right)\frac{\partial Y_{m,l}}{\partial \phi}, \tag{8.12}$$

where C and τ are defined in Eqs. (8.2). The initial and the boundary conditions for the temperature remain the same as in Eqs. (8.3). For the mass concentration, the initial and the boundary conditions are

$$\text{at } \tau = 0, \quad Y_{m,l} = Y_{m,l0},$$

$$\text{at } \phi = 0, \quad \frac{\partial Y_{m,l}}{\partial \phi} = \mathrm{Le}\frac{\partial Y_{m,l}}{\partial \tau},$$

$$\text{at } \phi = 1, \quad \frac{\partial Y_{m,l}}{\partial \phi} = f_m. \tag{8.13}$$

Here f_m is related to the liquid-phase concentration gradient $\partial Y_{m,l}/\partial y)_s$ as given by Eq. (8.11) and $Y_{m,l0}$ is the initial liquid-phase concentration of species m.

At any given time instant, with known droplet-surface temperature T_s and liquid-phase species mass fractions $Y_{m,l0}$ at the surface, the gas-phase species mass fractions at the droplet surface $Y_{m,gs}$ can be obtained by means of Raoult's Law [see Eq. (4.3a)].

Subsequently, B and ε_m can be obtained from relations presented just before Eq. (4.4), and the Blasius function discussed in Subsection 3.2.3 can be calculated. The vaporization rate ρv can be obtained as in the single-component droplet case.

Equations (8.10) and (8.11) then give the liquid-phase heat flux and species mass flux at the droplet surface. The coefficient $C(\tau)$ can be calculated from the gas-phase solution through Eqs. (8.2). Next the interior droplet temperature and species-concentration distribution can be obtained by the solution of the diffusion equations, Eqs. (8.1) and (8.12), in conjunction with the boundary conditions, Eqs. (8.3) and (8.13), respectively. The new droplet-surface temperature and the new liquid-phase mass fractions at the droplet surface are used for the gas-phase solution for the next time step. The same solution scheme is repeated at each time increment, and the solution is marched forward in time.

As in the single-component droplet formulation, the cases $C = 0$ and $C = C_0 =$ constant are considered here for the bicomponent droplet. The series solutions for the droplet temperature and the ODE formulations derived for the single-component droplet remain valid for the bicomponent droplet.

For the mass diffusion equation, it can be shown that with a time transformation the series solution for the temperature can be extended to yield a series solution for the corresponding mass diffusion equation. Again, when these series solutions are used as a Green's function in a superposition scheme, solutions for the case of a time-varying boundary condition can be obtained in integral equation form. Subsequently, when some new functions are defined, the integral equation can be transformed into a system of ODEs. This procedure, which is similar to that in the single-component droplet case, is subsequently detailed.

For $C = C_0$, the mass diffusion equation is, from Eq. (8.12), given by

$$\frac{\partial Y_{m,l}}{\partial \tau} = \frac{\phi}{\text{Le}} \frac{\partial^2 Y_{m,l}}{\partial \phi^2} + \left(\frac{1}{\text{Le}} + C_0 \phi\right) \frac{\partial Y_{m,l}}{\partial \phi} \tag{8.14}$$

together with the initial and the boundary conditions given by Eqs. (8.13).

By defining

$$\tau^* = (\tau/\text{Le}), \tag{8.15}$$

$$Y_{m,l}^* = Y_{m,l0} - Y_{m,l}, \tag{8.16}$$

$$C_0^* = C_0 \text{Le}, \tag{8.17}$$

we find that Eqs. (8.14) and (8.13) become

$$\frac{\partial Y_{m,l}^*}{\partial \tau^*} = \phi \frac{\partial^2 Y_{m,l}^*}{\partial \phi^2} + \left(1 + C_0^* \phi\right) \frac{\partial Y_{m,l}^*}{\partial \phi}, \tag{8.18}$$

with

(i) at $\tau^* = 0$, $Y_{m,l}^* = 0$,

(ii) at $\phi = 0$, $\dfrac{\partial Y_{m,l}^*}{\partial \tau^*} = \dfrac{\partial Y_{m,l}^*}{\partial \phi}$,

(iii) at $\phi = 1$, $\dfrac{\partial Y_{m,l}^*}{\partial \phi} = f_m^*(\tau^*)$, \tag{8.19}

where

$$f_m^*(\tau^*) = f_m(\tau). \tag{8.20}$$

Equations (8.18) and (8.19) resemble Eqs. (3.48) and (8.3), except for the last boundary condition. The solution $K^*(\tau, \phi)$ for a unity species gradient boundary condition for which $\partial Y^*_{m,l}/\partial \phi = 1$ can be obtained from Eq. (8.1) when τ and C_0 are replaced with τ^* and C_0^*, respectively. Thus,

$$K^*(\tau, \phi) = \frac{\tau^* + F_s(0, C_0^*, \phi)}{G_s(0, C_0^*, 1)} - \frac{1}{2}\frac{G_{ss}(0, C_0^*, 1)}{G_s^2(0, C_0^*, 1)} + \sum_{n=1}^{\infty} \frac{F(\lambda_n, C_0^*, \phi)}{\lambda_n G_s(\lambda_n, C_0^*, 1)}\exp(\lambda_n \tau^*),$$
(8.21)

which becomes

$$K(\tau, \phi) = \frac{\tau/\mathrm{Le} + F_s(0, C_0^*, \phi)}{G_s(0, C_0^*, 1)} - \frac{1}{2}\frac{G_{ss}(0, C_0^*, 1)}{G_s^2(0, C_0^*, 1)}$$

$$+ \sum_{n=1}^{\infty} \frac{F(\lambda_n, C_0^*, \phi)}{\lambda_n G_s(\lambda_n, C_0^*, 1)}\exp\left(\frac{\lambda_n \tau}{\mathrm{Le}}\right).$$
(8.22)

Here the λ_n's are the roots of

$$G(\lambda_n, C_0^*, 1) = 0.$$
(8.23)

Then, the solution for the case in which $\partial Y_{m,l}/\partial \phi = f_m$ at the surface is given by the superposition method as

$$Y^*_{m,l}(\tau, \phi) = -\int_0^{\tau} f_m(x)\frac{\partial K(\tau - x, \phi)}{\partial \tau}\,dx.$$
(8.24)

Substituting Eq. (8.22) into Eq. (8.24) gives

$$Y^*_{m,l}(\tau, \phi) = -C_1 f_m(\tau) - \int_0^{\tau} \frac{f_m(x)}{\mathrm{Le}}\left\{\frac{1}{G_s(0, C_0^*, 1)}\right.$$

$$\left. + \sum_{n=1}^{\infty} \frac{F(\lambda_n, C_0^*, \phi)}{G_s(\lambda_n, C_0^*, 1)}\exp\left[\frac{\lambda_n(\tau - x)}{\mathrm{Le}}\right]\right\}\,dx,$$
(8.25)

Setting $\phi = 1$ in Eq. (8.25) yields the surface-species mass fraction

$$Y^*_{m,ls}(\tau) = \gamma \int_0^{\tau} f_m(x)\,dx + \sum_{n=1}^{\infty}\left\{\sigma_n \int_0^{\tau} f_m(x)\exp\left[\frac{\lambda_n(\tau - x)}{\mathrm{Le}}\right]dx\right\},$$
(8.26)

where

$$\gamma = \frac{-1}{\mathrm{Le}\,G_s(0, C_0^*, 1)},$$
(8.27)

$$\sigma_n = \frac{-F(\lambda_n, C_0^*, 1)}{\mathrm{Le}\,G_s(\lambda_n, C_0^*, 1)}, \quad n \geq 1.$$
(8.28)

With the definitions

$$g_0(\tau) \equiv \gamma \int_0^{\tau} f_m(x)\,dx,$$
(8.29)

$$g_{nm}(\tau) \equiv \sigma_n \int_0^{\tau} f_m(x)\exp\left[\frac{\lambda_n(\tau - x)}{\mathrm{Le}}\right]dx, \quad n \geq 1,$$
(8.30)

Eq. (8.24) becomes

$$Y^*_{m,ls}(\tau) = \sum_{n=0}^{\infty} g_{nm}(\tau),$$ (8.31)

where, at $\tau = 0$,

$$g_n(0) = 0 \quad \text{for all } n \geq 1.$$ (8.32)

Taking the τ derivatives of the g_i's gives

$$\frac{dg_0}{d\tau} = \gamma f(\tau),$$ (8.33)

$$\frac{dg_n}{d\tau} = \frac{\lambda_n}{\text{Le}} g_n(\tau) + \sigma_n f(\tau), \quad n \geq 1.$$ (8.34)

Equations (8.32)–(8.34) now form a system of ODEs that replaces integral equation (8.26).

For the special case $C_0 = 0$ we have the series solution in terms of Bessel functions. It can be shown from Eq. (8.4) that

$$K(\tau, \phi) = \phi - \frac{1}{2} + \frac{\tau}{\text{Le}} - 4 \sum_{n=1}^{\infty} \frac{J_0(\lambda_n \phi^{1/2})}{\lambda_n^2 J_1'(\lambda_n)},$$ (8.35)

where the prime denotes a derivative and

$$J_1(\lambda_n) = 0.$$ (8.36)

Substituting Eq. (8.35) into Eq. (8.24) gives

$$Y^*_{m,l}(\tau, \phi) = -\left[\phi - \frac{1}{2} + \frac{\tau}{\text{Le}} - 4 \sum_{n=1}^{\infty} \frac{J_0(\lambda_n \phi^{1/2})}{\lambda_n^2 J_1'(\lambda_n)} \right] f_m(\tau)$$

$$- \int_0^{\tau} \frac{f_m(x)}{\text{Le}} \left\{ 1 + \sum_{n=1}^{\infty} \frac{J_0(\lambda_n \phi^{1/2})}{J_0(\lambda_n)} \exp\left[\frac{-\lambda_n^2(\tau - x)}{4\text{Le}} \right] \right\} dx.$$ (8.37)

Setting $\phi = 1$ gives

$$Y^*_{m,ls}(\tau) = -\frac{1}{\text{Le}} \int_0^{\tau} f_m(x) \left\{ 1 + \sum_{n=1}^{\infty} \exp\left[\frac{-\lambda_n^2(\tau - x)}{4\text{Le}} \right] \right\} dx.$$ (8.38)

Defining

$$g_{om}(\tau) \equiv \frac{-1}{\text{Le}} \int_0^{\tau} f_m(x) \, dx,$$ (8.39)

$$g_{nm}(\tau) \equiv \frac{-1}{\text{Le}} \int_0^{\tau} f_m(x) \exp\left[\frac{-\lambda_n^2(\tau - x)}{4\text{Le}} \right] dx,$$ (8.40)

we have

$$\frac{dg_{om}}{d\tau} = -\frac{f_m(\tau)}{\text{Le}},$$ (8.41)

$$\frac{dg_{nm}}{d\tau} = -\frac{f_m(\tau)}{\text{Le}} - \frac{\lambda_n^2}{4\text{Le}} g_{nm}(\tau), \quad n \geq 1,$$ (8.42)

as well as Eqs. (8.31) and (8.32).

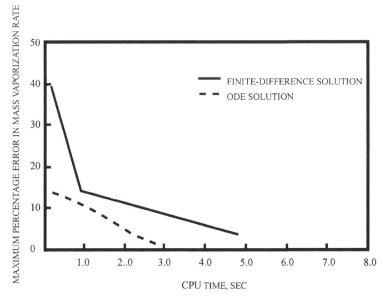

Figure 8.1. Maximum percentage error in mass vaporization rate versus CPU time. (Tong and Sirignano, 1986b, with permission of *Numerical Heat Transfer.*)

Results of the computations can be found in Subsection 4.2.1. As in the single-component droplet case, the results for piecewise constant C are almost identical to those for exact C. The two results are not distinguishable within the accuracy of the plots. The results for $C = \text{constant} = -0.6$ are in closer agreement than those for $C = 0$. The results for $C = 0$ have a substantial error, which suggests that the surface-regression term cannot be neglected in the diffusion equation, particularly for the high Lewis number case.

As the Lewis number increases, the system becomes stiffer and more equations are needed in the ODE solution. The series solutions converge rather slowly for large eigenvalues, and asymptotic series are needed. Although there are no foreseeable difficulties in obtaining the asymptotic series for the large Lewis number case, the results for Le = 10 and 30 were obtained by the finite-difference method. The objective here is to examine the various C approximations.

The computational efficiencies of the finite-difference solution and the ODE solution were studied. The case $C = 0$ and Le = 30 is considered. Recall that the solutions for $C = 0$ are in terms of Bessel functions, whose asymptotic expansions are known. The number of equations needed for accurate representation in the ODE solution depends mainly on the stiffness of the problem (governed by the Lewis number) but only very weakly on C, the surface-regression parameter. In other words, for a given Lewis number one should expect to use the same number of equations with the various C approximation schemes for the same accuracy. The finite-difference solution with 100 grid points is considered to be exact and is used as reference in calculating the percentage numerical error. Figure 8.1 shows the maximum percentage numerical error in overall mass vaporization rate versus CPU time.

The ODE solution curve appears to lie below the finite-difference solution curve, which indicates that the former solution scheme is more efficient.

As previously mentioned, only the droplet-surface temperature and the liquid-surface species concentration are needed for determining the vaporization rate. The interior droplet temperature and species concentration are not needed for determining the heat and mass fluxes to the droplet and the vaporization rate. The advantage of the ODE formulation is that it eliminates the spatial dependence from the problem and results in reduced computation time.

There are certain shortcomings with the approaches of Tong and Sirignano (1982a, 1982b, 1983, 1986a, 1986b), Prakash and Sirignano (1978, 1980), and Lara-Urbaneja and Sirignano (1981), as previously discussed. First, constant transport and thermal properties are used. Second, the analysis is strictly correct only when the Reynolds number based on relative velocity is sufficiently large to have a laminar boundary layer. Abramzon and Sirignano (1989) have an approach intended to cover a wider Reynolds number range and to treat variable properties. A discussion of that method for calculating droplet-vaporization rates is given in Chapter 3.

8.2 Numerical Schemes and Optimization for Spray Computations

Certain studies have been performed to address the issue of optimizing the numerical calculations for sprays. In this section we examine axisymmetric sprays. Laminar, nonreacting situations are considered. Turbulence and chemical reactions would modify characteristic length and time scales, thereby affecting the optimal numerical scheme. Two problems are considered: (i) an axisymmetric, steady jet flow and (ii) a model unsteady, axisymmetric confined flow.

In both cases, an Eulerian mesh is used for the parabolic gas-phase equations. Note that, in the model problem, the spatial differential operator is elliptic. A Lagrangian scheme is used for the hyperbolic equations describing the vaporizing-liquid phase. All hydrodynamic and thermal interactions between the two phases are considered. Generally, with one exception, consistent second-order-accurate numerical schemes are considered.

In the first problem, the subset of gas-phase equations has been attacked by four numerical methods: a predictor–corrector-explicit method, a sequential-implicit method, a block-implicit method, and a symmetric operator-splitting method. The computations predict the hydrodynamics and transport in the gas phase as well as the droplet trajectories. The size, temperature, and velocity of each droplet group are also predicted. At low error tolerances, the sequential-implicit method gives the best results; at large error tolerances, the explicit and the operator-splitting methods give better results. The block-implicit scheme is least effective at all accuracies. Details are discussed in Subsection 8.2.1. Further information is found in the work of Aggarwal et al. (1985).

In the second problem, a set of model equations is studied but the technique applies very well to a more general and more physically accurate set of equations as well. The integration scheme and scheme for interpolation between the Lagrangian and the Eulerian mesh are demonstrated to be second-order accurate. Effects of mesh size, number of droplet characteristics, time step, and the injection pulse time are determined by means of a parameter study. The results indicate slightly more sensitivity to grid spacing than to the number of droplet characteristics. More detailed information is given in Subsections 8.2.2. Still further information is given by Aggarwal et al. (1983).

8.2.1 Two-Phase Laminar Axisymmetric Jet Flow

There are two basic objectives for the present study. First, it is intended to develop a set of consistent second-order-accurate hybrid numerical schemes for a flow of an air, fuel-vapor, and fuel-spray mixture. In this scheme, the gas-phase properties are computed by an Eulerian approach, whereas the condensed-phase properties are obtained by a Lagrangian (or method of characteristics) approach. The second objective is to compare the various finite-difference schemes in terms of stability, accuracy, and efficiency. For the gas-phase equations, a predictor–corrector-explicit method (MacCormack, 1969), a sequential-implicit method, a block-implicit method (Briley and McDonald, 1977), and a symmetric operator-splitting method are examined. For the condensed phase, a second-order Runge–Kutta scheme is used.

The second part of this study deals with the comparison of four finite-difference schemes for the gas phase, namely the predictor–corrector-explicit, the sequential-implicit, the block-implicit, and the symmetric operator-splitting schemes. The predictor–corrector-explicit method is a variation of the two-step scheme of Mac-Cormack (1969), who used a conservative form of the governing equations. In this section, the method is used for a set of equations in the nonconservative (or convective) form. Note that, in the presence of mass and momentum exchange between the phases, the concept of conservation loses its usual relevance. The sequential-implicit method is based on the Crank–Nicolson implicit scheme (Ames, 1977). In the present case, which involves a set of coupled nonlinear PDEs, the equations are solved sequentially, and the nonlinearities are handled through iterations. The block-implicit method uses a linearization, as suggested by Briley and McDonald (1977). The second-order accuracy is maintained by use of the Crank–Nicolson differencing.

Because of the different time scales present in the problem, it was originally thought that a split-operator approach (Yanenko, 1971) would be effective. A second-order-accurate symmetric two-step formulation was used. This is similar to the one used by Rizzi and Bailey (1975) for chemically reacting three-dimensional flows and by Kee and Miller (1978) for one-phase diffusion flames.

The hexane fuel was used in these calculations because it is quite volatile and thus gives the maximum amount of stiffness that would be encountered in practice. Accuracy was weighed against efficiency. At the beginning of the study it was thought that the split-operator method would be the preferred approach. However, the results indicate that this is the case only at relatively large error limits. As the error criterion is reduced, the sequential-implicit scheme becomes more efficient. Reasons for this behavior are subsequently discussed in this subsection.

The physical model, shown schematically in Fig. 8.2, consists of an axisymmetric flow of two unconfined concentric jets having uniform but different initial

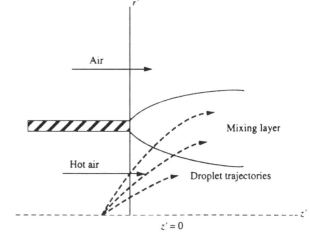

Figure 8.2. Schematic of the two-phase flow. (Aggarwal et al., 1985, with permission of *Numerical Methods in Partial Differential Equations.*)

properties. Thus the initial ($z = 0$) profiles of gas velocity, temperature, and species mass fractions have a discontinuity at the boundary of the inner jet. The radius of the inner jet is taken to be 0.5 cm and that of the outer jet to be 2.0 cm. The initial velocity and temperature in the inner jet are assumed to be 50 cm/s and 1000 K, respectively. The injection of liquid fuel in the form of groups of droplets is simulated by the assumption that a conical droplet flow originates from a point source located at a point ($z < 0$) on the axis of symmetry. Initially all the characteristics, each representing a group of droplets, are positioned at $z = 0$. The radial and the axial components of droplet velocities are obtained from the assumption of conical flow and that, in each conical section associated with given characteristics, the liquid mass flow is the same.

The equations that govern the conservation of mass, momentum, and energy for the gas phase are written in axisymmetric coordinates (r–z). Instead of the Navier–Stokes equations, the boundary-layer equations with an assumption of uniform pressure are used. The equations for a compressible one-phase jet flow are quite standard and can be found in Hornbeck (1973). The presence of the liquid phase introduces the phenomena of exchange of mass, momentum, and energy between the phases. Specifically, the source terms representing these exchange processes need to be included in the gas-phase equations.

The major assumptions used for the gas-phase equations are as follows: The gas-phase flow is assumed to be an axisymmetric steady laminar flow in an unconfined space. The gas-volume displacement effect that is due to the presence of droplets is neglected.

The diffusion of mass, momentum, and heat is governed by Fick's law, Newton's law for viscous stresses, and Fourier's law of heat conduction, respectively. The gas viscosity is inversely proportional to the gas density. The radiative heat transfer is neglected.

The Prandtl number, Schmidt number, and Lewis number are assumed to be unity. The modification to consider nonunity (but order unity) values of these numbers is trivial.

The axial diffusion of mass, momentum, and heat is neglected compared with that in the radial direction. This has been proven to be a realistic assumption in jet flows. The resulting simplification is significant when the solution is developed by a finite-difference approximation (Roache, 1976).

The liquid-phase equations are written following a Lagrangian description. As a result, the initial position and the other droplet properties are prescribed and the ODEs are written to describe the changes of these properties as one follows the droplet trajectories. The assumptions used in writing these equations are as follows: The droplet interaction is neglected, and the droplets are assumed to be spherical. The relationships used to describe the interphase heat and mass transfer rates are based on a quasi-steady analysis of the gas film surrounding the droplet. A vaporizing droplet exists in a hot surrounding without any chemical reaction. The assumption is made that the heat is transferred by conduction from the hot surroundings to the droplet surface and fuel vapor diffuses by molecular diffusion from the droplet surface to the surrounding. The effect of convection is taken into account by a Ranz–Marshall-type correction.

The momentum exchange between the droplets is assumed to be described by a drag law, as suggested by Ingebo (1956).

The temperature in the interior of the droplet is assumed to be uniform, which implies a very high liquid thermal conductivity. This is based on the assumption that the characteristic time for the heat transfer in the droplet interior is much smaller than the characteristic time for the temporal increase of droplet temperature. Thus, for a droplet temperature below the wet-bulb temperature, the heat transfer from the gas phase is used to heat the droplet and to vaporize it.

Solution Procedure

This subsection summarizes the methods of solving the governing equations. The four numerical schemes, which are examined herein, have been selected after substantial numerical experiments with several numerical methods. For example, the numerical results indicated that a Crank–Nicolson implicit formulation is superior to the standard implicit method. Thus the implicit methods with only the Crank–Nicolson differencing are included in the comparison. Similarly, the standard one-step explicit method is inferior to the predictor–corrector-explicit method and is therefore not included in the comparison. Also, the one-step operator-splitting method is discarded in favor of the two-step operator-splitting method.

The four numerical schemes for solving the gas-phase equations, which are discussed, are (1) predictor–corrector-explicit scheme, (2) sequential-implicit scheme, (3) block-implicit scheme, and (4) operator-splitting scheme. The liquid-phase equations are solved by a second-order Runge–Kutta method. Because we have a subset of coupled parabolic (in the z direction) equations and a subset of Lagrangian equations, a marching procedure is used to advance the solution in the z direction. In the discussion that follows, a grid point means a field point in the finite-difference approximation of Eulerian gas-phase equations. A characteristic point means a point along the trajectory of a group of droplets, but at a given axial (z) location. A subscript i indicates a radial location, whereas a superscript k indicates an axial location.

For the discussion of the numerical schemes, the gas-phase equation is written in the form

$$\frac{\partial \phi}{\partial z} = \left(-\frac{V}{U} + \frac{D}{rU} \right) \frac{\partial \phi}{\partial r} + \frac{D}{U} \frac{\partial^2 \phi}{\partial r^2} + \frac{S_\phi}{U}. \tag{8.43}$$

Initial conditions are prescribed (at $z = 0$) as

$$\phi = \phi_A \quad \text{for } 0 < r < 1$$
$$= \phi_B \quad \text{for } 1 \leq r \leq R. \tag{8.44}$$

Boundary conditions were given in the prior discussion.

Explicit Predictor–Corrector Scheme

A predictor–corrector scheme, based on the MacCormack (1969) method, is used. The governing equation can be rewritten as

$$\frac{\partial \phi}{\partial z} = A(\phi, r, z, V) \frac{\partial \phi}{\partial r} + B(\phi, r, s) \frac{\partial^2 \phi}{\partial r^2} + \frac{S_\phi}{U}, \tag{8.45}$$

$$\bar{\phi}_i - \phi_i^k = \alpha_i^k \left[\phi_i^k - \phi_{i-1}^k \right] + \beta_i^k \left[\phi_{i+1}^k - 2\phi_i^k + \phi_{i-1}^k \right] + \left(\frac{S_\phi}{U} \right)_i^k \Delta z, \tag{8.46}$$

where

$$\alpha = A \frac{z}{\Delta r},$$

$$\beta = B \frac{z}{\Delta r}. \tag{8.47}$$

β and S_ϕ / U are updated by use of $\bar{\phi}$ and α, and

$$\phi_i^{k+1} = \frac{1}{2} \left[\phi_i^k + \bar{\phi}_i + \bar{\alpha}_i(\bar{\phi}_{i+1} - \bar{\phi}_i) + \bar{\beta}_i(\bar{\phi}_{i+1} - 2\bar{\phi}_i + \bar{\phi}_{i-1}) + \left(\frac{\bar{S}_\phi}{U} \right)_i \Delta z \right], \tag{8.48}$$

where $\bar{\alpha}$ and $\bar{\beta}$ are the updated values. The general procedure in the predictor–corrector scheme is as follows. Obtain gas-phase properties at the droplet locations by using a linear interpolation. Using these gas-phase properties, evaluate the gas-phase source terms $\{S_{\phi i}^k\}$ at the droplet locations. The source terms are distributed to the neighboring gas-phase locations by a linear approximation. The predicted values of the gas-phase properties are obtained by the solution of the equations for $\bar{\phi}$. These values are used to obtain gas-phase radial velocity from the integral form of the continuity equation by use of the latest values; the gas-phase properties are obtained at the droplet locations. A second-order Runge–Kutta scheme calculates the droplet properties, and the gas-phase source terms are obtained. Now the corrector step gives the corrected values at the forward z location. The droplet properties and the gas-phase source terms are reevaluated. These source terms are used in the next cycle.

Sequential-Implicit Scheme

A standard Crank–Nicolson implicit scheme and an iterative procedure are used to treat the nonlinear terms in the finite-difference equations. The Crank–Nicolson scheme gives

$$\frac{(U_i^{k+1} + U_i^k)\phi_i^{k+1}}{\Delta z} = \{L(\phi)\}_i^{k+1} + \{L(\phi)\}_i^k + \{S_\phi\}_i^{k+1} + \{S_\phi\}_i^k, \qquad (8.49)$$

where

$$L(\phi) = \left(-V + \frac{D}{r}\right)\frac{\partial \phi}{\partial r} + D\frac{\partial^2 \phi}{\partial r^2}. \qquad (8.50)$$

Further simplification gives the finite-difference equation in the tridiagonal matrix form:

$$A_i^l \phi_{i-1}^{k+1} + B_i^l \phi_i^{k+1} + C_i^l \phi_{i+1}^{k+1} = D_i^l, \qquad (8.51)$$

where

$$A_i^l = \left(-V_i^l - 2\frac{D_i^l}{\Delta r} + \frac{D_i^l}{r_i}\right)\frac{\Delta z}{\Delta r},$$

$$B_i^l = 2U_i^l + 2U_i^k + 4D_i^l \frac{\Delta z}{\Delta r},$$

$$C_i^l = \left(V_i^l - 2 - 2\frac{D_i^l}{\Delta r} - \frac{D_i^l}{r_i}\right)\frac{\Delta z}{\Delta r},$$

$$D_i^l = 2(U_i^l + U_i^k)\phi_i^k + 2\Delta z\big[\{L(\phi)\}_i^k + \{S_\phi\}_i^k + \{S_\phi\}_i^l\big]. \qquad (8.52)$$

Note that the superscript l indicates an iteration within the step z^{k+1}–z^k. The sequence of steps in the implicit method is as follows. The gas-phase properties are interpolated from the grid points to the characteristic point. $\{S_\phi\}_i^k$ is calculated, and the iteration is started by use of $\phi_i^l = \phi_i^k$. The droplet properties are calculated at a new z location $(k+1)$. $\{S_\phi\}_i^k$ is then calculated, and the finite-difference equations are used to calculate ϕ_i^{l+1}. The values of ρ, D, and V are now updated, and the convergence criterion is checked. After convergence, droplet properties are recalculated. Note that the liquid-phase properties are updated by the solution of the droplet equations in each iteration.

Block-Implicit Scheme

The block-implicit method is used in a form suggested by Briley and McDonald (1977). In this method, the nonlinear terms at the new z location are linearized by a Taylor series expansion about the old z location. The result is a set of coupled, linear difference equations for the dependent variables at the new z level. The second-order accuracy in the z direction is obtained by a Crank–Nicolson differencing, which gives (at radial location i and axial location $k + 1/2$) the following relation:

$$\frac{\phi^{k+1} - \phi^k}{\Delta z} = \left\{ -\frac{V}{U}\phi_r + \frac{T}{r}\frac{1}{U}\phi_r + \frac{T}{U}\phi_{rr} \right\}^{k+1/2} + \left(\frac{S_\phi}{U} \right)^{k+1/2}. \tag{8.53}$$

Note that in Eq. (8.53) the subscript i has been omitted for convenience. Also, the subscript r indicates a derivative in the r direction. The linearization process can be explained by the linearization of one term. For the term $(T/U)\phi_r$, we get

$$\left\{ \frac{T}{U}\phi_r \right\}^{k+1/2} = \left\{ \frac{T}{U}\phi_r \right\}^k + \frac{1}{2}\left\{ \left(\frac{T}{U} \right)^k (\phi_r^{k+1} - \phi_r^k) + \left(\frac{\phi_r}{U} \right)^k (T^{k+1} - T^k) \right.$$
$$\left. + (T\phi)^k \left[-\frac{1}{U^2} \right]^k (U^{k+1} - U^k) \right\}.$$

After linearization of the other terms in the same fashion, Eq. (8.53) becomes

$$U^k \frac{\phi^{k+1} - \phi^k}{\Delta z} = -\frac{1}{2}[V^{k+1} + V^l]\left[\phi_r^k + \frac{1}{2}\phi_r^{k+1} - \frac{1}{2}\frac{U^{k+1}}{U^k}\phi_r^k \right]$$
$$+ \frac{1}{2r}\left[(\phi_r^k + \phi_r^{k+1})T^k + \phi_r^k T^{k+1} - T^k \frac{U^{k+1}}{U^k}\phi_r^k \right]$$
$$+ \frac{1}{2}\left[(\phi_{rr}^k + \phi_{rr}^{k+1})T^k + \phi_{rr}^k T^{k+1} - T^k \frac{U^{k+1}}{U^k}\phi_{rr}^k \right]$$
$$+ \frac{1}{2}\left[\left(\frac{S_\phi}{U} \right)^k + \left(\frac{S_\phi}{U} \right)^l \right]U^k. \tag{8.54}$$

The superscript l indicates an iteration within a z step. It should be further noted that the preceding linearization is not used for the radial gas velocity, because this velocity is obtained from an integral form of the continuity equation. The source terms are also not linearized, as these involve droplet properties. Thus an iterative procedure becomes necessary, even for the block-implicit method in the present case. Equation (8.54) can be written in block tridiagonal form as

$$\bar{A}_i \phi_{i-1}^{k+1} + \bar{B}_i \phi_i^{k+1} + \bar{C}_i \phi_{i+1}^{k+1} = D_i, \tag{8.55}$$

where $\bar{A}_i$, $\bar{B}_i$, and $\bar{C}_i$ are 3×3 square matrices (as there are three dependent variables). ϕ and D_i are 3×1 matrices.

Operator-Splitting Method

In the present case, the two-phase source terms are split from the spatial derivative terms, i.e., two operators are used. The first operator, called the transport operator (L_t), gives the fluid-mechanical contribution, whereas the second operator, called the source operator (L_s), gives the source-term contribution to the overall solution. A symmetric operator, which advances the solution by two z steps, has been used. In the second z step, the order of splitting is reversed. The symmetric two-step operator is demonstrated to have second-order accuracy (Rizzi and Bailey, 1975), whereas a single-step operator is only first-order accurate. The procedure can be explained by

$$\frac{\partial \phi}{\partial z} = L_t(\phi) + L_s(\phi), \tag{8.56}$$

where the transport operator is

$$L_t(\phi) = \left(-\frac{V}{U} + \frac{D}{rU}\right)\phi_r + \frac{D}{U}\phi_{rr}$$

and the source operator is

$$L_s(\phi) = \frac{S_\phi}{U}.$$

Splitting these operators, we obtain

$$\frac{d\phi}{dz} = L_s(\phi), \tag{8.57}$$

$$\frac{\partial \phi}{\partial z} = L_t(\phi). \tag{8.58}$$

Equation (8.57) is solved by a second-order Runge–Kutta scheme. This involves the solution of the droplet equations, which are also solved by a second-order Runge–Kutta scheme. If the source operator is much faster than the transport operator [Eq. (8.56) is stiff], a smaller step size is used in solving the source operator. This is really the justification for using an operator-splitting method. The solution of Eq. (8.58) is used in transport-operator equation (8.57), which is solved by a sequential-implicit method. In the next z step, the order of the operators is reversed. Thus the solution after two z steps can be written as

$$\phi^{k+2} = L_s L_t L_t L_s \phi^k. \tag{8.59}$$

A few additional details of the solution procedure follow. First, the source operator $(L_s\phi^k)$ is advanced by a Δz step. During a substep of this operator, the droplet properties are assumed to be frozen. Here the Δz step is composed of several substeps. After a gas-phase substep is completed, we update the droplet properties by keeping the newly calculated gas-phase properties frozen. Completion of the Δz step for L_s gives the solution $\bar{\phi}$, which is used in the transport step. A Crank–Nicolson implicit differencing is used to solve the transport step. An iterative procedure is used to treat the nonlinear terms. The converged solution represents the completion of one full Δz step. In the second Δt step, the order of operators is reversed; the transport operator is solved first. Thus one full cycle advances the solution by two axial steps.

Results

The basic features of the two-phase laminar axisymmetric jet flow were discussed by Aggarwal et al. (1985). The outer jet is gaseous, whereas the inner jet consists of hot air and liquid fuel in the form of spray. Because the intention in the present study is to use a system of stiff PDEs, the calculations should be performed for a relatively volatile fuel. Therefore hexane fuel is used in the present calculations.

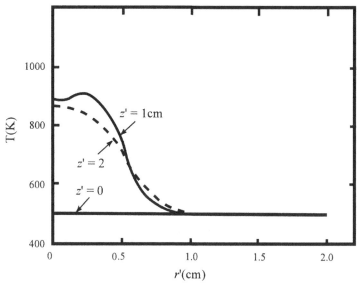

Figure 8.3. Gas-temperature profiles in the radial direction. (Aggarwal et al., 1985, with permission of *Numerical Methods in Partial Differential Equations.*)

Figures 8.3 and 8.4 show the results of the two-phase computations obtained with the sequential-implicit method. The step sizes in the radial and the axial directions are $\Delta r = 0.01$ cm and $\Delta z = 0.001$ cm, respectively. The initial droplet radius is 100 μm, and the initial droplet temperature is 300 K. Figure 8.3 gives the gas-temperature profiles in the radial direction at axial locations $z = 0$, 1.0, and 2.0 cm. The reason for giving the profiles at $z = 1$ and 2 cm is that the comparison of the various numerical schemes is based on the computations between these axial locations. The corresponding fuel-vapor mass fraction profiles are shown in Fig. 8.4. These figures demonstrate the development of the gas-phase mixing layers with two-phase interactions. For example, the mixing-layer thickness is $\sim$0.6 cm at $z = 1$ cm and increases to $\sim$0.9 cm at $z = 2$ cm. It is also noteworthy that the droplet radii

Figure 8.4. Fuel-vapor mass fraction profiles in the radial direction. (Aggarwal et al., 1985, with permission of *Numerical Methods in Partial Differential Equations.*)

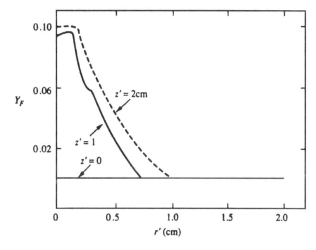

Table 8.1. *Examples of data for one*
point on the effectiveness curve for the
sequential-implicit method ($\Delta t = 0.05$)

Δz	Error	CPU (s)
0.5	0.0300	0.40
0.2	0.0128	0.69
0.05	0.0084	1.05
0.04	0.0069	2.43
0.02	0.0067	5.05
0.01	0.0064	10.03

range between 84.8 and 91.51 μm at $z = 1$ cm. The corresponding range is 74.1–75.1 μm at $z = 2$ cm. This means that, at $z = 2$ cm, more than half of the droplet mass has vaporized.

We now turn to a comparison of the relative effectiveness of the four different schemes previously described for the gas-phase equations. In particular, for each of these schemes, errors versus CPU time required were plotted. The effectiveness curves were constructed by the selection of discrete values for Δt and then determination of the optimal value for Δz. The latter has the property that any other choice of Δz would lead to a point above or to the right of the effectiveness curve. As an example, the data that yield one point on the effectiveness curve for the sequential-implicit method are shown in Table 8.1. Because this problem does not have a known analytical exact solution, we used a numerical solution on very fine grids as our standard. In particular, numerical approximation for the block-implicit scheme was used with $\Delta z = 0.001$, $\Delta r = 0.01$ (these variables are dimensionless here). The error associated with any approximation with grid parameters Δz and Δr is defined by

$$\frac{100}{R}\left[\sum_i (\phi_{ei} - \phi_i)^2 r_i \Delta r \right]^{1/2}. \tag{8.60}$$

Here ϕ_{ei} denotes the standard solution at the ith grid point, ϕ_i is the numerical approximation, and the sum is over all radial grid points r_i at a given value of z. In the results subsequently discussed, $z = 2$ was chosen as the point where the errors were compared.

The effectiveness curves are shown in Fig. 8.5. Observe that, for the larger error tolerances, both the explicit and the operator-splitting methods give the best results. However, as the error criteria are reduced, the sequential-implicit scheme becomes the more effective method. The relative performance of the explicit approach deteriorates with decreasing error requirements because of its subquadratic convergence rate. This is shown by Table 8.1 and results from nonconservative terms in the PDE. The operator-splitting scheme, on the other hand, is second-order accurate (see Table 8.2). However, it suffers from severe combinatorial (or bookkeeping) costs as the grid is refined. The second-order block-implicit scheme suffers from the same defects and is the least effective scheme at all accuracy requirements.

Table 8.2. *Examples of accuracy and cost trade-offs*

Method	Δr (cm)	Δz (cm)	L_2	CPU (s)	Iterations
Sequential implicit	0.05	0.01	0.0064	10.23	200
	0.1	0.02	0.0238	3.64	100
Block implicit	0.05	0.01	0.0064	25.14	244
	0.1	0.02	0.0238	8.46	146
Operator splitting	0.05	0.01	0.0060	8.58	202
	0.10	0.02	0.0230	2.64	102
Predictor–corrector	0.05	0.01	0.008	5.17	100
	0.10	0.02	0.024	1.97	50

8.2.2 Axisymmetric Unsteady Sprays

In the study discussed here, the numerical experimentation is performed by consideration of a system of model equations. In selecting a model problem, it is highly desirable to choose one that is as simple as possible, consistent with retaining the essential features that affect the computational efficiency. Thus the model equations in the present study retain the mathematical character of the parent equations but are considerably simplified otherwise. It must be understood that the intent in this chapter is to discuss a methodology for solving spray equations and not to develop an improved model of spray phenomena. For that purpose, the use of model equations is convenient. Improved physical models of spray behavior are discussed in Chapters 9 and 10.

The nonlinearity of the parent equations is retained by consideration of nonlinear source terms in the equations. These source terms express the exchange rates between the phases. The gas-phase properties are assumed to be represented

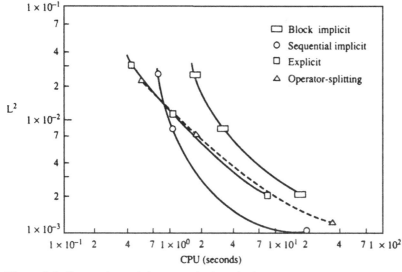

Figure 8.5. Comparison of the numerical methods: accuracy versus computer time. (Aggarwal et al., 1985, with permission of *Numerical Methods in Partial Differential Equations*.)

by a parabolic heat-diffusion-type equation. Three equations are considered to represent the liquid-phase properties, i.e., the droplet size, the droplet velocity, and the droplet number density. The numerical experiments on these equations indicate that, for the parabolic equation, an efficient finite-difference approach is based on an alternating-direction-implicit (ADI) scheme involving the solution of a set of tridiagonal matrices. For droplet equations in Eulerian form, the finite-difference techniques appear inadequate in certain situations that are important in spray combustion. For example, the droplet radius and the liquid velocities often become multivalued functions, as droplet paths cross as the flow develops. This can be treated in a natural way by the method of characteristics, but it is difficult to handle by finite differences with an Eulerian mesh. Realize that the droplet or particle path and the characteristic lines for the hyperbolic equations are identical. (See Chapter 7 for a more complete discussion of the characteristics.)

Note that the multivaluedness of the solutions can also occur whenever the initial droplet size or droplet-velocity distribution is polydisperse. This type of multivaluedness is usually treated by considering the polydisperse spray to be the sum of a finite number of superimposed monodisperse sprays. The type of multivaluedness that is emphasized here, however, can occur with a monodisperse spray. In particular, it happens whenever particle paths cross. The first type of multivaluedness appears first in the inflow boundary conditions and therefore is known a priori to occur, allowing it to be treated readily in the previosly mentioned manner. The second type of multivaluedness first appears in the interior of the calculation domain and cannot be predicted before calculation.

The gas-phase properties are governed by parabolic PDEs, whereas the droplet properties are governed by hyperbolic equations. We treat the latter by the method of characteristics and reduce them to ODEs. These two sets of equations are non-linearly coupled because of the mass, momentum, and heat transfer between the phases. In the present study, the model system contains five equations: one parabolic equation for the gas-phase scalar θ (temperature) and four ODEs for four unknowns [defined before Eq. (8.63)], n, $\vec{X}$, s, and $\vec{U}_l$. It may be noted that, in a real system containing many gas species, the mass densities of various species will be given by similar parabolic equations. The gas-phase velocity as well as the droplet-surface temperature are assumed to be known. Thus the momentum coupling is neglected for the gas phase; for the liquid phase it is properly taken into account. The set of model equations is nondimensionalized by use of characteristic values of length, velocity, temperature, and droplet size. The characteristic values used are, respectively,

$$Z_c = 10 \text{ cm}, \qquad U_c = 100 \text{ cm/s},$$

$$T_c = 500 \text{ K}, \qquad R_c = 100 \text{ } \mu\text{m}.$$

The nondimensional equation for θ in axisymmetric coordinates (r, z) can be written as

$$\frac{\partial T}{\partial t} = -\frac{\partial}{\partial z}(UT) + \frac{\alpha}{r}\frac{\partial T}{\partial r} + \alpha\left(\frac{\partial^2 T}{\partial r^2} + \frac{\partial^2 T}{\partial z^2}\right) - S_T, \qquad (8.61)$$

where

$$S_T = \frac{K_1 n}{2\pi \Delta r \Delta z}[1 + K_5(T - T_l)](1 + K_4\{S[(U - U_{lz})^2 + U_{lr}^2]\}^{1/4})S^{1/2}$$

$$\times \ln[1 + K_5(T - T_l)]. \tag{8.62}$$

The quantity $2\pi r \Delta r \Delta z$ represents the volume of a computational cell in an axisymmetric cylindrical geometry, and the subscript l represents a Lagrangian variable associated with any computational droplet. In the preceding equations, U and T_l are assumed to be known constants. Thus the gas velocity is assumed to be uniform and in the axial direction. Consequently, the radial convection term does not appear in Eq. (8.61). Here, S_T is the nondimensional heat transfer rate between the phases. The variable K_1 represents the heat transfer time constant, and K_4 and K_5 are assumed to be constants. In physically realistic two-phase situations, K_4 would represent the ratio of specific heat at constant pressure and the latent heat of vaporization and K_5 would be the coefficient of the Reynolds number correction, where the Reynolds number is based on the droplet size and droplet velocity relative to the gas (see Aggarwal et al., 1983).

Liquid-Phase Equations

The liquid-phase properties of interest are $\vec{U}_l$ (the liquid velocity), $\vec{X}$ (the droplet position), $S = R^2$ (where R is the droplet radius), and n (the number of droplets associated with a given mass of liquid). The governing equations are of the hyperbolic type, and are usually written in the following Lagrangian form:

$$\frac{dn}{dt} = 0, \tag{8.63}$$

$$\frac{d\vec{X}}{dt} = \vec{U}_l, \tag{8.64}$$

$$\frac{dS}{dt} = -K_2(1 + K_4\{S[(U - U_{lz})^2 + U_{lr}^2]\}^{1/4})\ln[1 + K_5(T - T_l)], \tag{8.65}$$

$$\frac{d\vec{U}_l}{dt} = \frac{K_3}{S^{1/2}}(\vec{U} - \vec{U}_l). \tag{8.66}$$

At the inflow ($z = 0$), the adiabatic boundary condition for T is prescribed as

$$U_\infty \rho_\infty c_p T_\infty = U_0 \rho_0 c_p T - \lambda(\partial T/\partial z)_{z=0},$$

which in the nondimensional form reduces to

$$U_\infty = T_0 - \alpha(\partial T/\partial z)_0, \qquad \alpha = \lambda/\rho c_p U_0 Z_0,$$

where T_∞ is prescribed.

An outflow boundary condition is needed to make the computational domain finite. We now use the standard outflow conditions at these points (see Fix and Gunzburger, 1977). Mathematically, it takes the form

$$\partial^2 T/\partial z^2 = 0 \quad \text{at } z = 1.$$

At the r boundary, the boundary conditions are

$$\partial T/\partial r = 0 \quad \text{for } r = 0 \quad \text{and} \quad r = 1.$$

Initially at time $= 0$, T is assumed to be the same as T_∞.

The droplet flow is assumed to be initially conical and flowing from a point source at a point on the axis of symmetry where $z < 0$. Initially, all the characteristics are positioned at $z = 0$. The initial values of S and n are assumed to be uniform, and U_l is obtained from the assumption of a conical flow.

The inflow boundary conditions in our example have been chosen to be single valued (monodisperse spray) in order to study only multivaluedness that is due to the crossing of particle paths (or equivalently, characteristics). Such multivaluedness will occur in some of the cases studied.

Numerical Aspects and Solution Procedure

For the parabolic equation, an ADI scheme is used. Thus a full-time advancement takes place in two steps. First, time is advanced by a half step in the z orientation and then it is advanced by another half step in the r orientation. The difference equations in the z orientation are

$$\frac{T^{n+1/2} - T^n}{\Delta t/2} = -\frac{\partial}{\partial z}\left(U_z T^{n+1/2}\right) + \frac{\alpha}{r}\frac{\partial}{\partial r}T^n + \alpha\frac{\partial^2}{\partial r^2}T^n + \alpha\frac{\partial^2}{\partial z^2}T^{n+1/2} - (S_T)^n,$$

and then in the r orientation, we have

$$\frac{T^{n+1} - T^{n+1/2}}{\Delta t/2} = -\frac{\partial}{\partial z}\left(U_z T^{n+1/2}\right) + \frac{\alpha}{r}\frac{\partial}{\partial r}T^{n+1} + \alpha\frac{\partial^2}{\partial r^2}T^{n+1} + \alpha\frac{\partial^2}{\partial z^2}T^{n+1/2} - (S_T)^{n+1},$$

which is second-order accurate in Δr, Δz, and Δt. Because S_T^{n+1} is nonlinear, it needs to be evaluated either by an iterative or a quasi-linearization procedure. One crude approximation is to take S_T^{n+1} to be the same as S_T^n. Then the results are less than second-order accurate in time. To obtain second-order accuracy in time, an iterative scheme is used, where S_T^n is used as the initial approximation and is updated until the desired accuracy is obtained. In another method involving the quasi-linearized procedure, S_T^{n+1} is written as

$$S_T^{n+1} = S_T^n + (\partial S/\partial T)^n (T^{n+1} - T^n).$$

In the first approximation T^{n+1} is assumed to be the same as T^n. Because of coupling from liquid-phase equations, it is updated by iteration until desired accuracy is achieved.

ODEs (8.63)–(8.66) are integrated over the same time step by a standard second-order predictor–corrector scheme (Gear, 1971). The local values of temperature for Eq. (8.65) at each point on the trajectory are obtained from linear interpolation of the four surrounding values of the gas-phase solution in the computation cell through which the droplet is passing.

The energy interchanges occurring as the droplets traverse each grid cell are evaluated by superimposing to the four surrounding grid points, as shown in

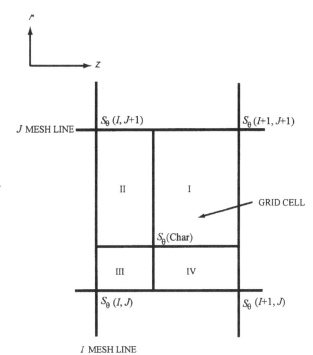

Figure 8.6. Second-order interpolation scheme. (Aggarwal et al., 1983, with permission of the *Journal of Computational Physics.*)

Fig. 8.6. Therefore,

$$S_T(I, J) = S_T(\text{char.})^* \frac{\text{Volume I}}{\text{Total Cell Volume}},$$

and $S_T(I + 1, J)$, $S_T(I + 1, J + 1)$, and $S_T(I, J + 1)$ are readily determined in analogous fashion.

Because of the linear interpolation, this maintains the second-order accuracy in our finite-difference scheme, as opposed to the first-order accuracy that would result if we assumed that the coefficients are constant in each cell (Dukowicz, 1980; Gosman and Johns, 1980). In the iteration cycle, the solutions for the liquid properties are first advanced one full-time step by means of Lagrangian calculations, and then the θ solution is advanced two half-time steps in the ADI subcycle.

Numerical Results

In this subsection we discuss the results from selected numerical experiments, as well as the specific integration schemes used in these simulations. As previously noted, two ways were considered for treating the nonlinearity S_T in the T equation, namely, iteration with respect to the source term S_T and quasi-linearization. The former has the advantage in that the distribution scheme just discussed for the source term S_T is physically clear. We found, however, that the number of iterations grew very rapidly with Δt, and the time-step restrictions were far more severe than those required for reasonable engineering accuracy. The distribution scheme used in quasi-linearization distributed S_T and $\partial S_T / \partial T$ in exactly the same way; S_T was distributed in the iterative scheme. From a physical point of view this is somewhat ad hoc; however, in those cases in which the iteration converged, the difference in the answers between the two approximations occurred in the fourth or fifth decimal place. Moreover, the CPU time required for the interactive scheme was considerably higher than that required for the quasi-linearization. Thus all of the results subsequently presented used the latter scheme.

The computer code developed for these simulations can use a variety of schemes for integrating ODEs (8.63)–(8.66) arising from the Lagrangian approach. To maintain a second-order approximation in our scheme, a predictor–corrector second-order Runge–Kutta method was used to integrate those equations.

Three grids were used in the calculations. The parameters varied were the number of characteristics N, the time step Δt, the grid spacings Δr and Δz, and finally the injection pulse time τ_p. (The physically continuous injection process is represented in discretized fashion by injection pulses with a period τ_p). Data for the individual grids are as listed below:

$$\begin{array}{lllll}
\text{coarse grid:} & N = 3, & \Delta t = 0.02, & \Delta r = \Delta z = 0.1, & \tau_p = 0.04; \\
\text{base grid:} & N = 6, & \Delta t = 0.01, & \Delta r = \Delta z = 0.05, & \tau_p = 0.02; \\
\text{fine grid:} & N = 24, & \Delta t = 0.005, & \Delta r = \Delta z = 0.025, & \tau_p = 0.01.
\end{array}$$

To get an estimate of the order of accuracy of our approximations for the temperature field T, we measured the discrete L_2 error at steady state, assuming that the fine-grid approximation was exact. More precisely, letting T_i denote the value of approximate temperature at the ith grid point (as computed on the coarse or the base grid) and letting T_i^E denote the analogous value for the fine grid, we then define

$$E = \left(\sum_i |T_i - T_i^E|^2 \Delta z \Delta r \right)^{1/2}$$

as the measure of error, where the sum is over all grid points. On the coarse grid, this error was 3.6×10^{-2}, whereas on the base grid it was 4×10^{-3}. This demonstrates a quadratic convergence in our scheme in the sense that E is reduced by at least a factor of 4 (in this case, it is 9) when the grid size is halved.

The results obtained from the base grid were selected to illustrate various features of the flow. The results of those calculations are shown in Figs. 8.7–8.10 for gas temperature, droplet trajectories, and droplet size. The phenomenon modelled is unsteady but reaches a steady state after an initial transient period. At $t = 0.5$ (50 time steps), for example, the solution is in the midst of the transient while by $t = 3$ (300 time steps), a steady state has been well established. Figure 8.7 displays the gas temperature θ during the transient period ($t = 0.5$). Note that the gas temperature initially (at $t = 0$) was equal to 2.0; the effect of the vaporizing droplets is to cool the gas because energy is required for vaporization. The injection occurs in the neighborhood of the origin (see Figs. 8.9 and 8.10); there, the droplet number is greatest and the cooling effect is greatest. On account of large local gradients in space and time, sensitivities to mesh size and time step are most severe in this region. The gas temperature in this neighborhood decreases with time, so that the cooling effect becomes less severe as time proceeds. There seems to be a potential

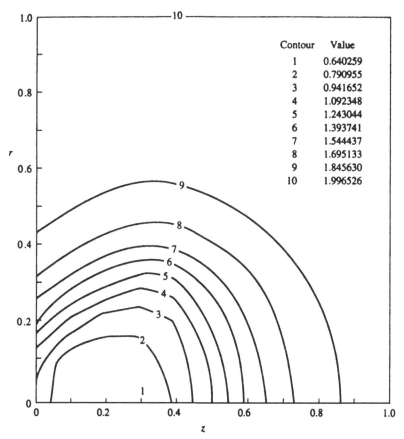

Contour	Value
1	0.640259
2	0.790955
3	0.941652
4	1.092348
5	1.243044
6	1.393741
7	1.544437
8	1.695133
9	1.845630
10	1.996526

Figure 8.7. Temperature contours (base case, time = 0.5). (Aggarwal et al., 1983, with permission of the *Journal of Computational Physics.*)

for benefit from a nonuniform grid and variable time steps, but this possibility has not yet been explored.

Figure 8.8 demonstrates the steady-state ($t = 3$) gas-temperature profile. The coldest region is along the axis of symmetry, which is essentially the center of the spray cone. Again, even in the steady state, the largest gradients occur in the neighborhood of the origin.

Figures 8.9 and 8.10 show steady-state results for droplet size and droplet trajectories. Note that, in Fig. 8.9, the circle radii represent droplet volume $S^{3/2}$ and not droplet radius $S^{1/2}$. Because a computer plots those circles with some discretization of diameters, only the roughest inferences should be made from such graphs. The droplet size is definitely seen to decrease substantially as it moves through the hot gas. Of course, the droplets at the edge of the spray vaporize much more rapidly than the droplets in the spray center. The gas velocity is greater than the axial component of the initial droplet velocity so that drag causes the droplets to accelerate downstream. The trajectories are clearly seen to involve a turning of the droplets in that direction.

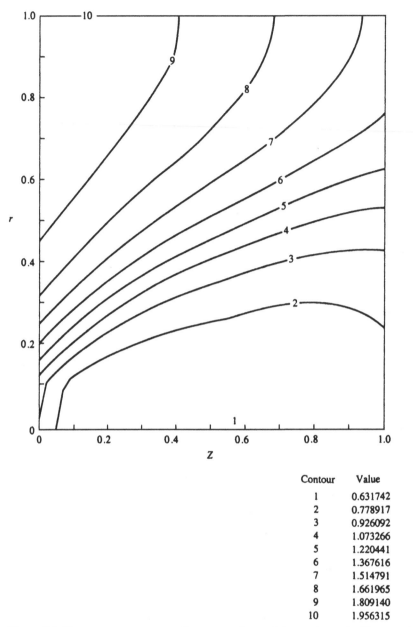

Contour	Value
1	0.631742
2	0.778917
3	0.926092
4	1.073266
5	1.220441
6	1.367616
7	1.514791
8	1.661965
9	1.809140
10	1.956315

Figure 8.8. Temperature contours (base case, time $= 3.0$, steady rate). (Aggarwal et al., 1983, with permission of the *Journal of Computational Physics*.)

Note that the cell Reynolds number varied between 1.35 and 5.0 for all cases considered and, in particular, was 2.5 for the 20×20 mesh base case. The next set of results is given in Figs. 8.11 and 8.12 and displays the sensitivity of the approximation to changes in the number N of characteristics used and to the grid spacing Δz and Δr. In these figures, comparisons are made with the results obtained from the base grid when one or more of these parameters was varied. Figure 8.11 deals with the case in which the number of characteristics on the base grid was

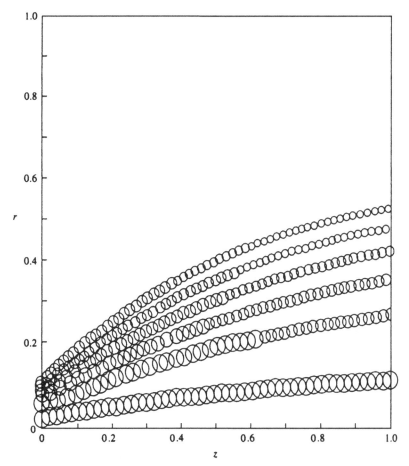

Figure 8.9. Droplet volume (base case, time = 0.5) (Aggarwal et al., 1983).

changed from $N = 6$ to $N = 3$. Plotted are contours of constant values of

$$[T^{(6)} - T^{(3)}]/T^{(6)},$$

where $T^{(6)}$ is the temperature field obtained from the base grid and $T^{(3)}$ is the temperature field obtained from the base grid except where N has been reduced to 3.

Figure 8.12 has analogous contours except where the mesh spacing in the base grid has been changed from $\Delta r = \Delta z = 0.05$ to $\Delta r = \Delta z = 0.1$. These figures tend to indicate that the two-phase flow is satisfactorily resolved on the base grid and, in fact for most purposes, even the coarse grid may be satisfactory. The approximation is slightly more sensitive to the grid spacing than to the number of characteristics used.

Changes in the time step produce more delicate effects. First, the nature of the initial condition places definite restrictions on the size of Δt, at least for small times t. The reason is that the sharp gradients produced near $t = 0$ may cause the calculated temperature to go negative if Δt is too large. At this point, the calculations must be terminated because of the nature of the source terms. This, for example, was the case when Δt was increased in base-grid size from 0.02 to 0.05. It is

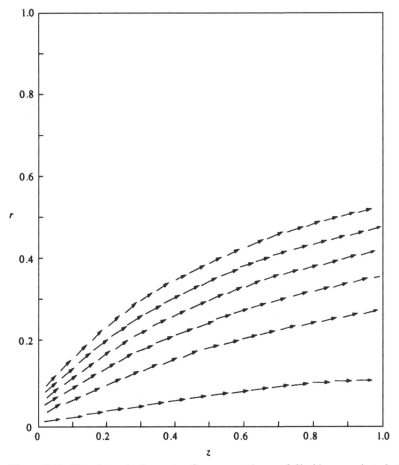

Figure 8.10. Droplet-velocity vector (base case, time = 3.0). (Aggarwal et al., 1983, with permission of the *Journal of Computational Physics*.)

emphasized that this restriction is far less severe than the restriction on Δt found in the iterative scheme.

Similar considerations show that the ratio $\tau_p/\Delta t$ must be sufficiently small. For example, if in the base grid (where $\Delta t = 0.01$, $\tau_p = 0.02$) the pulse time τ_p were increased to 0.05, then negative temperature would occur.

Probably the most important feature of the Lagrangian formulation is demonstrated in Figs. 8.13 and 8.14. In this axisymmetric case, injection is considered to occur from a circular line source at $z = 0$. This may be viewed as the limit of a ring of injection orifices whereby both distances between orifices and orifice diameter go to zero. In this case, many intersections of droplet trajectories will occur (see Fig. 8.13) so that solutions for all droplet properties are multivalued. Note that the gas temperature, as shown in Fig. 8.14, is still single valued. The crossing of the characteristic would not be allowed by a finite-difference scheme. Numerical diffusion would merge and smear the characteristics. In the case of a compressive wave in a gas dynamic field, such merging could give respectable global representation to shock-wave formulation; however, in the present droplet study, such a result would be

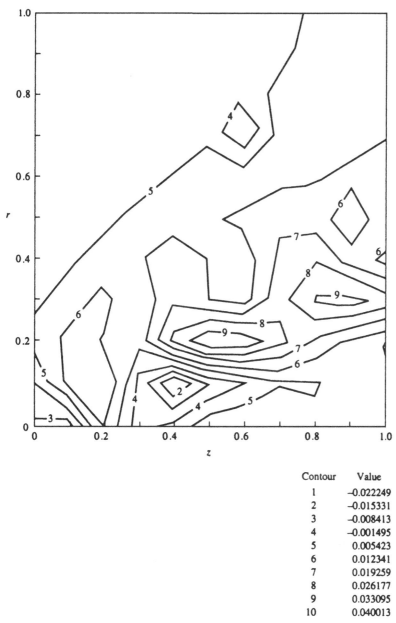

Contour	Value
1	-0.022249
2	-0.015331
3	-0.008413
4	-0.001495
5	0.005423
6	0.012341
7	0.019259
8	0.026177
9	0.033095
10	0.040013

Figure 8.11. Temperature-difference contours (normalized change, three characteristics versus six characteristics; time = 3.0). (Aggarwal et al., 1983, with permission of the *Journal of Computational Physics.*)

nonsensical. Again, with typical number densities, a negligibly small fraction of intersecting droplets will actually collide. These issues are more thoroughly discussed in Chapter 7.

Finally, a comparison was made between the second-order scheme proposed by Gosman and Johns (1980) and that of Dukowicz (1980). To do this, we retained all features of our discretization (ADI, method of characteristics, etc.) except for the replacement of the second-order distribution scheme with the first-order version.

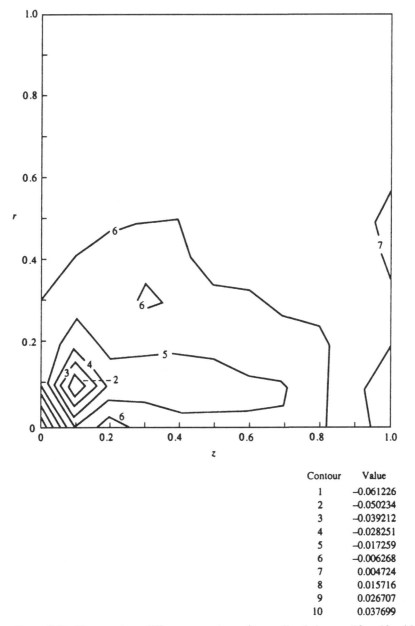

Contour	Value
1	−0.061226
2	−0.050234
3	−0.039212
4	−0.028251
5	−0.017259
6	−0.006268
7	0.004724
8	0.015716
9	0.026707
10	0.037699

Figure 8.12. Temperature-difference contours (normalized change, 10×10 grid versus 20×20 grid; time $= 3.0$). (Aggarwal et al., 1983, with permission of the *Journal of Computational Physics.*)

Therefore the comparison is with the model equations by use of their proposed interpolation scheme. (It is not a comparison between their calculations and our calculations because different equations were used.) The latter had L_2 errors (with the fine grid as exact) of 3.6×10^{-2} and 1×10^{-2} for the coarse grid and the base grid, respectively. This convergence is superlinear because everything but the source term S_T is treated with second-order accuracy and the coupling through S_T is rather

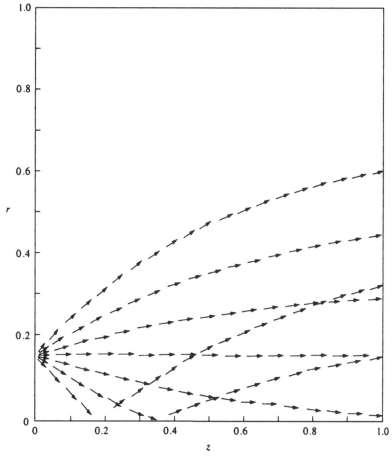

Figure 8.13. Droplet-velocity vector (circular-line-source case; time = 3.5, steady rate). (Aggarwal et al., 1983, with permission of the *Journal of Computational Physics.*)

weak in this particular model. However, this convergence is definitely subquadratic and theoretically should actually become linear as the grid spacings approach zero because of the first-order treatment of S_T. Further, it is noteworthy that Crowe et al. (1977) have used a source-distribution scheme that is similar to that of Gosman and Johns (1980) to solve steady-state spray equations. Because they are solving the steady-state two-dimensional planar equations compared with the time-dependent axisymmetric equation used here, the present results cannot be compared with those of Crowe et al.

A system of model equations that retains the essential mathematical and numerical character of the parent equations for treating a typical two-phase spray flow is used. Through numerical experimentation on these equations, it is recommended that an Eulerian representation for the gas-phase properties and a Lagrangian representation for the liquid-phase properties be used with any spray model. See, for example, the model by Aggarwal et al. (1981). Indeed, for certain flow situations involving multivalued droplet properties, this is the most appropriate approach. From this approach, an efficient numerical algorithm is developed

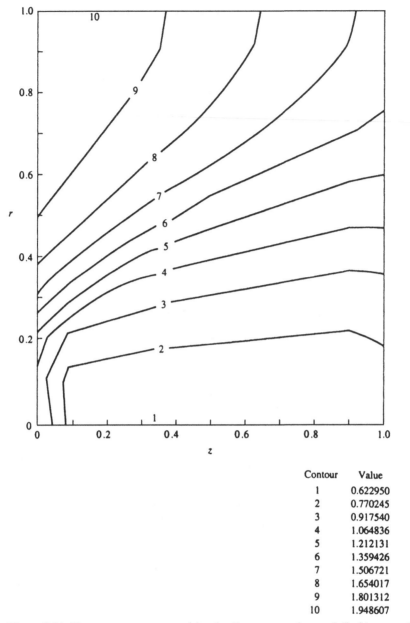

Contour	Value
1	0.622950
2	0.770245
3	0.917540
4	1.064836
5	1.212131
6	1.359426
7	1.506721
8	1.654017
9	1.801312
10	1.948607

Figure 8.14. Temperature contours (circular line source, time = 3.5). (Aggarwal et al., 1983, with permission of the *Journal of Computational Physics.*)

that is consistently second-order accurate. In this algorithm, the unsteady axisymmetric gas-phase equations are solved by a second-order-accurate-implicit ADI scheme, whereas the Lagrangian equations are solved by a second-order Runge–Kutta scheme. The treatment of exchange laws between the phases is also made second-order accurate by use of two-dimensional linear interpolation and volume-weighted distribution. The results of two different numerical experiments are presented. In the first experiment, the sensitivity of those results to the changes in time step, droplet pulse time, grid size, and number of groups of droplet characteristics used to describe the droplet injector is examined. All these results confirm a quadratic convergence as the grid size or the time step is varied. It may also be indicated that the results are slightly more sensitive with respect to the grid spacing than to the number of characteristics.

The purpose of the second numerical experiment was to demonstrate a physical situation that has multivalued droplet properties. For this case, a finite-difference solution of droplet equations based on a Eulerian description is inadequate. On the other hand, the Lagrangian description becomes a natural method for this type of flow. In addition, a comparison was made between the interpolation scheme presented in this chapter and the interpolation scheme proposed by Gosman and Johns (1980) and Dukowicz (1980). With the latter scheme, which gives superlinear (but subquadratic) convergence, results are slightly inferior to those presented in this subsection. Future computations of more realistic flows with relatively stronger two-phase coupling might reveal more prominent differences between first- and second-order source-function distributions.

8.2.3 Solution for Pressure

The previous discussion in this section does not address situations in which the solution for pressure is coupled with calculations for other gas-phase and liquid-phase properties. The iterations required for pressure solvers can be the controlling factor in computational speed; in fact, the relative merits of various computational schemes might change when a pressure solution is introduced.

One-dimensional spray calculations in a closed volume require pressure solutions; see Seth et al. (1978), Aggarwal and Sirignano (1984, 1985b), for example. Aggarwal and Sirignano (1984) tested a range of numerical schemes. Pressure calculations require small time steps for convergence. As a result, the explicit scheme for the gas-phase calculations is more computationally efficient than the implicit or the operator-splitting schemes since large time steps cannot be taken.

There has not been a thorough study of competing numerical schemes for strong elliptic multidimensional flows (that is, for confined flows with recirculation). Derivatives of TEACH codes are commonly used (Molavi and Sirignano, 1988; Raju and Sirignano, 1989, 1990a) with implicit calculations for the gas-phase·properties. Other schemes could very well be superior.

8.3 Point-Source Approximation in Spray Calculations

In large-scale spray computations, the coupling of the continuous and the dispersed phases is achieved by the introduction of source terms in the mass, momentum, and energy equations for the continuous gas phase, and the effect of the gas phase is introduced in the form of boundary conditions on the gas film surrounding the droplet. This section presents an evaluation of the point-source approximation used in the gas–liquid coupling of these spray calculations. From the mathematical point of view, the introduction of mass, momentum, and energy sources into the gas-phase equations amounts to representing these effects as point sources (or sinks) of mass, momentum, and energy, respectively. Numerical resolution is limited by the size of the gas cell used in the calculation and the point-source approximation is most accurate if the droplet is much smaller than the numerical cell. Although this is certainly the case if a large combustion chamber is being resolved with a cost-effective grid, a number of fundamental situations exist in which resolution on the scale of a few droplet spacings or even a few droplet diameters is required. Examples are ignition by a hot wall (Aggarwal and Sirignano, 1985a, 1986) or hot gases, interaction between a reaction zone and a stream of droplets (Rangel and Sirignano, 1988a, 1988b, 1989a) and droplet–droplet and droplet–wall interactions. These phenomena are strongly dependent on the spray initial conditions, geometry, relative velocities, composition, and temperatures.

As the gas-phase grid size is refined to improve the resolution of the gas-phase calculations, the point-source approximation becomes increasingly deficient because the ratio of the droplet volume to the gas-cell volume increases. This situation imposes a limitation on the use of the point-source approximation. The error occurs in two ways. On one hand, the evaluation of the gas-phase properties used as boundary conditions for the droplet calculations becomes inappropriate, as is subsequently demonstrated. On the other hand, the representation of the droplet as a point disturbance of the flow field also becomes incorrect. An evaluation of these errors is presented in two different configurations. One configuration is a zero Reynolds number droplet treated as spherically symmetric. Another configuration is the axisymmetric flow past a droplet at Reynolds numbers between 1 and 100.

Interpolation for Phase-Exchange Source Terms

In both the multicontinua method and the discrete-particle method, a phase-exchange source-term value for the gas phase is generally not immediately given at a gas-phase mesh point. Rather, it is presented at the instantaneous position of the droplet. Its value must be extrapolated to the mesh points. The lowest-order approximation involves simply transferring the value from the droplet location to the nearest mesh point. This produces numerical errors of the same first order as numerical diffusion errors that are avoided for the liquid phase by the Lagrangian method. It is superior therefore to distribute the source term in a weighted manner to the neighboring mesh points. There will be two, four, or eight neighboring mesh points, depending on whether we have a one-, two-, or three-dimensional calculation. The droplet or average droplet under consideration is within a rectangle defined by the corners at the four mesh points (p, n), $(p + 1, n)$, $(p + 1, n + 1)$, and $(p, n + 1)$. Consider that the nearest mesh point is (p, n). The lowest-order approximation merely transfers the source term S to the mesh point (p, n) with resulting first-order accuracy. The second-order scheme divides the source term S into four parts, each associated with a mesh point and weighted inversely to the proximity of that mesh point to the droplet position, namely

$$S_{p,n} = \frac{A_{p,n}}{A} S, \qquad S_{p+1,n} = \frac{A_{p+1,n}}{A} S,$$

$$S_{p+1,n+1} = \frac{A_{p+1,n+1}}{A} S, \qquad S_{p,n+1} = \frac{A_{p,n+1}}{A} S,$$

where A is the area of the original rectangle and $A_{p,n}$, etc., are the areas of the subdivided rectangles shown in Fig. 8.6. Note that the extension to three dimensions involves eight subdivided rectangular volumes.

Note that the source point is taken as the center of the droplet. Furthermore, in the multicontinua approach, many droplets can be represented by each average droplet. Also, any given mesh point can be a corner point for several cells (four in two dimensions), each of which can contain droplets and transfer portions of the source terms to the mesh point. These source contributions from various droplets are simply additive. Finally, note that the same geometrical weighting factors should be used to evaluate gas properties at the droplet position (for input as source terms to the liquid-phase equations), given the properties at the neighboring mesh points.

Spherically Symmetric Case

In the discrete-particle method, we must present ambient-gas conditions for the droplet in order to determine the droplet-vaporization rate, heating rate, drag, and trajectory. In a finite-difference scheme, the gas properties at immediately neighboring mesh points to the instantaneous droplet position are used as the ambient conditions on the subgrid droplet model. Theoretically, these ambient conditions should exist at the edge of the gas film (boundary layer and wake) surrounding the droplet; in practice, however, the neighboring mesh points might be in the gas film, therefore leading to errors in the matching of the subgrid droplet model to the gas phase. The surprising result is that, as the gas-phase grid is refined, the error increases; the neighboring mesh points to the droplet actually move closer to the droplet and into the surrounding gas film. In the case in which the droplet moves with the surrounding gas, we can assume that the gas film is spherically symmetric with respect to the droplet center. In that case, Rangel and Sirignano (1989b) have shown that the mass vaporization rate given by Eqs. (2.10) should be multiplied by the factor $1 + e$ to correct for the error in application of the ambient boundary condition on the droplet model. For the case in which the gas-phase grid size is smaller than the droplet radius, it is found that

$$e = -\frac{1}{1 + \delta},$$

where δ is the ratio of the distance between the mesh point (where ambient conditions are applied) and the droplet center to the droplet radius. This analytical solution of Rangel and Sirignano considers the ambient-gas conditions as given and addresses only errors associated with the location of the application of these conditions. It does not address the compounding of this error in a coupled gas–liquid calculation by the modification of the ambient values because of the modified vaporization rate; it is assumed that, as one problem is corrected, the other problem is automatically corrected in the coupled calculation.

Note that, for this case in which the relative velocity between the droplet and the ambient gas is zero, the droplet can be accurately considered as a point source from the gas perspective, that is, the point-source approximation has been shown by Rangel and Sirignano (1989b) to predict exactly the gas-film properties surrounding the droplet. Whereas the droplet size affects the vaporization rate, the ambient gas is affected only by the vaporization rate, and, in other ways, it is not affected by the droplet size. This simplification does not apply exactly to the case in which a finite relative velocity exists.

In spite of its limited applicability, the spherically symmetric case is useful for the purpose of this numerical evaluation because it admits an analytical solution. Consider a monocomponent fuel droplet in an otherwise undisturbed gas. The detailed analysis is presented in the work of Rangel and Sirignano (1989b).

The solution of the governing equations, assuming that c_p, ρD, and $\rho \alpha$ are constants, is the classical result for the Stefan flow (Williams, 1985; Glassman, 1987). In terms of the fuel mass fraction,

$$\frac{\rho_s u_s R}{\rho_s D_s} = \ln(1 + B_M), \tag{8.67}$$

or, in terms of the temperature,

$$\frac{\rho_s u_s R}{\rho_s \alpha_s} = \ln(1 + B_T), \tag{8.68}$$

where the transfer numbers are

$$B_M = \frac{Y_{Fs} - Y_{F\infty}}{1 - Y_{Fs}}, \qquad B_T = \frac{c_p(T_\infty - T_s)}{L}. \tag{8.69}$$

The fuel mass fraction and temperature profiles in the gas are given by

$$\frac{Y_F - 1}{Y_{F\infty} - 1} = \left(1 + \frac{Y_{Fs} - Y_{F\infty}}{1 - Y_{Fs}}\right)^{-R/r}, \tag{8.70}$$

$$\frac{c_p(T - T_s)/L + 1}{c_p(T_\infty - T_s)/L + 1} = \left[1 + \frac{c_p(T_\infty - T_s)}{L}\right]^{-R/r}. \tag{8.71}$$

In numerical computations, the droplets are represented by point sources of mass such that the instantaneous rate of mass production per droplet is $\dot{m}$. The solution in spherical coordinates with only radial dependence yields the fuel mass fraction and temperature profiles as

$$\frac{Y_F - 1}{Y_{F\infty} - 1} = \exp\left(-\frac{\dot{m}}{4\pi \rho D r}\right), \tag{8.72}$$

$$\frac{c_p(T - T_s)/L + 1}{c_p(T_\infty - T_s)/L + 1} = \exp\left(-\frac{\dot{m}}{4\pi \rho D r}\right). \tag{8.73}$$

As opposed to Eqs. (8.70) and (8.71), which give the fuel mass fraction and temperature profiles for $r > R$, Eqs. (8.72) and (8.73) give the fuel mass fraction and temperature profiles for $r > 0$. Note, however, that Eqs. (8.72) and (8.73) are singular at the origin, as expected from the physical character of the point-source approximation. It is interesting to note that if the vaporization or mass production rate appearing in Eqs. (8.72) and (8.73) is explicitly prescribed as

$$\dot{m} = 4\pi \rho D R \ln\left(1 + \frac{Y_{Fs} - Y_{F\infty}}{1 - Y_{Fs}}\right) \tag{8.74}$$

or equivalently by

$$\dot{m} = 4\pi\rho\alpha R \ln\left[1 + \frac{c_p(T_\infty - T_s)}{L}\right], \tag{8.75}$$

then Eqs. (8.72) and (8.73) become identical to Eqs. (8.70) and (8.71). This means that a spherically symmetric vaporizing droplet is equivalent to a point source located at the droplet center.

In principle, there is no error introduced by use of the point-source approximation; the inaccuracy in the numerical computations, however, occurs in the evaluation of the transfer number. This evaluation is done with the gas-phase properties in an appropriate neighborhood of the droplet. The number of gas cells or grid points involved depends on whether one-, two-, or three-dimensional calculations are being performed. In a control-volume approach, the properties of the gas cell containing the droplet are used (Rangel and Sirignano, 1988a, 1988b). A second-order-accurate scheme utilizes a weighted average of the most immediate grid points (Aggarwal et al., 1983).

In the simple spherically symmetric case considered here, the grid extends from the point source so that the first gas node is coincident with the source and successive nodes are located at intervals Δt. The gas grid points are labeled starting at the origin, $j = 0, 1, 2, 3$, etc. Consider the point-source representation of a droplet of radius R so that this radius is smaller than the grid spacing Δt. Equation (8.72) may be used to evaluate the fuel mass fraction at any distance r from the droplet center. In standard numerical calculations, the transfer number is evaluated with the properties at the closest gas grid point as if they were the conditions at infinity. To generalize this procedure, the transfer number is evaluated with the use of the gas properties at any desired grid point. Note that, when a gas node is selected sufficiently far from the droplet, the exact solution is approached. Note also that the first node cannot be used because it is located inside the droplet and is also a singular point. The new transfer number B_M is different from the transfer number B_M obtained with the exact solution with the conditions at infinity. The calculated mass vaporization rate is then

$$\dot{m}^* = 4\pi\rho D R \ln(1 + B_M^*), \tag{8.76}$$

where $B_M^* = (Y_{Fs} - Y_{F\infty}^*)/(1 - Y_{Fs})$ and the analytical solution is used to calculate the fuel mass fraction Y_∞^* at the selected gas grid point.

With the use of Eqs. (8.72) and (8.76) and the definitions of B_M and B_M^*, the relation for the numerical vaporization rate is immediately found in terms of the exact transfer number as

$$\dot{m}^* = 4\pi\rho D R\left(1 - \frac{1}{j\Delta r'}\right)\ln(1 + B_M), \tag{8.77}$$

and the error in the vaporization rate is

$$e = \frac{\dot{m}^* - \dot{m}}{\dot{m}} = -\frac{1}{j\Delta r'}. \tag{8.78}$$

Because the numerical vaporization rate is less than the exact vaporization rate, it follows that the numerical fuel mass fraction at any gas node should be lower than

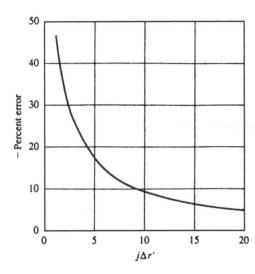

Figure 8.15. Percent error in the vaporization rate as a function of grid size: spherically symmetric droplet. (Rangel and Sirignano, 1989b, with permission of *Numerical Heat Transfer.*)

the exact value. The numerically predicted fuel mass fraction at location $r = j\Delta r'$ may be directly obtained from Eq. (8.70) as

$$\frac{Y_{F\infty}^{*(n)} - 1}{Y_{F\infty} - 1} = \left(1 + B_M^*\right)^{-1/j\Delta r'}, \tag{8.79}$$

where the superscript (n) indicates a new correction to the fuel mass fraction because the numerical transfer number B_M^* has been used on the right-hand side instead of the actual value B_M. The left-hand side of Eq. (8.79) may be conveniently expressed in terms of transfer numbers so that

$$\frac{1 + B_M^{*(n)}}{1 + B_M} = \left(1 + B_M^*\right)^{-1/j\Delta r'}. \tag{8.80}$$

To compute the numerical vaporization rate and the error to any order of accuracy, Eq. (8.80) is used recursively to increase the order of the numerical factor. The following expressions are obtained for the numerical vaporization rate and the error:

$$\dot{m}^* = 4\pi\rho D R F_{\Delta r} \ln(1 + B_M), \tag{8.81}$$

where

$$e = -\frac{1}{j\Delta r'} + \frac{1}{j^2\Delta r'^2} - \cdots = \sum_{n=1}^{\infty} \left(-\frac{1}{j\Delta r'}\right)^n = -\frac{1}{1 + j\Delta r'} \tag{8.82}$$

and $F_{\Delta r} = 1 + e$.

As Eq. (8.82) indicates, the error is dependent on only the ratio for the grid size to the droplet size times the index number of the node used to represent the ambient conditions. This product is simply the dimensionless distance from the selected node to the droplet center. The vaporization rate is always underpredicted. The most important conclusion is that a reduction in the gas-phase mesh size, presumably in order to improve the gas-phase resolution, results in a larger error in the calculation of the vaporization rate. The results are presented in Fig. 8.15, which shows the error in the numerical calculation of the vaporization rate when the gas-phase conditions are used at the selected gas nodes for the calculation of the transfer number. The comparisons are made with the exact solution.

Convective Case

Rangel and Sirignano (1989b) have shown for the convective case that a point-source approximation plus a free stream does not provide an accurate velocity field around the droplet (even in the inviscid limit). A point-source, doublet, and free-stream combination provides a much more accurate description. In this case, the flow field around the droplet depends on both the vaporization rate and the droplet diameter. There are still inaccuracies besides the viscous effects on the velocity field. For example, with the point-source approximation, the vaporization mass flux is assumed to be uniform over the droplet surface, which is not accurate.

We consider the case of steady, axisymmetric vaporization of an isolated droplet in an otherwise uniform flow of velocity u_∞. Because there is no analytical solution for the full Navier–Stokes and passive scalar equations, we explore the behavior of the numerical solutions. We are interested in the numerical coupling between a simplified droplet model and the gas-phase numerical calculation. To simplify the analysis further, consider a constant-density flow. A variable-density calculation would be more appropriate if large temperature gradients were involved, but such a consideration is not required for the purpose of our evaluation. Under the previous assumptions, the velocity field may be directly obtained from the continuity equation. The simplest solution is a point source of strength $\dot{m}/\rho$ located at the center of the droplet. The combined velocity potential for the source and uniform flow is found by superposition:

$$\phi = u_\infty x + \frac{\dot{m}}{4\pi\rho} \frac{1}{(x^2 + y^2)^{1/2}}, \tag{8.83}$$

where ϕ is the velocity potential so that $v = -\nabla\phi$. x is a coordinate along the axis of symmetry, and y is the radial coordinate normal to x. Because the strength of the point source depends on the vaporization rate, the velocity field is still coupled to the fuel mass fraction field and the latter must be obtained from the solution of the species equation. This is the most common form of the point-source approximation used in spray computations. Figure 8.16 shows the streamlines for a case in which the Stefan velocity $\dot{m}/4\pi R^2$ is equal to the uniform flow velocity u_∞. The contour of the droplet is also shown in the figure, but note that the boundary conditions on the droplet surface are not satisfied. Therefore the front stagnation point occurs at the surface of the droplet and would in fact occur inside the droplet if the uniform-flow velocity were higher than the Stefan velocity (weak vaporization). An additional violation of the physics exists because the blowing velocity is larger in the rear of the droplet than it is in the front. These shortcomings of the point-source approximation are expected to become negligible as the gas-phase grid size is increased with respect to the droplet size because the far field is more properly represented.

An improvement of the near-velocity field can be obtained if the velocity potential for a doublet is superposed on that of the point source and the uniform flow.

Contour interval: 2.42 E – 01 From –1.00E + 00 To 3.83E + 00

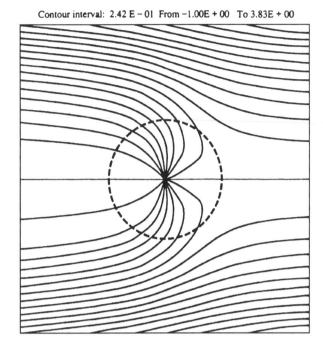

Figure 8.16. Streamlines for a point source in a uniform flow. (Rangel and Sirignano, 1989b, with permission of *Numerical Heat Transfer.*)

The new velocity potential then is

$$\phi = u_\infty x + \frac{\dot{m}}{4\pi\rho} \frac{1}{(x^2 + y^2)^{1/2}} + \frac{u_\infty R^3}{2} \frac{x}{(x^2 + y^2)^{3/2}}. \tag{8.84}$$

The new velocity potential satisfies the condition of zero normal velocity at the droplet surface for the case of zero vaporization. For the vaporizing case, the forward stagnation point is correctly located upstream of the droplet surface. A rear stagnation point falls inside the droplet but the solution for $x^2 + y^2 < R^2$ (inside the droplet) is irrelevant. The new Stefan velocity is uniform along the droplet surface, which is still physically incorrect but that may be sufficient when only an average vaporization rate is needed. The streamlines for the system of point source, doublet, and uniform flow when the blowing velocity is equal to the uniform-flow velocity are shown in Fig. 8.17.

For the sake of simplicity, we assume that the fuel mass fraction at the droplet surface is known or, equivalently, that the droplet-surface temperature is known by means of an equilibrium relation. In practical situations, the droplet-surface temperature is not known, and therefore the species equation and the liquid and the gas energy equations are all coupled. The species equation is solved by standard finite differences with a control-volume approach (Patankar, 1980). The gas-phase numerical cells are hollow cylinders of inner radius $(j - 1/2)\Delta y$, height Δx, and thickness Δy. However, the cells located at the axis of symmetry ($y = 0$) are solid cylinders of radius $1/2\Delta y$ and height Δx.

The velocity field is given by the velocity potential [Eq. (8.83) or Eq. (8.84)], and because the vaporization rate is not known a priori, the solution must be iterated until convergence is achieved. The vaporization model used is basically the stagnant-film model (Abramzon and Sirignano, 1989) with a correction to account

Contour interval: 2.38E − 01 From −1.00E + 00 To 3.76E + 00

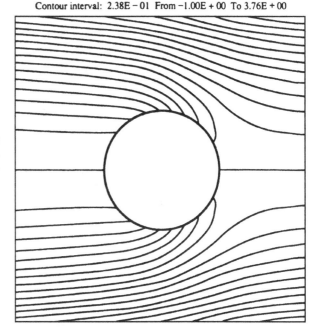

Figure 8.17. Streamlines for a doublet and a point source in uniform flow. (Rangel and Sirignano, 1989b, with permission of *Numerical Heat Transfer.*)

for convective effects. In terms of the mass transfer number, the vaporization rate is given by

$$\dot{m} = 4\pi \rho D C_{\mathrm{Re}} \ln\left(1 + \frac{Y_{Fs} - Y_{F\infty}^*}{1 - Y_{Fs}}\right), \tag{8.85}$$

where $Y_{F\infty}^*$ is the value of the fuel mass fraction used as the ambient condition (because this value is an average of the fuel mass fraction in the vicinity of the droplet, it differs from the actual fuel mass fraction at infinity $Y_{F\infty}$). The convective correction is that suggested by Clift et al. (1978):

$$\mathrm{Re} \geq 1, \quad C_{\mathrm{Re}} = 0.5\left[1 + (1 + \mathrm{Re})^{1/3}\mathrm{Re}^{0.077}\right],$$

$$\mathrm{Re} < 1, \quad C_{\mathrm{Re}} = 0.5\left[1 + (1 + \mathrm{Re})^{1/3}\right]. \tag{8.86}$$

Here, the Reynolds number is based on droplet diameter.

To isolate the effect of the flow model from that of the grid-to-drop ratio, the following results are obtained with the use of a very fine mesh. No significant differences were observed between results obtained with a grid-to-drop ratio of 0.2 versus 0.1. The grid-to-drop ratio is defined as the grid size $\Delta y = \Delta x$ divided by the droplet radius R. Results are presented for the finer mesh in Fig. 8.18, where the fuel mass fraction contours, normalized by Y_{Fs}, are shown. This case corresponds to a droplet with $Y_{Fs} = 0.5$, which is reported by Conner (1984). The source and doublet combination produces a better approximation to the detailed numerical solution than does the point source alone. The point-source model implies a local vaporization rate (Stefan flow) that is larger in the rear of the droplet than in the front of it. On the other hand, the point source plus doublet model implies a uniform vaporization rate along the droplet boundary. As a result, the two models result in a thinner mass

Contour interval: 1.00E−01 From 0.00E+00 To 1.00E+00

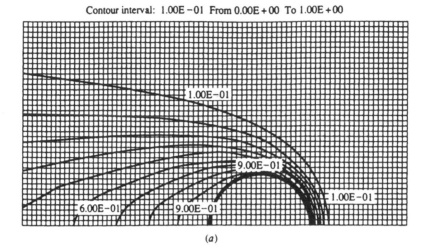

(a)

Contour interval: 1.00E−01 From 0.00E+00 To 1.00E+00

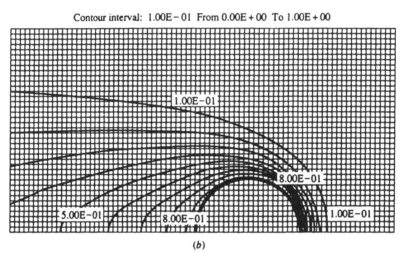

(b)

Figure 8.18. Fuel mass fraction contours for a droplet in an axisymmetric convective flow: (a) point source in uniform flow, (b) point source and doublet in uniform flow. (Rangel and Sirignano, 1989b, with permission of *Numerical Heat Transfer.*)

fraction boundary layer near the front stagnation point and a longer mass fraction wake than the exact calculation, but this effect is greater when the point source is used alone. Because no viscous effects are considered, neither model can simulate the separated wakes reported by Conner and Elghobashi (1987).

For each of the two velocity fields considered, i.e., point source in uniform flow and point source plus doublet in uniform flow, we examine the numerical solution of Reynolds numbers between 2.5 and 100, and for different gas-grid-size to droplet-size ratios. The vaporization rate obtained with the fuel mass fraction in the vicinity of the droplet $Y_{F\infty}^*$ is compared against the one obtained with the actual fuel mass fraction at infinity $Y_{F\infty}$. Whenever possible, a comparison is also made with the finite-difference results of Conner and Elghobashi (1987), who performed detailed calculations of the fuel mass fraction field around a droplet ith specified surface fuel mass fraction.

Contour interval: 1.00E−02 From 0.00E+00 To 1.00E−01

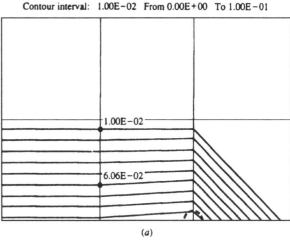

(a)

Contour interval: 1.01E−02 From 0.00E+00 To 1.01E−01

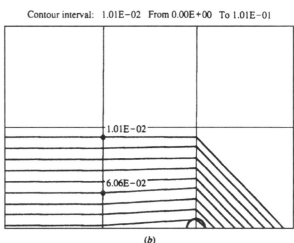

(b)

Figure 8.19. Fuel mass fraction contours for a grid-to-drop ratio of 10: (a) point source in uniform flow, (b) point source and doublet in uniform flow. (Rangel and Sirignano, 1989b, with permission of *Numerical Heat Transfer.*)

Figures 8.19–8.21 show the fuel mass fraction profile for grid-to-drop ratios of 10, 3, and 1 and for Re = 10. In all cases, the fuel mass fraction at the droplet surface is prescribed as $Y_{Fs} = 0.85$. The fuel mass fraction at infinity is zero ($Y_{F\infty} = 0$), and the mass transfer number B_F is formed with this actual value of the fuel mass fraction at infinity. This last step fully an couples the mass fraction solution from the velocity solution because the vaporization rate is determined by Eq. (8.85) with $Y_{F\infty}^* = Y_{F\infty}$. The mass transfer number is $B_F = 5.67$, and the dimensionless vaporization is $\dot{m} = 3.47$. This last result is within 5% of the value of 3.30 reported by Conner and Elghobashi. This value is obtained from extrapolation of the Sherwood number correlation reported by those authors in Fig. 13 of their paper.

In a numerical calculation, the value of $Y_{F\infty}$ is not readily available. It is therefore approximated by $Y_{F\infty}^*$, which is a suitable average in the neighborhood of the droplet. The next set of results, summarized in Table 8.3, illustrates the error in the vaporization rate when the gas properties of the nearest gas node are used for the calculation of the transfer number. Note that, in this case, the nearest gas node coincides with the droplet. The error increases as the grid-to-drop ratio decreases

Table 8.3. *Dimensionless vaporization rate and the percent error[a] for several values of the Reynolds number and different grid-to-drop ratios*

Re	$\Delta x' = \infty$	10	5	3	2
2.5	2.494	2.193	1.642	n.c.[b]	n.c.
	(0)	(12.1)	(34.2)		
10	3.467	3.283	2.818	1.942	n.c.
	(0)	(5.3)	(18.7)	(44.0)	
25	4.549	4.415	4.045	3.185	n.c.
	(0)	(2.9)	(11.1)	(30.0)	
100	7.246	7.158	6.894	6.140	n.c.
	(0)	(1.2)	(4.9)	(15.3)	

[a] The percent error is given in parentheses. The error is always negative.

[b] n.c. indicates that the solution does not converge.

because the transfer number is increasingly underpredicted. This occurs because as the grid-to-drop ratio decreases, the fuel mass fraction at the gas cell that coincides with the droplet increases, regardless of the value of Y_{Fs}. Moreover, for grid-to-drop ratios less than or equal to 2 (3 for Re = 2.5), the solution does not converge (indicated by n.c. in Table 8.3) because the fuel mass fraction at the gas node coinciding with the droplet is greater than Y_{Fs}. For a given grid-to-drop ratio, the error in the vaporization rate decreases as the Reynolds number increases because the higher

Contour interval: 6.41E−02 From 0.00E+00 To 6.41E−01

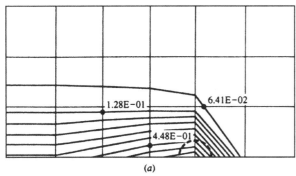

(a)

Contour interval: 7.57E−02 From 0.00E+00 To 7.57E−01

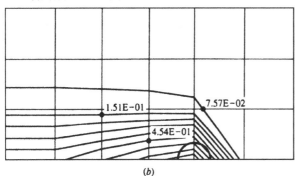

(b)

Figure 8.20. Fuel mass fraction contours for a grid-to-drop ratio of 3: (a) point source in uniform flow, (b) point source and doublet in uniform flow. (Rangel and Sirignano, 1989b, with permission of *Numerical Heat Transfer.*)

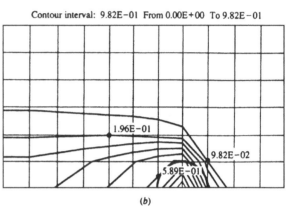

Figure 8.21. Fuel mass fraction contours for a grid-to-drop ratio of 1: (a) point source in uniform flow, (b) point source and doublet in uniform flow. (Rangel and Sirignano, 1989b, with permission of *Numerical Heat Transfer.*)

relative velocity reduces the fuel concentration at the location of the droplet. This implies that, in most practical situations, the error will be larger in the early portion of the droplet lifetime and will decrease as the droplet decelerates in the carrier gas.

The increased error in the vaporization rate as the grid-to-drop ratio decreases resides, for the most part, in the way in which the transfer number is evaluated. An average value of the fuel mass fraction in the neighborhood of the droplet may lead to the errors just described. Because it is unlikely that the droplet will coincide with a gas node in a real situation, it is useful to evaluate the error for cases in which the droplet is between two gas nodes rather than coinciding with one of them. Details of the analysis can be found in the work of Rangel and Sirignano (1989b).

Table 8.4 contains the errors in the vaporization rate for the case of $Re = 10$ for several grid-to-drop ratios and three different relative locations of the droplet with respect to the grid. When $f < 0.5$, the droplet is closer to the downstream-gas node, whereas for $f > 0.5$, it is closer to the upstream-gas node. The case of $f = 0$ or $f = 1$ corresponds to the droplet's coinciding with a gas node. To calculate the vaporization rate, the transfer number is evaluated with the average conditions at the location of the droplet. The fuel mass fraction $Y_{F\infty}^*$ is calculated by the average of the values at the two nearest nodes. The errors appearing in Table 8.4 show the same trend as those included in Table 8.3, namely, the error increases as the grid-to-drop ratio decreases. Some differences are observed, depending on the relative location of the droplet with respect to the gas mesh. These differences are due to the variations in the computed velocity field and are insignificant for our purposes.

Table 8.4. *Percent error in the vaporization rate for*
Re = 10 *and several values of the grid-to-drop ratio*

f	$\Delta x' = 10$	5	3	2
0	5.3	18.7	44.0	n.c.
	[3.4]	[5.6]	[0.0]	[5.3]
0.25	4.6	26.7	40.0	42.6
	[3.1]	[6.8]	[8.3]	[0.0]
0.75	4.2	14.6	31.2	49.9
	[3.6]	[7.8]	[2.0]	[0.0]

Note: The values inside the brackets indicate the error
when the transfer number is evaluated with the aid of
Eq. (8.87). The error is always negative. The abbreviation
n.c. means *not converged.*

As it was previously noted, it makes more sense to evaluate the transfer number
by using the conditions in the approaching flow, upstream of the fuel mass fraction
boundary layer. Even if the Reynolds number is low and there is no viscous boundary layer, one can define a fuel mass fraction layer with a characteristic thickness so
that the ambient conditions are evaluated just beyond this layer in the approaching
flow. Let δ_{99} be the point, measured from the droplet center along the axis of symmetry, where the fuel mass fraction profile has achieved 99% of its total variation
across the layer. Rangel and Sirignano (1989b) provide here a correlation that fits
the data of Conner (1984) within 5%:

$$\frac{\delta_{99}}{R} = \left(1 + \frac{5}{\text{Re}^{0.6}}\right)(1 + B_F)^{0.17}, \tag{8.87}$$

as can be seen in Fig. 8.22. When the mass transfer number B_F^* is evaluated with
the conditions δ_{99}, the error in the vaporization rate is greatly reduced, as shown
by the entries in square brackets in Table 8.4. The errors are all less than 9%, and
convergence is achieved for any droplet relative location down to a grid-to-drop
ratio of 2. Table 8.5 contains the error in the vaporization rate when the transfer
number is calculated in the manner just described for the same Reynolds numbers

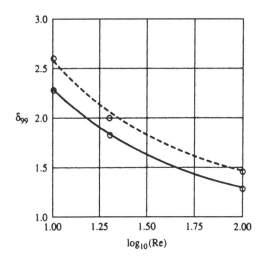

Figure 8.22. Thickness of the fuel mass fraction
layer at the front stagnation point of a vaporizing droplet. Solid curve for $B = 1$. Open circles
are from the numerical data of Dukowicz (1980).
(Rangel and Sirignano, 1989b, with permission of
Numerical Heat Transfer.)

Table 8.5. *Percent error in the vaporization rate for several values of the Reynolds number and different grid-to-drop ratios when the conditions upstream of the droplet are used to evaluate the transfer number*

Re	$\Delta x' = 10$	5	3	2
2.5	5.3	0.0	0.8	4.6
10	3.4	5.6	0.0	5.3
25	2.1	5.1	2.5	10.0
100	1.0	2.9	4.9	6.9

Note: The error is always negative.

Figure 8.23. Schematic of the transfer number evaluation method: (a) with conditions given at the droplet location, (b) with conditions given upstream of the droplet. (Rangel and Sirignano, 1989b, with permission of *Numerical Heat Transfer.*)

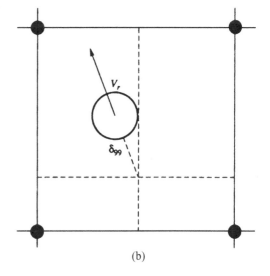

(a)

(b)

and grid-to-drop ratios used in Table 8.3. To evaluate the fuel mass fraction at the location defined by δ_{99}, linear interpolation between the two neighboring gas nodes is used. The errors appearing in Table 8.5 are all substantially lower than those of Table 6.3 and never exceed 10%. This error is acceptable, considering the order of magnitude of other errors and uncertainties associated with a numerical spray computation (e.g., physical properties, drag correlations, chemical kinetics). The procedure just described for the evaluation of the appropriate transfer number may be readily generalized to situations in which there is no alignment between the approaching flow and the gas-phase numerical grid. In the two-dimensional case illustrated in Fig. 8.23, a droplet interacts with four neighboring gas nodes represented by the filled circles at the corners of the grid shown in Fig. 8.23(a). The size of the droplet is also indicated in the figure, although the droplet is represented as a point source. At any given time, the point source divides the rectangular area between the nodes into four rectangular subareas, as indicated by the dashed lines. The standard method for the evaluation of the transfer number consists of averaging the gas properties at the location of the droplet by means of the formula

$$Y_{F\infty}^{*} = f_{1}Y_{F1} + f_{2}Y_{F2} + f_{3}Y_{F3} + f_{4}Y_{F4}, \qquad (8.88)$$

where the subscripts 1, 2, 3, and 4 denote the four gas nodes surrounding the droplet and f_{i} is the ratio of the rectangular area opposite node i to the total area in Fig. 8.23. As previously shown, large errors may be incurred when the properties at the location of the droplet are used if the grid-to-drop ratio is less than 10. To evaluate the transfer number by use of the condition of the approaching flow, the following procedure is recommended [Fig. 8.23(b)]. The relative velocity between the gas and the droplet can be calculated. (This does not represent an increase in computational time since the relative velocity is already calculated for the solution of the droplet trajectory.) The magnitude of the relative velocity is V_{r}, and it has components u_{r} and v_{r} in an x–y coordinate system aligned with the grid of Fig. 8.23(b). Next, the thickness of the fuel mass fraction layer at the front stagnation point of the droplet is calculated with Eq. (8.87), and this value determines the location at which the fuel mass fraction is evaluated, i.e.,

$$\delta_{x} = -\frac{u_{r}}{V_{r}}\delta_{99}, \qquad \delta_{y} = \frac{v_{r}}{V_{r}}\delta_{99}, \qquad (8.89)$$

measured from the location of the point source (the droplet center). The point determined with the aid of Eq. (8.89) divides the rectangular grid area into the four subareas delineated by the dashed lines in Fig. 8.23(b). The fuel mass fraction at this location is then determined with the use of Eq. (8.88) with the area ratios of Fig. 8.23(b). The value of the fuel mass fraction is the one used to calculate the mass transfer number of Eq. (8.87).

9 Spray Applications

The spray equations have been studied and solved for many applications: single-component and multicomponent liquids, high-temperature and low-temperature gas environments, monodisperse and polydisperse droplet-size distributions, steady and unsteady flows, one-dimensional and multidimensional flows, laminar and turbulent regimes, subcritical and supercritical thermodynamic regimes, and recirculating (strongly elliptical) and nonrecirculating (hyperbolic, parabolic, or weakly elliptic) flows. The analyses discussed here will not be totally inclusive of all of the interesting analyses that have been performed; rather, only a selection is presented.

Spray flows can be classified in various ways. One important issue concerns whether the gas is turbulent or laminar. In this chapter, only laminar flows are considered; the turbulent situation is discussed in Chapter 10. Another issue concerns whether thermodynamic conditions are subcritical, on the one hand, or near critical to supercritical, on the other hand.

In the most general spray case, the gas and the droplets are not in thermal and kinematic equilibria, that is, the droplet temperature and the droplet velocity differ from those properties of the surrounding gas. Of course, heat transfer and drag forces result in the tendency to move toward equilibrium. The equilibrium case is sometimes described as a locally homogeneous flow. It is possible to have thermal equilibrium or kinematic equilibrium without the other. When thermal equilibrium exists, the analysis described in Chapter 7 is simplified because the droplet temperature T_l can be set equal to the gas temperature T and Eq. (7.82) or its alternative forms, Eq. (7.83) or Eq. (7.87), can be avoided. For the kinematic-equilibrium case, the droplet velocity u_l can be set equal to the gas velocity u and Eq. (7.76a) or its other forms, Eqs. (7.76b), (7.77), (7.78), or (7.86), can be bypassed. The bypassed equations can be used after the temperature and the velocity properties are resolved to determine the droplet-heating rates and the drag forces.

In this chapter, our analytical examples do not contain any assumptions about thermal or kinematic equilibrium, that is, the so-called separated spray flows are studied whereby the droplet and the gas properties are separated in their values. The relaxation time to reach equilibrium is considered comparable to or longer than the droplet lifetime or residence time.

The analyses in this chapter are divided into three groups: one-dimensional spray phenomena, counterflow (planar and axisymmetric) spray behavior, and two-dimensional (steady and unsteady) parallel droplet-stream behavior. The fourth group of analyses addresses recirculating spray flows. These particular computations with recirculating flow were intended to simulate practical liquid-fueled combustors; so turbulent behavior is included in the model. This fourth class will therefore be discussed in Chapter 10.

Some of the earlier research used an Eulerian description of the two-continua formulation. For example, Seth et al. (1978) and Aggarwal and Sirignano (1982, 1984) performed one-dimensional analyses of flames propagating through fuel sprays. They did not resolve flame structure to a scale smaller than the average spacing between droplets and suffered some artificial diffusion associated with the Eulerian formulation. Sirignano (1985b) considered a simple vaporizing spray to develop integral solutions. Other research addressed the two-continua problem with a Lagrangian description. Aggarwal et al. (1981, 1983, 1985) studied axisymmetric idealized two-continua droplet-laden-jet-flow problems to establish some of the computational foundations and performance analyses for the Lagrangian method. They determined the trade-offs between computational time and accuracy. Among other issues, some guidance is given concerning the number of characteristics (or, equivalently, average particles) that should be selected to yield consistent accuracy with the Eulerian mesh selection for the gas-phase equations. The multicontinua formulation was used by Raju and Sirignano (1989, 1990a) and Molavi and Sirignano (1988). They utilized the Lagrangian approach to describe the droplet behavior. Raju and Sirignano and Molavi and Sirignano considered axisymmetric and planar two-dimensional idealized configurations related to center-body and dump combustors. Single-component and multicomponent fuels were considered.

The subgrid vaporization models used in these spray studies have evolved over the years. Originally, spherically symmetric models with Ranz–Marshall corrections were used. During the middle 1980s, the Tong and Sirignano model came into use. More recently, the Abramzon and Sirignano model has been used. Also, experimental drag coefficients, such as those of Ingebo (1956) and Renksizbulut and Yuen (1983), have been replaced with computational correlations, such as those of Haywood et al. (1989) and Chiang et al. (1992) for input to spray models.

In Sections 9.1, 9.2, and 9.6 we discuss spray combustion phenomena by using the two-continua method, and in Sections 9.3, 9.4, and 9.5, we use the discrete-particle method. The general conclusions of the one-dimensional planar, two-dimensional planar, and spherically symmetric studies are that (i) resolution on the scale of the spacing between droplets is important in the determination of ignition phenomena and flame structure, (ii) more than one flame zone can exist at any instant with spray combustion, (iii) an inherent unsteadiness results for the flame structure, (iv) diffusionlike or premixedlike flames can occur, (v) ignition and flame propagation for sprays can sometimes be faster than for gaseous mixtures of identical stoichimetry, and (vi) droplets tend to burn in a cloud or group rather than individually.

The collective results to come from various sections of this chapter strongly indicate that there is a serious flaw in the simple concept of a flame sweeping a heterogeneous combustible mixture and resulting in the conversion of liquid fuel to gaseous products in a short distance. Except for very small droplets or very volatile liquids, the flame structure will be much more complex. Diffusion flame structures will be common. Transient droplet heating, droplet-motion modification that is due to drag, and internal liquid circulation have been shown to be important factors in the predictions.

Note that differences exist in some of the spray modelling details among the six problems discussed in this chapter. These variations occurred primarily because the problems were solved during different periods of time while advances in droplet modelling were occurring simultaneously. Sometimes, a particular preference of a coauthor prevailed.

Some noteworthy contributions to spray theory that are not discussed in detail in this book are the works of Borghi and Loison (1992), Bracco (1974), Chen et al. (1988), Chigier (1976), Gupta and Bracco (1978), Kerstein (1984), and Sirignano (1985a, 1986).

9.1 Spherically Symmetric Spray Phenomena

One-dimensional, planar unsteady spray configurations have been studied extensively. Various subgrid vaporization models were studied by Aggarwal et al. (1984), who found that substantial global differences in the two-phase flow can result from different vaporization models; it is clear that accurate subgrid modelling of vaporization is required. Other one-dimensional planar studies based on the discrete-particle formulation include those of Aggarwal and Sirignano (1984, 1985a, 1985b, 1986). Those planar one-dimensional studies are discussed in Section 7.3. Continillo and Sirignano (1988) extended the study to a spherically symmetric spray configuration. Aggarwal (1987, 1989) and Continillo and Sirignano (1991) considered multicomponent liquid sprays in one-dimensional, planar, and spherically symmetric configurations. Generally vaporization rates, ignition delays, and flame-propagation rates were predicted.

In this section, an outline of the analysis and results of Continillo and Sirignano (1988, 1991) is given. The case of unsteady, spherically symmetric flame propagation through a multicomponent fuel spray is considered. Ignition occurs at the spherical center with subsequent outward radical propagation of the flame. A two-continua, Eulerian–Lagrangian formulation is used. A modified Abramzon–Sirignano droplet-vaporization and -heating model is used for the calculations.

Internal droplet circulation, viscous dissipation, spatial variation of pressure, and radiative heating is neglected. One-step chemical kinetics is considered. Although the polydisperse spray case was analyzed by Continillo and Sirignano (1988), only monodisperse spray calculations are made. Unitary Lewis number with constant ρD is assumed. The void volume fraction is set to unity. The field is unbounded and quiescent at infinity, as prescribed in the work of Continillo and Sirignano (1991). Note that the earlier work did consider a closed-volume calculation. See Continillo and Sirignano (1988).

Equations (7.62), (7.63), and (7.81) can be recast in spherically symmetric form by use of the previously mentioned assumptions. In particular, we obtain the following equations for the gas-phase properties:

$$\frac{\partial \bar{\rho}}{\partial t} + \frac{1}{r^2}\frac{\partial (r^2 \bar{\rho} u)}{\partial r} = \dot{M}, \tag{9.1}$$

$$\bar{\rho}\frac{\partial Y_m}{\partial t} + \bar{\rho} u\frac{\partial Y_m}{\partial r} - \frac{1}{r^2}\frac{\partial}{\partial r}\left(\rho D r^2 \frac{\partial Y_m}{\partial r}\right) = (\varepsilon_m - Y_m)\dot{M} + \bar{\rho}\dot{w}_m, \tag{9.2}$$

$$\bar{\rho} c_p\frac{\partial T}{\partial t} + \bar{\rho} c_p u\frac{\partial T}{\partial r} - \frac{1}{r^2}\left(\rho D r^2 c_p\frac{\partial T}{\partial r}\right) = \bar{\rho}\sum \dot{w}_m Q_m - \dot{M}(h - h_s + L_{\text{eff}}). \tag{9.3}$$

Note that the energy equation has been simplified by the neglect of certain transport terms related to variations in mass fraction. The five boundary conditions on the gas mass fraction and the gas temperature are

$$\frac{\partial Y_m}{\partial r}(0, t) = \frac{\partial T}{\partial r}(0, t) = \frac{\partial Y_m}{\partial r}(\infty, t) = \frac{\partial T}{\partial r}(\infty, t) = u(0, t) = 0. \tag{9.4}$$

The initial conditions are that pressure is uniform and temperature and mass fraction are specified.

The Abramzon–Sirignano droplet model for multicomponent fuels, as discussed in Chapter 4, is used for the spray calculations. It is modified when the effective thermal and mass diffusivities are set to the actual diffusivities, that is, $\chi = \chi_d = 1$ in Eqs. (3.51) and (4.8a). Hexane–decane liquid mixtures are considered. One-step oxidation kinetics (with different preexponential constants for each component) is considered; decane kinetics is faster by a factor of 1.5.

As the flame propagates, some vaporization occurs ahead of the flame. This fuel vapor has a higher fraction of the more volatile component than the original liquid fraction. If sufficient prevaporization and mixing with oxygen has occurred, a premixed character of the flame exists. Some droplets are swept by the propagating flame; vaporization of these droplets occurs in a high-temperature region behind the flame, with the fuel vapor diffusing to the flame from behind. The flame then has a diffusion flame character with oxygen and fuel vapor diffusing to the flame from opposite sides. A flame can have both a diffusionlike character and a premixedlike character. As initial droplet diameter increases or as the initial mass fraction of the volatile component decreases, the flame behavior moves from a premixed flame character to a diffusion flame character. When the diffusion flame character is dominant, vaporization is the slowest process and is rate controlling. For the case in which the premixed character is dominant, chemical kinetics is slower than the vaporization and becomes the rate-controlling factor. Therefore, when vaporization is rate controlling, an increase in the initial fraction of hexane, the more volatile component, results in a faster flame-propagation rate. On the other hand, when chemical kinetics is rate controlling, an increase in the initial hexane mass fraction results in a decrease in the propagation rate. Of course, an increase in the initial droplet diameter or a decrease in the initial hexane mass fraction yields a decrease in the vaporization rate, making the likelihood of vaporization-rate control greater. These intuitive predictions are reflected in the computational results shown in Figs. 9.1 and 9.2 for initial droplet diameters of 50 and 17.5 μm, respectively. Note that, for the

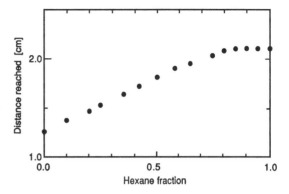

Figure 9.1. Distance reached by the flame after 10 ms as a function of fuel composition. Initial droplet diameter, 50 μm. (Continillo and Sirignano, 1991, with permission of Springer-Verlag.)

17.5-μm case, the pure hexane limit results in a lower speed than the pure decane limit.

Figures 9.3 and 9.4 show temperature, hexane mass fraction, decane mass fraction, and oxygen mass fraction profiles for two different initial droplet diameters: 50 and 6 μm, respectively. In Fig. 9.3 for the larger droplets, the vaporization is rate controlling and a fuel-rich region exists behind the flame front at the various times indicated. A different behavior is seen in Fig. 9.4, in which the smaller droplets vaporize faster, creating the zone of maximum fuel-vapor concentration in front of the flame. For the large-droplet case, there is a diffusion flame with some premixed flame character as well. For the smaller droplets, the character is primarily like that of a premixed flame. The more volatile component tends to vaporize earlier and further ahead of the flame than the less volatile component.

9.2 Counterflow Spray Flows

It is well known that two opposing jet flows result in a classical stagnation-point flow. Studies have been made for reacting flows in which one of the jets is laden with fuel droplets. Continillo and Sirignano (1990) performed a similarity analysis for a planar configuration by using one-step chemical kinetics; n-octane fuel was considered. Gutheil and Sirignano (1998) extended the similarity analysis, considering both planar and axisymmetric configurations with detailed chemical kinetics

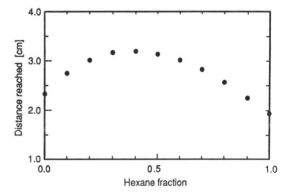

Figure 9.2. Distance reached by the flame after 5 ms as a function of fuel composition. Initial droplet diameter, 17.5 μm. (Continillo and Sirignano, 1991, with permission of Springer-Verlag.)

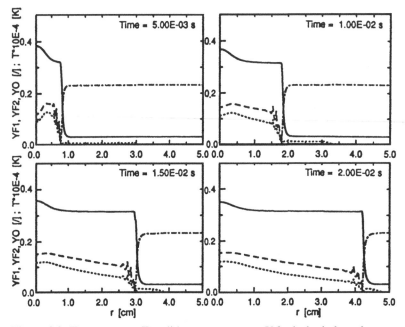

Figure 9.3. Temperature T, solid curves; oxygen YO, dashed–dotted curves; hexane $YF1$, dotted curves; and decane $YF2$, dashed curves, as functions of the radial coordinate at successive times. Base case. Initial droplet diameter, 50 μm; fuel composition: 50%–50%. (Continillo and Sirignano, 1991, with permission of Springer-Verlag.)

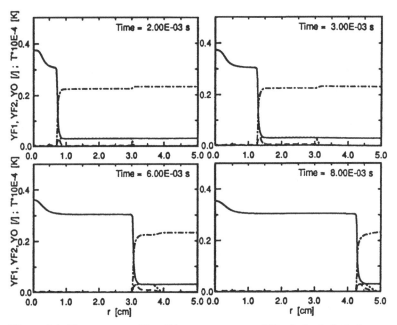

Figure 9.4. Temperature T, solid curves; oxygen YO, dashed–dotted curves; hexane $YF1$, dotted curves; and decane $YF2$, dashed curves, as functions of the radial coordinate at successive times. Initial droplet diameter, 6 μm; fuel composition, 50%–50%. (Continillo and Sirignano, 1991, with permission of Springer-Verlag.)

for both n-heptane and methanol fuels. Other spray stagnation-point analyses were performed by Li et al. (1992), Chen et al. (1992), Gao et al. (1996), and Greenberg and Sarig (1996). Puri and Libby (1989) considered a single stream of droplets in a counterflow. See the review of spray stagnation flows by Li (1997).

Consider now the formulation of the governing equations, following from the works of Continillo and Sirignano and Gutheil and Sirignano. A two-continua method is used for the monodisperse two-phase flow. The gas-phase properties are described in an Eulerian frame, and a Lagrangian frame is used for the droplet properties. y is taken as the original flow direction of the jets; x is the transverse direction (the radial direction in the axisymmetric configuration). The variables are nondimensionalized by the ambient conditions at $y = -\infty$, v_0^-, ρ_0^-, T_0^-, and $p_0^- = \rho_0^- R T_0^-$, as the reference values for gas kinematic viscosity, density, temperature, and pressure. The two outer flows are potential flows described by

$$u_0^+ = K^+ x, \quad u_0^- = K^- x, \tag{9.5a}$$

for the x component of velocity and by

$$v_0^+ = -K^+ y, \quad v_0^- = -K^- y, \tag{9.5b}$$

for the y component of velocity, where K^+ and K^- are the strain rates. The balances of static pressure and dynamic pressure require that

$$K \equiv K^- = \sqrt{\rho_0^+/\rho_0^-}\, K^+. \tag{9.5c}$$

The time is nondimensionalized by the reciprocal of K, and the velocity nondimensionalization uses $\sqrt{v_0^-}\, K$. The reference length is given by the quantity $\sqrt{v_0^-/K}$. The standard similarity transformation yields

$$\eta \equiv \int_0^y \rho\, dy', \tag{9.5d}$$

$$f \equiv \int_0^\eta \frac{u}{x}\, d\eta', \tag{9.5e}$$

where u is the transverse or radial velocity component. Also, we have for the axial-velocity component,

$$v = \left[(\alpha+1) f - \int_0^\eta (\dot{M} T)\, d\eta' \right] T$$
$$\equiv -[(\alpha+1) f + f_v] T. \tag{9.6}$$

Here $\alpha = 0$ for planar counterflow and $\alpha = 1$ for axisymmetric counterflow. The steady form of continuity equation (7.62) is automatically satisfied.

The classical boundary-layer approximation is made on gas-phase momentum equation (7.75) after the neglect of gravity and the assumption that the void volume is unity. The y-momentum equation has been reduced to a statement of negligible change in pressure in the y direction by means of the boundary-layer approximation. The x-momentum equation for a monodisperse spray is represented as

$$\frac{d}{d\eta}\left(\frac{d^2 f}{d\eta^2} \right) + [(\alpha+1) f + f_v]\frac{d^2 f}{d\eta^2} = \left(\frac{df}{d\eta} \right)^2 - T - \frac{[\dot{M}(u_l - u) - F_{Dx}]T}{x}, \tag{9.7}$$

where the product of gas density and dynamic viscosity is considered to be constant.

Under the previously mentioned approximations, gas-species-conservation equation (7.73) and gas energy equation (7.81) become

$$\frac{1}{\mathrm{Sc}}\frac{\mathrm{d}^2 Y_m}{\mathrm{d}\eta^2} + [(\alpha+1)f + f_v]\frac{\mathrm{d}Y_m}{\mathrm{d}\eta} = -T\dot{w}_m - (\varepsilon_m - Y_m)\dot{M}T, \tag{9.8}$$

$$\frac{1}{P_r}\frac{\mathrm{d}^2 T}{\mathrm{d}\eta^2} + [(\alpha+1)f + f_v]\frac{\mathrm{d}T}{\mathrm{d}\eta} = T\sum Q_m\dot{w}_m - T\dot{M}(h - h_s + L_{\mathrm{eff}}). \tag{9.9}$$

Note that Gutheil and Sirignano (1998) generalized the treatment of the species conservation by considering detailed transport. That extension is not reflected here. Constant specific heat and equal binary diffusion coefficients are assumed here.

The gas-phase boundary conditions are prescribed at plus and minus infinity:

$$\frac{\mathrm{d}f}{\mathrm{d}\eta}(-\infty) = 1, \qquad Y_i(-\infty) = Y_{i-\infty},$$

$$\frac{\mathrm{d}f}{\mathrm{d}\eta}(\infty) = -\sqrt{\frac{\rho_{-\infty}}{\rho_\infty}}\frac{\mathrm{d}f}{\mathrm{d}z}(-\infty) = -\sqrt{T_\infty}, \tag{9.10}$$

$$T(\infty) = T_\infty, \qquad Y_i(\infty) = Y_{i\infty}. \tag{9.11}$$

We consider that all variables are continuous in η so that a formal matching procedure is not required at the interface between the two counterflows. This continuity is a result of the viscosity, conduction, and diffusion. The definition for $f(\eta)$ implies that $f(0) = 0$, which provides a third condition on the solution of the third-order differential equation.

The Abramzon–Sirignano droplet-heating and -vaporization model was utilized by Continillo and Sirignano and by Gutheil and Sirignano. One important modification was that the effective thermal diffusivity for the liquid was set equal to the actual thermal diffusivity. $\chi = 1$ was taken in Eq. (3.51) before it was solved with conditions (3.52a)–(3.52h).

The droplet-motion and -size variations were determined through a Lagrangian calculation with Eqs. (7.84)–(7.86). The droplet number density was determined locally by the steady-state form of Eq. (7.71). We have therefore for the u_l and the v_l components of the droplet velocity that

$$\frac{1}{x^\alpha}\frac{\partial(nx^\alpha u_l)}{\partial x} + \frac{\partial(nv_l)}{\partial y} = 0$$

or

$$\frac{\partial(nu_l)}{\partial x} + \alpha x^{-1}nu_l + \frac{\partial(nv_l)}{\partial y} = 0.$$

So if $u_l = K_l(\eta)x$ and n is independent of x, we have

$$\frac{1}{n}\frac{\partial(nv_l)}{\partial y} = -(\alpha+1)K_l(\eta). \tag{9.12}$$

This linear dependence on x will apply for low droplet Reynolds numbers. Note that, in the free stream outside the boundary layer, K_l will be a constant $K_{l,-\infty}$ independent of η. From Eqs. (3.54) and (7.90), a linear relation results for the droplet velocity. (Although Continillo and Sirignano did not consider the effect of Stefan convection on the droplet drag, its effect is included here.) Dividing the vector equation into two components and neglecting gravity, we have

$$\frac{du_l}{dt} = \frac{9\mu}{2\rho_l R^2 (1 + B_M)} (u - u_l),$$ (9.13a)

$$\frac{dv_l}{dt} = \frac{9\mu}{2\rho_l R^2 (1 + B_M)} (v - v_l).$$ (9.13b)

Equation (7.13a) can be transformed to

$$\frac{d(u_l/x)}{dt} = \frac{9\mu}{2\rho_l R^2 (1 + B_M)} \left(\frac{u}{x} - \frac{u_l}{x} \right) - \left(\frac{u_l}{x} \right)^2.$$ (9.13c)

For the self-similar solution, u/x, v, R, and B_M will not depend on x. Therefore u_l/x and v_l are independent of x. It follows that

$$\frac{1}{v_l} \frac{dv_l}{dt} + \frac{1}{n} \frac{dn}{dt} = -(\alpha + 1)K_l(\eta) = -\frac{1}{s}\frac{ds}{dt}$$ (9.14a)

if we define

$$s = e^{(\alpha+1) \int_{-\infty}^{y} \frac{K_l}{v_l} dy'}.$$ (9.14b)

Then

$$v_l n s = \text{constant} = (v_l n s)_{-\infty} = (v_l n)_{-\infty}.$$ (9.14c)

This droplet continuity equation allows the determination of the droplet number density given the droplet-velocity field and the initial upstream number density.

One might infer that the number density in Eq. (9.14c) goes to infinity as the droplet velocity goes to zero. This was suggested and discussed by Chen et al. (1992). However, the correct interpretation is more complex. Here, we consider two different situations. Under certain conditions, droplets move toward the opposed gas stream but never cross the stagnation plane. Other situations occur where the droplets cross the stagnation plane, then reverse direction, and move backward toward the stagnation plane.

First, consider the case in which the droplet never crosses the stagnation plane and no reversal of droplet direction occurs. Here there is no problem because the y component of droplet velocity goes to zero only as the x position of the droplet becomes infinite, that is, $s \to \infty$ in Eq. (9.14c) as $v_l \to 0$. So here n does not become infinite for the same reason that gas density at the stagnation plane remains finite.

The case in which the droplet reverses direction is more challenging because x is finite where $v_l = 0$. Here the original formulation in the work of Continillo and Sirignano (1990) determines $\text{div}(n\vec{V}_l) = 0$, where $\vec{V}_l$ is the droplet-velocity vector (u_l, v_l). Multivalued solutions exist when the droplet reverses; there will be distinct solutions for forward-moving and backward-moving droplets. So each solution applies only on its own sheet, that is, the drops are on one sheet before they reverse; after the first reversal, the backward-moving drops are on a second sheet. Then, after a second reversal, the drops are on a third sheet. Only by distinguishing the sheets can the divergence analysis be applied. Therefore a different Eq. (9.14c) applies for each sheet, yielding a different droplet number density n for each sheet. The source terms in the gas-phase equations account for the contributions from all of the sheets. Now consider the edge of the sheet where the first reversal occurs. This is a sink of droplets for the first sheet and source for the second sheet. The magnitudes of the source and the sink are equal because of the conservation of droplets. Note that the divergence becomes infinite there ($\mathrm{d}v_l/\mathrm{d}y \to \infty$ as $v_l \to 0$). The flux at the edge is not zero, even though the normal velocity component $v_l \to 0$ there. So $n \to \infty$ there and the flux remains finite (a mathematical abstraction because the volume where n is infinite has zero dimension); any finite volume has a finite number of droplets inside that volume. That point can be proved as follows. From the work of Continillo and Sirignano it may be seen that $(1/v_l)(\mathrm{d}v_l/\mathrm{d}t) = \mathrm{d}v_l/\mathrm{d}y$ and $(1/n)(\mathrm{d}n/\mathrm{d}t) = (v_l/n)(\mathrm{d}n/\mathrm{d}y)$. Define the y value at reversal as y_c. Near $y = y_c$, we have the approximation $v_l = k(y - y_c)^p$, where k is a constant and the exponent is bounded with $0 < p < 1$. Then $\mathrm{d}v_l/\mathrm{d}y$ is proportional to $(y - y_c)^{p-1}$, which becomes infinite at $y = y_c$. It follows that $(1/n)(\mathrm{d}n/\mathrm{d}y) = -(1/v_l)(\mathrm{d}v_l/\mathrm{d}y)$ in the limiting behavior; this assumes that s remains finite. So n is proportional to $(y - y_c)^q$, where q must be negative and nonzero. By continuity, nv_l must remain finite and nonzero so that $p = -q$. Now $\int_{y_c}^{y} n \, \mathrm{d}y'$ is proportional to $(y - y_c)^{1-p}$. Because $1 - p > 0$, the total number of droplets in the volume between $y = y_c$ and any other larger y value is finite, not infinite. The point therefore is that, although n becomes infinite in the limit, it remains integrable.

Continillo and Sirignano (1990) considered two opposed air streams with normal octane liquid droplets injected with the left-hand-side stream. Figure 9.5 shows that two flames occur in the stagnation region. They are fed primarily by fuel vapor diffusing from vaporizing droplets that have penetrated to the zone between the two flames. Some vaporization occurs before the droplets penetrate the left-hand flame; therefore that flame has a combined premixed and diffusion character. Figure 9.5 also shows the effect of varying strain on the behavior of the flow. Increasing the strain results in narrowing the two reaction zones and causing them to approach each other and the stagnation plane. Ultimately the increasing strain rate causes the two reaction zones to merge, and only one temperature peak is seen in Fig. 9.5(c). The study by Li et al. (1992) predicts the locations of the vaporization zone and the reaction zone as functions of the strain rate for heptane and methanol sprays transported in a nitrogen stream and mixing with a counterflowing oxygen–nitrogen stream. Extinction conditions are also predicted.

Gutheil and Sirignano (1998) also performed calculations for the counterflow sprays by using detailed kinetics and transport. Otherwise the approach was identical to that of Continillo and Sirignano. Although Continillo and Sirignano did not find that their n-octane droplets crossed the stagnation plane, Chen et al. (1992) and Puri and Libby (1989) did show that possibility. Gutheil and Sirignano also found that result for high strain rates, as shown in Fig. 9.6. Both the flame and the vaporization zones move to the air side of the counterflow, where they overlap with the droplets penetrating the flame. With the high strain rates, or equivalently with the low residence times, the droplets pass the stagnation plane and then turn back.

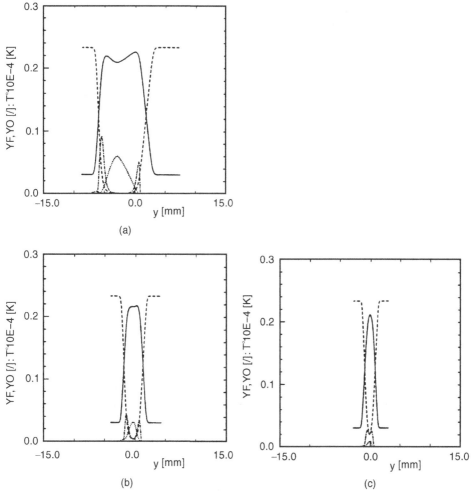

Figure 9.5. Influence of the strain rate. Profiles of fuel-vapor and oxygen mass fractions, temperature, and log of the reaction rate. Initial droplet diameter, $D_0 = 50\,\mu m$. (a) $K = 100\,\mathrm{s}^{-1}$, (b) $K = 300\,\mathrm{s}^{-1}$, (c) $K = 500\,\mathrm{s}^{-1}$. (Continillo and Sirignano, 1990, with permission of *Combustion and Flame*.)

They pass the stagnation plane again but now from the opposite direction. Then they turn once more, moving asymptotically toward the stagnation plane. The spray is no longer locally monodisperse because three sheets of solutions exist at the region near the stagnation plane.

Gutheil (2001) extended the work to examine structure and extinction for monodisperse and bidisperse counterflow ethanol spray flames. There were significant differences between the monodisperse and bidisperse cases. At high strain rates, the large droplets were found to control flame width and stability. The small droplets dominated the reaction-zone structure. Lacas et al. (1992) and Schlotz and Gutheil (2000) considered the counterflow configuration for oxygen droplets with a hydrogen gas. Lacas et al. considered only low (subcritical) pressure whereas Schlotz

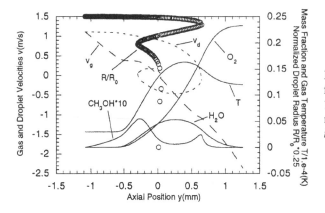

Figure 9.6. Outer structure of the flame for the conditions of Fig. 9.2 with $a = 1400/s$ (extinction). (Gutheil and Sirignano, 1998, with permission of *Combustion and Flame*.)

and Gutheil examined conditions that were at high pressure but still subcritical; both monodisperse and bidisperse cases were examined numerically. At low strain rates, the Sauter mean diameter could be used to represent the behavior.

9.3 One-Dimensional Planar Spray Ignition and Flame Propagation

Aggarwal and Sirignano (1984, 1985a, 1986) considered the unsteady, planar one-dimensional fuel spray–air ignition and flame propagation in a closed volume. They used an Eulerian formulation for the gas phase and a Lagrangian analysis for the liquid phase. In these studies, the importance of the gas-phase behavior on a microscale smaller than the distance between neighboring droplets was considered. It necessitated the use of the discrete-particle approach rather than the two-continua method. We shall see that certain important and interesting findings require resolution; in particular, the nonlinearities in the phenomena couple with nonuniformities on the microscale to yield a minimum ignition delay (or ignition energy) and a maximum flame-propagation speed at a finite initial droplet size rather than at the premixed limit.

The governing equations follow from a one-dimensional version of Eqs. (7.62), (7.73), and (7.81) for the gas phase and Eqs. (7.88)–(7.90) for the liquid phase. The gas-phase equations are written for the void volume equal to unity, the Lewis number equal to unity, and constant specific heats. The pressure is assumed to be uniform but time varying as a result of the closed volume.

The gas-phase equations for a perfect gas can be recast as

$$\frac{\partial \rho}{\partial t} + \frac{\partial}{\partial x}(\rho u) = \dot{M}, \tag{9.15}$$

$$\frac{\partial Y_m}{\partial t} + u\frac{\partial Y_m}{\partial x} - D\frac{\partial^2 Y_m}{\partial x^2} = (\varepsilon_m - Y_m)\dot{M}/\rho + \dot{w}_m, \tag{9.16}$$

$$\frac{\partial \phi}{\partial t} + u\frac{\partial \phi}{\partial x} - D\frac{\partial^2 \phi}{\partial x^2} = \left(\frac{p}{p_r}\right)^{\frac{1-\gamma}{\gamma}} \dot{w}_F \frac{Q}{c_p T_r} - \frac{\dot{M}}{\rho}\left[\phi - \phi_s + \frac{L_{\text{eff}}/c_p T_r}{(p/p_r)^{\frac{\gamma-1}{\gamma}}}\right], \tag{9.17}$$

where ϕ, which can be regarded as a transformed entropy variable, is defined as

$$\phi = (T/T_r)(p/p_r)^{(1-\gamma)/\gamma}. \tag{9.18}$$

The reference quantities T_r and p_r are taken as the temperature and the pressure values at the initial time for the constant-volume problems. (In other problems, they might be taken as the inflow conditions.)

The one-dimensional liquid-phase equations for a polydisperse spray can be written as

$$\frac{dx^{(k)}}{dt} = u_l^{(k)}, \tag{9.19}$$

$$\frac{du_l^{(k)}}{dt} = \frac{1}{2}\rho\pi R^2 C_D\left[u - u_l^{(k)}\right]\left|u - u_l^{(k)}\right|, \tag{9.20}$$

$$\frac{dR^{(k)}}{dt} = \frac{3\dot{m}^{(k)}}{4\pi\left[R^{(k)}\right]^2\rho_l}. \tag{9.21}$$

A liquid-phase heat diffusion equation can be selected from the choices considered in Chapters 2 and 3. The method for proper transformation of the source and sink terms from the Lagrangian grid to the Eulerian grid and vice versa is discussed in Section 8.2.

Aggarwal et al. (1984) used a system of equations equivalent to the preceding equations to evaluate spray vaporization models in situations without chemical reaction. They concluded that significant differences can result from the different droplet-heating and -vaporization models that they studied [types (ii), (iii), and (v)]. In high-temperature environments, the spatial variation, as well as the temporal variation, of the temperature within the droplet should be resolved. When spatial variation of the liquid temperature is important, the internal circulation will have an effect; therefore it should be considered for accuracy. Aggarwal et al. did show that significant variations in composition and temperature could occur on the scale of the spacing between neighboring droplets. For example, see Fig. 9.7, in which mass fraction versus axial position at certain instances of time are plotted. These microscale variations were later shown to be important.

Aggarwal and Sirignano (1985a) considered an initially monodisperse fuel–air spray in contact with a hot wall at one end; this wall was sufficiently hot to serve as an ignition source. The air and the droplets initially were not in motion; only the expanding hot gases cause a relative motion and drag force on the droplets. On account of the low-speed motion, internal circulation is not considered to be important, and the droplet-heating and -vaporization model is spherically symmetric. The parameters varied in this study included initial droplet size, hot-wall temperature, equivalence ratio, fuel type (volatility), and the distance from the hot wall to the nearest droplet.

As expected, increases in the hot-wall temperature resulted in a decrease for the heating time required before ignition (ignition delay). An increase in the fuel volatility also resulted in a decrease for ignition delay and for the accumulated energy required for ignition to occur (ignition energy). Volatility influence was

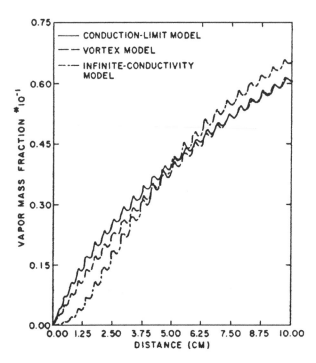

Figure 9.7. Fuel-vapor mass fraction versus distance at 16 ms for different liquid-phase models. (Aggarwal et al., 1984, with permission of *AIAA Journal*.)

especially pronounced for larger initial droplet sizes. Equivalence ratio also had an influence; too little fuel vapor or too much fuel vapor could inhibit ignition.

The effects of the droplet-heating and -vaporization model choice was shown to be significant for less-volatile fuels or for larger initial droplet sizes. A comparison was made between finite-conductivity and infinite-conductivity models here. The differences were much less important for volatile fuels or small initial droplet size. The infinite-conductivity model overpredicts ignition delay and ignition energy for the same reasons that it underpredicts vaporization rate during the early lifetime (see Chapter 2). For example, a hexane droplet with a 52.5-μm initial radius has a 2% difference in the ignition-delay prediction between the two models, but a hexane droplet with a 105-μm radius shows a 35% disagreement.

One interesting result is that a strong dependence on the distance between the nearest droplet and the ignition source exists for the ignition delay and for the ignition energy (Fig. 9.8). Note that an optimal distance occurs whereby delay and energy are minimized. This implies that, in a practical spray, in which distance to the ignition source is not controlled precisely or known, we should not expect a reproducible ignition delay or ignition energy. It is a statistical quality, and a range of variation will occur.

Another interesting result is that, for a fixed equivalence ratio and a fixed distance from the ignition source, the dependence of ignition delay and ignition energy on the initial droplet radius is not monotonic. In particular, minimum ignition delay and energy occur for a finite droplet size rather than at the premixed limit (zero initial droplet radius). This optimal initial droplet size increases as fixed volatility increases.

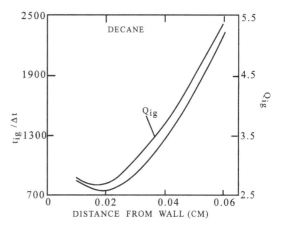

Figure 9.8. Variation of ignition-delay time and ignition energy with the location of droplets nearest to the hot wall. (Aggarwal and Sirignano, 1985a, with permission of The Combustion Institute.)

These findings of an optimal distance from the ignition source and an optimal initial droplet size require the resolution of the phenomena on the microscale and depend on the nonlinearity of reaction rate. The temperature, fuel-vapor mass fraction, and oxygen mass fraction are highly nonuniform in the gas surrounding a droplet. The reaction rate will reach a maximum at some position, depending on these values. For a given overall equivalence ratio (OER), the global reaction rate in the heterogeneous (nonuniform) case could be greater than in the homogeneous (uniform) case. This is a consequence of the fact that the average of a nonlinear function can differ significantly from the nonlinear function of an averaged quantity.

Aggarwal and Sirignano (1986) extended their analysis to consider polydisperse sprays. Otherwise, the same configuration was studied. Again, the minimal ignition delay could be found at a finite initial droplet radius. Figure 9.9 shows the results for a bidisperse hexane spray in which the two initial sizes in the distribution differ by a factor of 2 with equal mass at each initial size. The smaller of the two diameters d_s is plotted on the abscissa.

Various average diameters have been used to represent the distribution in a spray. If $f(R)$ is the size-distribution function for the spray, we calculate an average

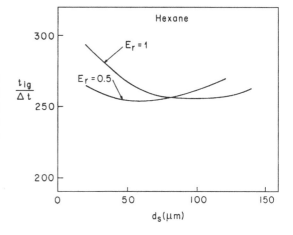

Figure 9.9. Ignition-delay time versus the smaller droplet diameter for a bidisperse spray at different overall equivalence ratios: hexane. (Aggarwal and Sirignano, 1986, with permission of *Combustion Science and Technology*.)

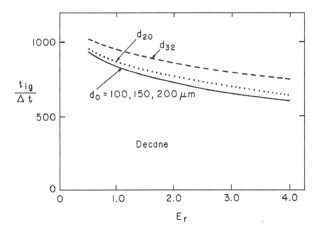

Figure 9.10. Ignition-delay time versus OER ratio for polydisperse spray with three initial drop sizes with equal mass distribution (solid curve), monodisperse with d_{20} (dotted curve), and monodisperse with Sauter mean diameter d_{32} (dashed curve). (Aggarwal and Sirignano, 1986, with permission of *Combustion Science and Technology*.)

diameter by

$$d_{nm} = 2 R_{nm} = 2 \left[\frac{\int_0^\infty f(R) R^n \, dR}{\int_0^\infty f(R) R^m \, dR} \right]^{\frac{1}{n-m}}. \tag{9.22}$$

For the particular bidisperse spray configured in Fig. 9.9, d_{nm} will scale linearly with d_s.

For the purpose of representing the ignition process with an average droplet size, the results correlate better with d_{20} rather than with the more commonly used d_{32}. This indicates that total surface area of the droplets is the important factor. See Fig. 9.10 for a comparison of a disperse spray and surrogate monodisperse sprays for decane; the ignition energy correlates with the value of a monodisperse spray at d_{20} over a wide range of equivalence ratios. A plot of ignition delay versus equivalence ratio at given initial size distribution leads to the same conclusion.

It was also found for the bidisperse spray that the ignition process was controlled by the smaller-size droplets. Figure 9.11 shows there is a substantial sensitivity to the distance of the smaller droplets from the ignition source while there is no sensitivity to the location of the larger droplets.

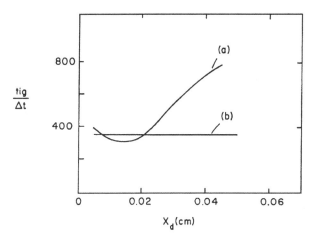

Figure 9.11. Variation of ignition-delay time with the location of nearest droplets in each size group for a bidisperse spray. Initial drop diameters d_0, 50 and 100 μm with equal mass distribution, $E_r = 1.0$, decane, (a) location of smaller droplets varied, (b) location of larger droplets varied. (Aggarwal and Sirignano, 1986, with permission of *Combustion Science and Technology*.)

Aggarwal and Sirignano (1985a) conjectured that the finding of an optimal size for ignition delay indicated the existence of an optimal droplet size for flame-propagation speeds. Aggarwal and Sirignano (1985b) demonstrated indeed that flame speeds for laminar heterogeneous mixtures could be greater than the established laminar premixed flame speed at the same OER and for the same fuel. The same configuration as previously discussed for the monodisperse spray was used. However, the calculation was continued after the ignition. For example, it was shown that, at an equivalence ratio of 0.5 for hexane, the flame travelled further (with a greater pressure rise in the closed volume) for the case with an initial droplet radius of $20\,\mu$m than for the premixed case. At an equivalence ratio of unity, the flame speed increased as droplet size decreased, reaching a maximum value for the premixed limit.

Other interesting findings for the flame propagation were that the flame exhibited both diffusionlike and premixedlike behaviors; flame-propagation rates and pressure-rise rates increased as the fuel volatility increased; transient droplet heating was an important factor affecting the speed; the flame propagation was unsteady in the closed volume with time-varying conditions ahead of the flame; and the difference in motion between the droplet and the gas resulted in a stratification of the composition that was not originally present. Also, the chemical-reaction zone thickness was found to be of the order of magnitude of the average distance between droplets. This implied that variations in fuel-vapor concentration could be significant, resulting in comparable qualitative effects on the average reaction rates that were found in the ignition study. The nonlinearity in the reaction-rate function combined with this nonuniformity can explain the finding of the optimal droplet size again. This comparable length scale between the reaction-zone thickness and the droplet spacing indicates an important and inherent intermittency in the phenomenon. The review by Aggarwal (1998) on spray ignition is recommended to the reader.

9.4 Vaporization and Combustion of Droplet Streams

Studies have been performed on two-dimensional planar flows with parallel streams of droplets. The basic configuration with the moving array of droplets is displayed in Fig. 9.12. Steady-state calculations that approximate the droplet streams to be continuous liquid streams were performed by Rangel and Sirignano (1986, 1988b, 1989a). The continuity and the momentum equations for the gas flow are not solved in these analyses; rather, the pressure and the gas velocity are assumed to be uniform. The droplet velocities are determined by conservation-of-momentum principles. These studies actually lie between two-continua formulations and discrete-particle formulations. They resolve the flow in one transverse direction on a scale smaller than the spacing between droplets; however, the properties are not resolved on the scale of the droplet spacing in the liquid-stream direction. Averaging of liquid properties occurs in one of the two spatial directions, thereby yielding a steady-flow result that integrates implicitly over time to remove the effect of the inherent intermittency associated with the spacing between the droplets in the stream. (Averaging in the third direction is implicit because the problem *ab initio* was

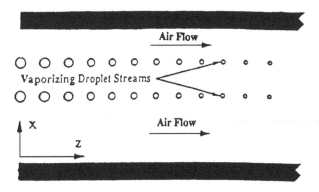

Idealized combustor

reduced to a two-dimensional statement.) Certain important phenomena can still be identified, even though properties are smoothed to eliminate the intermittency. Figure 9.13 shows fuel-vapor mass fraction in a two-droplet-stream configuration. A multiple flame is shown because regions of large fuel-vapor mass fractions are separated. Early vaporization and mixing of the fuel vapor with the air together with the introduction of an ignition source allow a premixed flame to occur. The droplets penetrate the premixed flame before complete vaporization has occurred; therefore continued vaporization in the presence of a hot gas with excess air is sustained. This causes diffusion flames to be established around each stream of droplets. Delplanque and Rangel (1991) performed a related analysis for one stream of droplets flowing through the gaseous viscous boundary layer near a hot wall. They included the important effect of gaseous thermal expansion that created a gas motion and droplet drag that moved the droplet stream away from the wall as it flowed downstream.

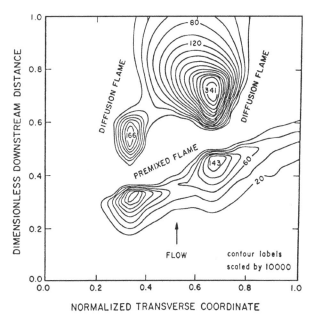

Figure 9.12. Two-dimensional, planar, parallel droplet-stream configuration. (Delplanque et al., 1991, with permission of the American Institute of Aeronautics and Astronautics.)

Figure 9.13. Fuel-vapor mass fractions in the parallel two-droplet-stream configuration. (Rangel and Sirignano, 1989a, with permission of *Combustion and Flame*.)

The intermittent effects were addressed in the unsteady analyses of Delplanque et al. (1990, 1991) and Rangel and Sirignano (1988a, 1991a). These studies are fully in the domain of discrete-particle formulations (with the understanding that any two-dimensional representation involves averaging in the third dimension). The inclusion of the intermittent effects yields results that support the general conclusions of the steady-state analyses.

The unsteady droplet-stream studies of Rangel and Sirignano and of Delplanque et al. considered two types of unsteadiness: one is due to the transient process of ignition and flame establishment; the other is due to the inherent intermittency of the flow of discrete particles. (Later, Delplanque and Sirignano, 1996, considered oscillatory combustion and vaporization. See Section 9.8.) The gas velocity and the droplet velocity are considered to be oriented in only the axial direction with no transverse convection or advection. The gas velocity is taken to be uniform and constant with time. Specific heat is constant and the Lewis number is unity. One-step chemical reaction is considered. The gas-phase equations are

$$L(T) \equiv \frac{\partial T}{\partial t} + u \frac{\partial T}{\partial z} - \frac{1}{\text{Pe}} \left(\frac{\partial^2 T}{\partial x^2} + \frac{\partial^2 T}{\partial z^2} \right)$$

$$= -\sum_{j=1}^{J} \frac{n\dot{m}_j}{\rho} \left(T - T_s + \frac{L_{\text{eff}}}{c_p} \right) \delta(x - x_j) \delta(z - z_j) + \sum_{m=1}^{M} (Q_m/c_p)\dot{w}_m, \quad (9.23)$$

where N is the number of droplets per unit length in the transverse y direction, δ is a delta function identifying the instantaneous x_j and z_j coordinates for the droplet, J is the total number of droplets in the two-dimensional array, and M is the total number of species in the gas. For the mass species equations, we have

$$L(Y_m) = \sum_{j=1}^{J} \frac{N}{\rho} \varepsilon_m \dot{m}_j \delta(x - x_j) \delta(z - z_j) + \dot{w}_m. \quad (9.24)$$

The assumptions about uniform and constant pressure and velocity in the gas allow us to bypass the momentum and the continuity equations.

The channel walls were considered to be adiabatic except for a portion that served as the ignition (heat) source. Conditions at inflow were specified, and zero gradients were imposed at the outflow. A droplet model essentially identical to the model discussed in Section 9.3 was used; a slightly modified drag coefficient was used, and a correction factor dependent on the Reynolds number was used for the vaporization. The equations were solved by finite-difference methods.

Figure 9.14 shows that both diffusion flame and premixed flame structures exist. Furthermore, it is seen that the diffusion flame can envelop more than one droplet. The studies of Delplanque et al. extended the parallel stream model to multicomponent fuels. Delplanque et al. also studied the effect of the point-source approximation correction discussed in Section 4.3. Figure 9.15 shows the importance of the correction for the prediction of droplet temperature and droplet mass in the regions where those quantities are rapidly varying with time. See also the work of Leiroz and Rangel (1995, 1997, 2007) on the vaporization and combustion of droplet streams at zero-valued Reynolds number and with transient gas-phase interactions.

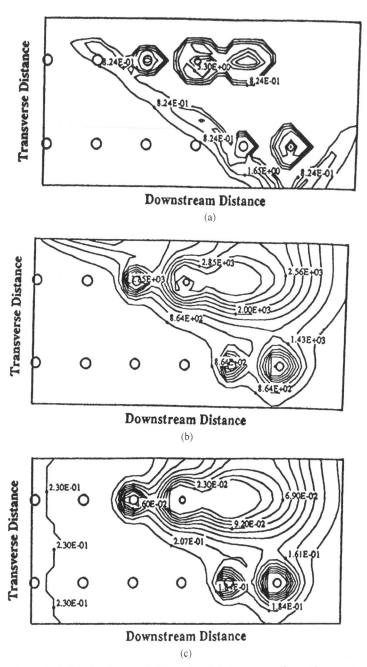

Figure 9.14. Unsteady, parallel two-droplet-stream configuration contour plots at nondimensional time = 0.016: (a) reaction rate with contour intervals of 0.824, (b) gas temperature with contour intervals of 283 K, (c) oxygen mass fraction with contour intervals of 0.023.

A variety of interesting experimental studies have occurred on single- and double-droplet streams. See Sangiovanni and Labowsky (1982), Queiroz and Yao (1989), Connon and Dunn-Rankin (1996), Shaw et al. (2002), and Khau et al. (2007). In these papers, important interactions among the droplets have been quantified,

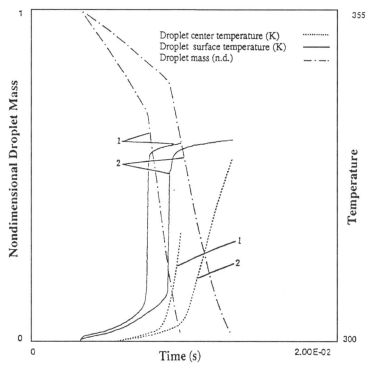

Figure 9.15. Droplet temperature and mass versus time for the second droplet in the stream nearest the ignition source in a parallel stream configuration. Point-source approximation correction for curve 1. No correction for curve 2. (Delplanque et al., 1991, with permission of the American Institute of Aeronautics and Astronautics.)

showing consistent trends with theory; instabilities of the geometrical array in certain domains have been noted; and extensions to bi-component liquids have occurred. See the discussion in Subsection 12.3.4 about the formation of a linear stream of uniform-sized droplets.

9.5 Flame Propagation Through Metal-Slurry Sprays

In Section 4.3, we discussed the vaporization and the combustion of individual metal-slurry droplets. Surface regression, shell formation, bubble formation, pore drying, coalescence, and metal-particle burning were discussed. In this section, those droplet models are engaged in order to study a two-dimensional, steady flame in a flowing metal-slurry spray. In particular, the effort of Bhatia and Sirignano (1995) is discussed.

The gas-phase analysis followed the analysis of Rangel and Sirignano (1986). A flow in a rectangular channel of length L_1 and width L_2 is considered with adiabatic sidewalls except for a portion that provides a uniform, steady heat flux to the gas and serves thereby as a flame holder. The flow is in the z direction; dependence on the transverse x direction is considered whereas integration over the y direction removes that dependence and yields a two-dimensional problem. Cases with single and multiple parallel streams of slurry droplets flowing in the z direction were considered. The droplet streams were approximated, from the work of Rangel and Sirignano, as line sources of mass and line sinks of heat in the gas-phase analysis; this assumption allowed the steady formulation of the problem, neglecting the intermittent effects discussed in Section 9.4. Furthermore, the transient behavior before and following ignition is not considered in the gas-phase analysis. Of course, an individual droplet still experiences transient behavior in Lagrangian time as it flows downstream.

Simplifications for the gas-phase equations include neglect of the transverse velocity in the x direction, neglect of streamwise (z-direction) heat conduction and mass diffusion, unitary Lewis number, inviscid and isobaric flow, neglect of the reaction force to droplet drag, thereby allowing the bypass of the gas momentum equation, constant specific heat and thermal conductivity, an equation of state whereby the product of gas density and temperature remains constant in the isobaric state (neglect of molecular-weight variations in a perfect gas), one-step oxidation kinetics for the hydrocarbon vapors, and neglect of the liquid volume.

The gas-phase equations will have a sink term that is due to oxygen consumption at the droplets on account of oxidation at the surface of the dry-metal agglomerate and metal–vapor oxidation in a gas film surrounding the remaining metal particle following complete vaporization of the hydrocarbon liquid in the slurry droplet. This sink has a magnitude of

$$\sum_{j=1}^{N} n_j \dot{m}_{O,j},$$

where j is an integer index indicating the particular droplet stream.

Equations (7.62), (7.73), and (7.82) can now be rewritten as

$$\frac{\partial}{\partial z}(\rho u) = \sum_{j=1}^{N} n_j \dot{m}_j - \sum_{j=1}^{N} n_j \dot{m}_{O,j}, \tag{9.25}$$

$$L(Y_F) \equiv \rho u \frac{\partial Y_F}{\partial z} - \rho D \frac{\partial^2 Y_F}{\partial x^2} = \sum_{j=1}^{N} n_j [\dot{m}_j (1 - Y_F) + \dot{m}_{O,j} Y_F] - \rho \dot{w}_F, \tag{9.26}$$

$$L(Y_O) = -\sum_{j=1}^{N} n_j [\dot{m}_j Y_O + \dot{m}_{O,j}(1 - Y_O)] - \rho \dot{w}_O, \tag{9.27}$$

$$c_p L(T) = -\sum_{j=1}^{N} n_j \{ \dot{m}_j [c_p(T - T_s) + L_{\text{eff}}] - \dot{m}_{O,j} [c_p(T - T_s) - L_{O,j}] \} + \rho \dot{w}_F Q. \tag{9.28}$$

$L_{O,j}$ is a result of the sink of energy associated with surface oxidation of the metal agglomerate and of the energy source that is due to vapor-phase burning of the metal.

The boundary conditions are adiabatic walls and impermeable walls, yielding

$$\frac{\partial T}{\partial x} = \frac{\partial Y_F}{\partial x} = \frac{\partial Y_O}{\partial x} = 0$$

at the walls except for a region where heat is applied and the heat flux $\dot{q}w$ (or normal temperature gradient) is prescribed. At $z = 0$, the inlet conditions for Y_O, Y_F, and T are given.

After nondimensionalization of Eqs. (9.25)–(9.28) and placement of Eqs. (9.26)–(9.28) into a canonical form, a Green's function can be applied to yield the solution. The canonical form is

$$\frac{\partial J_k}{\partial \zeta} - P_1 \frac{\partial^2 J_k}{\partial \zeta^2} = \sum_{j=1}^{N} P_2 S + P_5 S_c, \tag{9.29}$$

where $k = 1, 2, 3$ corresponds to Eqs. (9.26), (9.27), and (9.28), respectively, so that J_k is proportional to Y_F, Y_O, or T. $\zeta = x/L_2$ and

$$\zeta = \frac{4\pi \rho D z}{(m u_l / R)_{z=0}}.$$

The source and sink terms S and S_c will be functions (generally nonlinear) of J_1, J_2, and J_3.

The Green's function,

$$G(\zeta, \xi; \zeta', \xi') = 1 + 2 \sum_{n=1}^{\infty} e^{-P_1 n^2 \pi^2 (\zeta - \xi')} \cos n\pi \zeta \cos n\pi \zeta', \tag{9.30}$$

was used to obtain the solution

$$J_k(\zeta, \xi) = 1 + 2 \sum_{n=1}^{\infty} J_{kn}(\xi) \cos n\pi \zeta, \tag{9.31}$$

where J_k is governed by the nonlinear integrodifferential equation

$$\frac{dJ_{kn}}{d\zeta} = \sum_{J=1}^{N} P_2 S \cos n\pi \zeta_j + \int_0^1 P_5 S_c \cos n\pi \zeta' d\zeta' + P_8 \dot{q}_w - P_1 n^2 \pi^2 J_{kn}. \tag{9.32a}$$

The initial condition at $\xi = 0$ is that

$$J_{kn}(0) = \int_0^1 J_k(\zeta, 0) \cos n\pi \zeta \, d\zeta. \tag{9.32b}$$

Equation (9.32a) is solved by numerical integration. The coupling with the droplet model discussed in Section 4.3 occurs through the source and sink terms. Simultaneous integration of those differential equations modelling the droplet behavior is required. P_1, P_2, P_5, and P_8 are nondimensional groupings defined by Bhatia and Sirignano (1995).

The hydrocarbon–air combustion yielded both a premixedlike flame and a diffusion flame of the types discussed in Section 9.4; farther downstream another diffusion flame existed because of the combustion of the aluminum vapor. Better mixing is found with multiple liquid-stream arrangements but little difference in ignition delay or burning time occurs between single-stream and multiple-stream configurations. As shown in Figs. 9.16 and 9.17, the time required for burning the metal is the longest and therefore rate controlling. That time increases with an increase in the metal loading in the slurry and with an increase in the metal-particle radius to droplet radius ratio.

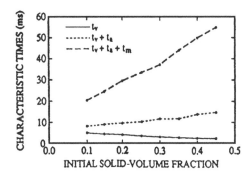

Figure 9.16. Comparison of liquid-hydrocarbon-vaporization time (t_v), dry-agglomerate heating time (t_a), and the metal-burning time (t_m), for $\phi_1 = 0.5 = T_\infty = 1250\,\text{K}$ and $r_m/r_{1,i} = 0.04$. (Bhatia and Sirignano, 1995, with permission of *Combustion and Flame*.)

9.6 Liquid-Fueled Combustion Instability

In Section 3.4, we discussed the response of the droplet-vaporization rate to the oscillations of ambient conditions caused by an acoustic wave. Combustion instability of liquid-fueled and liquid-propellant engines was explained to be a motivating factor. Bhatia and Sirignano (1991) performed a one-dimensional computational analysis of liquid-fueled combustion instability. Their work differed from the studies of Tong and Sirignano (1989) and Duvvur et al. (1996) in certain key aspects. Bhatia and Sirignano considered a total system with upstream and downstream boundary conditions and explicit chemical reaction. They did not impose acoustic oscillations or vaporization-rate control of combustion; rather, they predicted the oscillations and demonstrated that the vaporization process was a controlling factor in the combustion process.

Bhatia and Sirignano used a one-dimensional, unsteady version of the spray equations, as discussed in Chapter 7. Unlike the analyses of Sections 9.1, 9.3, 9.4, and 9.5, the momentum equation for the gas phase was fully exercised. This equation is important in governing the oscillatory behavior; the pressure will be highly nonuniform in the chamber. The chemical kinetics (one-step) scheme of Westbrook and Dryer (1984) was used for the *n*-decane fuel. Turbulent mixing was neglected. The effects of any transport in the transverse direction were, of course, neglected in the one-dimensional analysis.

The gas-film and effective liquid-conductivity model of Abramzon and Sirignano (1989) was used to describe the unsteady heating and vaporization of the

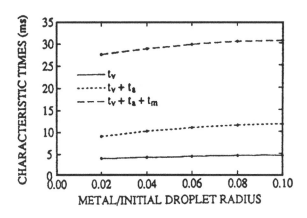

Figure 9.17. Comparison of liquid-hydrocarbon vaporization time (t_v), dry-agglomerate heating time (t_a), and the metal-burning time (t_m), for $\phi_1 = 0.5$, $T_\infty = 1250\,\text{K}$ and initial metal loading $\phi_{m,i} = 20\%$ by volume. (Bhatia and Sirignano, 1995, with permission of *Combustion and Flame*.)

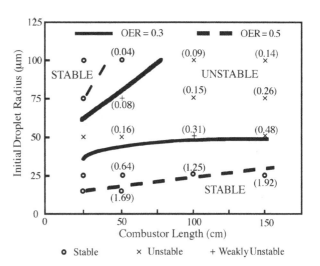

Figure 9.18. Domain of stable–unstable operation: droplet radius and combustor length. Points shown correspond to OER = 0.3. Numbers in parenthesis indicate τ_p/τ_{dh} (Bhatia and Sirignano, 1993b).

liquid-decane fuel. The initial droplet diameter or equivalently the characteristic droplet-heating time was found to be an important parameter.

The initial conditions for gas temperature, pressure, species concentration, and velocity and for liquid-droplet temperature, velocity, location, and radius are specified. For the inflowing liquid at the upstream end, the stagnation pressure is specified but the mass flow fluctuates as the pressure at the upstream end oscillates. The short-nozzle approximation of Crocco and Sirignano (1966) is used for the downstream boundary condition. It results in a constant value of the Mach number at the exit plane as the other conditions, including velocity and sound speed, do oscillate. In the computation, the inlet velocity is forced for one cycle of oscillation; then a free oscillation is allowed. It grows in amplitude for some conditions, giving a limit-cycle behavior. In other cases, it decays. The phenomenon was studied for a range of initial droplet diameter, chamber lengths, and OERs.

Figure 9.18 shows the stability map on a plot of initial droplet diameter versus combustor length for two different OERs. In the figure, the most unstable region approximately centers on a value of 0.15 for the ratio of characteristic time for the oscillation to the characteristic heating time. Accounting for differences in the normalization for Tong and Sirignano (1989) versus Bhatia and Sirignano (1991), this value of 0.15 coincides with the most unstable value of 0.6 found by Tong and Sirignano.

Chemical kinetics is very fast and clearly not rate controlling. With the neglect of transverse mixing (which is assumed instantaneous in a one-dimensional approximation), vaporization rate is found to be the controlling process. The characteristic droplet-heating time is clearly the most important combustion parameter in determining the combustor stability.

The combustion process can occur in either the stable or the unstable mode, depending on various parameters, as indicated in Fig. 9.18. The unstable oscillations result in a limit-cycle behavior. Figure 9.19 shows a limit-cycle oscillatory behavior for the 50-μm droplet radius, 50-cm chamber length, and OER of 0.3. For

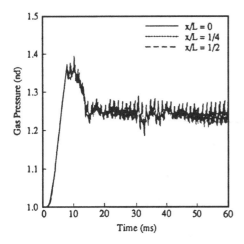

Figure 9.19. Pressure for $L = 50$ cm, OER $= 0.3$, $r_i = 50\,\mu$m (Bhatia and Sirignano, 1993b).

the 100-μm droplet diameter, marginal instability with much smaller amplitudes is found.

High-frequency oscillations involving a standing mode in the combustor with near-velocity nodes at the two ends occur when the ratio of oscillation time period to droplet-heating time is ~ 0.15. Low-frequency oscillations involving a uniform but pulsating pressure field in the combustor result when the droplet-heating time is close to the value of the gas-residence time in the combustor.

For the high-frequency oscillations at lower values of the equivalence ratio, the frequency is essentially the fundamental resonant frequency determined by the combustion-chamber length. At higher equivalence ratios, overtones can dominate with the frequencies again determined by the length. The limit-cycle behavior does not depend on the frequency of the initial disturbance. The oscillation moves to a natural frequency within a few cycles after the initial forcing function ceases.

An increase in equivalence ratio enlarges the domain of unstable operation. The instability domain generally occurs at intermediate values of initial droplet diameters and intermediate values of combustor length, as shown in Fig. 9.18.

Some discussion of combustion instability in a liquid-oxygen–liquid-hydrogen propellant rocket motor at supercritical or near-critical conditions is presented in Section 9.8.

9.7 Spray Behavior in Near-Critical and Supercritical Domains

Ryan et al. (1990) showed in the subcritical case that the corrections yielded by the group-combustion approach and by the droplet-array approach do not differ significantly. Because flames in jet and rocket engines are usually anchored in region of high spray density, the influence of neighboring droplets on high-pressure and supercritical droplet combustion behavior must be evaluated. Jiang and Chiang (1994a, 1994b, 1996) recently contributed an approach based on the model developed by Bellan and Cuffel (1983), who considered a cloud of droplets surrounded by a hotter homogeneous gas region. Convective effects are neglected (except for

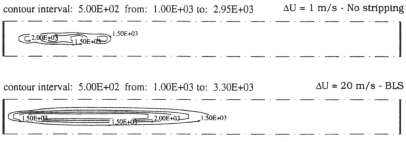

Isotherms at t = 2.00 ms

Figure 9.20. Influence of boundary-layer stripping on the combustion of LOX droplet streams at supercritical conditions. Isotherms surround the droplet array. (Delplanque and Sirignano, 1995, with permission of *Combustion Science and Technology*.)

Stefan flow). Multiple-droplet interactions are modeled with the sphere-of-influence concept. The simulations of Jiang and Chiang (1996) indicate that a droplet is less likely to reach the critical mixing conditions if it is in a cloud than if it is isolated. Furthermore, they found that such droplet clouds do not follow the d^2 law at any pressure.

An analysis of the effect of species and temperature nonuniformities induced by neighboring droplets on the transcritical combustion behavior of LOX droplets in a convective environment was contributed by Delplanque and Sirignano (1995), who used the droplet-array approach (parallel streams), with the droplets followed in a Lagrangian manner. The model that predicts the droplet behavior is described in Section 5.2 and allows secondary atomization in the stripping mode. Despite the hindering effects of droplet drag and accelerating gas flow resulting from the mass flux contribution of the vaporizing droplets, the gasification rate is still controlled by secondary atomization (Fig. 9.20). The characteristic time is 1 order of magnitude smaller than that associated with primary vaporization. However, the reaction zone created by the combustion of preceding droplets enables the droplet surface to reach the critical mixing conditions.

In summary, conditions of practical interest generally involve numerous droplets in a convective environment. The results obtained so far indicate that the presence of neighbors precludes droplets in clouds from reaching the critical mixing state. Furthermore, the d^2 law was found to be invalid at any pressures for droplets in clouds.

9.8 Influence of Supercritical Droplet Behavior on Combustion Instability

In Section 5.2 it was shown that supercritical ambient conditions and their influence on physical properties such as surface tension significantly affect droplet lifetime. The goal now is to show how this result can affect the predicted overall performance of propulsion systems by using the example of one of the most challenging research problems in rocket engine technology: combustion instability.

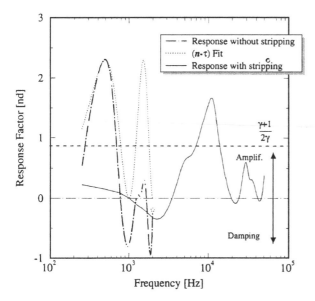

Figure 9.21. Response factor for an isolated LOX droplet with and without stripping. $T_\infty = 1000$ K, $P_\infty = 100$ atm, $\Delta U = 20$ m/s (Delplanque and Sirignano, 1996).

The open-loop response of LOX droplets to prescribed oscillatory ambient conditions consistent with liquid rocket engines was investigated by Delplanque and Sirignano (1996), who used the supercritical droplet-combustion models described in the previous sections both for isolated droplets and droplet arrays. This study evaluated the stability of the combustion chamber assuming concentrated combustion at the injector end, short nozzle, and isentropic flow downstream of the combustion zone, following the work of Crocco and Cheng (1956). A response factor G was computed to quantify the Rayleigh criterion, which states that an initially small-pressure perturbation will grow if the considered process adds energy in phase (or with a small enough phase lag) with pressure:

$$G = \iint (\dot{w}' p') \, dt \, dz \Big/ \iint (p')^2 \, dt \, dz. \tag{9.33}$$

The primes denote fluctuations with respect to the nonoscillatory values. Delplanque and Sirignano note that an underlying assumption to this definition of G is that the gasification rate provides a good approximation of the energy-release rate.

The frequency at which the peak response factor occurs is mainly correlated to the droplet lifetime. Therefore, because secondary atomization in the stripping mode results in a 1-order-of-magnitude reduction in droplet lifetime, it causes a corresponding shift in the peak frequency, as shown in Fig. 9.21. Consequently, when stripping occurs, the peak frequency is significantly larger than the acoustic frequencies of the common modes for standard cryogenic rocket engine chambers. Delplanque and Sirignano argued that, because in these engines droplets are likely to undergo secondary atomization in the stripping regime for most of their lifetime, this phenomenon could explain the observed better stability of such engines compared with that of storable propellant engines. See Harrje and Reardon (1972).

Furthermore, under the modelling assumptions, the droplet-gasification process can drive combustion instabilities for the longitudinal mode (with or without stripping).

Estimates of the influence of neighboring droplets on the droplet response to an oscillatory field obtained with the droplet-stream model previously described indicate that the isolated droplet configuration underestimates the driving potential of the gasification process.

10 Spray Interactions with Turbulence and Vortical Structures

The interactions of a spray with a turbulent gas flow is important in many applications (e.g., most power and propulsion applications). Two general types of studies exist. In one type, the global and statistical properties associated with a cloud or spray within a turbulent field are considered. In the other type, detailed attention is given to how individual particles behave in a turbulent or vortical field. Some studies consider both perspectives. Most of the research work in the field has been performed on the former type of study. Faeth (1987), Crowe et al. (1988), and Crowe et al. (1996) give helpful reviews of this type of research.

The interactive turbulent fields can be separated into homogeneous turbulent fields and free-shear flows (e.g., jets and mixing layers). In some theoretical studies, two-dimensional vortical structures interacting with a spray have been examined. Most of the studies deal with situations in which the contribution of the spray to the generation of the turbulence field is secondary, that is, there is a forced gas flow whose mass flux and kinetic-energy flux substantially exceed the flux values for the liquid component of the dilute flow. The turbulent kinetic-energy flux of the gas flow is much less than the mean kinetic-energy flux of the gas flow and is comparable to the mean kinetic energy of the liquid flow. Therefore the turbulent field is much more likely in this situation to receive kinetic energy transferred from the mean gas flow than kinetic energy transferred from the mean liquid flow. An exception to this situation would be a liquid-propellant rocket motor in which all of the forced flow is initially in liquid form. In this case, the turbulent kinetic energy appears directly or indirectly through transfer from the mean kinetic energy of the liquid. The direct transfer is defined to be the type in which turbulent fluctuations in the liquid flow or boundary layer and wake instabilities in the gas flow past the liquid (e.g., vortex shedding over droplets) cause the gas-phase turbulent fluctuations. Indirect transfer is the case whereby vaporization causes the mean kinetic energy of the liquid to be transformed into mean kinetic energy of the vapor, which in turn gets partially transferred to the turbulent kinetic energy of the gas.

Faeth (1987) compared three types of two-phase models for turbulent flows: locally homogeneous flow (LHF), deterministic separated flow (DSF), and stochastic separated flow (SSF). He prefers SSF for practical dilute sprays; that model

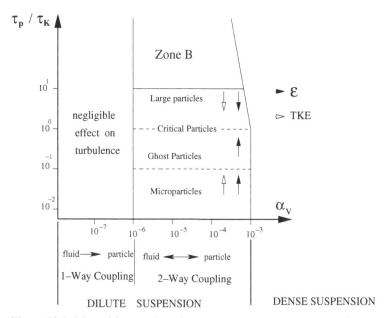

Figure 10.1. Map of flow regimes in turbulent particle-laden flows (Elghobashi, 2006).

considers finite interphase transport rates and uses random-walk computations to simulate turbulent dispersion for the dispersed phase. Crowe et al. (1988) review both time-averaged and time-dependent free-shear flows of two phases. The time-dependent methods capture the instantaneous flow and can better calculate particle trajectories. In the time-averaged method, a steady flow with gradient diffusion is usually considered. Crowe et al. (1996) review computational approaches to two-phase flows.

In this chapter, we examine relevant work for particle-laden and bubble-laden turbulent flows as well as the works for turbulent spray flows. The term "discrete particle" is used to include drops, bubbles, and solid particles. The "carrier phase" or "continuous phase" means the gas for spray flow, the liquid for a bubbly flow, and the gas or liquid for particle-laden flows. Furthermore, we consider related works that address transitional or other vortical flows.

Figure 10.1, from Elghobashi (2006) is an update of the work of Elghobashi (1991, 1994) and shows the domains of various types of coupling between the phases. The two parameters in the map are the ratio of the droplet dynamic response time τ_p to the Kolmogorov time scale for turbulence τ_K, and the volume fraction of particles (equal to $1 - \theta$ in Chapter 7). The Kolmogorov scale is the shortest time scale in the turbulence-kinetic-energy spectrum. The time ratio is plotted on the ordinate, and the volume fraction is plotted on the abscissa. Volume fraction is immediately determined from the cube of the ratio of average distance between neighboring particles to the particle diameter (for spheres) or, equivalently, the product of the number of particles times particle volume divided by the fluid volume. Small normalized distances between neighboring particles indicate a dense spray or suspension whereas larger distances result in a dilute condition. For a very dilute spray whereby $1 - \theta = O(10^{-7} \text{ to } 10^{-6})$, momentum transfer between the gas and the droplets is important from the droplet perspective; the impact of the momentum transfer on the gas is negligible. This situation is described as a one-way coupling. As the spray becomes less dilute, momentum transfer between the phases is significant for both phases, and we describe this domain as two-way coupling. In Zone B of the figure, higher Reynolds number and slow droplet or particle response (large τ_p) occur and turbulent production is enhanced by the droplets. Higher Reynolds number can result in vortex shedding in the droplet wake. Note though that the higher Reynolds number can be accompanied by a higher Weber number, which results in droplet breakup, leading to smaller droplets and with lower Reynolds number.

The domain below Zone B of the figure is divided into four portions by Elghobashi according to the results of certain direct numerical simulations (DNSs): large particles $\tau_p/\tau_K > 1$; critical particles $\tau_p/\tau_K = O(1)$; ghost particles $\tau_p/\tau_K < 1$; and microparticles $\tau_p/\tau_K \ll 1$. In the large-particle regime, both turbulent kinetic energy k and turbulent dissipation rate ε are reduced relative to the single-phase flow. In the critical-particle domain, k is reduced but ε remains about the same relative to a particle-free flow. For the ghost-particle domain, k remains about the same but ε is increased relative to a particle-free flow. Both k and ε are increased relative to a particle-free flow in the microparticle domain.

As the figure shows, very dense sprays $[1 - \theta \geq O(10^{-3})]$ yield a four-way coupling situation in which, in addition to the two-way coupling characteristics, the droplets or particles exchange momentum directly with neighboring droplets or particles because of collisions (including near collisions that result in the interaction of boundary layers and wakes).

Some measurements were made by Birouk and Gökalp (2002, 2006) for suspended droplets of five different alkanes vaporizing in homogeneous isotropic turbulence with zero mean relative velocity between the gas and droplets. Bicomponent liquids were also examined. A correlation was developed showing that vaporization rate increased as $\text{Re}_{tL}^{2/3}$, where the turbulent Reynolds number $\text{Re}_{tL} = q^{0.5} L/v$; q, L, and v are the turbulent kinetic energy, integral scale of turbulence, and kinematic viscosity, respectively. For the zero mean flow, a d^2 law was obtained but the vaporization-rate constant (i.e., slope of the d^2 versus time curve) was increased by the factor $1 + 0.0063 \text{Re}_{tL}^{2/3} \text{Sc}$. So, clearly, as the turbulent Reynolds increases above a value of 100, the correction becomes significant.

There are two ways that turbulent fluctuations can affect the rates of exchanges of mass, momentum, and energy with the dispersed phase. Both large and small eddies can change the ambience of the discrete particle. For example, the relative velocity between the continuous phase and the particle and the ambient temperature and composition at the edge of the fluid film surrounding the particle will fluctuate because of the turbulence. This changes the exchange rates between the phases but, by itself, does not necessarily change the nondimensional rates for a discrete particle as given by the Nusselt number Nu, Sherwood number Sh, drag coefficient C_D, and the lift coefficient C_L. The second type of influence can come from particle interactions with turbulent scales of the same scales as the particle or its surrounding fluid film. Here, the local configuration for transport and momentum exchange can be changed so that Nu, Sh, C_D, and C_L are modified as discussed in Section 3.4.

This second manner has not been yet incorporated into the computations that are subsequently discussed. It remains a topic for future exploitation.

In most practical spray problems, it is not possible with existing computational resources to resolve the behaviors internal to each droplet, in the fluid film (boundary layer and wake) surrounding each droplet, and sometimes in the spaces between neighboring droplets. Also, the smallest turbulent eddies are impractical to resolve in most practical situations. So averaging over at least the smallest scales for the spray and the turbulence is commonly required, and models are needed to close the system of equations.

In the approaches to the turbulent spray computation, a formal averaging process is made first over at least some, if not all, of the turbulence and spray length scales. For Reynolds-averaged Navier–Stokes (RANS) simulations, averaging is done over all turbulent scales, whereas, for large-eddy simulations (LESs), the averaging is done over the smaller scales only. Actually, even so-called DNSs, which aim to simulate all turbulent scales, begin the analysis with a system of equations that resulted from averaging and modelling at the scale of the droplet size and, sometimes, also at the scale of average distance between neighboring droplets. In Chapter 7, we identified an approach that simultaneously averages over droplet scales and turbulence scales; however, that has not yet been used in computations.

In any of the existing approaches, there is a clear need for improved modelling of droplet vaporization in a turbulent, reacting flow as commonly found in many practical applications. For example, in the latest designs of many liquid-fueled, continuous-flow combustors, a rate-controlling process is droplet vaporization. This process has critical impacts on required volume for complete combustion and spatial variations of mixture ratio and gas temperature. In turn, these characteristics have critical impacts on energy efficiency, formation of pollutants, and turbomachinery design for gas-turbine engines. The rate of vaporization controls the ultimate rates of energy release and species mass conversions. These vaporization rates are very sensitive to instantaneous droplet diameter; relative velocity between the droplet and the averaged gas environment; the turbulent velocity, concentration, and temperature fluctuations in the droplet vicinity; the proximity of neighboring droplets (i.e., droplet number density); the transient droplet heating; and the effect of internal liquid circulation on the transient heating.

Although these parameter sensitivities of the vaporization rate were previously modelled in some way, there are three major deficiencies that must be corrected before a useful, predictive computational code can be developed. First, these effects must be embodied simultaneously in one model. Current vaporization models typically focus on the physics associated with some subset of these parameters. Second, the vaporization model must be able to handle a wide and realistic range of turbulent scales for length and time. For example, there is interest in subgrid-size turbulent eddies that range from a size comparable to the droplets to sizes that are at least an order of magnitude greater than the droplet. Third, to perform either time-averaged (i.e., RANS simulations) or LESs of turbulent sprays, a rational two-way coupling must be prescribed between the subgrid physics and chemistry and the supergrid physics (i.e., large-eddy fluctuations and mean flow) for both phases. In either case, it is important to have some understanding about the dynamic interactions between vortical structures in the surrounding gas and the droplets. This would provide a foundation for understanding the turbulence–spray interaction. Current knowledge about the interaction of a droplet with unsteady gaseous flows, including interactions with vortical structures, was discussed in Section 3.4.

In developing a strategy for subgrid modelling, it is important to know the average droplet and spray scales (e.g., droplet diameter and spacing between neighboring droplets) and the smallest turbulent eddy scales. The comparison of spray, droplet, and turbulence scales is necessary to make a sound judgement about methodology. If the droplet and spray scales are smaller than the smallest turbulent scales, the droplets can be modelled as if they were in a laminar, unsteady fluid wherein gradients in the droplet far field are small compared with the droplet scale. Here, standard droplet and droplet group models discussed in Chapters 2–6 become reasonable to use. Therefore, the sequential averaging process becomes acceptable; in that process the two-phase flow equations are first developed with some explicit or implicit averaging over droplet and spray scales, followed by filtering (averaging by a different name) of the turbulent fluctuations using a larger scale for this turbulence filtering than the droplet and spray scales. While there is an acceptability here, some mismatch can still occur in computational-resource commitment because many droplets might appear in a particular Eulerian computational cell. Because these droplets in that cell are now all subject to the same averaged environmental conditions, there can be some waste in treating so many droplets.

The challenge for the subgrid modelling strategy occurs, as discussed in Chapter 7, in the situation in which the smallest turbulent scales are comparable to or smaller than the droplet and spray scales. Then the sequential averaging makes no sense. If the averaging for the two-phase flow and the droplet modelling were done first, the smallest scales of turbulence would have been averaged away and should have been included in the droplet model. In this situation, the sequential averaging processes should be replaced with one averaging process that addresses in the subgrid modelling the droplet behavior, the smallest turbulent scales, and the interaction between the droplets and the turbulence at those subgrid scales. Although such a fundamental approach is outlined in Chapter 7, it has not been fully used. So existing computations for particle-laden, bubble-laden, and spray flows are valid only for situations in which the discrete-particle scales are much smaller than the smallest turbulent scales.

In the works discussed in the following sections of this chapter, the carrier-phase equations are solved numerically by use of the traditional Eulerian gridding. The discrete phase is resolved typically with a Lagrangian tracking method. In some works, all particles are tracked whereas in others, only average or representative drops are tracked to reduce computational costs.

In this chapter, we examine the interaction of sprays with vortices in the next section, and we discuss Reynolds-averaged turbulence models, DNSs, and LESs in the remaining sections.

10.1 Vortex–Spray Interactions

Some theoretical studies have been made on time-dependent behavior within mixing layers and jets laden with particles or droplets. Generally, inviscid vortex methods for the dispersed phase are used. Coupling between the hydrodynamics of the two phases is assumed to be one way only. Chein and Chung (1987) emphasized the effects of vortex pairing in a mixing layer. The pairing was found to enhance the entrainment and the dispersion of the particles. They found optimal dispersion in the midrange of the Stokes number (ratio of particle aerodynamic response time to the flow characteristic time). Chung and Troutt (1988) extended the method to jet flows. They noted the importance of the large-scale component of turbulence in the dispersion process.

Interesting experimental work on particle dispersion in mixing layers was performed by Lazaro and Lasheras (1989, 1992a, 1992b) for flows with and without acoustic forcing. In the unforced case, a similarity for particle dispersion independent of particle size is found if the coordinates are normalized by a length proportioned to a density ratio times a Reynolds number times a diameter. Both experimental and theoretical evidence from Lazaro and Lasheras (1989), Chung and Troutt (1988), Chein and Chung (1988), and others indicates that the dispersion of particles with high inertia can exceed that of passive scalars. This agrees with the previously cited DNS results for homogeneous turbulence.

Rangel (1990, 1992) extended the vortex method to consider heat transfer and vaporization of the droplets as well as dispersion. Vortex pairing was also considered in a planar, temporal mixing layer with one-way coupling. The larger droplets tended to be less sensitive to the vortical structure on account of their higher inertia. The smaller droplets were more easily entrained but tended to vaporize completely before vortex pairing had a significant effect. The gas temperature and vapor mass fraction fields were determined by finite-difference Eulerian computations with the droplets serving as sources and sinks. Vortex-dynamical methods were used to calculate gas velocity, and the droplet properties were calculated by Lagrangian discrete-particle methodology. The effects of droplet kinematic inertia were carefully examined. Figure 10.2 demonstrates the droplet motion and the vapor mass fraction contours in a case in which three initially parallel droplet streams are moving through a temporally developing, two-dimensional mixing layer. Droplet and gas properties are averaged in the third dimension. Rangel and Continillo (1992) considered vaporization and ignition for the two-dimensional interaction of a viscous line vortex with a fuel-droplet ring or cloud. The effects of chemical kinetic and vaporization parameters on the ignition-delay time were determined. Bellan and Harstad (1992) modelled a cluster of droplets embedded in a vortical structure extending previous studies (Bellan and Harstad, 1987, 1988). The cluster and the vortex were assumed to convect together, which differs from the assumption of previous studies. Centrifugal effects caused fuel vapor to accumulate in the vortex core. Harstad and Bellan (1997) extended those works to vaporizing polydisperse droplets in an inviscid vortex structure. They found that that each initial size class of droplets results in a range of sizes because of the nonuniform environment for the droplets after experiencing centrifuging. So, even in an initially monodisperse spray, a polydisperse spray would result with time.

Park et al. (1994) considered the axisymmetric unsteady jet flame with n-heptane droplets carried by a nitrogen jet that is coflowing with surrounding air. Gravity effects and two-way coupling were included. Strong interactions between the droplets and the vortex structures were discovered. These interactions and

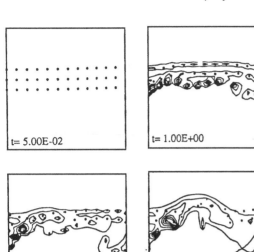

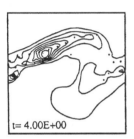

Figure 10.2. Temporal development of vapor mass fraction contours and droplet position with three droplet streams moving through a temporally developing two-dimensional mixing layer. (Rangel, 1992, with permission of Taylor & Francis, Inc.)

gravity affected the flame structure. Aggarwal et al. (1996) considered a droplet-laden axisymmetric jet with two-way hydrodynamic coupling. The dispersed phase had a significant effect on the vortex dynamics. Vaporization also affected the dynamics. Park et al. (1996, 1998) examined a droplet-laden jet undergoing transition. Both evaporating and nonevaporating droplets were considered. The effect of swirl and the flow was examined. Consistent with previous results of others, they found that the spray was affected by the vortex structures. Interestingly, they also found that, at high mass loading of the droplets, substantial modification to the vortex dynamics could occur.

Aggarwal et al. (1996) considered a particle-laden two-dimensional, unsteady, spatially developing shear layer; they confirmed the prior experimental evidence by Hishida et al. (1992) of a correlation between the Stokes number and particle dispersion by the transitional shear layer. Ling et al. (1998) used a pseudo-spectral method to analyze a particle-laden three-dimensional temporal mixing layer. There was only a one-way coupling in their calculations. Particle dispersion is still dominated by the two-dimensional large-scale structure motion. Particles tend to concentrate on the circumference of the structure, especially if their Stokes number is of the order of unity. The Stokes number is defined as the ratio of the particle response time (at low Reynolds number) to the unsteady flow-field characteristic time. There are three-dimensional effects, however; for example, counterrotating "rib" vortices can transport particles to or from the large structures, modifying the distribution of the particles.

In general, much remains to be determined about the modulation of both the mean flow and the turbulent fluctuations or vortical structures by the spray of droplets or cloud of particles. In different situations, the presence of droplets or particles can either enhance or reduce turbulence. More analyses with two-way coupling are necessary.

10.2 Time-Averaged Turbulence Models

Turbulent flows are a computational challenge because fluctuations occur over a wide spectrum of length scales. Therefore it is difficult and usually practically impossible to resolve details over a large physical domain. Equations governing statistical properties of the flow are often attractive because they reduce the complexity required in the solution. However, they do introduce the closure problem, which requires modelling of some terms to relate some unknown quantities to other unknowns; otherwise the number of equations is not sufficient. Often the modelling of terms is guided by scanty evidence so that controversy is attached to this general approach.

Elghobashi and Abou-Arab (1983) developed a two-equation turbulence model for two-phase flows that eliminated much of the ad hoc character of simulating the character of the interaction between the two phases in a turbulent flow. Although the formulation is derived from continuity and momentum equations, modelling at third order is necessary to achieve closure. The resulting equations describe the turbulent kinetic energy and the dissipation of the turbulent kinetic energy for the continuous phase. Elghobashi and Abou-Arab (1983) extended for the two-phase flow case the classical $k - \varepsilon$ model for single-phase fluids. The approach was based on a two-continua representation (as described in Chapter 7). Their closure scheme for the averaged equations was based on certain key assumptions: (i) neglect of fourth-order correlations; (ii) neglect of the pressure–velocity correlation's contribution to the diffusion of turbulence; (iii) modelling of the averaged divergence of the pressure–velocity product, following the work of Launder et al. (1975); (iv) gradient transport modelling for various correlations of velocity with other quantities; and (v) modelling of the droplet-velocity–gas-velocity correlation by means of the Chao (1964) linearized Lagrangian equation of particle motion. The analysis introduced several additional production terms and dissipation terms into the turbulent-kinetic-energy equation and the dissipation of turbulent-kinetic-energy rate equation; these new terms disappear as the droplet number density goes to zero.

The Elghobashi and Abou-Arab model involved many terms to be modelled because both liquid and gas properties and void fractions were involved. Although the modelling was performed in general tensor form, the final modelled equations were presented for the boundary-layer approximation in cylindrically symmetric form. Several new constants were introduced requiring empirical determination. Comparisons of calculations with experimental results for particle-laden round jets were quite favorable.

Elghobashi et al. (1984) applied the two-equation model to a particle-laden jet calculation and demonstrated respectable agreement with experiment. They showed that additional dissipation of the flow resulted in this two-phase case. Mostafa and Elghobashi (1985a, 1985b) extended the model to account for vaporization. Rizk and Elghobashi (1985) studied the motion of a spherical particle near a wall, accounting for the effects of lift and the modification of drag. Rizk and Elghobashi (1989) extended the two-phase, two-equation turbulence model to account for confined flows with wall effects.

In the previously mentioned research, Elghobashi and co-workers used Eulerian formulations for both the dispersed and the continuous phases. Gosman and Ioannides (1981), Mostafa and Mongia (1988), Mostafa et al. (1989), and Berlemont et al. (1991) used a stochastic Lagrangian method for the dispersed phase, together with a two-equation turbulence model for the continuous phase. The first study did not account for the modification of the two-equation turbulence model because of the presence of the second phase. MacInnes and Bracco (1992a) examined stochastic dispersion models for nonvaporizing sprays. They considered several cases of homogeneous turbulence and of shear flows. Concerns were raised about many existing models that predicted eventual nonuniform distribution of particles in a flow that initially possessed a uniform distribution. An approximate correction was derived. The work was extended to include the effects of finite-particle response time by MacInnes and Bracco (1992b).

An axisymmetric, turbulent reacting spray flow in a center-body injection configuration was analyzed by Raju and Sirignano (1989, 1990a). A steady-state solution was found asymptotically in time by means of finite-difference techniques. Monodisperse and polydisperse sprays and single-component and multicomponent liquid fuels were studied. Transient droplet heating and vaporization were considered following the methods outlined in Chapters 2 and 3. A multicontinua approach of the type discussed in Chapter 7 was used. The gas-phase properties were calculated on an Eulerian grid whereas the liquid properties were calculated by a Lagrangian method. A $k - \varepsilon$ model for the gas-phase turbulence was used with one-way coupling only with regard to the turbulence fluctuations (still a two-way coupling for the mean flow properties was utilized).

Figure 10.3 shows droplet trajectories and size for a polydisperse spray calculation, and Fig. 10.4 indicates gas-velocity magnitude and direction at the grid points of a nonuniform computational mesh. A double vortex results; the outer vortex is driven by the incoming air while the inner vortex is driven by the injected liquid. The smaller droplets in the spray tend to become bound in the recirculation zone whereas the larger droplets penetrate beyond the recirculation zone. The recirculation zone was predicted to be fuel rich, and the turbulent transport of fuel vapors from that zone to the incoming air was predicted to be important as a rate-controlling factor in the combustion rate. The calculations used solid-sphere drag data that overpredicted the drag on the vaporizing droplets and subsequently underpredicted the amount of liquid fuel that emerged from the recirculation zone. Improved calculations with more accurate drag coefficients should increase the amount of liquid fuel emerging from the zone, increasing the relative importance of the vaporization rate and decreasing the relative importance of the turbulent transport.

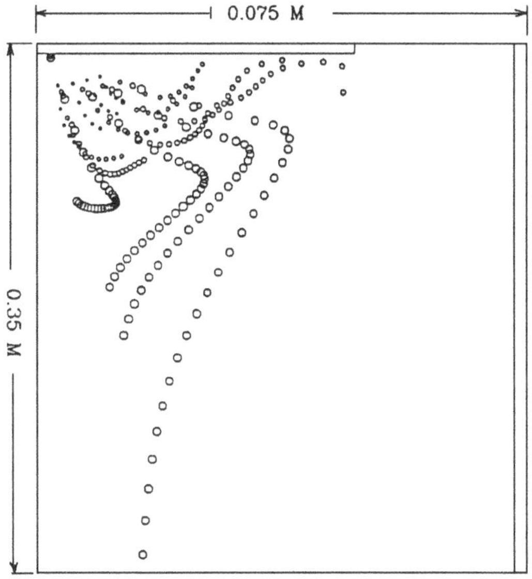

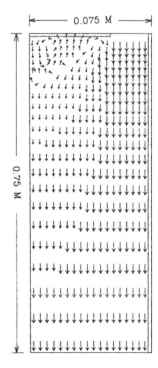

Figure 10.3. Droplet trajectories and sizes for a polydisperse spray with a center-body injector configuration (Raju and Sirignano, 1989).

Hollmann and Gutheil (1996, 1998) developed a spray flamelet model to be used for turbulent spray diffusion flames. Detailed chemistry for methanol–air reactions was included. The approach uses a computational library for laminar-flamelet structure that is characterized by mixture fraction and scalar dissipation rate. A Lagrangian liquid-phase treatment was coupled with an Eulerian gas-phase calculation. Extension to ethanol sprays was made by Düwel et al. (2007). Ge and Gutheil (2006, 2008) improved the probability density function (PDF) representation of the

Figure 10.4. Gas-velocity vectors for center-body injector configuration (Raju and Sirignano, 1989).

mixture fraction statistics by using Monte Carlo methods and also introduced a joint mixture ratio–enthalpy PDF. An overview is presented by Ge et al. (2006).

10.3 Direct Numerical Simulation

A major problem with time-averaged equations is the uncertainty about the modelling of terms to achieve closure. For single-phase fluids, advancement in computational power has encouraged the development of the method of DNS. Here the unsteady Navier–Stokes equation are simulated and solved over the full range of length scales. However, there are still limitations on the size of the physical domain and the range of length scales that can be solved; specifically, the allowable magnitude of the Reynolds number based on the largest length and velocity scales is limited to values generally smaller than practical values. However, the method can potentially be helpful in evaluating closure methods for averaged equations and in providing physical insights to the small-scale physics. Also, there is the expectation that the limit on the Reynolds number will increase as computational capabilities grow.

The DNS methods as applied to two-phase flows have generally not been truly without modelling in a fashion analogous to the single-phase fluid treatment. Applications have been confined to situations in which the smallest scale of turbulence is considerably larger than the droplet or the particle size. A few exceptions for flows laden with nonvaporizing solid particles are subsequently discussed. The interaction between the two phases is modelled by some simplifying equations of the types discussed in earlier chapters. Because velocity gradients on the scales of the droplet diameter and boundary-layer thickness are not resolved, the vorticity generated by means of the droplet–gas interaction are not determined. So current two-phase DNS is useful only for special classes of turbulent two-phase flows. It cannot be helpful, for example, in evaluating turbulent flow created primarily by a dense spray interacting with a gas. A liquid-propellant rocket engine is an example in which an accurate simulation of the details of the smallest scales of turbulence requires DNSs of both phases; DNS for one phase without DNS for the other phase would not yield a DNS for either phase or for the interactions between the phases.

Both DNS and LES analyses often begin with equations that treat the continuous- or carrier-fluid properties as continuous variables and, through a Lagrangian tracking scheme, follow and describe the behavior of representative particles, droplets, or bubbles. So an implicit averaging was made on the droplet scale because the carrier-fluid properties are constructed to exist everywhere, including within discrete-particle volumes, and some average discrete-phase behavior is described. Even if the DNS considers all particles, the continuous carrier-fluid properties and the absence of resolution of the behavior in the film surrounding the particle implies averaging has been performed. Furthermore, for LES, another averaging (or filtering) is performed explicitly to avoid resolution of the smallest scales of turbulence. The modelling of the droplet behavior has neglected any direct interaction of small eddies with the droplets. Rather, the resulting equations are based implicitly on the assumption that the smallest eddies of the turbulence are larger than the largest droplet scales. In Chapter 7, the droplet averaging process and the LES length-scale filtering process were unified into one process. This is especially power-ful in the situation in which the smallest eddy scales are comparable to the droplet scales.

First, we discuss works for nonvaporizing situations; in particular, particle-laden and bubble-laden flows are discussed. Important developments in LES and DNS for bubble-laden and particle-laden turbulent flows have been made in re-cent years. For a DNS of particle- or bubble-laden flows with homogeneous turbu-lence, see Elghobashi and Truesdell (1992, 1993), Boivin et al. (1998), and Druzhinin and Elghobashi (1998). Ling et al. (1998) considered temporal mixing layers, and Druzhinin and Elghobashi (2001) analyzed spatially developing mixing layers. See also Ferrante and Elghobashi (2003) and Squires and Eaton (1991).

Squires and Eaton (1990, 1991), Elghobashi (1991), Elghobashi and Truesdell (1991, 1992, 1993), and Ferrante and Elghobashi (2003) analyzed two-phase flows by DNS of isotropic homogeneous turbulent flows. Squires and Eaton used only Stokes drag as the force on the particle and considered both one-way and two-way cou-plings of the two phases with regard to the turbulent field. Elghobashi and Truesdell considered viscous and pressure drag, pressure-gradient and viscous-stress forces on the particle, Basset correction, gravity, and virtual mass. One-way coupling was considered by Elghobashi and Truesdell (1992) whereas two-way coupling was con-sidered in their other papers (1991, 1993); the particles did not modify the turbu-lent field substantially for dilute flows. Lagrangian methods were used to calcu-late particle trajectories. At zero gravity and short dispersion times, inertia caused particle diffusivity to exceed the fluid diffusivity. Both gravity and inertia reduced lateral dispersion. At low frequencies, particle energy remained higher than the fluid energy. Particle energy was lower than fluid energy in the mid-to-high-frequency range. Drag and gravity tended to be the dominant forces. The Basset correction was the next largest force term but generally an order of magnitude smaller.

Elghobashi and Truesdell (1993) considered solid particles in a homogeneous turbulent flow. The particles were points of mass whose velocities were individually described by Eq. (3.63). A reversed force on the incompressible carrier fluid was described according to Eq. (7.74) for constant density, no vaporization, $\theta = 1$, and viscous stress linearly proportional to velocity gradient. Particles sizes were smaller than the smallest scale of turbulence (Kolmogorov scale). Calculations were per-formed for cases with gravity and for cases without gravity.

The small particles in the absence of gravity transferred their momentum and kinetic energy to the high-wave-number end of the turbulence spectrum for the gas, increasing somewhat the energy content at the high-wave-number end beyond the amount normally found for a single-phase fluid. The viscous dissipation rate was also found to be higher by Elghobashi and Truesdell. Eventually this enhances the decay rate of the turbulence energy, decreases the Kolmogorov scale, and increases the growth rate of the integral scale.

DNS calculations from Druzhinin and Elghobashi (2001) for a spatially devel-oping mixing layer that is laden with bubbles are shown in Figures 10.5 and 10.6. Of course, the same principles and equations apply to droplet-laden and bubble-laden flows. The mixing layer is originally two dimensional but becomes unstable in three dimensions. Figure 10.5 shows the *y* component of vorticity that focuses or

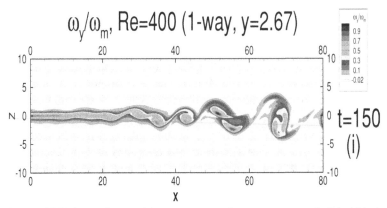

Figure 10.5. Spanwise vorticity component for one-way-coupled, bubble-laden, spatially developing mixing-layer flow. Clustering of vorticity and pairing of vortex structures (Druzhinin and Elghobashi, 2001). (Also, see color plate.)

clusters during the instability. Then the vortex-structure pair causing the mixing-layer growth. The concentration of the bubbles was originally uniform but is modified significantly by the vortical motion, as shown in Figure 10.6.

Ferrante and Elghobashi (2003) showed that, in the presence of gravity, the particles transfer momentum and kinetic energy to the smallest scales of turbulence in an anisotropic manner to a direction aligned with the gravity vector. Then energy is transferred at the same wave number to the other two directions. This tends to reduce the turbulent-energy-decay rate and the integral scale growth rate. Consequently, both turbulent-kinetic-energy levels and viscous dissipation are increased and the Kolmogorov scale is decreased here.

Ferrante and Elghobashi (2004a, 2004b) considered spatially developing boundary layers, using an Eulerian–Lagrangian approach. A method for developing inflow conditions for DNS of the turbulent boundary layer was first developed by rescaling downstream velocity and entering the rescaled conditions for inflow in an iterative manner. Then the flow with submillimeter bubbles was examined. For

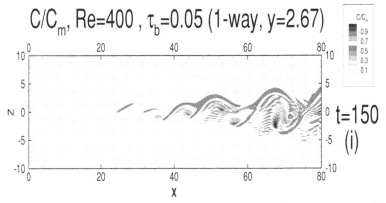

Figure 10.6. Spatial variation of initially uniform bubble concentration (Druzhinin and Elghobashi, 2001). (Also, see color plate.)

bubble-volume fractions in the 0.001–0.02 range, the velocity vector field had a positive divergence value. This created a positive mean velocity normal to and away from the wall that pushes streamwise vortical structures away from the wall, thereby increasing the displacement effect and reducing both the streamwise velocity and the wall skin friction. Turbulent-kinetic-energy production is reduced by the movement of the maximum Reynolds stress region away from the wall to a zone of smaller velocity gradient. Ferrante and Elghobashi (2005) showed that an increase in Reynolds number brings the vortical structures closer to the wall, reducing the decrease in drag.

Ferrante and Elghobashi (2007a) used an Eulerian–Eulerian approach to consider a periodic array of vortex structures (bubble-laden Taylor–Green vortex flow) and showed that the positive velocity divergence created in the vortex core by bubble clustering resulted in decay of the vorticity. Ferrante and Elghobashi (2007b) showed that the Eulerian–Lagrangian method is more accurate than the Eulerian–Eulerian approach, which does not present the accurate bubble concentration that produces the velocity divergence. In their Lagrangian calculations, Ferrante and Elghobashi examined the trajectory and behavior of every bubble, differing from the typical previous study that followed only representative or average bubbles or particles. In some regions where bubbles collected, the number of bubbles in an Eulerian computational cell became large.

A few true DNS calculations for nonvaporizing solid particles in a fluid with decaying isotropic turbulence have been performed. Zhang and Prosperetti (2005) developed a numerical method for resolving flows around solid spheres freely moving in a turbulent fluid. The Stokes model was used in the region near the solid surfaces where the no-slip condition applies in order to transfer that condition to the fixed mesh; this avoids the remeshing to account for moving boundaries and gives a substantial benefit in terms of demand for computational resources. Their field contained 100 particles with particle density only 1.02 times the fluid density. The volume fraction was 0.1 and the particle diameter was 8.32 times the initial Kolomogorov scale. Uhlmann (2005, 2007), using an immersed boundary method that applies a force on the fluid at the Lagrangian position of the particle, considered as many as 4096 particles. Using that immersed boundary method, Ferrante and Elghobashi (2007c) and Lucci et al. (2008) examined various mass loadings with the solid material density varying from 2.56 to 10 times the fluid density. Up to 6400 particles were considered. An artificial repulsive force field was created to represent particle collisions. They showed that the particles increase the dissipation rate and reduce the turbulent kinetic energy, especially at the low-wave-number end of the spectrum because of the large particle sizes.

Let us now consider DNS for vaporizing droplets. Mashayek (1998a, 1998b) examined spray flows with vaporizing droplets in two flows: a gas flow with stationary (i.e., forced) isotropic turbulence and a turbulent shear flow. The turbulence was artificially forced in the one case so that dissipation would not cause a decrease in intensity. Variable density was allowed at low Mach number but the vapor and surrounding gas were assumed to have the same properties. For the shear flow, vaporization was largest in the region of highest strain rate. Generally, the qualitative behaviors were the expected results; vaporization rate decreased as initial droplet diameter, latent heat of vaporization, boiling temperature, or initial mass loading increased and the rate increased as ambient gas temperature increased. See the review on turbulent droplet- and particle-laden flows by Masheyak and Pandya (2003).

Bellan and co-workers used DNS methodology that addressed jet flows and temporal mixing layers through a range from laminar flows to transitional flows with vaporizing droplets. Miller and Bellan (1999) considered a three-dimensional temporally developing mixing layer between two streams, one of which was droplet laden. Almost 10^6 droplets were tracked. The nonequilibrium Langmuir–Knudsen vaporization-rate law was used. Vortex structures formed and the droplets were centrifuged out of the high-vorticity regions and migrated to the high-strain regions. Abdel-Hameed and Bellan (2002) considered laminar jet flows of different cross sections. The presence of drops increases the jet entrainment rate and shortens the core length by an order of magnitude compared with the single-phase jet. The droplet interaction with the flow creates a streamwise vorticity that modifies mixing rates. The circular jet cross section results in a lower entrainment rate than elliptical, rectangular, and triangular cross sections, similar to the known results for single-phase jets. Other papers by Bellan and co-workers that focused on the generation of databases using DNS that could guide subgrid modelling for LES are reviewed in the next section.

The method of continuous thermodynamics for multicomponent liquid droplets (discussed in Chapter 4) was extended to the case of DNS for transitional flows by Bellan and co-workers. See LeClercq and Bellan (2004, 2005a, 2005b) and Selle and Bellan (2007a, 2007b, 2007c). The same group extended their work on supercritical vaporization to configuration undergoing transition to turbulence. See Bellan (2000a, 2006), Okong'o and Bellan (2000b, 2002a, 2002b, 2004a), Miller et al. (2001), and Selle et al. (2007).

The current DNS approaches for bubble-laden, particle-laden, and spray flows are generally not truly DNSs. Certainly no studies have true DNS for vaporizing or nonvaporizing spray flows, although we can expect some to be developed in the near future. The studies typically do involve modelling at the droplet (meaning here droplet, solid particle, or bubble) level. That is, the droplet-liquid interior and the gas film surrounding each droplet are not resolved. Rather, each droplet is considered as a point with mass and inertia. The aerodynamic forces and heating rates and vaporization rates are treated through the use of sources and sinks, using algorithms for the source and sink strengths that have been determined through separate analytical or empirical studies; so, they are essentially subgrid models. To represent interactions between the phases, these source and sink terms are distributed to an Eulerian grid whose scale is larger than the discrete-particle size. So, effectively, some subgrid modelling and averaging exist in all two-phase "DNS" efforts that consider more than a few discrete-phase particles, droplets, or bubbles and rely on source–sink representations. Furthermore, as noted earlier, these subgrid models have neglected modification of Nu, Sh, C_D, and C_L through particle interactions with the smallest scales of turbulence.

Generally no attempt has been made modify the description of the surrounding gas-film behavior and the subgrid model used in order to account for the interactions (e.g., collisions) with vortical structures of dimensions comparable to the droplet size. So the issue raised in Chapter 7 can be raised here. That is, if the average spacing between particles (bubbles, or drops) is larger than the smallest turbulence scale, the two-fluid formulation involves averaging over a scale larger than that smallest turbulence scale and the DNS cannot be a true simulation of the turbulence.

A common practice in this field of particle-laden flows is to describe as turbulence all unsteady-flow fluctuations with small length and time scales compared with the largest scales in the particular flow. However, turbulence is classically defined as vortical fluctuations. For example, acoustical fluctuations, no matter how random they might be, are not turbulent. The particles moving through a fluid can cause fluctuations that are not totally due to viscous or vortical causes. Consider the theoretical inviscid limit wherein only a potential flow occurs around a moving particle or droplet; in a frame of reference moving relative to the particle, unsteady-flow fluctuations in velocity and pressure can be observed. If there are many particles moving through the fluid at random spacing and time intervals, random fluctuations in the fluid flow will be observed. However, there is no production of vorticity and therefore no turbulence in this limit. For a practical viscous fluid, we can expect a viscous boundary layer to surround each particle and a viscous wake to trail each particle. For a higher particle Reynolds number, the flow outside of the viscous layer will approximate a potential flow. Details are provided in Section 3.2 or in the text by White (1991). Only the deviations from potential flow within the viscous layers and wakes are vortical and can meet the definition of turbulence. The distinction between vortical and nonvortical flow fluctuations is important when we consider the interactions between fluctuations of different scales. More research is required for understanding this complex behavior and for correcting some current confusion in the field.

10.4 Large-Eddy Simulations

LESs involve resolution of the unsteady turbulent phenomena above a certain length scale. Filtering or averaging occurs for the phenomena that occur below a certain length scale. The scale at which filtering occurs is the numerical grid size; the physics, mechanics, and chemistry below this scale are modelled. It is known as subgrid modelling.

Useful overviews of LES methods for single-phase flows are given in Piomelli (1999) and Givi (2003). Bellan (2000b) gives a helpful review of LES for vaporizing droplet-laden flows. The LES approach was used for turbulent particle-laden channel flow by Wang and Squires (1996). Okong'o and Bellan (2000a, 2004b), Miller and Bellan (2000), Leboissetier et al. (2005), and Okong'o et al. (2008) used DNS results to guide development of LES for vaporizing sprays. Some applications of LES for spray combustion were described by Sankaran and Menon (2002a, 2002b), Menon (2004), Menon et al. (2004), and Menon and Patel (2005, 2006).

Figures 10.7 and 10.8 show recent results for nonreacting results of LES computations from Sankaran and Menon (2002a) for a spray combustor with swirl. In Figure 10.7, for a nonreacting case, droplet positions and azimuthal vorticity are shown for a low-swirl case, and Figure 10.8 shows the high-swirl case. The increased swirl forces the droplets to larger radial positions. For the reacting cases, the azimuthal

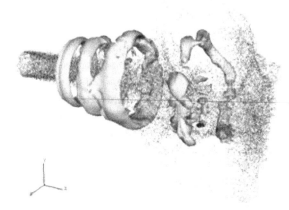

Figure 10.7. Azimuthal vorticity contour (light) and droplet positions (dark) with low-swirl conditions (Sankaran and Menon, 2002a, with permission of the *Journal of Turbulence*). (Also, see color plate.)

vorticity leads to the creation of a core recirculation zone that stabilizes the flame. The flame and therefore the high temperature regions are moved radially outward by the increased swirl.

The need for better physics-based subgrid closure for two-phase flows is indicated by the apparent limitations of existing models in characterizing what is

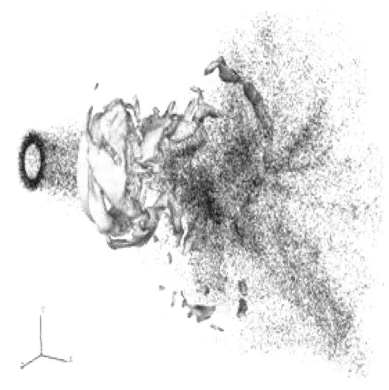

Figure 10.8. Azimuthal vorticity contour (light) and droplet positions (dark) with high-swirl conditions (Sankaran and Menon, 2002a, with permission of the *Journal of Turbulence*). (Also, see color plate.)

observed in actual applications. Issues of interest are the behavior of the spray breakup, droplet dispersion in swirling flow, and mixing under different operating conditions. An inherent requirement is that turbulent fine-scale mixing in a two-phase flow be accurately predicted over a wide range of operating conditions without changing the closure models. This was addressed by Menon (2004), Menon et al. (2004), and Menon and Patel (2005, 2006). Attempts to validate existing models have been only partly successful for dilute sprays as well as for dense sprays. For example, in gas-turbine combustors (Lefebvre, 1989) fuel is injected in a manner to create a rapidly atomized spray downstream of the injector. Thus, a few millimeters downstream of the fuel injector, the droplets are being rapidly dispersed into a hot swirling air flow, and efficient mixing is the essential goal for the design. Current design models based on RANS simulations are unable to deal with this requirement with high fidelity. Even existing LES models based on state-of-the-art subgrid models are only partly successful in properly capturing, for example, the spray spreading and penetration, the distribution of the mixed and unmixed fuel–air mixture, and the impact of this on pollutant formation; see Menon et al. (2004) and Menon and Patel (2005, 2006).

Recent simulations and experiments by Menon (2004), Menon and Patel (2005), and Colby et al. (2005) have shown that changes in the initial droplet-size distribution and/or turbulent intensity in the mixing region can have a profound effect on the mixing and combustion process farther downstream and can also directly impact pollutant emission. Clearly, models used to account for turbulence–droplet interaction play an important role in this effect.

Okong'o and Bellan (2000a, 2004b), Leboissetier et al. (2005), and Okong'o et al. (2008) examined transitional, temporal mixing layers carrying vaporizing droplets. DNS results are generated to provide a database for evaluation of models for LES approaches. Okong'o and Bellan introduced corrections to the filtered variables based on the filtered standard deviation in order to represent the unfiltered variables at the droplet locations for proper calculation of droplet source terms. The LES and DNS compared favorably for global growth of the mixing layer and mixing rates. However, differences were found in dissipation rates and the spatial distribution of the droplets. The filtered droplet source terms were overestimated by the LES. Of course, the LES was shown to be substantially less computationally expensive than DNS. Various subgrid models were compared; see the following subsection on closure.

Vaporization subgrid models used in spray combustion and in nonreacting spray flows have not yet accounted for interactions of vortices and vaporizing droplets of comparable dimensions. Some codes have accounted for large eddies interacting with much smaller droplets. Note that the turbulent Reynolds number for flow through a typical turbine combustor can be of the order of 10,000 with an integral scale that is tens of centimeters (Menon, 2004; Menon et al., 2004; Menon and Patel, 2006). Consequently the Kolmogorov scale is of the order of 100 μm, which is not much larger than the droplet diameter. DNS of turbulent two-phase flows (Ferrante and Elghobashi, 2003) has indicated that the presence of the discrete phase can generate small-scale turbulence. Energy at the end of the turbulent inertial range spectrum has been shown to transfer both to larger particle sizes and to smaller sizes because of the presence of droplets. Only one paper (Masoudi and Sirignano, 2000) accounted for interactions between an isolated vaporizing droplet and comparably sized vortices. See Section 3.4.

Sirignano (1972) predicted that the droplet aerodynamic forces and the momentum exchange associated with mass exchange (e.g., vaporization) caused the production of vorticity in a two-phase flow. This occurs independently of whether or not unsteady boundary layers or wakes are developed around the droplets. The existing DNS literature clearly indicates that the presence of particles or droplets affects the dissipation of turbulent kinetic energy. There is also evidence that the turbulence–particle interactions can create a less homogeneous situation at the microstructure. Thus an initially uniform distribution of particles (in terms of interparticle spacing) can be made highly nonuniform by the turbulence. The role of stratification within the microstructure is worthy of further examination as well because these local nonuniformities can impact fuel–air mixing, flame stability, and pollutant emission.

There are certain special features of droplet vaporization in spray combustion that must be reflected in the modelling: the vaporization time scale in the high-temperature environment can be as small as or smaller than the times for the liquid droplets to reach kinematic or thermal equilibrium with the gas; the Reynolds number (and Peclet number) based on relative gas–droplet velocity and droplet dimension can be $O(10)$ to $O(100)$, causing substantial stratification and large gradients within the gaseous microstructure surrounding droplets, thereby requiring more information than mere averages of the microstructure field; and droplet-vaporization and -heating rates can depend on internal spatial and temporal variations of liquid temperature, composition, and velocity, thereby requiring more information than simply average properties.

10.4.1 Proper Two-Way Coupling for LES Closure

In spray combustion, the coupling between the two phases is critical in both directions and must be properly modelled. The rates for liquid-phase heating, vaporization, and acceleration are totally dependent on the coupling with the gas phase. The gas phase is strongly affected by liquid-phase behavior because the rate of energy conversion and mass species conversion depends critically on droplet heating and vaporization. Furthermore, the mass and momentum of the liquid (discrete phase) can be within an order of magnitude of the values for the gas (continuous phase) although the volume fraction of liquid is orders of magnitude smaller. So the mass, momentum, and energy exchange has significant quantitative impact on conservation of these quantities for both phases.

Another issue concerns the proper coupling between the filtered LES equations and the subgrid models, as discussed in Chapter 7. The two-phase-flow equations are the result of averaging over length scales that are larger than distances between neighboring droplets so that the gas-phase and liquid-phase properties are represented by two superimposed continua. In practical turbulent spray LES computations (Sankaran and Menon, 2002a; Menon et al., 2004; Menon and Patel, 2005, 2006; Patel and Mennon, 2008; Patel et al., 2006), droplet groups that are representative of droplets in some neighborhood with similar properties (e.g., size, velocity, temperature) are tracked (primarily to reduce computational cost). Although the volume defining the neighborhood for droplet-property averaging is arbitrary, it is rational to make the averaging volume of the order of the computational-cell volume. The length scale for the averaging volume should be orders of magnitude larger than the droplet diameter or the average spacing between droplets. If the droplet scales and mesh size are comparable, very few droplets are involved in the averaging process and very large fluctuations in the average quantities can be expected.

Currently, these averaged two-phase-flow equations are further spatially filtered to obtain two-phase LES equations. Of course, "filtering" is averaging by another name; this means that some of the short wavelengths were already filtered in the two-phase flow equations. If the subgrid model for the two-phase flow did not reflect the presence of small vortices, there is danger of serious error. Recently, Sirignano (2005a) has proposed how to do LES filtering and the two-phase averaging in one consistent process; see Chapter 7. This study clearly identified the source terms and the flux terms that require subgrid modelling. Several of these terms had not previously been identified or modelled in spray combustion analysis. This new understanding has not been fully exploited at this time.

10.4.2 Gas-Phase Equations

By using an appropriate spatial Favre filter (Erlebacher et al., 1992; Sirignano, 2005a) to remove the high-frequency component, we obtain the filtered conservation equations for mass, species, momentum, and energy for the gas phase in a compressible two-phase mixture in the following general form:

$$\frac{\partial(\overline{\rho}\,\langle\Phi\rangle)}{\partial t} + \frac{\partial(\overline{\rho}\,\langle\Phi\rangle\,\langle u_j\rangle)}{\partial x_j} - \frac{\partial\langle Q_j\rangle}{\partial x_j} - \langle S\rangle = \frac{\partial Q_{sg,j}}{\partial x_j} + S_{sg}. \qquad (10.1)$$

Favre integration over a scale at least as large as the subgrid scales is implied through the use of the angled brackets, and the subscript *sg* implies a quantity that is obtained through subgrid models and or correlations from direct numerical simulations (DNSs). Q is a flux whereas S is a source or sink. Details about the fluxes, sources, and sinks for each conservation equation are given in Sirignano (2005a) and Chapter 7.

For LES, the flux and source terms

$$\alpha_{n,i} \stackrel{\text{def}}{=} \langle Y_n\rangle\,\langle u_i\rangle - \langle Y_n\,u_i\rangle\,;$$

$$\beta_{n,i} \stackrel{\text{def}}{=} \langle Y_n\rangle\,\langle V_{n,i}\rangle - \langle Y_n V_{n,i}\rangle;$$

$$\Gamma_{ij} \stackrel{\text{def}}{=} \langle u_i\rangle\,\langle u_j\rangle - \langle u_i u_j\rangle;$$

$$\Delta \stackrel{\text{def}}{=} \langle u_j\rangle\,\overline{\theta}\,\frac{\partial\langle\widehat{p}\rangle}{\partial x_j} - \overline{u_j\frac{\partial p}{\partial x_j}};$$

$$E_i \stackrel{\text{def}}{=} \langle u_i\rangle\,\langle h\rangle - \langle u_i h\rangle$$

are the subgrid terms that require closure. Also requiring closure is $\tilde{\omega}_n$, the *n*th filtered species reaction rate that is usually very problematic to achieve. Past studies have established accurate closure for Γ_{ij}, Δ, and E_i by using a localized dynamic closure approach based on the transport modelling of the subgrid kinetic energy. See the review for turbulent two-phase LES by Bellan (2000b). Also, see Menon and Kim (1996); Menon et al., (1996); Kim and Menon (1999, 2000); and Kim et al. (1999). Application to complex single- and two-phase nonreacting and reacting LES has demonstrated the robustness and accuracy of this approach.

The most common representation for the Reynolds stress subgrid modelling is the Smagorinsky model:

$$\Gamma_{ij} \stackrel{\text{def}}{=} \langle u_i \rangle \langle u_j \rangle - \langle u_i u_j \rangle = -2\nu_t(\langle S_{ij} \rangle - \frac{1}{3}\delta_{ij} \langle S_{kk} \rangle) + \frac{2}{3}k^{sgs}\delta_{ij}, \qquad (10.2)$$

where the rate of strain and the turbulent eddy viscosity are given by

$$S_{ij} = \frac{1}{2}\left[\frac{\partial u_i}{\partial x_j} + \frac{\partial u_j}{\partial x_i}\right]; \quad \nu_t = C_\nu \sqrt{k^{sgs}}\Delta_G, \qquad (10.3)$$

k^{sgs} is the subgrid kinetic energy per unit mass associated with the subgrid velocity fluctuations, Δ_G is the filter length used in the function G of Chapter 7, and C_ν is a "universal" constant. Note, of course, that the Smagorinsky model was developed for single-phase flow turbulence, which is known to have a different energy spectrum and cascade process. In particular, we know that the discrete phase affects turbulence generation at small scales that is not represented by the Smagorinsky model or any existing subgrid model. Note also that the application of the Smagorinsky model to two-phase flows in practice has not distinguished in the rate-of-strain term for the difference between the gradient of the average velocity and the average of the velocity gradient as indicated by Eq. (7.19).

In the Smagorinsky model, the energy transport term has been modelled as

$$E_i \stackrel{\text{def}}{=} \langle u_i \rangle \langle h \rangle - \langle u_i h \rangle = -\frac{C\Delta_G^2}{c_p Pr}\sqrt{\langle S_{kl} \rangle \langle S_{kl} \rangle}\frac{\partial \langle T \rangle}{\partial x_i}. \qquad (10.4)$$

See, for example, Bellan (2000b) here.

Models for $\alpha_{n,i}$, $\beta_{n,i}$, and $\bar{\omega}_n$ were proposed in the past primarily for gas-phase LES (Moin et al., 1991; Cook and Riley, 1994, 1998; DesJardin and Frankel, 1998; Fureby, 2000) with various levels of success; however, from a physical point of view, molecular diffusion, reactions, and heat release all occur in the small scales that are not resolved in LES. In the earlier linear-eddy mixing– (LEM–) LES approaches (Chakravarthy and Menon, 2000; Sankaran et al., 2003; Menon, 2004; Menon et al., 2004; Menon and Patel, 2005, 2006), closure for these terms were provided in the subgrid but with coupling to the large scales.

An alternative to the Smagorinsky model is the gradient model in which the velocity gradients are used to establish the eddy viscosity and the eddy diffusivity. The Reynolds stress is modelled to be an eddy viscosity multiplied by the gradient of the filtered velocity. Similarly, the energy or species mass-flux terms are modelled to be the products of an eddy diffusivity with the gradient of filtered temperature or species mass fraction. The product of the gradients $\partial \langle u_i \rangle/\partial x_k$ and $\partial \langle \phi \rangle/\partial x_k$ can be shown by an expansion to approximate a constant multiplied by the difference between $\langle u_i \rangle \langle \phi \rangle$ and $\langle u_i \phi \rangle$, where ϕ can be the vector or scalar of interest. In particular, we have

$$\Gamma_{ij} \stackrel{\text{def}}{=} \langle u_i \rangle \, \langle u_j \rangle - \langle u_i u_j \rangle = C\Delta^2 \frac{\partial \langle u_i \rangle}{\partial x_k} \frac{\partial \langle u_j \rangle}{\partial x_k}, \qquad (10.5)$$

$$E_i \stackrel{\text{def}}{=} \langle u_i \rangle \, \langle h \rangle - \langle u_i h \rangle = C\Delta^2 \frac{\partial \langle u_i \rangle}{\partial x_k} \frac{\partial \langle T \rangle}{\partial x_k}. \qquad (10.6)$$

Clearly, the Smagorinsky approach uses more terms in the models.

A third approach is the similar-scale concept. Here, the smaller turbulent scales are assumed to have a self-similar form and a second filtering is performed at a larger scale than the primary filter (typically, a few times larger). Because of self-similarity, the fluxes at the primary filter level are taken to be equal to those calculated from the second filtering at the larger scale.

Generally, these models do not consider directly subgrid interactions with the discrete phase. However, Menon and co-workers did include a coupling. Specifically, a dynamic relation (i.e., an evolution equation in the form of a partial differential equation) was introduced for the subgrid turbulent kinetic energy k^{sgs}. This equation had a source term related to the generation of turbulent kinetic energy that is due to droplet interactions.

In comparing of DNS results against LES results with various subgrid models, Okong'o and Bellan (2000a, 2004b), Leboissetier et al. (2005), and Okong'o et al. (2008) found that the case with the Smagorinsky model did not compare as favorably as the cases with gradient and scale-similarity models. The Smagorinsky model showed deficiencies in the predictions of dissipation rate and spatial distribution of droplets.

10.4.3 Liquid-Phase Equations

In developing the Lagrangian form of the equations, we define the time derivative following the liquid $d(\ )/dt = \partial(\ )/\partial t + \overline{u}_{l,j} \partial(\ )/\partial x_j$. The liquid density is constant: $\overline{\rho}_l = [1 - \overline{\theta}\,]\rho_l$, where θ is unity at a gas location in the subgrid and zero at a liquid location. The overbar implies volume-weighted integration over a scale at least as large as the subgrid. In the following discussion, we focus primarily on the dilute spray regime without attention to the capture of droplet–droplet interaction effects. The droplet location equation is $dx_i/dt = \overline{u_{l,i}}$ and the continuity equation can be written in Lagrangian form, respectively, as $d\overline{\theta}/dt = \dot{\overline{M}}/\rho_l$, where $\dot{M}$ is the averaged vaporization rate per unit volume. The momentum equation can be written as $\overline{\rho}_l\,(d\overline{u}_{l,i}/dt) = F(\overline{\theta},\, \tilde{u}_i,\, \text{sources, sinks, fluxes})$.

In LES, the Eulerian gas-phase equations and the Lagrangian liquid-phase equations are strongly coupled through the source terms and are solved simultaneously. An equation for averaged liquid energy can be written in a similar manner to that of the preceding equations. However, the averaging method does not allow spatial resolution of interior droplet temperatures. For a spray with rapidly vaporizing droplets in a high-temperature gaseous environment, the use of the averaged energy equation would give a uniform droplet temperature, thereby yielding inaccuracies in the surface gradient. So a different approach should be followed in order to resolve both spatial and temporal variations of liquid temperature within the droplets. This subgrid-modelling approach, by accounting for spatial variation of internal liquid temperature, will allow a more accurate determination of the liquid-droplet surface temperature and the heat flux at the droplet surface as functions of time for the average droplet at each location in the LES mesh. These surface-temperature and surface heat flux quantities are specifically what are needed to couple properly the gas-phase and liquid-phase energy calculations.

As one possible future approach, the droplet-temperature field can be solved by the Abramzon and Sirignano (1989) model that accounts for internal liquid circulation over a wide range of droplet Reynolds number; see Section 3.1. Essentially, this model combines a solution of a one-dimensional, unsteady, heat diffusion equation for droplet interior (using an effective conductivity to account for the change in length scale caused by internal liquid motion) with a quasi-steady, convective, gas-film (boundary-layer and near-wake) analysis. This model does not assume unity Lewis number for the gas film and has compared favorably with exact Navier–Stokes computations. It applies when the length scales for free-stream gradients are large compared with the droplet size. So it can be used when the vortex-core size is much larger than the droplet size. Also, a correction should be developed for the case of smaller vortices, as subsequently discussed.

10.4.4 Vortex–Droplet Interactions

Scalar transport, mixing, and combustion occur because of four distinct processes (which occur at their respective temporal and spatial scales): large-scale advection and interface stretching that are due to large-scale (energy-containing) eddies or coherent structures, small-scale turbulent mixing that is due to (isotropic) eddies in the inertial range of turbulence, molecular diffusion that brings the species into molecular contact, and finally, chemical kinetics. These smaller-scale phenomena can be described by certain subgrid models.

Computations of vortex–spray interactions were discussed earlier in this chapter. However, for subgrid modelling in the case in which the vortex is larger than the droplet size and the spacing between neighboring droplets but smaller than an averaging volume, some simpler algorithms are required but do not yet exist. Two models that have been used for gaseous mixing and chemical reaction in vortical flows are candidates for extension to turbulent spray flows: linear-eddy mixing (LEM) model and asymptotic vortex mixing (AVM) model.

Previous studies on mixing of species and chemical reaction within a vortical structure by Marble (1985), Karagozian and Marble (1986), and Cetegen and Sirignano (1988, 1990) can provide the useful foundation for developing a spray–vortex interaction model, which we identified as the AVM model. If the subgrid vortex is much larger than the droplet sizes, then, as shown in the cited single-phase-flow publications, the time-varying mixing rate and chemical-conversion rate within the eddy can be determined by solving an ODE. An advantage of this AVM model is that the subgrid diffusion length is continually and naturally determined by first-principles representation of fluid rotation, whereas the LEM model has an ad hoc, discontinuous process without any modelling of fluid rotation. The proposed research involves the extension of this model to two-phase flows.

A subgrid-simulation approach was developed based on the LEM model of Kerstein (1989), in which the reaction–diffusion processes (which occur at scales smaller than the grid and therefore can never be resolved in a LES) are simulated by a subgrid direct simulation within *each* LES cell (Menon et al., 1996, 2004; Menon and Calhoon, 1996; Smith and Menon, 1996; Chakravarthy and Menon, 2000, 2001; Menon, 2004; Menon and Patel, 2005). This method uses a "grid-within-grid" concept in which the LES filtered equations for mass, momentum, and energy are solved on the LES grid and the scalar transport equations are solved in a two-step procedure: a local one-dimensional (1D) direct simulation of the reaction–diffusion equations within the LES cell, and a three-dimensional (3D) Lagrangian transport of the scalar fields across the LES cell faces based on the resolved flux transport. Turbulent mixing by eddies smaller than the LES cell size are implemented within the 1D domain by a mapping procedure that recovers 3D inertial range turbulence dynamics. The 3D transport of the scalar fields (and its gradients) across the LES cell faces is based on the resolved fluxes. Reduction of the local subgrid domain to a 1D state allows this simulation approach to be feasible within the context of a LES.

In two-phase flow, although the droplets are tracked in the Lagrangian manner, the phase change is included within the subgrid LEM.

Although the LEM model possesses many of the characteristics of the mixing process, a fundamental limitation of this model is that the effect of turbulent stirring by eddies smaller than the grid is implemented as a stochastic instantaneous process. Therefore this stirring model cannot account for the actual time required for vortex roll-up. Thus it has been established to be accurate in only a stochastic sense rather than in a temporal sense (i.e., in high Re flows, many subgrid stirring events occur between each LES time step. If the effect of subgrid vortices on the scalar mixing needs to be included in a "deterministic" sense, then a new model has to be developed. Furthermore, for droplet–vortex interaction, the time-scale for interaction is very important because vaporization process is highly deterministic. In the following discussion, a new subgrid model that has the potential for addressing these specific issues is described.

It should be noted that the LEM–LES is computationally very expensive, especially when multispecies finite-rate kinetics is included. However, as past studies have demonstrated, a highly parallel solver can mitigate some of this cost overhead.

The AVM subgrid model for gaseous mixing and chemical oxidation reaction will take the earlier results of Marble (1985), Karagozian and Marble (1986), and Cetegen and Sirignano (1988, 1990) and place them into a form appropriate for subgrid-turbulence modelling. The vortex-mixing problem has been treated for single-phase flows with and without chemical reaction. The model uses an asymptotic approach to solve the partial differential equations governing heat and mass diffusion in a flow field that is straining and turning because of a vortex. For example, in these earlier studies, a 2D gaseous flame-vortex model was developed in a Lagrangian frame of reference. An analytical viscous line vortex model of the form

$$V_\theta = \frac{\Gamma}{2\pi r}(1 - e^{(r^2/4vt)}), \ V_r = 0 \tag{10.7}$$

is used, where Γ is the vortex circulation at large radii and ν is the kinematic viscosity. The scalar field is modelled as a material interface undergoing stretching under the action of this vortex. Using a material transformation and some simplifying assumptions (Cetegen and Sirignano, 1990), it was shown that, for a reacting vortex, the mass conservation for the fuel species Y_f can be reduced to the following form:

$$\frac{\partial Y_f}{\partial \tau} = D\frac{\partial^2 Y_f}{\partial \xi^2} + h(\tau)\frac{\omega_f M_f}{\rho}. \tag{10.8}$$

Here, $h(\tau) = dt/d\tau$ is related to stretching and M_f and ρ are the fuel molecular weight and mixture density, respectively. The transformation from (t, y) to (τ, ξ) as shown by Cetegen and Sirignano (1990) accounts for vortex stretching in the interface normal coordinate ξ. The extension of this work to exothermic multispecies reactions is straightforward by adding the energy equation for temperature and will be considered.

In past studies, the time-dependent results were integrated over the vortex volume to obtain useful subgrid results. For example, it is found that the data correlate so that the degree of mixedness f satisfies $f = 0.593\mathrm{Sc}^{-1/2}[\frac{4\nu t}{y_m x_m}]^{1/2} + 0.093\mathrm{Sc}^{0.06}$ $\mathrm{Re}[\frac{4\nu t}{y_m x_m}]$, where Re, Sc, ν, t, y_m, and x_m are the vortex Reynolds number, Schmidt number, kinematic viscosity, time, and dimensions for domain of integration over the vortex. Here, f is proportional to the product of the variance of two species concentrations or mass fractions and varies between zero for complete unmixedness and unity for complete mixedness. Furthermore, the scalar dissipation rate is directly proportional to df/dt; thereby, the correlation provides a useful subgrid model.

This approach simplified the effect of shear and normal straining by focusing primarily on the normal strain effect with the added reduction in the spatial dimensionality to one dimension. In some respects, this model is similar to the LEM model approach described earlier, in which the spatial dimensionality is reduced to one dimension by considering the domain in the direction of the scalar gradient. The key difference between this approach and the earlier LEM model is that in the AVM model the effect of a physical vortex size, lifetime, and turnover rate on the scalar field can be included whereas, in the LEM model, turbulent stirring is modelled with a stochastic mapping that implies instantaneous mixing of a scalar field by a vortex.

In the other limit, in which the smallest vortex structure is of the scale of the droplet diameter or spacing, new issues have to be addressed. Past studies by Kim et al. (1995, 1997) and Masoudi and Sirignano (1997, 1998, 2000) show that significant modifications of mass, momentum, and heat exchange can occur when a vortex collides with a particle or droplet. The existing models for droplet drag, heating rate, and vaporization rate are built for laminar microstructures and are inaccurate in a highly turbulent flow. Kim et al. found that significant modifications in the droplet drag and the creation of lift and torque on the droplet can occur due to the unsteady vortex–droplet interaction. Masoudi and Sirignano (2000) demonstrated significant modifications of the Sherwood number for a vaporizing droplet by means of interactions with vortices. The average vaporization rate can be increased or decreased by a 10% or more factor; in particular, the maximum fractional perturbation is correlated as $0.013(\Gamma_0/2\pi)^{0.85}\mathrm{Re}^{0.75}\tanh(0.75d_0/\sigma_0^{0.80})$, where Γ_0, Re, d_0, and σ_0 are the vortex circulation, droplet Reynolds number, initial displacement of the vortex from the droplet axis of symmetry, and the vortex-core size. Furthermore, Masoudi and Sirignano (1998) showed that vortex–solid particle interactions can significantly modify the Nusselt number when the gas temperature is stratified. So it can be strongly suspected that, for a vaporizing droplet, the stratified vapor concentration in the gas surrounding a droplet in synergism with a vortex (or vortices) will substantially increase the magnitude of the difference between vaporization rate for axisymmetric droplet situations and the rate for a droplet interacting with a vortex (or vortices).

This work on the collisions of small vortices with droplets should be extended in various ways: cases in which the vorticity vector and free-stream velocity vector are not orthogonal must be considered. The change in the eddy resulting from a collision with a droplet must be characterized. The temporal behavior and vaporization rate of an individual droplet undergoing repeated collisions with an inflowing train of vortices must be evaluated. A sufficiently wide range of parameters should be considered so that a good sense of the subgrid interactions of the smallest eddies with the spray can be determined.

The gas-phase flux terms Γ_{ij} and E_i have available LES models. However, it is expected that, for two-phase flows, the turbulent microstructure will be different from that of single-phase-flow models. It is expected that the model for viscous dissipation $\overline{\Phi}$ will differ for two-phase flows as Ferrante and Elghobashi (2003) and others have shown that the high-wave-number end of the energy spectrum is affected by the presence of particles. As previously noted, several researchers have found that, contrary to intuition about mixing, turbulence can make the distribution of interdroplet spacing highly nonuniform. This can increase the degree of stratification in the microstructure and cause terms related to the differences between averages of products and products of averages to be quite significant quantitatively.

Film Vaporization

11.1 Introduction

Although the emphasis in this book is on the dynamics of vaporization of liquids in the form of drops and sprays, it is important to note when liquids might better be applied in a form other a spray. Such a situation can develop when miniature devices are of interest. The use of a wall film rather than a spray might provide sufficient surface area of liquid to vaporize at desired rates. Also, other benefits might arise. In this chapter, we discuss a concept of liquid-film combustors that are superior for miniaturization.

   Combustion has the potential to provide simultaneously high-power density and high-energy density; these parameters make it more attractive than batteries and fuel cells for applications for which weight is an issue, e.g., flight or mobile power sources. So it is important to study this method of power generation on a small scale. The microgas turbine (combustor volume 0.04 cc), the mini (0.078-cc displacement) and micro (0.0017-cc displacement) rotary engine, the microrocket (0.1-cc combustion chamber), and the micro Swiss-roll burner are examples of such studies. See Dunn-Rankin et al. (2006), Waitz et al. (1998), Fu et al. (2001), Micci and Ketsdever (2000), Lindsay et al. (2001), and Sitzki et al. (2001). These devices are not yet sufficiently efficient to compete with the best batteries; however, the feasibility of internal combustion as a miniature power source has been shown. The major challenge for all miniature-combustor designs is the increasing surface-to-volume (S/V) ratio with decreasing size. Combustor-wall temperatures are kept fairly low because of material considerations, and high S/V ratio usually produces flame quenching, attracting researchers to quench-resistant fuels (e.g., hydrogen), high-preheat concepts (as with the Swiss-roll burner), or catalytic surfaces; see Chao et al. (2001). Kyritsis et al. (2002, 2004a,b) have an interesting approach using catalysts and electrospraying. Much of the earlier research in compact energy and power sources has derived from U.S. Defense Advanced Research Projects Agency programs in palm power and portable power. These programs are often device based and constrained by specific target applications, making it difficult to study the fundamentals of potentially new combustion regimes governed by small physical scales and high surface-to-volume ratios. A useful overview of those studies is given by Fernandez-Pello (2002) and by Walther and Fernandez-Pello (2001).

Recent studies indicate that the strategy of vaporizing liquid fuel from a film on the combustor wall offers significant advantages over spray vaporization for miniature combustors; see Sirignano et al. (2001), Sirignano et al. (2003, 2005), Pham et al. (2006), and Mattioli et al. (2009). The liquid-fuel film minimizes heat loss to the wall, inhibits flame quenching, and keeps the temperature of the wall material low. These findings are based on dimensional analysis, order-of-magnitude analysis, and experiments. The experimental observations and measurements provided the proof of concept for a University of California patent. The concept can be applied to rockets, ramjets, turbojets, (reciprocating and rotary) internal combustion engines, heating furnaces, kilns, boilers, and to any other combustor for which heat losses must be reduced and/or quenching must be prevented. Various geometrical combustor shapes can be used for the continuous or intermittent combustion application of this application. Here we discuss the application of the concept to continuous cylindrical combustors.

Current technology for larger combustors does not rely on any significant portion of the liquid-fuel filming on the combustor walls. Rather, to keep the ratio of liquid-surface area to liquid volume large enough to sustain the fuel-vaporization rate at the large required value, the fuel is injected as a spray. The intention is to vaporize the liquid as a spray before very much liquid deposits on the walls or solid surfaces of the combustor. That is, the droplets are suspended in the gas or at least the gas affects their motion so that they do not reach a solid surface and adhere to it before vaporization is completed. If the fuel were filmed in these larger engines, the surface area of the liquid would not be large enough to sustain the needed vaporization rate. However, the surface-to-volume ratio of the combustor and therefore for any wall film will grow as the volume of the combustors decreases. As a consequence, in the subcentimeter range under discussion here, the liquid film can offer as high a liquid-surface area for vaporization as a vaporizing spray can. Film-vaporization rates in miniature combustors will be adequate compared with spray-vaporization rates on account of the large surface area per unit volume. So film combustion is as competitive as spray combustion in miniature combustors from this perspective. Film vaporization would not be competitive, of course, in larger combustors. Film vaporization or combustion gains over spray combustion in terms of limiting wall heat losses that can cause a reduction in efficiency and a potential for flame quenching. With a spray, we have heat transferred from the gas to the walls plus from the gas to the droplets. Energy transferred to the liquid is regained by the gas phase upon vaporization, but wall losses are real losses. With the wall film, the heat cannot be transferred to the wall directly but rather goes to the liquid. Nearly all of this energy returns to the gas phase upon vaporization. The wall receives very little heat passed through the liquid and never achieves a temperature above the liquid boiling point. These temperatures are substantially lower than current technology would allow without some other mechanism for cooling the walls.

11.2 Miniature Film-Combustor Concept

When one is burning liquid fuels in miniature combustors, the logical choice is to inject all or a portion of the fuel directly as a film on the solid surfaces where high heat transfer from the combustion products occurs. In cases in which not enough liquid fuel is available to cover all critical surfaces with a fuel film, inert (noncombusting) liquids or film cooling with air can be used to augment the liquid fuel. Water would be one example here. When nonliquid fuels are involved in miniature combustors, it can still be sensible to augment the combustion with liquid fuel that is filmed on internal solid surfaces. It can also be sensible to use inert liquids for filming in some cases.

The liquid fuel (or inert) can be applied as a film by several means. It can be sprayed onto a chosen surface so as to avoid both substantial vaporization before striking the surface and rebound. It can be injected through an orifice, multiple orifices, or porous materials to flow tangentially along the surface. The liquid can be spread over the surface on account of its own momentum and surface tension, the friction forces on it caused by the neighboring flowing gases, and/or the designed use of certain force fields (e.g., electric field on charged liquid or gravity).

A small burner, such as a laboratory Bunsen burner, will not have the flame entering the tube supplying the combustible gas; rather, it sits above the exit of the tube. Technologies such as vortex generators or swirl generators can be used to have the flame move upstream and enter the tube. However, then the heat losses to the tube walls can result in melting of the tube. Even with a cooling mechanism to prevent tube destruction, thermal inefficiency and quenching of the flame are possible consequences. Many applications require that the combustion occur in a confined chamber (i.e., a combustor). Safety is one motivation. Another reason is that engine efficiencies and therefore engine power levels are increased substantially when combustion occurs at elevated pressures. The most practical way to maintain elevated pressures is through confinement of the fluids.

To enhance the likelihood of combusting in a confined chamber at high flow rates, this new technology can be combined with other well-known technologies such as the use of swirl generators and vortex generators that will enhance vaporization and mixing rates. Furthermore, swirling of the liquid as it is injected into the film, surface-tension modification, or both, can, in some cases, help to stabilize the film on a surface via centrifugal effects. Figure 11.1 shows the particular configuration that was used to validate the concept. Liquid fuel enters tangentially along the walls of a cylindrical chamber that serves as the combustor. Swirling air flows axially through the combustor tube and past the vaporizing wall film. The cylindrical combustor is chosen for our discussion here and for the laboratory demonstration and research. However, the film-combustion concept will extend to many other configurations.

Standard engineering analysis based on principles of conservation of mass, momentum, and energy has been performed for some examples of cylindrical combustors with diameters in the range of a few millimeters to 1 cm; see Sirignano et al. (2001, 2003). Kerosene and alcohol fuels have been examined. Air-flow velocities up to 10 m/s (and thereby air volume flow rates up to 1000 cc/s) have been

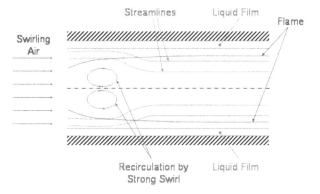

Figure 11.1. A schematic of the liquid-film combustor: the swirl air enters a combustion chamber with liquid fuel flowing in a thin wall film. The fuel vapor mixes and reacts, forming a flame.

considered at atmospheric pressure. Liquid volume flow rates have been maintained in stoichiometric proportion. Calculations indicate that, with all of the liquid fuel injected as a thin film on the internal side of the cylindrical wall, the vaporization rate and the gaseous mixing rate can be sufficiently high to sustain the combustion. Residence times can be sufficiently small to burn the fuel in a combustor that is no longer than a few centimeters. The film thickness is measurable in tens of micrometers but still the Reynolds number is of the order of or larger than unity so that viscous forces do not prevent the movement of liquid along the solid surface. If this combustor is considered to be an element in an engine with a typical thermal-cycle efficiency of 30% to 40%, delivered power levels of 1 kW and above are possible. Realize here that the combustor can be the size of a partially depleted cigarette and the full engine might not be much larger.

The required gas residence time will affect the design of the chamber length. At 10 m/s, the maximum velocity that we are considering, a few milliseconds residence time is achievable in a length of a few centimeters. The gas boundary layer along the liquid film has the inherent characteristic that the time for transverse diffusion across the layer thickness is equal to the residence time measured from the point where the layer development begins; so mixing of the fuel vapor and air in a diffusion layer should be achievable within the residence time. Mixing rates will, of course, be enhanced by swirl and/or vortex generation. Strong swirl can also cause vortex breakdown and recirculation that would increase the effective residence time. Chemical-reaction times, once ignition occurs and a flame is established, should be less than a millisecond.

Film vaporization and combustion gain over spray combustion in terms of limiting wall heat losses that can cause a reduction in efficiency and a potential for flame quenching. With a spray, we have heat transferred from the gas to the walls plus from the gas to the droplets. Energy transferred to the liquid is regained by the gas phase upon vaporization but wall losses are real losses. With the wall film, the heat cannot be transferred to the wall directly but rather goes to the liquid. Nearly all of this energy returns to the gas phase upon vaporization. The wall receives very little heat passed through the liquid and never achieves a temperature above the liquid boiling point.

Even without surface capillary forces, weak swirl (tangential velocity smaller than axial velocity) should be sufficient to maintain the liquid on the wall. For example, the radial acceleration $v^2/(d/2)$ is almost equal to the earth-sea-level gravitational acceleration if the tangential velocity $v = 20$ cm/s and $d = 1$ cm or if $v = 15$ cm/s and $d = 5$ mm. So, for this purpose, swirl velocities can be an order of magnitude less than the axial velocity. Stronger swirl, however, might be required to enhance vaporization and mixing so that the combustion can occur during the residence time.

Let us make some rough quantitative estimates for the design and operation of a miniature combustor. Figure 11.1 gives the concept of flow details with liquid-film combustion in a swirling air flow. The ratio of air mass flow rate to the fuel mass flow rate at stoichiometric proportion is expected to be $O(10)$ for typical liquid hydrocarbon fuels. For example, the ratio is 14.71 for C_nH_{2n} and 6.435 for methyl alcohol. We proceed by considering stoichiometric or near-stoichiometric conditions; however, the concept and the analysis are extendable to rich or lean operations. If we consider a tube of internal diameter d between 5 and 10 mm and with an axial air velocity u_g of 1 to 10 m/s, the air volumetric flow rate V_g is 2.5×10^{-5} to 10^{-3} m^3/s. The density ratio of liquid fuel to air will vary from $O(10^3)$ at atmospheric pressure to $O(10^2)$ at 10 atm; so, with stoichiometric proportions, the volumetric flow rate ratio (air to liquid) is between $O(10^3)$ and $O(10^4)$. Therefore the liquid volume flow rate V_l must have a value between $O(10^{-9})$ and $O(10^{-6})$ m^3/s. With a liquid density ρ_l of $O(1 \text{ g/cm}^3)$, this implies a variation from about 1 mg of fuel per second (for a 5-mm-diameter combustor, 1 m/s for air velocity, and 1-atm-pressure operation) to about 1 g/s (with 10-mm diameter, 10-m/s axial air velocity, and 10 atm of pressure).

The power range for these fuel flow rates can be significant; chemical energy release rates with typical hydrocarbon fuels will vary between 10 and 10^4 cal/s. Even with a poor overall engine efficiency of 20% to 30%, the power produced will be between 10 W and 10 kW for the range of parameters considered here.

The liquid axial and swirl velocity components will be primarily determined by the action of viscous shear from the gas flow. We can expect the liquid velocity u_l on average through the film to be severalfold smaller than the gas velocity. Let us assume that the reduction factor is $O(10^{-1})$. Then the thickness t of the liquid film can be estimated from a continuity equation that yields $t = V_l/(\pi d u_l)$ if all of the liquid enters the chamber at the upstream end. (If the liquid enters at N positions along the axial coordinate, the value just calculated should be divided by N to estimate the film thickness.) This indicates that, for stoichiometric proportions, the film thickness t might vary from about 1 μm (at 1 atm and the lower diameter value) to about 25 μm at 10 atm and the larger diameter. Film thickness will increase (decrease) as we go to richer (leaner) mixtures, larger (smaller) diameter, and/or higher (lower) pressures. We want to avoid, for the moment, the challenge of a film thickness that is only a few micrometers; so we will avoid lower-diameter combustors, lean operations, and lower-pressure combination that result in that situation.

The Reynolds number Re$_d$ for the air flow based on average velocity and combustor diameter d can vary between $O(10^2)$ and $O(10^5)$ for our range of parameters. Realize that, of these three orders of magnitude of difference in Re$_d$, only two relate to configuration change and we depend on the temperature that we use to evaluate viscosity and density. Clearly this range extends from the laminar to the turbulent range. The Reynolds number Re$_l$ for the liquid flow based on its density, viscosity, average velocity, and film thickness will vary from $O(10^{-1})$ to $O(10^{2})$. Furthermore, an estimate of the gas-phase laminar sublayer thickness in the turbulent case indicates that it is greater than the estimated film thickness. So the liquid film is expected to always be in the laminar range, although not necessarily a Stokes flow.

The global mass vaporization rate must equal the fuel mass flow rate into the chamber for the device to operate properly. The volume of the combustion chamber required for this match should not be too large if the device is to be miniature. The global vaporization rate depends on two factors: the total liquid-surface area in the combustor and the local vaporization rate per unit area. For a spray, the ratio of total liquid-surface area to total liquid volume is the surface-to-volume ratio for the average droplet: $(S/V)_{\mathrm{drop}} = 4\pi R^{2}/(4\pi R^{3}/3) = 3/R$, where R is the radius of an average droplet. In typical fuel sprays, an average droplet size will be in the range of 10 m to 100 m. For a thin cylindrical wall film of axial length L, the exposed surface-area-to-volume ratio is approximately $(S/V)_{\mathrm{film}} = \pi d L/(\pi d L t) = 1/t$.

From the continuity relations for gas and liquid, the stoichiometric proportion for the mass flows, and the preceding results for surface-to-volume ratios, the approximation follows that $S/V)_{\mathrm{film}}/(S/V)_{\mathrm{drop}} = (40/3)(\rho_{l}u_{l}/\rho_{g}u_{g})R/d$. From this result, it is seen that the film will have as much or more surface area if the combustor diameter is sufficiently small and/or the pressure is sufficiently low. In particular, a linear relation of d versus R whose slope decreases as pressure increases gives the boundary where the two ratios are equal. Therefore the film combustor can be designed to have an advantage whereby, for the same volume of liquid fuel in the chamber at any moment, it has a greater liquid surface area than would a typical spray. The diameter d of the combustor must take a value below the value given by the linear relation with R.

It is seen therefore that film vaporization rates in miniature combustors will not be inadequate compared with spray vaporization rates on account of surface-area deficiency or vaporization rate per unit area. Rather, film combustion is as competitive as spray combustion in miniature combustors from this perspective. Film vaporization would not be competitive, of course, in larger combustors.

Another important factor concerns transport rates through the gas to the liquid. The rate of vaporization is determined by the rate at which heat reaches the liquid to overcome the threshold set by the latent heat of vaporization. Also, combustion rates are determined by the vaporization rate. Residence times and chamber lengths will be determined by the rate at which the fuel vapor mixes with air. Let us consider a case in which $\mathrm{Re}_{d} = 10^{3}$, $u_{g} = 10$ m/s, and $\rho_{l}/\rho_{g} = 10^{3}$. Laminar flow formulas are used here. Emmons (1956) considered vaporization from a liquid-fuel surface with laminar mixing and reaction with an air flow past it. The film mass vaporization rate per unit area is of the order of $(2/\mathrm{Re}_{L})^{1/2}\rho_{g}u_{g}$, where the Reynolds number Re_{L} is based on the downstream length L of the liquid surface and on the velocity and the high-temperature properties outside the boundary layer.

The same formula gives an order-of-magnitude estimate for the vaporization rate of an advecting droplet if u_{g} is replaced with a relative droplet–gas velocity that has a smaller value. Re_{L} should also be based on the relative velocity and on the droplet diameter; it would have a substantially smaller value for the droplet. As a consequence, the vaporization rate per unit area for a droplet can be somewhat larger than the film value if no correction is made. To maintain an advantage for the film, we advocate the use of swirl and vortex generation. The film gains in transport rates with effective velocity increase more than the droplet gains because the droplet–gas relative velocity will not grow as fast as the gas velocity.

The total mass vaporization rate from the liquid surface is approximately given by $\pi d L (2/\mathrm{Re}_L)^{1/2} \rho_g u_g$. This value divided by the mass flow rate of air must be of $O(10^{-1})$ to maintain the stoichiometric or near-stoichiometric proportion. This relation can yield the required length of the fuel film. In particular, the approximation follows that $L/d = 10^{-3} \mathrm{Re}_d$, where Re_d is based on the high-temperature properties outside of the boundary layer and, for our range of parameters, is between $O(10^2)$ and $O(10^4)$. So, for example, we estimate that the value of L will exceed the order of magnitude of the diameter d only if both the pressure and the air velocity are at the high ends of their ranges. Again, this estimate is based on a laminar flow analysis for the heating and vaporization of the liquid. With a turbulent flow that occurs at the upper end of our Re_d range and with some swirl, we can expect faster transport, higher vaporization rates, and shorter L. So liquid-film lengths not too much larger than the combustor diameter appear to be sufficient.

The first laboratory demonstrations were done at UCI, and a typical result (Pham et al., 2006) is indicated by the images shown in Figure 11.2, which shows the Pyrex glass cylindrical combustor that was fabricated with an internal diameter of 1 cm. The air flows past swirl vanes downstream into the combustion chamber where liquid fuel is introduced tangentially through eight 1-mm-diameter feed tubes, a pair at each of four axial positions. The transparent combustor allows determination of the fact that the liquid does form a film as it flows over the inside cylindrical surface. It was difficult to ignite the methanol liquid-fueled flame from a cold start, so both methane gas and liquid-methanol fuels were used simultaneously. (With heptane liquid fuel, the gas assist was not necessary.) The procedure was to feed a bit of liquid fuel into the base of the combustor, flow the fuel–air mixture, and ignite the flame at the exit of the tube. The flame was now fed by both some liquid picked up as the air flowed past the pool at the base of the combustor and by the gas. We would then decrease the gas fuel flow rate until, at a critical condition, the flame jumped into the tube, where it burned in a confined state, as shown in Fig. 11.2. It is important to reiterate that burning occurred inside the tube only when liquid fuel was provided. In fact, if the liquid flow was stopped, leaving only the inflow of gaseous fuel, the flame continued to burn inside the tube until it exhausted any remaining film, and it then jumped up to the top of the tube to burn in the low-swirl stabilized state. The experiments by Pham et al. (2006) and Mattioli et al. (2009) examined improved methods of swirl to mix the fuel vapor and air faster and allow shorter combustors. The experiments showed that fuel film combustors are capable of providing up to 600 W of thermal power for less than 300 g of weight with a heat loss estimated to be less than 5% of the total thermal power production. The introduction of secondary air injection allows more complete combustion and an augmentation of the stability. The stable operating range with secondary air injection ranges in equivalence ratio from 0.4 to 1. Under the proper lean operating condition, continuous burning times are achieved with a stable flame holding and with total flame confinement inside the chamber. Tests showed further that the combustor can operate at higher pressures

Figure 11.2. Confinement of flame with methanol liquid wall film. 1-cm-diameter Pyrex tube with tangential fuel injection at several locations. Swirler inflow of air at bottom and exit of products at top. The flame extends beyond the exit. (Also, see color plate.)

when an exit nozzle is added and in any orientation with respect to gravity. This encouraged further analysis, which is discussed in the next section.

11.3 Analysis of Liquid-Film Combustor

Steady-state axisymmetric combustion in a cylindrical chamber with the liquid fuel introduced through a wall film was analyzed by Sirignano et al. (2005). Although the film-combustor concept is not limited to continuous combustors or to any particular geometry, they considered steady continuous combustion in a circular-cylindrical chamber. For the miniature combustor of interest, the swirling core flow of air, fuel vapor, and products operates in the laminar range. Figure 11.1 presents a sketch of the studied flow. The mixing and reaction occurs therefore in a gas surrounded by a liquid-fuel film. Figure 11.1 does oversimplify what was found experimentally. That is, a triple-flame structure was found at the leading edge of the flame. The model here considers only the diffusion-flame branch and does not address the lean-premixed-flame and rich-premixed-flame branches.

Under the condition of unitary Lewis number, the analysis of the scalar fields takes advantage of the existence of a linear combination of the scalar properties, known as the superscalar, which is spatially uniform. The base flow is assumed to have constant density and to be either a fully developed flow or a plug flow. Perturbations account separately for swirl effects, Stefan flow, and density variations; so some linearization becomes helpful. The effects of each perturbation type on the velocity fields and scalar fields are determined. The motion in the liquid film is coupled to the core gas motion. The diffusion flame character is portrayed. Vaporization rates and burning rates are determined; the effects of radial velocity perturbations, which are due to swirl and Stefan flow, and of the density variations on the transport, vaporization, and burning rates are determined. The analysis yields closed-form solutions in terms of eigenfunction expansions. Series solutions are found to certain ordinary differential equations describing radial variations in the dependent variables. Large Damkohler number solutions are calculated as functions of Reynolds number and Peclet number, showing strong dependencies. Methanol and heptane fuels are considered. Limits of high Peclet number are also examined. The results allow for the prediction of required chamber length for given chamber diameter, inlet flow conditions, liquid-fuel physicochemical properties, initial fuel temperature, and overall mixture ratio. The feasibility of the liquid-film concept, which was demonstrated in the laboratory, is supported theoretically.

The analytical small-perturbation approach taken offers some advantage over the direct numerical integration of the Navier–Stokes equations for the reacting flow. A disadvantage is that the interesting practical range for a few of the parameters extends beyond the applicable domain for small-perturbation theory; as a consequence, some of the results should be viewed as having qualitative value rather than quantitative value.

The major assumptions and governing equations are presented in the following discussion. The superscalar approach of Appendix B can be used to determine the flame temperature (in the limit of infinite chemical-kinetic rate) and surface temperature (before the field equations are solved) as functions of the heat per unit mass transferred to the liquid surface. The liquid-phase analysis is briefly discussed, and it is shown that, for certain cases, the liquid-phase field can be determined before the gas phase is resolved in detail. Solutions for the axial-velocity field and gas scalar field are given.

11.3.1 Assumptions and Governing Equations

The liquid fuel is injected along the cylindrical wall of the combustion chamber and is spread in a wall film by the shearing action of the gas. The gas flows axially but is injected with a weak swirl component to spread the liquid film and to provide the centrifugal effect that keeps the film on the wall. The swirl is not strong enough in this model to create a recirculation zone.

The liquid is assumed to be a single compound (e.g., heptane or methanol). The same assumptions used in Appendix B are made here. The gas phase that is surrounded by the liquid-fuel film has a laminar multicomponent flow with viscosity, Fourier heat conduction, Fickian mass diffusion, and one-step oxidation kinetics. Diffusivities for all species are assumed identical; the Lewis number value is unity; radiation is neglected; and kinetic energy is neglected in comparison with thermal energy so that, with regard to the energetics, pressure is considered uniform over the space although the pressure gradient can be significant in the momentum balance (small Mach number). Both steady and unsteady scenarios are subsequently discussed. Later, the analysis focuses on steady flows.

The liquid-film thickness will be much smaller than the chamber radius because the fuel mass flow rate is an order of magnitude smaller than the air-flow rate and the liquid density is orders of magnitude larger than the gas density.

Two possibilities are allowed: (i) no fuel vapor exists on the oxygen (upstream) side of the flame so that $Y_{F\infty}$ and a pure diffusion flame occurs and (ii) fuel vapor exists on both sides of each flame in the configuration so that both a premixed character and a diffusion character exist for the flame(s). Consequently, Eq. (B.21) is modified to have

$$\frac{h_\infty - h_s + \nu Q Y_{O\infty}}{L_{\text{eff}}} = \frac{Y_{Fs} - Y_{F\infty} + \nu Y_{O\infty}}{1 - Y_{Fs}} = B. \tag{11.1}$$

Equation (11.1) defines the well-known Spalding transfer number B for the Le $= 1$ case. More important, Eq. (11.1) together with the enthalpy-temperature relationship $h(T)$ and the phase-equilibrium law $Y_{Fs}(T)$ serve as boundary conditions on (steady or unsteady) heat transfer in the liquid because the effective latent heat of vaporization, $L_{\text{eff}} = L + \dot{q}_l / \dot{m}_f$, contains the normal temperature gradient on the liquid side of the interface. Equation (11.1) determines the surface temperature T_s and the surface mass fraction Y_{Fs} as functions of fuel properties, inflow conditions, and L_{eff}. Coupled with the liquid-phase heat transfer problem, T_s and Y_{Fs} are absolutely determined as functions of time. These surface values are affected by the details of the configuration or of the hydrodynamics and transport phenomena only through L_{eff}.

In the limit of infinite reaction rate, the flame becomes a sheet of zero thickness at which $Y_O = Y_F = 0$. Then, with constant specific heat, the modifications of Eqs. (B.23) and (B.24) for the limiting flame temperature T_f yield

$$\frac{T_f}{T_\infty} = \nu \left[\frac{Q}{c_p T_\infty} - \frac{L_{\text{eff}} / c_p T_\infty}{1 - Y_{Fs}(T_s)} \right] Y_{O\infty} + \frac{L_{\text{eff}} / c_p T_\infty}{1 - Y_{Fs}(T_s)} Y_{F\infty} = \frac{T_s}{T_\infty} + \frac{L_{\text{eff}} / c_p T_\infty}{1 - Y_{Fs}(T_s)} Y_{Fs}(T_s).$$
$$\tag{11.2}$$

Flame temperature as well as liquid surface temperature depend on the dynamics or configuration only through L_{eff}. Particularly, they depend on the inflow conditions, the fuel properties, and the heat per unit mass transferred to the liquid.

11.3.2 Liquid-Phase Thermal Analysis

The uniformity of the superscalar S over the gas phase does require that L_{eff} be instantaneously uniform over liquid surface. The latent heat of vaporization L is a fuel property and therefore will be uniform and constant. The uniformity of the heat flux to the interior per unit mass flux $\dot{q}_l / \dot{m}$ is a more demanding requirement. It can be satisfied in various situations. In one situation in which the wet-bulb temperature has been reached, $\dot{q}_l = 0$. In a second situation in which a thin quasi-steady thermal layer exists in the liquid near the surface, $\dot{q}_l = c_l(T_s - T_0)$ where c_l is the liquid specific heat and T_0 is the liquid interior temperature. Because the liquid-phase Prandtl number is much larger than unity, we can expect the thermal layer at the liquid surface to be much thinner than the liquid viscous layer. Then, the super scalar S has a uniform value in either of those two situations and Eq. (11.1) together with an equation of state for h and a phase-equilibrium law immediately yield surface values. Then Eq. (11.2) can readily give the limiting flame temperature. Furthermore, the details of the liquid flow do not modify the magnitude of L_{eff} in these situations with uniform surface temperature and thin thermal layer.

In this analysis, it is assumed that a thin quasi-steady thermal liquid layer exists adjacent to the surface. In particular, the layer thickness is taken to be smaller than the liquid-film thickness, and then the film thickness is not an interesting parameter in the heat transfer process. Also, the details of the liquid motion are not important for heat transfer. The solid–liquid interface can be considered as adiabatic under this situation. Note that L_{eff} remains constant as the characteristic diffusion length varies with downstream distance because $\dot{q}_l$ varies in proportion to $\dot{m}$.

More information, beyond the scope of this analysis, about flames in gaseous diffusion layers adjacent to vaporizing liquid films or pools can be found in the literature. Emmons (1956) considered the similar solution of diffusion and reaction with forced gas motion above a flat liquid surface without liquid motion or liquid-phase temperature gradients. Smirnov (1985) extended the Emmons two-dimensional, steady analysis and considered liquid motion and heating in a thick liquid film in which the solid-wall interface played no role. The effects of unsteady flame development, finite liquid pool depth, surface-tension-driven flow, and buoyancy in both phases were considered in 2D (Schiller et al., 1996b; Sirignano and Schiller, 1996; Schiller and Sirignano, 1997; Kim and Sirignano, 2003), axisymmetric (Kim et al., 1998), and 3D (Cai et al., 2002, 2003) configurations.

11.3.3 Fluid-Dynamics Analysis

Consider steady, viscous, laminar axisymmetric flow using a cylindrical coordinate system. Swirl is allowed. u, v, and w are the x, r, and θ components of velocity, respectively. The chamber walls are at $r = R$ and the liquid–gas interface is at $r = R_i$. Bulk viscosity is set to zero. Although the velocity, normal stress, and shear stress are continuous across the liquid–gas interface, discontinuities in density and viscosity will occur there.

The flow field is considered to be a fully developed laminar flow with certain perturbations. The unperturbed base flow variables are denoted with the subscript 0. The perturbations result from the combustion process (variations in density and temperature), imposed swirl, Stefan convection that is due to liquid-film vaporization, and differences between initial conditions and the developed flow asymptote. Although the base flow without these perturbations is fully developed, it still differs from classical fully developed pipe flow on account of the liquid wall film. The assumption of low Mach number (negligible compressibility), weak swirl, and weak

Stefan convection is reasonable. The strongest assumption is the consideration of density variation that is due to temperature variation as a perturbation, allowing the first approximation of density ρ_0 to be a constant.

The pressure stress balances the viscous stress with a negligible inertial term in the momentum equation. The viscosity is assumed to be constant through each phase with a discontinuity at the interface. Both velocity and shear stress are continuous across the interface whereas the normal derivative of velocity is discontinuous. The radial profile of u_0 is a parabola in each phase. It also implies that the pressure gradient $\partial p_0/\partial x$ is a constant. The centerline velocity u_c is given by

$$u_c = u_0(r = 0) = \left[R_i^2 \left(1 - \frac{\mu_l}{\mu_g} \right) - R^2 \right] \frac{1}{4\mu_l} \frac{dp_0}{dx}. \tag{11.3}$$

The nondimensional velocity is given by

$$u_0 = 1 - \varepsilon^2 \tilde{r}^2 \tag{11.4}$$

for $0 \le \tilde{r} \le R_i/R$ and

$$\frac{u_0}{u_c} = \frac{1 - \tilde{r}^2}{1 - \left(1 - \frac{\mu_l}{\mu_g}\right)\left(\frac{R_i}{R}\right)^2} \tag{11.5}$$

for $R_i/R \le \tilde{r} \le 1$, where

$$\varepsilon^2 = \frac{\frac{\mu_l}{\mu_g}}{1 - \left(1 - \frac{\mu_l}{\mu_g}\right)\left(\frac{R_i}{R}\right)^2}. \tag{11.6}$$

The parabolic profile is characteristic of a fully developed laminar flow in the gas core. Sirignano et al. (2005) also examined a case in which the core gas flow is represented as a plug flow; this represents a situation in which the flow is just beginning to develop (or a less practical situation in which the liquid viscosity is negligible). It is characterized by $\varepsilon = 0$, which implies that the gaseous viscous layer at the liquid surface is very thin (i.e., large gas-phase Reynolds number for the developing flow) and/or the liquid viscosity is negligible (i.e., large liquid-phase Reynolds number). The momentum-balance approximation just used does not apply in this limit; the gas-phase pressure drop becomes negligible. The core flow in this plug-flow limit has a flat profile that matches the inflow profile. $\varepsilon = 1$ is the limit where the liquid-film thickness is negligible. For the calculations in this section, $\tilde{r}_i = 0.98$ is taken. The viscosity ratio is taken as $\mu_l/\mu_g = 2.21631$ so that, for the parabolic-velocity-profile calculations, $\varepsilon = 1.01105$ is obtained from Eq. (11.6).

11.3.4 Scalar Analysis

Consider now the solutions for gas temperature and mass fractions for the Le $= 1$ condition. The superscalar S can be used as previously indicated. Consistent with the neglect of Stefan convection, the factor $1 - Y_{Fs}$ is replaced with one. If we solve for the first Shvab–Zel'dovich variable β_1, knowledge of S will immediately yield β_2. Then, assuming an infinite chemical-kinetic rate, we can determine temperature and mass fractions. For convenience, we add a constant to β_1 and then normalize it so that it becomes zero at the liquid interface and has unitary value at the inflow:

$$\beta = \frac{[Y_{Fs} + \nu Y_O - Y_F]}{[\nu Y_{O\infty} + Y_{Fs}]}. \tag{11.7}$$

Recall that Y_{Fs} and T_s can be determined a priori as previously indicated. The value of β at the flame is given by setting the oxidizer and fuel-vapor mass fractions in Eq. (11.7) to zero. Then, $\beta_{\text{flame}} = Y_{Fs}/[\nu Y_{O\infty} + Y_{Fs}]$. So, although the line for any particular value of β for a contour plot in the r–z plane does not depend on the specific fuel or oxidizer, the particular line that gives the flame position in the limit of infinite kinetics does depend on the choice of fuel and oxidizer. The contours of the primitive variables such as temperature and mass fraction will also depend on the fuel and oxidizer.

u_c and R are used to nondimensionalize time and space coordinates. The diffusivity D is considered to be constant. The lowest-order approximation for β is given by

$$\text{Pe}(1 - \varepsilon^2 \tilde{r}^2)\frac{\partial \beta}{\partial \tilde{x}} = \frac{1}{\tilde{r}}\frac{\partial}{\partial \tilde{r}}\left(\tilde{r}\frac{\partial \beta}{\partial \tilde{r}}\right) + \frac{\partial^2 \beta}{\partial \tilde{x}^2}, \tag{11.8}$$

where the Peclet number is given by $\text{Pe} = u_c R / D$. Solutions are obtained by use of eigenfunction expansions. Separation of the variable β yields

$$\beta = \sum_{n=1}^{\infty} c_n \Theta_n(\tilde{r}) e^{-k_n \tilde{x}} \tag{11.9}$$

where $\Theta_n(\tilde{r})$ is governed by the ordinary differential equation

$$\frac{\partial^2 \Theta_n}{\partial \tilde{r}^2} + \frac{1}{\tilde{r}}\frac{\partial \Theta_n}{\partial \tilde{r}} + \lambda_n^2\left(1 - \varepsilon_*^2 \tilde{r}^2\right)\Theta_n = 0. \tag{11.10}$$

We have $\Theta(\tilde{r}_i) = 0$ and

$$\varepsilon_* = \frac{\varepsilon}{[1 + k_n/\text{Pe}]^{1/2}}; \quad k_n = \frac{1}{2}\left[\left(\text{Pe}^2 + 4\lambda_n^2\right)^{1/2} - \text{Pe}\right]. \tag{11.11}$$

Another condition on the solution is that it must remain finite everywhere. A Pe value that is large compared with unity implies that axial diffusion is dominated by both axial advection and radial diffusion. As $\text{Pe} \to \infty$ the finite ε_* becomes independent of Pe and Eq. (11.10) and its boundary conditions indicate that Θ_n and λ_n become independent of Pe. For $\varepsilon = 0$, Θ_n and λ_n are independent of Pe for any Pe value. Then, for $\text{Pe} \to \infty$ and for any ε value, $k_n \to \lambda_n^2/\text{Pe}$, and Eq. (11.9) indicates that the dependencies of β on $\tilde{x}$ and Pe collapse to an exponentially decaying dependence on $\tilde{x}/\text{Pe}$.

When $\varepsilon = 0$, Eq. (11.10) becomes Bessel's equation of zero order. This limit can be used when a plug flow is considered; the value of u_c in the nondimensionalization scheme should be replaced with the uniform velocity given by u_0. Graetz solved heat and mass transfer problems of this type for both fully developed single-phase flow, $\varepsilon = 1$, and plug flow, $\varepsilon = 0$. Review of the Graetz work and related works by Nusselt and others can be found in Shah and London (1978) and Drew (1931). The current work differs from Graetz's work primarily through the effect of the liquid wall film and second through the inclusion of diffusion in the main flow direction as well as in the lateral direction. Drew extended the Graetz problem to include multidirectional transport. $\varepsilon = 0$ with finite Pe implies that the viscous flow is completely undeveloped dynamically but the scalar diffusion layer is developing. Equation (11.10) has a regular singular point at $\tilde{r} = 0$. Two series solution can be found; one is analytic and the other has a logarithmic singularity at the origin. We discard the second solution because the solution is constrained to remain finite. The recursive relation for the coefficients in the analytic series connects three sequential coefficients. A simpler recursive relation connecting only two coefficients is found through transformation of the dependent variable by extracting an exponential term. Equation (11.10), with the conditions that the solution become zero at the liquid interface $\tilde{r} = R_i/R = \tilde{r}_i$ and remain finite in the domain, presents a Sturm–Liouville problem when ε_* has a uniform value for all n. See Titchmarsh (1962). So an infinite number of eigenfunctions and eigenvalues are produced, namely,

$$\Theta_n(\eta, \gamma_n) = e^{-\gamma_n \eta^2/4} \left[1 + \sum_{m=1}^{\infty} \left(-\frac{1}{4} \right)^m \frac{\Pi_{p=0}^{m-1}(1 - 2p\gamma_n)}{[m!]^2} \eta^{2m} \right], \qquad (11.12)$$

where the definitions are

$$\eta = \frac{2\varepsilon_*\sqrt{1 + \gamma_n}}{\gamma_n} \tilde{r}; \quad \gamma_n = \frac{2\varepsilon_*}{\sqrt{\text{Pe } k_n + k_n^2 - 2\varepsilon_*}}. \qquad (11.13)$$

The values of γ_n are determined by finding the points where the undulating function Θ_n becomes zero for $\tilde{r} = R_i/R$.

For the case in which Pe $\gg 1$, the dependence of ε_* on k_n is weak. Nevertheless, because k_n and γ_n are related, an iterative solution is required. See Sirignano et al. (2005). The Sturm–Liouville theory guarantees the generation of a complete orthogonal set of functions only when ε_* is independent of k_n. Equation (11.11) shows that, with finite ε, this occurs when Pe becomes infinite. We shall consider that Pe is at least an order of magnitude greater than any interesting nondimensional wave number. Consequently we assume that the expansion of a function in a series of the eigenfunctions given by Eq. (11.9) is a decent approximation for large Pe. When $\epsilon = 0, \Theta_n = J_0(\lambda_n \tilde{r})$, which is the Bessel function of the first kind and zero order. Equation (11.10) and its boundary conditions do form a Sturm–Liouville problem for any Pe value in this $\epsilon = 0$ limit.

Once the values of γ_n are known, the functions Θ_n can be determined from Eq. (11.12), and the general solution that satisfies the boundary conditions at the liquid interface is given by Eq. (11.9). Fifty-two eigenfunctions were substituted into Eq. (11.9) by Sirignano et al. (2005) for the calculations of the scalar variables; this gave a decent approximation to the infinite sum. The values of c_n can be determined from the inflow conditions. Classical Sturm–Liouville theory yields

$$c_n = \frac{\int_0^{\tilde{r}} (\tilde{r} - \varepsilon_*^2 \tilde{r}^3) \beta(\tilde{r}, 0) \Theta_n(\tilde{r}) \mathrm{d}\tilde{r}}{\int_0^{\tilde{r}} (\tilde{r} - \varepsilon_*^2 \tilde{r}^3) \Theta_n^2(\tilde{r}) \mathrm{d}\tilde{r}}. \qquad (11.14)$$

Sirignano et al. considered the case in which the inlet condition is uniform over the cross-sectional area of the inlet; i.e., $\beta(\tilde{r}, 0) = 1$.

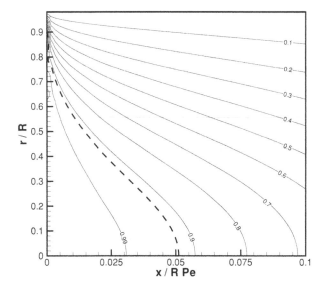

Figure 11.3. β contours for parabolic axial-velocity profile. Pe = 500. Air flows from left to right. Dashed curve gives heptane-flame contour.

The local nondimensional vaporization rate per unit area can be determined as a function of downstream position, with the neglect of Stefan convection, by relating it to the normal gradient of the scalar at the interface. The result is

$$\frac{\dot{m}R}{\rho D} = \frac{2}{\tilde{r}}[\nu Y_{O\infty} + Y_{Fs}] \sum_{n=1}^{\infty} \left[c_n e^{-k_n \tilde{x}} \eta_i \, e^{-\gamma_n \eta_i^2/4} \sum_{m=1}^{\infty} m \frac{\Pi_{p=0}^{m-1}(1 - 2p\beta_n)}{[m!]^2} \eta_i^{2m-1} \right]. \quad (11.15)$$

Integration over the axial coordinate and the circumference gives the total nondimensional vaporization rate over the liquid surface of length L:

$$\frac{R \int_0^{L/R} \dot{m} \, d\tilde{x}}{2\rho D[\nu Y_{O\infty} + Y_{Fs}]} = \sum_{n=1}^{\infty} \left[\frac{c_n}{k_n} \left(e^{-k_n L/R} - 1 \right) \eta_i \, e^{-\gamma_n \eta_i^2/4} \sum_{m=1}^{\infty} m \frac{\Pi_{p=0}^{m-1}(1 - 2p\beta_n)}{[m!]^2} \eta_i^{2m-1} \right]. \quad (11.16)$$

In the plug-flow limit of $\varepsilon = 0$, we have

$$\frac{\dot{m}R}{\rho D} = \frac{2}{\tilde{r}}[\nu Y_{O\infty} + Y_{Fs}] \sum_{n=1}^{\infty} e^{-k_n \tilde{x}}, \quad (11.17)$$

$$\frac{R \int_0^{L/R} \dot{m} \, d\tilde{x}}{2\rho D[\nu Y_{O\infty} + Y_{Fs}]} = \sum_{n=1}^{\infty} \frac{1 - e^{-k_n L/R}}{k_n}. \quad (11.18)$$

The major effect of fuel choice on vaporization effect comes through the parameter $\nu Y_{O\infty} + Y_{Fs}$. A very minor influence occurs for the case with a parabolic-velocity profile through the liquid viscosity that affects γ_n and k_n and thereby has the small influence on the vaporization rates.

11.3.5 Results

Figure 11.3 shows solutions of Eqs. (11.9) and (11.14) for Pe = 500 and for a base flow with parabolic axial-velocity profile (i.e., $\varepsilon \neq 0$). The solution for the scalar β does not depend on the choice of fuel; however, the contour value giving the

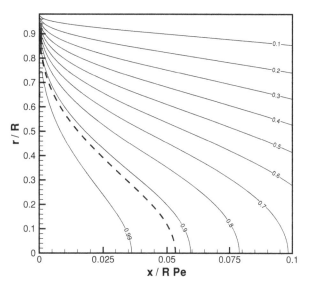

Figure 11.4. β contours for parabolic-velocity profile. Pe = 1000. Dashed curve gives heptane-flame contour.

thin-flame contour does depend on the fuel and oxidizer and the liquid temperature. The dashed curve shows the flame contour for the heptane film at a bulk temperature of 298 K vaporizing into the gas stream with air as the inflowing gas. The β value decreases with an increase in the value of the radial coordinate or of the axial coordinate. An increase of the Peclet number results in the relocation farther downstream of a contour curve representing a given value of the scalar. See Fig. 11.4 for the case in which Pe = 1000. In fact, for Peclet number of 750 and above, the results essentially collapse to a function of $\tilde{r}$ and $\tilde{x}/$Pe, as noted previously. More terms in the series solution of Eq. (11.9) are required for a converged solution as $\tilde{x}$ becomes smaller. So we find some error in the contours for values of $\tilde{x}/$Pe < 0.003. For this small region, especially at higher $\tilde{r}$ values, 52 terms in the summation of Eq. (11.9) did not give convergence. These contours do not depend on the particular fuel that is used or the liquid interface temperature because the scalar in Eq. (11.9) goes to zero value at the liquid interface. The value of β at the flame position will depend upon the choice of fuel and the liquid interface temperature. Figure 11.5 shows results for the plug-flow limit. Again, the results collapse to a function of $\tilde{r}$ and $\tilde{x}/$Pe only for large Peclet numbers; the results for Pe = 100 and 500 are essentially superimposed. The results are qualitatively similar to the parabolic profile results, but for a given value of β, the contour line extends farther downstream. This can be expected because, for the same Peclet number, the plug flow has a greater average velocity than the parabolic flow.

The flame position for heptane fuel at a bulk temperature of 298 K is shown in Fig. 11.6 at three different Peclet numbers under the assumption of infinite chemical kinetic rate so that a flame of zero thickness results. As expected, the flame moves farther downstream with increasing Pe. At Pe = 500, the combustion is completed at approximately 12.5-diameter length, as indicated by the flame collapsing to a point on the axis. The results for Pe = 750 and 1000 superimpose when plotted versus $\tilde{x}/$Pe. If relatively high mass flows and relatively short chamber lengths are desired,

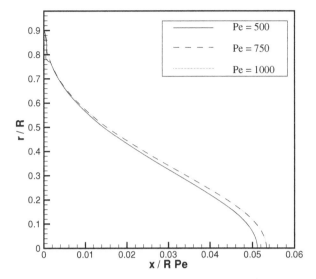

Figure 11.5. Flame contour, parabolic-velocity profile.

it is clear that the effective diffusivity must be increased (and the effective Peclet number must be reduced) through the generation of vortices or turbulence.

Temperature and mass fraction profiles for heptane with Pe = 500 are presented in Figs. 11.7 and 11.8. Peak temperature occurs at the flame with fuel vapor and oxidizer (air) each existing on only one side of the flame with both mass fractions decreasing with radial distance from the flame and going to zero values at the flame. For large Pe, the results for temperature and mass fraction will superimpose when plotted against $\tilde{x}/Pe$.

Figure 11.9 shows the changes in flame position that are due to a change in the initial fuel temperature or a change in fuel to methanol. Clearly the heptane

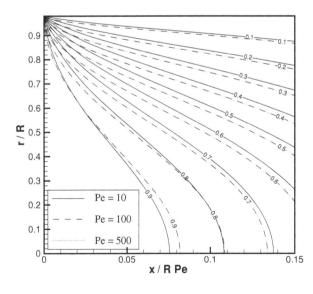

Figure 11.6. β contours for plug flow.

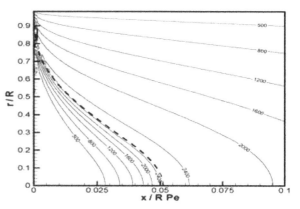

Figure 11.7. Calculated temperature contour in heptane–air liquid-film combustor. Pe = 500. Parabolic axial-velocity profile.

case leads to faster vaporization and shorter flame length, whereas a change in the initial fuel temperature has little effect. Figure 11.10 shows vaporization mass flux per unit area and integrated vaporization mass flux versus downstream distance for heptane with Pe = 500. The total (i.e., integrated) vaporization rate is roughly proportional to Pe because the film length is determined to maintain stoichiometric proportions whereas the air-inflow rate scales with Pe. Figure 11.11 shows an important implication of the integrated vaporization rate. L_{flame} is the downstream position at which combustion is completed; that is, it is value of the x position where the flame crosses the axis of symmetry. As expected, it increases roughly linearly with increasing Pe at larger Pe values. Furthermore, a greater distance is required for the less-volatile fuel, methanol, than for heptane. L^* is the fuel-film length required to vaporize at an integrated rate that matches, in stoichiometric proportion, the inflow of air. It is seen in Figure 11.11 that $L^* < L_{flame}$ for Pe = 10 or greater. A careful examination of the data indicates that the dependence of L^* on Pe is sublinear; that result is qualitatively in agreement with boundary-layer theory that predicts a square-root dependence. The implication of the linear and sublinear dependencies is that a crossing of the curves will occur but, apparently, it happens for a very small Pe value. The result implies that, to protect the chamber wall for the full length needed to complete combustion, the fuel must be supplied in excess of

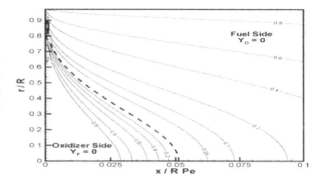

Figure 11.8. Calculated heptane fuel-vapor and oxygen mass fractions contours in liquid-film combustor. Lengths are normalized by cylinder radius. Pe = 500. Parabolic axial-velocity profile.

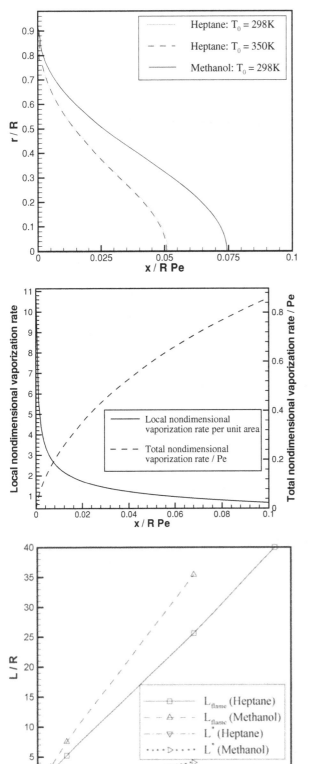

Figure 11.9. Nondimensional flame position for change in fuel or initial fuel temperature. Pe = 500. Parabolic axial-velocity profile.

Figure 11.10. Nondimensional (local and integrated) vaporization mass rates versus film length for heptane. Pe = 500. Parabolic axial-velocity profile.

Figure 11.11. Film length for stoichiometric flux proportion and flame length versus Peclet number for methanol and heptane fuels. Plug flow. Axial-velocity profile.

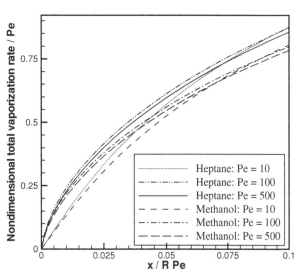

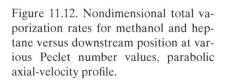

Figure 11.12. Nondimensional total vaporization rates for methanol and heptane versus downstream position at various Peclet number values, parabolic axial-velocity profile.

stoichiometric mixtures. An alternative option not explored here is to mix the fuel with a less volatile inert liquid (e.g., alcohol and water) to get the wall coverage for heat protection without exceeding stoichiometric proportions significantly. Or the inert liquid could be applied to the chamber walls just downstream of the end of the fuel film. Also, the addition of turbulence or vorticity in the gas can create a larger value of the effective diffusivity. Qualitatively, this is similar to decreasing the value of Pe, implying that the increased transport rates will shorten the required lengths for completion of vaporization and combustion. As noted earlier, an error in the use of the eigenfunction expansion increases as Pe becomes smaller.

Figure 11.12 extends the results of Fig. 11.10 for the integrated vaporization rate to other Peclet numbers and presents data for both heptane and methanol. Integrated vaporization rates increase with increasing Pe and methanol vaporizes at a slightly lower rate than heptane. The longer required lengths for film vaporization and completion of combustion for methanol appear to be primarily caused by the greater amount of methanol-vapor mass flux than heptane-vapor mass flux needed for a given air mass flow to maintain a stoichiometric proportion.

Generally, at this time, only qualitative comparisons with experiments can readily be made. Currently there are no experimental measurements of gas-phase and liquid-phase temperature and velocity and gas-phase concentrations inside a liquid-film combustor. There are some measurements by Sirignano et al. (2001, 2003) of wall temperatures and exit temperatures. The wall temperatures are below the liquid boiling point, in agreement with the theory, indicating that the liquid is insulating the wall. The exit-temperature measurements reinforce the clear visual observation that combustion of all of the fuel does not occur within the combustor; the flame is continued beyond the cylinder exit for the Peclet numbers tested. That is, for the higher values of Pe, $L^* < L_{\text{flame}}$, as predicted by the calculations shown in Fig. 11.11. Some unpublished results indicate that, in certain flow rate ranges, overall flame length can decrease as air and fuel flow rates increase. It is likely that vorticity generation is increasing sufficiently with increasing flow rate so that the effective diffusivity increases and the effective Pe decreases, thereby decreasing flame length as indicated by Fig. 11.11.

The results indicate that the liquid-film combustion is feasible, although, for this simple laminar flow, the required chamber length is an order of magnitude larger than the diameter. This indicates a need to impose swirl and/or vortex generation in order to increase transport rates and thereby reduce the required chamber length. In the next section, the effect of swirl imposed on the inflow is considered as a perturbation to the base flow. Similarly, perturbations to this base flow that are due to the Stefan velocity and the density variation are evaluated by Sirignano et al. (2005).

11.4 Concluding Remarks

The liquid-film combustor has been analyzed for steady, laminar, axisymmetric flow for the two cases of a fully developed parabolic velocity profile and a flat plug-flow velocity profile to describe the base flow. Linear perturbations have been used to treat the effects of swirl-induced flow, vaporization-induced flow, and gas expansion. A solution with a standing diffusion flame and cool walls is obtained, in qualitative agreement with experiments. So theoretical support is given to the experimental observation that, unlike premixed combustion or spray combustion, stable liquid-fuel film burning can occur in small-diameter chambers without quenching. In particular, the vaporizing liquid film allows the wall temperatures to remain below the liquid boiling point.

The flame length at higher Peclet number values is many combustor radii and is also substantially longer than the film length required for vaporizing a stoichiometric amount of fuel. This situation can result in dry walls exposed to high heat fluxes. This means that, for higher mass flows, it is desirable to increase significantly the effective diffusivity (so as to decrease the effective Peclet number and flame length) by means of vortex generation.

The perturbations to the velocity field that are due to vaporization and gas expansion have significant effects on the combustion process, extending the flame length and modifying the vaporization rate per unit surface area as a function of downstream position.

For the inviscid flow, swirl has no effect within this linear theory. This linearized analysis gives some indication of the important physics and of some length scales of the flow downstream of the air inlet; however, it cannot represent well the inflow and flame-holding details. The swirling inflow would have a 3D character and would typically involve jets from ports, wakes of flows through vanes, and/ or recirculation zones. Computational-fluid-dynamical methods are required for giving better resolution of all of the physics in the inflow and flame-holding region. So 3D solutions of the Navier–Stokes equation for the multicomponent reacting flow are required in the future for resolution of this upstream region.

Stability of Liquid Streams

12.1 Introduction

It is intended herein to present the current status of the fundamental understanding about the liquid-atomization processes for various injection configurations. The limitations of the theory and the need for future work will be made apparent. This chapter is not intended to be a guide for the practicing engineer; the current state of the art is based on empirical approaches that are discussed in Chapter 1. Rather, this chapter reviews theoretical research that should eventually lead to improved design methodology and design tools for liquid-atomization systems. Other overviews of the theory can be found in Lefebvre (1989), Bayvel and Orzechowski (1993), Sirignano and Mehring (2000, 2005), and Lin (2003).

The atomization problem could be divided according to three subdomains of the fluid mechanical field. The upstream subdomain lies within the liquid-supply piping, plenum chamber, and orifice (nozzle) of the injector hardware. More than the mass flow and average velocity from the orifice into the combustion chamber are important here; velocity and pressure fluctuations in the liquid that are due to turbulence, collapse of bubbles formed through cavitation, supply-pressure unsteadiness, and/or active-control devices are critical in affecting the temporal and spatial variation of the liquid flow over the orifice exit cross section. A small amount of research has been performed on this subject.

The second subdomain involves the liquid stream from the orifice exit to the downstream point where disintegration of the stream begins. The neighboring gas flow (or gas and droplet flow) is part of this subdomain. This region involves severe distortion of the liquid stream that typically occurs in a wavelike manner. We focus our discussion in this chapter on this subdomain. This is the subdomain where most of the research work has been performed. The nonequal treatment of the three subdomains is a reflection of the amount of research performed rather than of their relative importance.

The third subdomain in the spray formation phenomenon involves the cascading of the ligaments that occur in the earliest disintegration to smaller and smaller elements, resulting eventually in the essentially spherical droplets of the spray. This is the most challenging portion of the theoretical problem. No theory has been developed to explain the details of the dynamics in this cascade process. Some statistical theories have attempted to describe the droplets that are yielded from this process. The final breakup of liquid elements into smaller elements can be caused by capillary necking or disjoining pressure forcing the liquid to zero transverse thickness at some localities or by a fracturing of the liquid in which a viscous tensile stress causes the total normal stress to fall below the thermodynamic saturation pressure.

There are many types of injectors (sometimes called atomization systems) that inject liquid streams with sufficient energy to provide for breakup or atomization into droplets. Energy can be provided through various mechanisms: liquid pressure, air pressure, rotation of cups or disks, vibration or acoustics, and electric fields. Discussions of practical injectors can be found in Lefebvre (1989), and Bayvel and Orzechowski (1993).

There are some important geometrical characteristics of injected streams: (i) curvature of the gas–liquid interface and (ii) divergence of the liquid stream. The interface can have a mean or average curvature that is due to basic geometry; round jets, annular jets, coaxial jets, and conical jets share this character. A planar jet does not have a mean curvature. Any liquid jet or stream with a free surface can have curvature that is due to surface waves such as capillary waves. Curvature leads to a pressure jump across the surface that is due to surface tension. With wave motion at the surface, the pressure jump will vary with time and position along the surface thereby leading to pressure gradients parallel to the surface and subsequent accelerations. These induced motions can become unstable, causing the disintegration of the stream.

Divergence of the stream occurs in conical streams and fan jets (such as result from two impinging round jets). For hollow cone and fan jet injections, this results in a thinning of the stream and affects the eventual droplet size. Of course, the divergence increases the number of interesting dimensions for the phenomenon.

Important liquid properties for atomization are density ρ_l, surface tension coefficient σ, and dynamic viscosity μ_l, where the subscript l refers to the liquid. For common liquids, density is of the order of 1000 kg/m^3. The surface-tension coefficient is of the order of 10^{-2} N/m for many liquids. Dynamic viscosity varies from the order of 10^{-4} to the order of 1 kg/ms. Liquid density is very weakly dependent on pressure and temperature. Surface tension and viscosity are weakly dependent on pressure and will decrease as temperature increases.

There are four forces acting on the liquid that are typically important in atomization (not including electromagnetic forces acting on charged liquids):

gravity force,	$\rho_l L^3 g$;
inertia,	$\rho_l L^2 V^2$;
surface-tension force,	σL;
viscous force,	$\mu_l L V$;

where L and V are the characteristic length and velocity, respectively, and g is the acceleration that is due to gravity. Note that g can also represent any acceleration of the liquid that can be formulated as a reversed D'Alembert acceleration in a noninertial frame of reference. From these four forces, three independent well-known nondimensional groupings result:

Reynolds number, $\quad \mathrm{Re} = \frac{\rho_l L V}{\mu_l}$;

Weber number, $\quad \mathrm{We} = \frac{\rho_l L V^2}{\sigma}$;

Froude number, $\quad \mathrm{Fr} = \frac{V^2}{g L}$.

Another number in common use is the Ohnesorge number:

$$\mathrm{Oh} = \frac{\mathrm{We}^{0.5}}{\mathrm{Re}} = \frac{\mu_l}{(\rho_l \sigma L)^{0.5}}.$$

In some cases, the Bond number replaces the Froude number

$$\mathrm{Bo} = \frac{\rho_l L^2 g}{\sigma}.$$

When a periodic disturbance of frequency ω is imposed on the liquid, it may be convenient to use the Strouhal number:

$$\mathrm{St} = \frac{\omega L}{V}.$$

The characteristic velocity and length typically are based on the phenomenon to be observed. For discharging jets, the average stream velocity and the diameter (for a round cylindrical jet) or the thickness (for a liquid sheetlike flow) is typically used. For the temporal analysis of periodically disturbed streams, the disturbance wave number and a characteristic velocity other than the average sheet velocity might be used. The nondimensional groupings indicate that $\sqrt{gL}$ and $\sqrt{\sigma/\rho_l L}$ are natural characteristic velocities related to gravity waves and capillary waves, respectively. It is seen that gravity wave speeds dominate on large length scales whereas capillary wave speeds dominate on short length scales. With the small sizes typically associated with injected liquid streams, gravity waves are not interesting, but capillary waves are critical in the distortion and disintegration process.

In addition, there are viscous forces, gravity force, and inertia for the surrounding gas. This will result in four more nondimensional parameters: gas-phase Weber number, gas-phase Reynolds number, and the gas-to-liquid density and viscosity ratios. These Weber and Reynolds numbers are based on a gas characteristic velocity and gas properties.

An instability that involves only the liquid inertia, liquid viscous forces and body forces, and the surface tension are described as capillary instabilities. The inertia and the viscosity of the gas might modify the instability in a quantitative manner; however, they are not essential to the capillary mechanism. An instability that involves a parallel flow of gas with the liquid stream can also occur. If a continuous variation of the parallel velocity occurs, this hydrodynamic instability is named the Rayleigh instability in the inviscid case and the Orr–Sommerfeld instability in the viscous case (Drazin and Reid, 1981). (This Rayleigh hydrodynamic instability is distinct from the Rayleigh capillary instability discussed later.) When the variation of the parallel velocity involves a discontinuity of that velocity component at the liquid–gas interface forming a vortex sheet there, the phenomenon is identified as Kelvin–Helmholtz instability. When there is instability associated with an acceleration or body force in a normal direction to the gas–liquid interface, we identify the behavior as a Rayleigh–Taylor instability. Any of these instabilities can result in a disintegration of the liquid stream and the formation of smaller liquid ligaments and droplets.

Generally two types of instabilities for liquid streams have been analyzed: temporal instabilities and spatially developing instabilities, both of which are discussed here. Temporal stability analyses consider solutions that are periodic in space and can be oscillatory or exponential in time. The spatial periodicity implies infinite length for the liquid stream. Spatial stability analyses consider solutions for a semi-infinite stream flowing from a nozzle. Periodicity in time is considered, and the behavior in space can be oscillatory or exponential. The earliest analyses were temporal, but spatial stability is of greater practical interest. Note that perturbations of liquid jets in open-flow problems might evolve both in space and time and therefore temporal or spatial analyses need not provide the whole picture even in linear stability. Convective and absolute instabilities can become important here.

Two general modes of wave phenomena can be found for either temporal or spatially developing instabilities in most injection configurations. The "dilational" mode involves primarily a pulsing of the stream width of thickness. This mode has been also named a "sausage" mode. In a linearized treatment for a planar sheet, the pulsing is symmetric so that this mode is also identified as the "symmetric" mode. The other general mode is named the "sinuous" mode because the primary distortion involves a wavy ribbonlike motion of the stream with only secondary effects on the stream thickness. So the center of the stream cross section follows a wavy path in the sinuous distortion whereas it is not significantly affected in the dilational mode. The sinuous mode has an antisymmetric distortion in the linearized treatment of the two liquid–gas interfaces in the planar configuration and has also been named the "antisymmetric" mode.

The discussion of liquid stream stability is separated into four general categories based on the geometrical configuration: round jets, planar sheets, annular sheets, and "conical" sheets. They are addressed in Sections 12.3, 12.4, 12.5, and 12.6., respectively. The terms "sheet" and "free film" are used interchangeably to denote a liquid stream whose thickness is smaller than its other dimensions. In the atomization process, thin sheets or free films are often formed as a first step toward obtaining a spray with small droplets in a relatively fast manner. The term "conical" is used casually here following industrial practice; the "conical" surfaces discussed here are not necessarily generated by straight lines emerging from a common point. Rather, the generating lines are typically curved so that bell-shaped surfaces will be included in this category. The form categories are chosen arbitrary, and several of them can be portrayed as limiting cases of some other configurations. A common formulation of the governing equations is given in Section 12.2.

12.2 Formulation of Governing Equations

The general equations governing the motion of a liquid stream or bulk of liquid in another immiscible fluid and under the influence of gravity are presented for the specific case of an annular liquid stream. Three limiting cases are possible for this configuration. The round jet in an ambient-gas stream is recovered as the radius for the inner interface decreases to zero. As both the inner and outer annular radii grow infinitely large, the planar sheet configuration is recovered with the sheet thickness given by the difference of inner and outer annular radii. Finally, for finite values of the inner radius and infinitely large outer radius, we obtain the case of a submerged gas jet. Our discussion addresses the round jet, planar sheet, and annular sheet.

Boundary conditions must be applied at the liquid interface. One condition is the kinematic condition that a particle of fluid on the surface moves with the surface so as to remain on the surface. Or in other words, the velocity component normal to the interface is continuous across the interface, i.e., $\vec{v}_1 \cdot \vec{n} = v_2 \cdot \vec{n}$ where $\vec{v}_{1,2}$ are the local velocity vectors on both sides of the interface of fluid 1 and fluid 2, and $\vec{n}$ is the local unit normal vector of the interface.

The second condition considers the balance between the surface stresses on both sides of an interface between fluid 1 and fluid 2, including the pressure jump across the interface that is due to surface tension. The dynamic boundary condition is $(p_1 - p_2 + \sigma\kappa)\,\vec{n} = \left(\vec{\vec{\tau}}_1 - \vec{\vec{\tau}}_2\right) \cdot \vec{n}$, where σ is the surface-tension coefficient of fluid 1 in fluid 2 (in units of force per unit length), p_k is the pressure, $\vec{\vec{\tau}}_k$ is the viscous stress tensor, $\vec{n}$ is the unit normal (into fluid 2) at the interface, and κ is the local surface curvature, and $R_1^{-1} + R_2^{-1}$, where R_1 and R_2 are the principal radii of curvature of the surface.

We consider only Newtonian viscous incompressible fluids with gravity as the only body force. Let the subscript $k = 1, 2, 3$ now refer to either the fluid within the liquid annulus ($k = 1$), the fluid within the core of the annulus ($k = 2$), or the fluid surrounding the annulus ($k = 3$). Denoting the annular liquid jet's inner (subscript $-$) and outer (subscript $+$) interfaces by $r_-(t, z, \theta)$ and $r_+(t, z, \theta)$, respectively, the kinematic boundary conditions at these interfaces are given by

$$v_{r,k}\left(r_-, \theta, z, t\right) = \frac{\partial r_-}{\partial t} + v_{z,k}\left(r_-, \theta, z, t\right)\frac{\partial r_-}{\partial z} + \frac{1}{r_-}v_{\theta,k}\left(r_-, \theta, z, t\right)\frac{\partial r_-}{\partial \theta}, k = 1, 2,$$

$$(12.1)$$

$$v_{r,k}\left(r_+, \theta, z, t\right) = \frac{\partial r_+}{\partial t} + v_{z,k}\left(r_+, \theta, z, t\right)\frac{\partial r_+}{\partial z} + \frac{1}{r_+}v_{\theta,k}\left(r_+, \theta, z, t\right)\frac{\partial r_+}{\partial \theta}, k = 1, 3.$$

$$(12.2)$$

These equations establish that the jet's surfaces are material surfaces. The dynamic boundary conditions establish that the shear stress is continuous, and the jump in normal stresses across the interfaces is balanced by surface tension. Assuming inviscid fluids only, the viscous shear stress at both interfaces r_+ and r_- vanishes, and the dynamic boundary conditions at the two interfaces reduce to a condition for the normal stress only, i.e.,

$$p_1\left(r_-, \theta, z, t\right) = p_2\left(r_-, \theta, z, t\right) - \sigma_-\kappa_-, \qquad (12.3)$$

$$p_1\left(r_+, \theta, z, t\right) = p_3\left(r_+, \theta, z, t\right) + \sigma_+\kappa_+, \qquad (12.4)$$

where $\sigma_\pm$ denote the constant surface-tension coefficient between fluid 1 and fluid 2 (subscript $-$) or fluid 1 and fluid 3 (subscript $+$), respectively, and κ_- and κ_+ refer to the local curvature of the inner and outer interface, respectively.

The pressure jump across the interface that is due to surface tension is obtained from the local curvature $\kappa_{(\pm)}$, i.e.,

$$\kappa = \nabla \cdot \vec{n} = \frac{1}{R_1} + \frac{1}{R_2}, \tag{12.5}$$

where $\vec{n}$ is the local normal unit vector at the considered interface.

For an inviscid flow without vorticity, it is possible to set the velocity equal to the gradient of a scalar velocity potential:

$$\vec{v} = \nabla \phi.$$

Substitution of this relation into Equation (12.1) yields Laplace's equation $\nabla^2 \phi = 0$. This approach is not necessarily limited to inviscid flow; Joseph et al. (2007) showed how potential flow can be assumed for incompressible, constant-property viscous flow. The unsteady Bernoulli equation can be developed by integrating the inviscid vector form of the momentum equation along a path that is locally tangent to the velocity vector (not necessarily the particle path in an unsteady case). We can obtain

$$\rho_l \frac{\partial \phi}{\partial t} + \frac{\rho_l}{2} \nabla \phi \cdot \nabla \phi + p_l = p_{\text{ref}}, \tag{12.6}$$

where p_{ref} is a constant reference pressure. Equation (12.6) is especially useful when applied along the liquid interface where it can relate velocity potential to pressure. An important nonlinear effect is introduced through the Bernoulli equation.

In the limit of very large values for the radial locations of the two interfaces $r_+ (z, \theta, t)$ and $r_- (z, \theta, t)$, we recover the equations for the three-dimensional distortion of an incompressible inviscid planar fluid sheet separating two (different) inviscid incompressible fluids. Also, for finite values of r_+ but zero value for r_-, the governing equations for a round jet in an ambient fluid are obtained. In the limit of very large values for r_+, the equations for a submerged gas jet are obtained if $\rho_2 \ll \rho_1$.

12.3 Round Jet Analyses

The round liquid jet is the simplest injection configuration and has received the most attention. It is well known that four distinct regimes can be found on a plot of Weber number versus Reynolds number (or on a plot of Weber number versus Ohnesorge number.) These four regimes, shown in Figure 12.1, are separated on a log–log plot by three straight lines of negative slope (Sirignano and Mehring, 2000). At the lowest values of We and Re, the Rayleigh capillary mechanism prevails. Here aerodynamics interaction with the surrounding gas is not important and axisymmetric dilational oscillations occur. At higher values of We and Re, the first-wind-induced regime appears where aerodynamics becomes important and nonaxisymmetric sinuous oscillations appear. In these first two regimes, the resulting droplet diameters after breakup are of the same order of magnitude as that of the diameter of the original jet. At still higher We and Re, the second-wind-induced regime appears that is characterized by smaller droplets after breakup. At the highest We and Re values, the atomization regime appears wherein breakup occurs very close to the nozzle exit with the smallest droplets as a result.

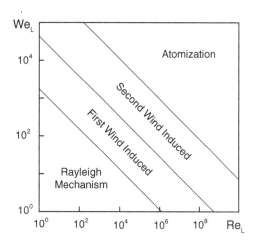

Figure 12.1. Modes of disintegration on a liquid Reynolds number and Weber number plane.

Consider an initially cylindrical liquid jet undergoing the capillary oscillation. Assume no gravitational or viscous effect. At low pressures, the density of the surrounding gas is negligible compared with the liquid density so that gas inertia can be neglected. Thereby the liquid jet is essentially flowing in a passive medium, implying very small or zero values for the gas-phase Weber number. For the round jet, only one free surface exists, i.e., $R = r_+ (z, \theta, t)$ and $r_- = 0$; and because the surrounding medium has been assumed to be passive, field equations for fluid 3 ($k = 3$) do not have to be solved.

Accordingly, the only kinematic and dynamic boundary conditions that apply here are Eq. (12.2) for $k = 1$ and Eq. (12.4) with $p_3 = 0$, respectively.

12.3.1 Temporal Stability Analysis

Rayleigh (Drazin and Reed, 1981) made an early contribution to the understanding of the formation of droplets from a liquid stream by capillary effects. In a temporal stability situation, a Galilean transformation can be made to the infinite stream so that stream velocity is eliminated. (In the spatial stability case, that transformation is not useful because the nozzle exit position should remain stationary in the frame of reference.) Small magnitudes of disturbances to the initially cylindrical column of liquid allow a linearized analysis. Two approaches are possible. One uses separation of variables to solve the governing Laplace partial differential equation constrained by certain interface conditions and predicts spatial and temporal behavior. Another approach seeks a final configuration with minimum surface energy.

Rayleigh sought solutions for the dispersive waves of the form

$$p' = P(r)e^{in\theta}e^{ikz}e^{i\omega t},$$

$$u'_r = U(r)e^{in\theta}e^{ikz}e^{i\omega t},$$

$$R' = ce^{in\theta}e^{ikz}e^{i\omega t},$$

for the perturbations of pressure, radial velocity, and jet radius, where n and k are wave numbers in the circumferential and axial directions, ω is an angular frequency, c is a constant, and primes denote perturbation quantities. The function P is governed by a modified Bessel equation. It is found that the growth or decay rate is given by the dispersion relation

$$\omega^2 = \frac{\sigma}{R_0^3 \rho_l} \frac{k R_0 I_n'(k R_0)}{I_n(k R_0)} \left[k^2 R_0^2 + n^2 - 1 \right], \tag{12.7}$$

where $I_n(kr)$ is a modified Bessel function of the first kind and R_0 is the undisturbed radius of the liquid cylinder. Because the right-hand side of Eq. (12.7) is real, ω^2 is real and, because the term outside the brackets is positive, ω^2 takes the sign of $\left[k^2 R_0^2 + n^2 - 1 \right]$. Because n is an integer or zero, ω^2 is always positive when n is a nonzero integer, resulting in a stable oscillatory behavior. When $n = 0$, oscillations with growth or decay can result when $k^2 R_0^2 - 1 < 0$. For $k R_0 > 1$, neutral oscillations can result. Note that $n = 0$ occurs for axisymmetric behavior. So only axisymmetric modes can grow in amplitude. Note that the condition for growth is that $k R_0 < 1$, which implies that the wavelength $\lambda = 2\pi/k$ is longer than the circumference of the jet. The nonaxisymmetric modes with $|n| \geq 1$ do not grow or decay. The maximum growth rate occurs when the wavelength-to-undisturbed-radius ratio is $\lambda/R_0 = 9.016$. The axisymmetric linear-theoretical behavior results in a dilational or varicose waviness of the interface. The amplitude growth continues until the stream is severed at positions one wavelength apart. Eventually one spherical droplet forms for each original wavelength along the cylindrical liquid. The droplet radius is given by

$$R_{\text{drop}} = R_0 \left(\frac{3\pi}{2 k R_0} \right)^{1/3} = R_0 \left(\frac{3\lambda}{4 R_0} \right)^{1/3}. \tag{12.8}$$

Rayleigh showed that the perturbations of surface area and of surface energy during temporal instability take the sign of

$$k^2 R_0^2 + n^2 - 1 ,$$

which is the factor that emerged from Eq. (12.7). So when $|n| \geq 1$, surface energy would increase. Because this is not allowed by the energy-conservation principle, nonaxisymmetric modes do not occur. Only with $n = 0$ (axisymmetric behavior) and $k R_0 < 1$ (wavelength longer than circumference) is a decreasing surface energy solution obtained. This implies that a more stable configuration is being found. The modal analysis and the surface energy analysis therefore lead to identical conclusions.

12.3.2 Surface Energy

For hydrostatic capillary phenomena in a mechanical system in equilibrium, the total system energy is unvaried under arbitrary virtual displacements that are consistent with the constraints of that system. The surface energy of the system is modified by the pressure work done on the surface. The pressure in the liquid at the surface is

$$p = p_a + \sigma \left(\frac{1}{R_1} + \frac{1}{R_2} \right), \tag{12.9}$$

where p_a is the gas pressure at the interface.

Several recent analyses of the final droplet diameter and droplet velocity joint distributions based on the maximum-entropy principle assume certain constraints on the surface energy. Some workers assume that the surface energy is conserved so that the original liquid stream and the final droplet collection have the same surface energy. Another group assumes that the sum of kinetic energy and surface energy is conserved. As will be subsequently shown, these constraints do not apply to the simplest configuration that we understand, namely the Rayleigh capillary breakup mechanism. Therefore we have no reason to accept these constraints for more complex situations. In the Rayleigh breakup, the velocities of the jet stream and of the droplets are identical so that the kinetic energy is constant there. (No aerodynamic interaction with the gas is considered.)

As the surface of a jet distorts to form droplets, the surface area will decrease during the process. For example, consider a sinusoidal disturbance of the surface of an initially cylindrical jet of radius R_0. As a higher-order effect, the mean radius R_m of the disturbed jet will be smaller than the original radius R_0. If the local radius of the surface of the round jet is given by $R = R_m + \varepsilon cos(2\pi x/\lambda)$, it follows that to conserve volume $R_0^2 = R_m^2 + \varepsilon^2/2$. Therefore $R_m < R_0$ if $|\varepsilon| > 0$. The surface area of the disturbed liquid is $S = 2\pi R_m \lambda$ for a segment that extends one wavelength. This area is smaller by a factor of R_m/R_0 than an area segment of the same length for the undisturbed liquid surface. As the amplitude ε increases, the disturbed surface area becomes still smaller. This can be counterintuitive because, if the change in mean radius is not considered, one might expect increased surface area. The decrease in surface area results in a proportional decrease in surface energy.

As the surface area decreases, work is done that converts surface energy into some other form (Sirignano and Mehring, 2005). As the surface moves, the liquid acquires a kinetic energy of oscillation that is distinct from the translational kinetic energy. The presence of viscous dissipation will cause the oscillation to cease. The dissipation will slightly increase the liquid temperature by a negligible amount. Although the prescribed kinetic energy and viscous-dissipative heating are insignificant compared with the internal energy, they are not negligible compared with the surface energy. For example, if we consider the fastest-growing temporal instability for the Rayleigh mechanism, the wavelength-to-initial-radius ratio of 9.016 together with Eq. (12.8) indicates that the surface area (and surface energy) ratio between the final drop and the initial cylinder is

$$\frac{S_{\text{drop}}}{S_0} = 2\left(\frac{R_{\text{drop}}}{R_0}\right)^2 \left(\frac{\lambda}{R_0}\right)^{-1} = 0.7933.$$

This means that about 21% of the original surface energy changed to kinetic energy of oscillatory motion and then dissipative heat. For ideal inviscid fluids, energy dissipation is absent and the droplets resulting from stream disintegration would be subject to continuous oscillation or distortion. Note that Eq. (12.9) can be used to demonstrate that the pressure difference between the ambient value and the liquid volume increases by a factor of

$$\frac{2R_0}{R_{\text{drop}}} = 2\left(\frac{4}{3}\frac{R_0}{\lambda}\right)^{1/3}. \tag{12.10}$$

For the fastest-growing Rayleigh wavelength, this factor is 1.058. This pressure increase indicates a slight enthalpy increase that can be significant for the energy balance in the flowing system in the spatially developing instability. For example, a cylindrical jet flowing at velocity u_0 has a surface energy flux of $2\pi\sigma R_0 u_0$ and a pressure work term exactly one-half of that value. So this 6% increase in the pressure contribution to the enthalpy requires only a 3% change in the surface energy. We expect therefore viscous dissipation to be a dominant factor for the spatially developing instability as well as for the temporal instability.

When aerodynamic effects are present, the kinetic energy associated with the relative motion of the surrounding gas can be transferred to the liquid. In this case, the resulting spray can have more energy than the original liquid jet. In the case in which the gas stream causes the liquid stream to disintegrate into many droplets with diameters smaller than the original transverse dimension of the liquid stream, the surface energy per unit volume of droplets will be much greater than that value for the initial stream. If the gas entrainment of the liquid accelerates the liquid, liquid kinetic energy increases. These transfers of energy from the gas to the liquid are not possible in the capillary limit in which gas density is neglected. In the aerodynamic case in which gas density is finite, meaningful energy-conservation principles can be written for the liquid and gas only as a combined system.

12.3.3 Spatial Stability Analysis

Consider now a round jet flowing at mean velocity u_0 from an orifice. Keller et al. (1973) found the dispersion relation

$$(\omega + k u_0)^2 = \frac{\sigma}{\rho_l R_0^3} \frac{k R_0 I_n'(k R_0)}{I_n(k R_0)}\left(k^2 R_0^2 + n^2 - 1\right) \tag{12.11}$$

from a three-dimensional inviscid analysis. Here we consider ω to be real and to be imposed through modulation at the orifice exit. The wave number k can be complex whereby the imaginary part implies exponential growth or decay. When the wave number n (which is only real and represents azimuthal waves) takes a nonzero integer value, k cannot become complex. So asymmetric modes are neutrally stable. For axisymmetric behavior (i.e., $n = 0$), neutral stability still occurs when the wavelength is shorter than the circumference $\left(k^2 R_0^2 - 1 > 0\right)$ whereas exponential behavior occurs when the wavelength exceeds the circumference $\left(k^2 R_0^2 - 1 < 0\right)$. The latter observation can also be made by use of simplified reduced one-dimensional model equations (Bechtel et al., 1995).

The Keller et al. dispersion relation yields four values of k, one of which shows a long wavelength oscillation with exponential growth of amplitude. However, that was never found experimentally. This inconsistency was later explained by Bogy (1978a, 1978b), who introduced the use of group velocities to the linear analysis of capillary instability problems on a one-dimensional inviscid jet.

GROUP VELOCITY. It is well known that, for conservation of wave numbers in multi-dimensional unsteady cases,

$$\frac{\partial k_i}{\partial t} + \frac{\partial \omega}{\partial x_i} = 0. \tag{12.12}$$

When $\omega = \omega(k_x, k_y, k_z) = \omega(\vec{k})$, as given by the dispersion relationship, we have

$$\frac{\partial k_i}{\partial t} + \frac{\partial \omega}{\partial k_j}\frac{\partial k_j}{\partial x_i} = \frac{\partial k_i}{\partial t} + C_j\left(\vec{k}\right)\frac{\partial k_j}{\partial x_i} = \frac{\partial k_i}{\partial t} + C_j(\vec{k})\frac{\partial k_i}{\partial x_j} = 0, \qquad (12.13)$$

where $C_i(\vec{k})$ is the group velocity that can be derived by differentiation from a dispersion relation such as Eq. (12.11). Because k_i is the gradient of a phase α, it follows that

$$\frac{\partial k_j}{\partial x_i} = \frac{\partial k_i}{\partial x_j} = \frac{\partial^2 \alpha}{\partial x_i \partial x_j}.$$

In a one-dimensional situation, the wave number k becomes a scalar and the gradient becomes a spatial derivative. For a linear system, the group velocity gives the rate at which information or energy is propagated, i.e., group velocity and energy propagation velocity are identical (Lighthill, 1978). Each of the four branches for k has its own group velocity. When C is positive, information for that branch is propagated in the downstream direction, so that a boundary condition is required at $x = 0$ to provide that information. A negative value of C means that information is propagated in the upstream direction. That branch requires a boundary condition at infinity. As a consequence, Bogy showed that all four boundary conditions should not be placed at the orifice exit as had been done by previous investigators. An unstable mode with nonzero-valued group velocity is named a convective instability whereas an unstable mode with zero-valued group velocity is an absolute or non-convective instability. With an absolute instability, the growth is confined to some fixed region where the group velocity is zero.

It can be assumed that no energy is added to the stream oscillation at infinity. All modulation occurs at the nozzle exit, $x = 0$. The branch that yields exponential growth in the downstream direction has negative group velocity so that in the absence of energy addition or modulation downstream, it is never excited. This explains why the exponential growth terms predicted by theory are not observed experimentally. Following this observation, Pimbley (1976) and Bogy (1978a, 1978b, 1979) correctly analyzed the jet problem for large Weber numbers. However, the authors failed to notice the absolute instability later found by Leib and Goldstein (1986) for small jet velocities. The latter authors also showed that neutral waves with upstream (i.e., negative) group velocity can be found on the jet even downstream from their source (i.e., the nozzle exit). Mehring and Sirignano also made the similar observation for spatially developing annular liquid sheets (Mehring and Sirignano, 2000b).

12.3.4 Nonlinear Effects

The nonlinear terms in the governing equations for the round jet have profound effects. They predict the deviation from sinusoidal wave distortion and the subsequent deviation from a monodisperse spray. For a simple example, under conditions in which the round jet would form a single stream of droplets, linear theory explains that equisized droplets are formed, one for every wavelength of a disturbance to the round stream. However, the experimental observation is that every wavelength can produce a large (parent) droplet and a smaller "satellite" droplet in an axisymmetric configuration. The satellite initially is attached to both the upstream and down-stream parent droplets but breaks first from one of them and then is accelerated by surface tension toward the other. The time for merger with the connected droplet in-creases with increasing wavelength and decreases with increasing amplitude of the disturbance. For a sufficiently long time for merger compared with the residence time, we can consider the satellite as stable.

The simplest way to explore nonlinear effects in jet breakup is to extend Rayleigh's linear analysis to higher order in the perturbation amplitude ε. Such an expansion was attempted by Yuen (1968), who used the method of strained co-ordinates in a third-order perturbation expansion. Yuen's analysis provided some indication of the nonuniform capillary breakup of an inviscid liquid jet and also pre-dicted a shift of the stability limit to larger wave numbers than observed within linear theory (i.e. $k = 1/R_0$). Nayfeh (1970) showed the invalidity of Yuen's result near the stability limit, $k = R_0$. Using the method of multiple scales, he obtained two second-order expansions: one valid away from $k = 1/R_0$ and one valid near $k = 1/R_0$. The latter has an even larger instability region than predicted by Yuen with a cutoff wave number of $k = \left[1 + 3\varepsilon^2/\left(4R_0^2\right)\right]/R_0$ {versus $k = \left[1 + 9\varepsilon^2/\left(16R_0^2\right)\right]/R_0$ given by Yuen, where ε denotes the amplitude of the imposed disturbance of wave-length k}. The validity of Nayfeh's result was demonstrated by Eggers (1997), who compared Nayfeh's solution with those obtained from nonlinear numerical simula-tions of the one-dimensional lubrication equations.

The spatially developing case must be examined rather than the temporally developing case to break a false symmetry. Pimbley and Lee (1977) performed a second-order analysis for the spatially developing case, although they used nozzle boundary conditions that were later recognized as inappropriate by Bogy (1978a). They also conducted an experiment. Some significant differences from the tempo-ral analysis were found. In the temporal case, symmetry existed around the position of maximum amplitude. This symmetry is no longer present in the spatial stability case. The relationship between a large droplet and the satellite droplet immediately downstream of it will, in general, be different from the relationship with the satel-lite droplet immediately upstream. In the flowing spatially unstable round jet cases, Pimbley and Lee found that the nonlinear effects produce higher harmonics in the diameter as a function of downstream position and time. The second-harmonic am-plitude is typically smaller than the fundamental amplitude but causes two local maxima per wavelength and leads to a satellite droplet. So, every wavelength pro-duces one parent droplet and one satellite droplet in this theory.

Various authors addressed shortcomings of the linear and nonlinear one-dimensional models by Bogy (1978b) and Pimbley and Lee (1977). Using a formal perturbation expansion, Schulkes (1993) showed that the inviscid one-dimensional equations of Lee (1974) are inconsistent because terms that have been retained in the boundary conditions should have been rejected according to the approximations made for the momentum equations.

A systematic derivation of inviscid and viscous one-dimensional equations start-ing from the full Navier–Stokes equations was performed by Bechtel et al. (1992), Eggers and Dupont (1994), and García and Castellanos (1994). The discussion on self-consistent higher-order one-dimensional models was continued by Bechtel et al.

(1995), identifying Eggers' "regularized" model, applied by Eggers and DuPont (1994), Shi et al. (1994), Brenner et al. (1994), and Eggers (1995), as consistent with exception of the surface-tension or curvature term. That term appears, to all orders, coupled to the leading-order approximations of all other physical effects. On the other hand, the same authors note that the models of Ting and Keller (1990), Forest and Wang (1990), and Bechtel et al. (1992) for the same physical geometry maintain the slenderness approximation consistently in all terms. From this understanding, the authors produced one-dimensional nonlinear slender Newtonian jet equations for leading-order behavior and higher-order corrections, incorporating the effects of surface tension, inertia, viscosity and gravity. The models by García and Castellanos (1994) and Bechtel and co-workers (1992, 1995) were contrasted with each other by Eggers (1997). Using essentially a Galerkin approximation of the equations of motion, Eggers (1997) illustrated the appearance of the full curvature term from surface-tension forces even at leading order and providing a consistent strong argument for the "ad hoc" consideration of the full curvature term in otherwise lower-order one-dimensional models. As Eggers notes, a consistent leading-order systems (i.e., without the full curvature term) is unstable against short-wavelength "noise," making it ill-suited for numerical simulations. Because the leading-order equations are unaffected by higher-order terms, this problem remains even if higher-order perturbation terms are included.

The use of the slender-jet approximation, used by many authors to predict nonlinear jet distortion away from breakup and until breakup, will lead to increasing errors as contributions of short-wavelength disturbances increase during jet pinch-off. Even more severe limitations apply to inviscid one-dimensional or slender-jet models, for which the development of finite time singularities is observed even before jet breakup. For a more detailed discussion on the limitations of one-dimensional models, we refer to Schulkes (1993) and Eggers (1997).

There can be difficulties caused by satellite formation in cases in which a uniform stream of droplets is desired. Chaudhary and Redekopp (1980) and Chaudhary and Maxworthy (1980a, 1980b) compensated for the generation of higher harmonics with the superposition of a third harmonic that, if properly modulated, produced a uniform droplet size. Orme and Muntz (1987) developed an injection scheme that modulated at lower frequency than the Rayleigh capillary instability frequency; so more than one droplet is formed in the modulated wavelength. Because of asymmetry in droplet velocity over the modulated wavelength, all N droplets in that wavelength coalesce into a larger droplet, forming a stream of uniform droplets. Whereas the Orme and Muntz scheme involved open-loop control with amplitude modulation, Strayer and Dunn-Rankin (2001) proposed using a control method that used deterministic transfer functions to calculate a required input signal. Rohani et al. (2009) extended the concept closer to a closed-loop-control capability by describing a process to produce this transfer function. See the discussion in Section 9.4 about the use of linear streams of uniform-sized droplets in theory and experiments.

NUMERICAL SIMULATION. The capillary temporal stability of a round viscous jet in a vacuum was studied by Ashgriz and Mashayek (1995) by a Galerkin finite-element numerical solution of the Navier–Stokes equations together with the free-surface conditions. A Reynolds number is defined to be

$$\text{Re} = \frac{1}{v}\left(\frac{\sigma R}{\rho_l}\right)^{1/2}, \tag{12.14}$$

which is actually a reciprocal Ohnesorge number. With $\text{Re} \leq O(1)$, imposed modulations are overdamped and no oscillations occur. Higher Re values result in oscillations. Above a certain value of Re dependent on the wave number, satellite droplets are formed with satellite size increasing with Re; clearly, viscosity is found to inhibit satellite formation and, for sufficiently large viscosity, any droplet formation is inhibited. As the modulation amplitude is increased, the threshold Re value for satellite formation increases and satellite size decreases. The results confirm that the second-harmonic component is responsible for the satellite formation.

Ashgriz and Mashayek (1995) found that growth rates agree well with linear theory for low values of Re, but linear theory overpredicts the growth rates at higher Re values. Breakup time decreases exponentially with increasing amplitude. The cutoff wave-number value (below which the jet is unstable) increases with initial amplitude of the oscillation.

Heister and co-workers (Hilbing et al., 1995; Spangler et al., 1995; Hilbing and Heister, 1996, 1997) and Lundgren and Mansour (1988) performed an interesting set of round-jet calculations employing a boundary element method (BEM). Hilbing and Heister (1996) showed that variations in Weber number, modulation frequency (and thereby wavelength), and amplitude can be used to control the eventual droplet size. See Fig. 12.2. Satellite droplet velocities are found to be less than the main droplet velocity, indicating the possibility of recombination. Only inviscid axisymmetric behavior was considered with the neglect of aerodynamic interactions. Mansour and Lundgren studied the Rayleigh capillary instability, neglecting viscosity and air inertia satellite droplets were predicted at all unstable wave numbers. A comparison of the nonlinear calculations with classical linear analysis indicated that the linear treatment predicts well the early jet deformation, although the formation of the satellites is not predicted. The technique represents the flow by singular dipole solutions of Laplace's equation distributed over the surface. The authors claim that this is equivalent to distributed vortex rings of varying strength determined by local surface distortion.

Spangler et al. (1995) extended the work of Lundgren and Mansour to consider the inertia of the surrounding air. Now the Kelvin–Helmholtz and capillary instability mechanisms can appear in an integrated fashion. Aerodynamic interactions were important even at low relative velocities, and the nonlinear effects were significant for initial deflections as small as 1% of the undisturbed radius. Transition from the Rayleigh regime to the first-wind-induced regime and then to the second-wind-induced domain occurred approximately at gas-phase Weber numbers of 1 and 2, respectively. Nonlinear and aerodynamic effects tended to cause broad troughs and narrow peaks in the waveform with two points of minimum radius for each wavelength. One satellite drop for each main drop was formed in the first-wind-induced regime. At higher Weber numbers in the second-wind-induced regime, a spiked peak occurs in the waveform with the pinching off of a fluid ring at the peak

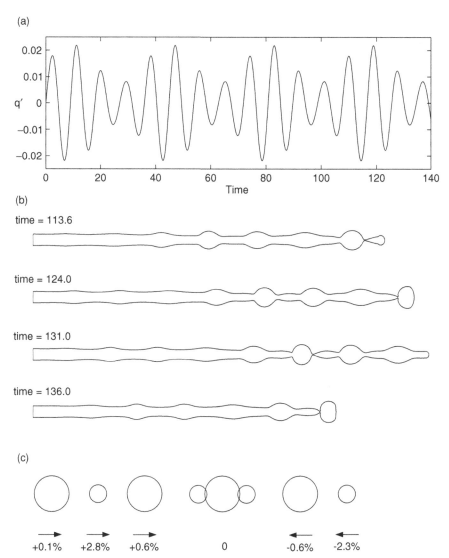

Figure 12.2. Jet breakup with an amplitude-modulated inflow velocity: (a) inflow disturbance waveform, (b) jet profiles at four successive pinch events, (c) relative droplet velocities
(Hilbing et al., 1995).

location for each wavelength. This indicates that small droplets will be formed when
this ring of fluid disintegrates. Of course, axisymmetric calculations cannot capture
this breakup.

Hilbing and Heister (1997) analyzed the unsteady nonlinear evolution of a high-
speed viscous liquid jet issuing from a circular orifice. The authors used an integral
method for a thin viscous region at the jet periphery, whereas they used a BE for
the inviscid "core" flow. Under certain conditions, jet "swelling" was observed, and
the boundary layer thinned out to a shear layer over a length of about half an orifice
radius.

12.3.5 Viscous Effects

Weber (1931) considered the viscous effects in a one-dimensional temporal instability analysis of the axisymmetric case. His results yielded that the growth rate is given by

$$\alpha = -\frac{3}{2}\frac{\mu}{\rho_l}k^2 \pm \left[\left(\frac{3}{2}\frac{\mu}{\rho_l}k^2 \right) + \frac{\sigma R_0^2 k^2}{2\rho_l R_0^3}\left(1 - k^2 R_0^2 \right) \right]^{1/2}. \tag{12.15}$$

When the viscosity $\mu = 0$, this agrees with the result of Bogy's one-dimensional analysis (1978a) in the temporal limit where $u_0 = 0$. Nondimensionalization of Eq. (12.15) yields an equation containing the Ohnesorge number Oh with $L = R_0$.

Viscosity tends to damp any oscillation causing an exponential decay. The decay rate increases (because of increasing velocity gradients) as the wavelength increases with increasing viscosity. The breakup length and time also increase.

STREAM PINCH-OFF. The local behavior of the liquid stream at the point in space and time where the thickness goes to zero thickness is an important matter of current research. The large fractional change in thickness results in large velocity gradients. Viscosity becomes important locally; by means of the viscous forces, singularities that can appear in the inviscid solution may not manifest.

Eggers (1997) presents a thorough review of the pinch-off behavior of a round capillary jet. He presents the local characteristic time and length as

$$t_c = \rho_l^2 v_l^3 / \sigma^2, \tag{12.16}$$

$$l_c = \rho_l \, v_l^2 / \sigma, \tag{12.17}$$

where v_l is the kinematic viscosity of the liquid. Nondimensionalization in terms of these quantities would aid in the development of self-similar solutions. These characteristic dimensions result from a balance of surface tension, viscous, and inertial forces. In particular, the Reynolds number and Weber number based on l_c can each be set to unity. The characteristic velocity in these two numbers can be defined as l_c/t_c. As the liquid stream comes closer to the zero-thickness point, we can expect that continuum theory will become invalid and intermolecular forces will increase in importance. The final stream pinch-off can be predicted only if intermolecular forces are properly taken into account. Although Eggers' analysis was developed for the round jet, the discussed concepts should be applicable to the liquid sheet as well.

12.3.6 Cavitation

Cavitation can be important in high-pressure atomization. Recent experimental studies by Tamaki et al. (1998, 2001), Hiroyasu (2000), and Payri et al. (2004) show that the occurrence of cavitation inside the nozzle makes a substantial contribution to the breakup of the exiting liquid jet. In the traditional criterion, cavitation occurs when the pressure drops below the breaking strength of liquid (P_c), which in an ideal case is the vapor pressure at local temperature. Winer and Bair (1987) and, independently, Joseph (1998) proposed that the important parameter in cavitation is the total stress that includes both the pressure and viscous stress. Kottke et al. (2005) conducted an experiment on cavitation in creeping shear flow where the reduction of hydrodynamic pressure does not occur. However, because of high shear stress they observed the appearance of cavitation bubbles at pressures much higher than vapor pressure. Archer et al. (1997) observed a drop in the shear stress in the start-up of a steady shearing flow of a low-molecular-weight polystyrene that was due to opening of bubbles within the flow at stress equal to 0.1 MPa.

To study the mechanisms by which cavitation enhances the breakup process, Dabiri et al. (2007, 2008) developed numerical models for aperture and orifice flows by using a finite-volume method for integration and a level-set formulation to track the interface and model the surface tension. The approach requires evaluation of the viscous stress because cavitation depends on molecular-related properties. In particular, averaged values for turbulence cannot be used.

Flow of a liquid through an aperture is a free-jet problem that offers an analytic solution for the flow field. It was shown by Dabiri et al. (2008) that, for a viscous potential flow (VPF), the constant-speed condition on the free surface leads to zero normal viscous stress on the free surface, and hence satisfies the boundary condition of viscous flow as well. The full Navier–Stokes equations for the aperture flow were solved numerically by Dabiri et al. for Reynolds numbers Re between 1 and 1000 and Weber number We between 10 and 1000. By use of the theory of viscous potential flow, the viscous stresses could be found, and by use of the total-stress criterion for cavitation, the regions that are vulnerable to cavitation are identified and the results are compared with the solution of the VPF. For high Reynolds numbers, solutions are similar except in boundary layers. An elaboration on VPF is given by Joseph et al. (2007).

Dabiri et al. (2007) conducted a numerical simulation of two-phase incompressible flow in an axisymmetric geometry of the orifice. The orifice has a rounded upstream corner with radius r and a sharp downstream corner with length-to-diameter ratio (L/D) between 0.1 and 5. Figure 12.3 shows an example of solution for the velocity and pressure fields for Re = 2000 and We = 1000. The cavitation number is defined as $K = \frac{p_u - p_d}{p_d - p_c}$ where p_u and p_d are upstream and downstream pressures, respectively, and p_c is the critical pressure. In our nondimensional formulation, the upstream and downstream pressure values, the critical pressure, and therefore the K value are not required for obtaining the Navier–Stokes solution. The total stress including viscous stress and pressure has been calculated in the flow field and, from there, the maximum principal stress is found (T_{11}). The total-stress criterion for cavitation is applied to find the regions where cavitation is likely to occur and compared with those of the traditional pressure criterion. Results are shown in Fig. 12.4. For each problem, K has a value representative of that problem, say $K = K^*$. In Fig. 12.4, the volume enclosed by the K^* contour is the volume where the cavitation criterion is met and the flow is therefore vulnerable to cavitation. It is observed that the viscous stress has significant effects on cavitation, especially for nozzles with larger length-to-diameter ratios. Namely, for any given value of K, the volume

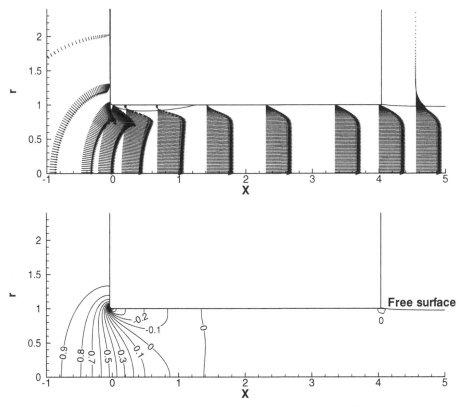

Figure 12.3. Orifice flow for Re $=$ 2000, We $=$ 1000: (a) Velocity profiles for flow, (b) pressure contours.

susceptible to cavitation is larger under the total-stress criterion. As explained, the method is not restricted to any value of K. The figure portrays only a sample of the infinite number of contours that can be obtained by postprocessing of the Navier–Stokes solutions.

Figure 12.5(a) shows the threshold value of cavitation number above which cavitation occurs in the nozzle. The effect of geometry and occurrence of hydraulic flip in the orifice on the total stress is also studied. Here, flow in nozzles with different radii of curvature at the inlet corner is considered. r/D is varied between 0.01 and 0.04 while other parameters of the flow and domain are kept constant. Figures 12.5(b) and 12.5(c) show the threshold value of K versus r/D. For both Reynolds numbers of 1000 and 2000, the K_{th} increases as the r/D increases. This is expected because the larger the radius of curvature, the smaller the increase in velocity and drop in pressure, so the less chance of cavitation.

BUBBLES IN THE FLOW. The growth and collapse of cavitation bubbles in the nozzle flow are simulated by a one-way-interaction model. That is, the liquid flow affects bubble growth and collapse but the bubble size change does not modify the liquid flow in this preliminary calculation. We expect that a future calculation with a two-way-interaction will result in moderation of the maximum bubble size significantly. This analysis provides the effects of different flow parameters such as Reynolds

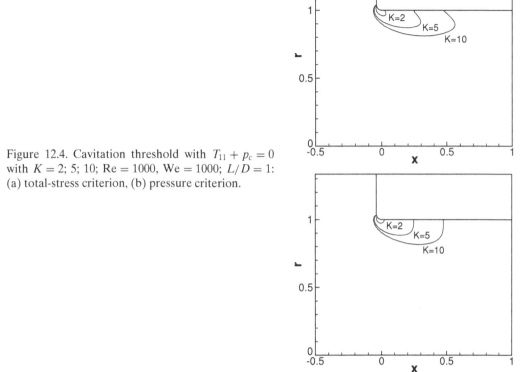

Figure 12.4. Cavitation threshold with $T_{11} + p_c = 0$ with $K = 2;\ 5;\ 10$; $\mathrm{Re} = 1000$, $\mathrm{We} = 1000$; $L/D = 1$: (a) total-stress criterion, (b) pressure criterion.

number, Weber numbers and size of nucleation sites on the cavitation character. In the one-way interaction it is assumed that a nucleation cite that could be represented as a submicrometer-size bubble is moving with the local velocity of flow, and as it reaches the low-pressure regions, near corners, it expands and creates a cavitation bubble, and later, by moving to higher-pressure regions, it will collapse. Also, it is assumed the existence of the bubble will not significantly change the flow field around it. In the future, this model will be expanded to consider a two-way-interaction between bubbles and the flow, where the change in the flow field that is due to bubble growth and collapse will also be modeled. Starting from a point in the flow, one can find the streamline and position of particle with time and pressure felt by particle. Then this pressure is used to solve the equation governing size of the bubble. Figure 12.6 shows the variation of the pressure coefficient and normalized bubble size through an orifice flow with $\mathrm{Re} = 8000$, $\mathrm{We} = 5620$, $r/D = 4.55 \times 10^{-3}$, and $(p_{\mathrm{supply}} - p_{\mathrm{exit}})/p_{\mathrm{exit}} = 2.727$. In the low-pressure region near the corner, bubbles grow and then collapse. The bubble first grows followed by collapse in size with some oscillation. A lag occurs in the bubble size variation; i.e., maximum size does not occur at minimum pressure because capillary pressure and viscous normal stress are considered in the balance at the bubble–liquid interface. Also, phase change is considered as the bubble remains saturated with vapor of the liquid in the isothermal flow.

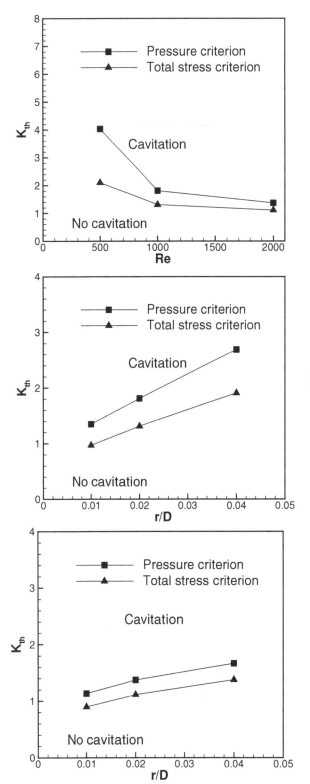

Figure 12.5. Threshold values of K above which orifice cavitation occurs; $L/D = 2$: (a) corner $r/D = 0.02$; (b) Re $= 1000$; (c) Re $= 2000$.

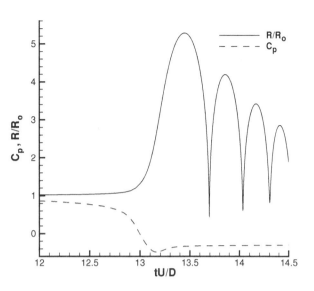

Figure 12.6. Pressure vs. the x component of bubbles position in the orifice; bubble size relative to its original size vs. the x component of bubbles position in the orifice. $Re = 8000$, $We = 5620$, $r/D = 4.55 \times 10^{-3}$, and $(p_{\text{supply}} - p_{\text{exit}})/p_{\text{exit}} = 2.727$.

12.4 Planar Sheet Analyses

The dynamics of sheets of fluid was studied as long ago as 1833 by Savart who produced and analyzed bell-like or flat axisymmetric sheets, created by a disk-shaped obstruction in the path of a cylindrical water jet, or by impingement of two jets. Dorman (1952), Fraser and Eisenklam (1953), and, later, Dombrowski and Fraser (1954) were the first to describe the breakup and drop formation of plane fan sheets. Three modes of sheet disintegration, referred to as *rim, wave,* and *perforated-sheet* disintegration were identified. In the rim mode, forces created by surface tension cause the free edge of a liquid sheet to contract into a thick rim, which then breaks up by a mechanism corresponding to the disintegration of a free jet. This mode of disintegration is most prominent where the viscosity and surface tension of the liquid are both high. In the perforated-sheet disintegration mode, holes appear in the sheet and are delineated by rims formed from the liquid that was initially included inside. These holes grow rapidly in size until the rims of adjacent holes coalesce to produce ligaments of irregular shape that finally break up into drops of varying size. Disintegration through the generation of wave motion on the sheet and in the absence of perforations is referred to as wave disintegration, whereby areas of the sheet, corresponding to half or full wavelengths of the oscillation, are torn away before the leading edge is reached. The relative importance of the different modes can greatly influence both the mean drop size and the drop size distribution (Lefebvre, 1989). Rim-sheet disintegration is found to be important at low relative velocities, whereas wave-sheet disintegration due to aerodynamic interaction with the surrounding gas dominates at most injection pressures used in practice. Our discussions will focus on the wave disintegration mode.

12.4.1 Linear Theory

Hagerty and Shea (1955) and Squire (1953) first considered temporal (spatially periodic) behavior on an infinite liquid sheet at a low gas-to-liquid-density ratio. Rangel and Sirignano (1991b) and Sirignano & Mehring (2000) extended the analysis to a high-density ratio. In the limit of large radius r ($r \to \infty$), the equations for an annular sheet yield the governing equations for a planar liquid sheet (fluid 1) in a surrounding ambient gas (fluid 2 = fluid 3) in Cartesian coordinates. Considering two-dimensional disturbances only (i.e., $\partial \dots /\partial\theta = 0$) and neglecting the effects of viscosity and gravity in both the liquid and the gas phase, linearized treatment (Sirignano and Mehring, 2000) yields the velocity potential:

$$\phi = (A \cosh ky + B \sinh ky)e^{i(kx-\omega t)} \tag{12.18}$$

for the liquid sheet of thickness $2a$, where $A = 0$ for the antisymmetric mode and $B = 0$ for the symmetric mode and y and x denote the directions perpendicular and parallel to the main flow direction of the sheet, respectively. Furthermore,

$$\phi = A_1 e^{\pm ky}e^{i(kx-\omega t)}, \tag{12.19}$$

where A_1 assumes a different value for the gas region $y < -h$ below the liquid than for the gas region $y > h$ above the liquid. The plus sign is used in the exponent if $y < -h$ whereas the minus sign applies if $y > h$.

The dispersion relation can be determined with the preceding relations plus the interface conditions:

$$\frac{\omega}{ku_0} = \frac{1}{2}\frac{\rho - C}{\rho + C} \pm \frac{[2\pi/\text{We}_1(C+\rho) - C\rho]^{1/2}}{\rho + C}$$

$$= \frac{1}{2}\frac{\rho - C}{\rho + C} \pm \frac{[kh/\text{We}_2(C+\rho) - C\rho]^{1/2}}{\rho + C}, \tag{12.20}$$

with $\text{We}_1 = \rho_l u_0^2 \lambda/\sigma$ and $\text{We}_2 = \rho_l u_0^2 h/\sigma$. $C = \coth kh$ for the symmetric mode, and $C = \tanh kh$ for the antisymmetric mode. Here u_0 is the relative velocity between the liquid and the gas. ω will have an imaginary part for sufficiently low σ, sufficiently high $\rho = \rho_g/\rho_l$, and sufficiently high relative velocity. When $\sigma = 0$, we have a pure Kelvin–Helmholtz behavior, whereas for $u_0 = 0$, or $\rho = \rho_g/\rho_l = 0$, we have pure capillary waves.

The two solutions obtained from Eq. (12.20) for ω/k denote two different waves travelling in opposite directions along the sheet. From Eq. (12.18), note the fluid sheet is unstable if ω/k in Eq. (12.20) becomes complex, i.e., when the Weber number We_1 or We_2 exceeds its critical value:

$$\text{We}_{1,c} = 2\pi\left[\frac{1}{\rho} + \frac{1}{C}\right], \quad \text{We}_{2,c} = kh\left[\frac{1}{\rho} + \frac{1}{C}\right]. \tag{12.21}$$

The growth rate of an initial disturbance is given by the imaginary part of the disturbance frequency ω provided by Eq. (12.20), i.e.

$$\text{Im}\left[\omega/(ku_0)\right] = \pm\sqrt{\frac{C\rho}{(C+\rho)^2} - \frac{kh}{\text{We}_2}\frac{1}{C+\rho}}, \tag{12.22}$$

where, according to Eq. (12.18), the "+" sign denotes the solution with unstable behavior and Im means "the imaginary part of." The results obtained from the prescribed linear analysis can be summarized as follows:

1. Only two "principal" modes of sheet distortion, i.e., sinuous or varicose (dilational) modes, develop on planar sheets. The varicose waves are dispersive whereas the sinuous waves are generally dispersive but become nondispersive, as kh and ρ both tend toward zero.

2. The stability and growth rates of any sinuous or varicose disturbance depend on the Weber number $\mathrm{We}_2 = \rho_l u_0^2 h / \sigma$ and the density ratio $\rho = \rho_g / \rho_l$. See the results for growth rate in Fig. 12.7.

3. For all density ratios, the growth rates for both sinuous and dilational waves increase as the Weber number We_2 is increased. The maximum growth rate for the sinuous disturbances does not significantly change with changes in the density ratio ρ. However, the maximum growth rate for the dilational case increases significantly as ρ is increased. For low-density ratios, the maximum growth rate for the sinuous case is always higher than for dilational waves. As ρ is increased beyond a certain value, the maximum growth rate for dilational waves eventually overcomes the value for sinuous growth.

4. For all density ratios, there exists a region of wave numbers in which dilational waves are more unstable than the sinuous ones; the latter might even be stable in that region.

5. The disturbance wavelength at the maximum growth rate decreases as the density ratio is increased for either sinuous or dilational waves.

6. For intermediate Weber numbers and small thickness-to-wavelength ratios, sinuous waves are more unstable than dilational ones with the instability occurring at Weber numbers larger than unity and with the most unstable wave number given by

$$k \approx \rho_g u_0^2 / 2\sigma,$$

with a growth rate of

$$\omega_{\max} = \frac{1}{2} \frac{\rho_g}{\rho_l} \frac{u_0}{h} \mathrm{We}^{1/2}. \tag{12.23}$$

At low Weber numbers (near or less than 1), the varicose or dilational mode is more unstable, and at very large Weber numbers, the instability of both modes is equally important. In fact, for $\mathrm{We} < 1$, sinuous waves even become neutrally stable, so that in that case only varicose waves can be aerodynamically unstable. The Weber number at which the instability of sinuous waves starts dominating the varicose instability depends on the gas-to-liquid density ratio ρ_g / ρ_l and increases with increasing values of ρ_g / ρ_l. At low gas density, the sinuous wave dominates over a wide range of Weber numbers. However, at high ambient-gas densities (e.g., $\rho_g \geq 0.25 \rho_l$), the dilational mode has the faster growth rate over a very wide range of Weber number.

VISCOUS EFFECTS. Temporal and spatial stability analyses show that liquid viscosity plays a dual role in the stability of liquid sheets. At low Weber numbers, viscosity introduces an additional mode of instability, which (under certain conditions) can grow faster than the aerodynamic instability. However, at high Weber numbers, linear theory shows that aerodynamic instability always dominates and

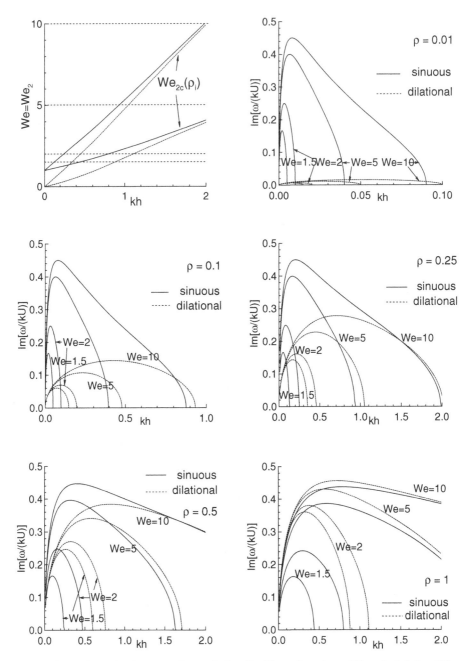

Figure 12.7. Dimensionless growth rate $\text{Im}[\omega/(ku_0)]$ as a function of $kh = 2\pi h/\lambda$ for different values of $\text{We} = \text{We}_2$ and $\rho = \rho_g/\rho_l$.

liquid viscosity always reduces the disturbance growth rates and shifts the dominant disturbances to longer wavelengths (Li and Tankin, 1991; Li, 1993). Differences between predictions for the most unstable wavelength, i.e., wavelength with the maximum growth rate, on two-dimensional viscous sheets and observed dominant waves on plane and fan sheets were explained by disturbances forced onto the sheet internally by flow disturbances even at subcritical Reynolds numbers (Hashimoto and Suzuki, 1991) or externally by nozzle vibrations at some natural frequency (Crapper et al., 1973; Crapper and Dombrowski, 1984). Sheet thinning in fan sheets and conical sheets and the magnitude of initial disturbance amplitudes of unstable waves were also used to argue that disturbances at the optimum wavelength, i.e., the wavelength with maximum growth rate, as predicted by linear two-dimensional theories, do not necessarily dominate the instability or breakup of practical liquid sheets.

12.4.2 Fan Sheets

A fan sheet or fan jet involves a thin liquid sheet that, in the undisturbed state, has a plane as the midsurface between the two gas–liquid interfaces of the sheet. Furthermore, the streamlines in this plane are straight lines diverging from a common point; the name "fan sheet" provides a description of the geometry. The velocity along these radial streamlines is found to be approximately constant with distance from the origin (Clark and Dombrowski, 1972). The curvature of the nearly planar gas–liquid interfaces is insufficient to cause pressure gradients and velocity change in the steady case. So it can be assumed from continuity that, in the steady state, the product of thickness and distance from the theoretical point of origin of the streamlines is a constant. Accordingly, the sheet thickness decreases as it flows downstream. The thickness variation of fan sheets can be described by $t = K/x$, where t is the local sheet thickness at distance x from the nozzle, and K is a thickness parameter depending on fluid viscosity, surface tension, and liquid injection pressure (Dombrowski et al., 1960).

In practice, of course, the fan sheet does not emerge from a point; it can emerge from a properly designed nozzle or be the result of two identical round jets impinging at an angle with the center plane of the fan sheet coinciding with the symmetry plane of the two-jet configuration. These impinging jets are common in liquid rocket engines. The antisymmetric mode only has been found experimentally at atmospheric pressure. So analysis focused on that mode. However, it would not be surprising to see the importance of the symmetric dilational mode increase with increases in pressure based on other experiences with planar sheets.

Dombrowski and Johns (1963) found that, for a viscous fluid in an attenuating sheet, the wave with maximum growth is not necessarily the one with the maximum growth rate, whereas, in the inviscid case, the wave of maximum growth has also the maximum growth rate through the sheet. Here, maximum growth was determined by integrating the growth rate equations over time.

12.4.3 Nonlinear Theory

More insight on the sheet breakup mechanism for the sinuous mode was provided by Clark and Dombrowski (1972). They considered a second-order temporal analysis of the aerodynamic growth of sinuous waves on nonattenuating inviscid sheets and predicted the appearance of the first harmonic dilational mode that is due to energy transfer from the fundamental sinuous mode, which subsequently leads to sheet-breakup at half wavelengths of the fundamental sinuous mode. The same observation was also made within the third-order stability analysis by Jazayeri and

Li (1996) and within the nonlinear discrete-vortex method simulations of the two-dimensional sheet by Rangel and Sirignano (1991b). In both analyses, the nonlinear sinuous growth rates were found to be less than predicted by linear theory. Also, as subsequently shown, the prescribed nonlinear coupling between sinuous and dilational capillary waves does not depend on the presence of a surrounding gas flow. Rangel and Sirignano, using a vortex-dynamics method, performed a two-dimensional, nonlinear, inviscid analysis of the temporal stability of planar sheets. Using the vortex-dynamics approach, Rangel and Sirignano (1988c) had earlier considered the nonlinear stability of thick liquid films. The liquid–gas interfaces are themselves vortex sheets in the inviscid limit. The strength of the vortex sheet can be modified because of surface tension and density discontinuity across the interface.

Rangel and Sirignano (1991b) assumed periodic spatial behavior on an infinitely long liquid stream and calculated the temporal behavior for both the sinuous and dilational modes. The method is powerful in that no restrictions on the magnitude of the gas-density-to-liquid-density ratio, on the sheet-thickness-to-wavelength ratio, or on the amplitude of the sheet distortion are required. However, the method cannot readily be extended to semi-infinite sheet configurations or to three-dimensional and axisymmetric cases. Also, the numerical method requires great care when the sheet thickness locally becomes comparable to the arc distance between discrete-vortex elements.

Figures 12.8 and 12.9 give results of Rangel and Sirignano for the sinuous and dilational modes, respectively, in the case in which gas and liquid densities are equal and the initial sheet thickness is one quarter of the wavelength of the disturbance. The sinuous case displays half-wave thinning. The theory cannot predict stream breakup because numerical errors occur as the sheet thickness approaches the length of discretization. However, the possibility of the stream breaking at half-wavelengths is indicated. At the larger gas-to-liquid density ratio, an oscillation in the waveform occurs; the sinuous shape disappears and reappears. The dilational waves can assume a "heart" shape with thinning at wavelength intervals.

THIN-SHEET ANALYSIS. A thin planar liquid sheet infinitely or semi-infinitely long in the flow direction (x) is considered. The liquid sheet is initially injected into a gas of negligible density compared with the liquid density with the undisturbed velocity u_0 and the undisturbed semithickness a. Following Kim and Sirignano (2000), the liquid is assumed inviscid, incompressible, and free of gravity force.

Define the sheet thickness,

$$\tilde{y}(x, z, t) = y_+ - y_-, \quad \bar{y}(x, z, t) = (y_+ + y_-)/2, \tag{12.24}$$

where the subscripts $+$ and $-$ denote values at the upper and lower sheet surfaces, respectively. Also, describe pressure difference across the sheet Δp and average pressure $\bar{p}$ in the sheet in relation to $\tilde{y}$ and $\bar{y}$ by using

$$\Delta p = p_+ - p_- = -\sigma \left[(f_{1+} + f_{1-}) \frac{\partial^2 \bar{y}}{\partial x^2} + \frac{1}{2} (f_{1+} - f_{1-}) \frac{\partial^2 \tilde{y}}{\partial x^2} + (f_{2+} + f_{2-}) \frac{\partial^2 \bar{y}}{\partial z^2} \right.$$

$$\left. + \frac{1}{2} (f_{2+} - f_{2-}) \frac{\partial^2 \tilde{y}}{\partial z^2} + (f_{3+} + f_{3-}) \frac{\partial^2 \bar{y}}{\partial x \partial z} + \frac{1}{2} (f_{3+} - f_{3-}) \frac{\partial^2 \tilde{y}}{\partial x \partial z}, \right.$$

$$\tag{12.25}$$

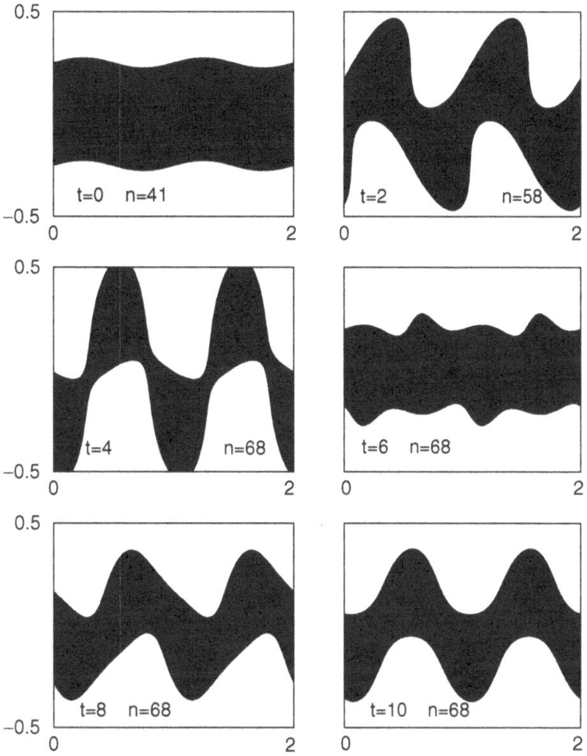

Figure 12.8. Time evolution of sinuous sheet disturbance for $\rho = 1$, $W = 0.67$, and $h/\lambda = 0.25$ (Rangel and Sirignano, 1991b).

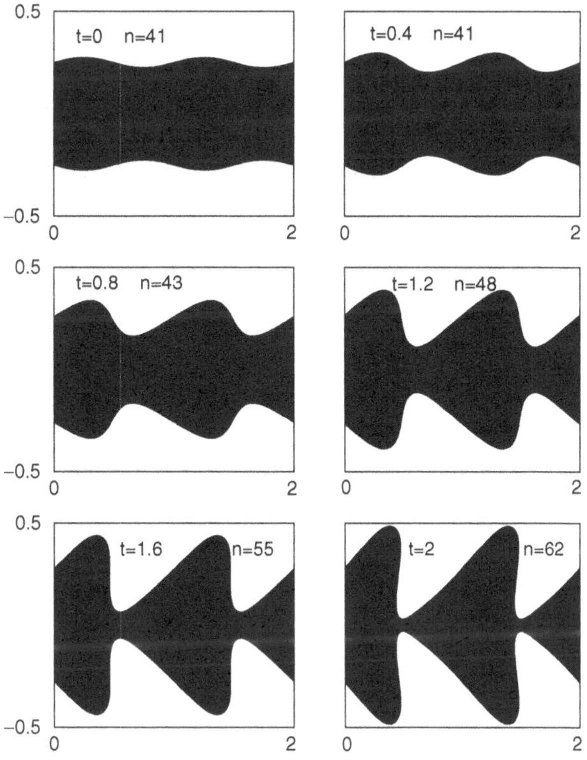

Figure 12.9. Time evolution of dilational sheet disturbance for $\rho = 1$, $W = 0.67$, and $h/\lambda = 0.25$ (Rangel and Sirignano, 1991b).

$$\bar{p} = (p_+ + p_-)/2 = -\frac{\sigma}{2}\left[(f_{1+} - f_{1-})\frac{\partial^2 \bar{y}}{\partial x^2} + \frac{1}{2}(f_{1+} + f_{1-})\frac{\partial^2 \tilde{y}}{\partial x^2} + (f_{2+} - f_{2-})\frac{\partial^2 \bar{y}}{\partial z^2} \right.$$

$$\left. + \frac{1}{2}(f_{2+} + f_{2-})\frac{\partial^2 \tilde{y}}{\partial z^2} + (f_{3+} - f_{3-})\frac{\partial^2 \bar{y}}{\partial x \partial z} + \frac{1}{2}(f_{3+} + f_{3-})\frac{\partial^2 \tilde{y}}{\partial x \partial z}, \right. \tag{12.26}$$

where f_{1+}, f_{1-}, f_{2+}, f_{2-}, f_{3+}, and f_{3-} are defined as

$$f_{1\pm} = \frac{1 + \left(\bar{y}_z \pm \frac{1}{2}\tilde{y}_z\right)^2}{\left[1 + \left(\bar{y}_x \pm \frac{1}{2}\tilde{y}_x\right)^2 + \left(\bar{y}_z \pm \frac{1}{2}\tilde{y}_z\right)^2\right]^{3/2}},$$

$$f_{2\pm} = \frac{1 + \left(\bar{y}_x \pm \frac{1}{2}\tilde{y}_x\right)^2}{\left[1 + \left(\bar{y}_x \pm \frac{1}{2}\tilde{y}_x\right)^2 + \left(\bar{y}_z \pm \frac{1}{2}\tilde{y}_z\right)^2\right]^{3/2}},$$

$$f_{3\pm} = \frac{-2\left(\bar{y}_x \pm \frac{1}{2}\tilde{y}_x\right)\left(\bar{y}_z \pm \frac{1}{2}\tilde{y}_z\right)}{\left[1 + \left(\bar{y}_x \pm \frac{1}{2}\tilde{y}_x\right)^2 + \left(\bar{y}_z \pm \frac{1}{2}\tilde{y}_z\right)^2\right]^{3/2}}. \tag{12.27}$$

For a sheet whose thickness is small compared with the wavelength of a disturbance, we consider u, $\partial v/\partial y$, w and $\partial p/\partial y$ to be nearly constant with variation of y. For the two-dimensional disturbance, Mehring and Sirignano (1999) showed that these behaviors were predicted as the leading behavior in an asymptotic representation for long wavelengths. The problem can therefore be reduced to a two-dimensional, unsteady formulation. We define average velocities $\bar{u}(x, z, t)$, $\bar{v}(x, z, t)$, and $\bar{w}(x, z, t)$ to be

$$\bar{u}(x, z, t) = \frac{1}{\bar{y}}\int_{y_-}^{y_+} u dy, \quad \bar{v}(x, z, t) = \frac{1}{\bar{y}}\int_{y_-}^{y_+} v dy, \quad \bar{w}(x, z, t) = \frac{1}{\bar{y}}\int_{y_-}^{y_+} w dy \quad (12.28)$$

Average pressure $\bar{p}(x, z, t)$ is defined in a similar manner. The conservation equations can be integrated term-by-term from y_- to y_+ and incorporated with the kinematic and dynamic boundary conditions and the preceding definitions. The results are

$$\frac{\partial \bar{y}}{\partial t} + \frac{\partial \bar{y}\bar{u}}{\partial x} + \frac{\partial \bar{y}\bar{w}}{\partial z} = 0, \tag{12.29}$$

$$\frac{\partial \bar{u}}{\partial t} + \bar{u}\frac{\partial \bar{u}}{\partial x} + \bar{w}\frac{\partial \bar{u}}{\partial z} = -\frac{1}{\rho}\left(\frac{\partial \bar{p}}{\partial x} - \frac{\Delta p}{\bar{y}}\frac{\partial \bar{y}}{\partial x}\right), \tag{12.30}$$

$$\frac{\partial \bar{w}}{\partial t} + \bar{u}\frac{\partial \bar{w}}{\partial x} + \bar{w}\frac{\partial \bar{w}}{\partial z} = -\frac{1}{\rho}\left(\frac{\partial \bar{p}}{\partial z} - \frac{\Delta p}{\bar{y}}\frac{\partial \bar{y}}{\partial z}\right), \tag{12.31}$$

$$\frac{\partial \bar{v}}{\partial t} + \bar{u}\frac{\partial \bar{v}}{\partial x} + \bar{w}\frac{\partial \bar{v}}{\partial z} = -\frac{1}{\rho}\frac{\Delta p}{\bar{y}}. \tag{12.32}$$

Equations (12.29)–(12.32) show that the number of unknowns is five ($\bar{y}, \tilde{y}, \bar{u}, \bar{v}$, and $\bar{w}$) but the number of equations is four. An additional equation is obtained by combining the kinematic boundary conditions for v_+ and v_- and by using $\bar{v} = (v_+ + v_-)/2$. Mehring and Sirignano (1999) showed that v can be expressed by a polynomial expansion in terms of y or $(y - \bar{y})$. As a consequence, v can be expressed as a linear function of y by the first-order approximation. Thus the expression $\bar{v} = (v_+ + v_-)/2$ is consistent with (12.28) by the first-order approximation:

$$\bar{v} = \frac{\partial \bar{y}}{\partial t} + \bar{u}\frac{\partial \bar{y}}{\partial x} + \bar{w}\frac{\partial \bar{y}}{\partial z}. \tag{12.33}$$

Equations (12.29)–(12.33) represent the "lubrication approximation" or "slender jet" assumption. For liquid sheets (or jets) with large amplitude distortions, such as observed close to pinch-off, this assumption needs careful reevaluation, e.g., comparison with solutions to the full system of equations (Mehring and Sirignano, 1999). For two-dimensional planar sheet distortion, (12.29)–(12.33) agree with the results of Mehring and Sirignano whereby $\partial(\ )/\partial z = 0$ and $\bar{w} = 0$. The system of equations has been solved by finite-difference computations using the Richtmyer splitting of the Lax–Wendroff method.

Considering only two-dimensional disturbances, the reduced-dimension approach yields, after linearization, two partial differential equations governing small-amplitude dilational and sinuous capillary waves on thin two-dimensional inviscid ($\eta_1 = 0$) planar sheets, i.e.,

$$\frac{\partial^2 h}{\partial \tau^2} + \frac{\sigma a}{\rho_1}\frac{\partial^4 h}{\partial \xi^4} = 0, \tag{12.34}$$

$$\frac{\partial^2 Y}{\partial \tau^2} - \frac{\sigma}{\rho_1 a}\frac{\partial^2 Y}{\partial \xi^2} = 0. \tag{12.35}$$

Equation (12.34) describes thickness variations h (dilational waves) and Eq. (12.35) describes variations in the sheet-centerline position Y whereas the sheet thickness remains constant (sinuous waves). τ and ξ denote time and space variables in a coordinate system moving with the undisturbed liquid stream. The previous equations readily show that linear sinuous and dilational capillary waves are decoupled. A modal analysis also quickly reveals that dilational waves are dispersive, whereas sinuous waves are nondispersive. Equations (12.34) and (12.35) are both well-known equations in mechanics, governing the transverse vibration of a uniform beam and vibrations on a taut string, respectively (Graff, 1975). Within the analysis of thin liquid sheets, Taylor (1959) first derived and analyzed Eqs. (12.34) and (12.35), the latter in its steady form and for a radially expanding planar sheet.

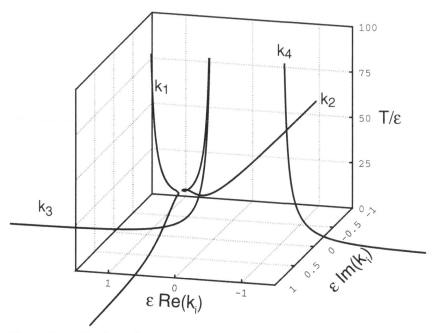

Figure 12.10. The dependence of wave numbers k_i for modulated symmetric distorting semi-infinite sheets on forcing period $T = 2\pi/\omega$ (Mehring and Sirignano, 1999).

A similar linear analysis of spatially developing dilational and sinuous capillary waves on thin planar sheets shows that there are four dilational ($k_{1,...,4}$) and two sinuous ($l_{1,2}$) waves generated if a harmonic disturbance is locally forced onto the moving sheet at a frequency ω, i.e.,

$$k_{1,2} = \frac{1 \pm (1 - 8a\,\epsilon\,\omega/u_o)^{1/2}}{4a\epsilon},$$

$$k_{3,4} = \frac{-1 \pm (1 + 8a\,\epsilon\,\omega/u_o)^{1/2}}{4a\epsilon},$$

$$l_{1,2} = \frac{\omega}{2a\,(1 \pm 2\epsilon)}, \tag{12.36}$$

with $\epsilon = 1/\sqrt{2\text{We}}$ and $\text{We} = 2\rho a u_o^2/\sigma$, where $2a$ is the undisturbed sheet thickness and u_o is the undisturbed base flow velocity of the sheet exiting from the orifice. Figure 12.10 illustrates the dilational-mode wave numbers k_i ($i = 1, \ldots, 4$) as a function of the time period $T = 2\pi/\omega$ of the imposed sheet modulation.

Group-velocity arguments analogous to those used by Bogy (1978a, 1979) for slender cylindrical liquid jets show that, of the four dilational waves generated by a harmonic dilational forcing at the nozzle exit, only two (i.e., k_1 and k_3) will appear downstream from the nozzle, resulting in the superposition of two waves of similar wavelengths or a single travelling wave superimposed onto a wave with an exponentially decaying envelope in the downstream direction. For sinuous sheet modulations, only two wave numbers are generated, both of which will have positive group velocities relative to the nozzle exit if $\epsilon < 0.5$ or We > 2, resulting in a beat in the envelope of two sinuous waves travelling in the downstream direction. However, for We < 2, only wave number $l_1 = \omega / [2a(1 + 2\epsilon)]$ is expected to appear downstream from the nozzle. The appearance of dilational and sinuous capillary waves on planar sheets is independent of the imposed modulation frequency, because the group velocity for each wave number has either positive or negative values for any forcing frequency. Linear analysis also shows that there exists no forcing frequency, resulting in exponential growth of sinuous or dilational sheet disturbances forced at the nozzle exit.

The general nonlinear dimensionally reduced system of equations for sinuous and dilational capillary waves on thin planar two-dimensional (inviscid) liquid sheets in a passive ambient gas or void is obtained as a special case of Eqs. (12.29)–(12.33). Temporal analysis reveals that the transverse oscillation of nonlinear sinuous travelling waves already observed by Rangel and Sirignano (1991(b)) for the case of finite-density air also appears for the zero-ambient-density case. Both vortex-dynamics simulations (Rangel and Sirignano, 1991(b)) and reduced-dimension analysis (Mehring and Sirignano, 1999; Kim and Sirignano, 2000) for thin periodically disturbed sheets show that, for the nonlinear dilational travelling wave, a temporal steepening and desteepening of the wave ("wobbling") occurs. The latter analysis also reveals the possibility of sheet instability (due to capillary effects) triggered by a small-amplitude dilational sheet disturbance with wavelength λ_s superimposed onto a dilational travelling wave with shorter wavelengths and, in particular, for $\lambda = \lambda_s / n$, where n has an integer value. This type of instability was first observed by Matsuuchi (1974, 1976).

These equations can also be used to analyze the nonlinear distortion of thin planar liquid sheets, which are modulated at the nozzle exit. See Figs. 12.11 and 12.12 for modulations of the transverse and axial-velocity components, respectively. The characteristic envelope in variations of the sheet-thickness or the sheet-centerline location in the downstream direction along the dilational or sinuous modulated sheet predicted by linear theory and described earlier is altered somewhat by nonlinear effects. As within the temporal analysis, the modulated sinuous sheet distortion excites the dilational mode as the sheet propagates downstream, leading to fluid agglomeration in the maximum deflection region of the sinuous distorting sheet, as already observed experimentally by Hashimoto and Suzuki (1991).

For both sinuous and dilational modulations, nonlinear effects result in the accumulation of fluid into lumps, connected by threads of fluid that show increased thinning and eventually break up at points close to these blobs of fluid. This observation is analogous to the case of a cylindrical jet, in which nonlinear effects are responsible for the satellite droplets between the larger droplets. Drop-size predictions from these nonlinear two-dimensional results can again be obtained by use of the simplified assumption that the larger two-dimensional liquid columns and the thin threads between them will disintegrate according to Rayleigh's theory, assuming that no further breakup of these ligaments occurs in the longitudinal direction, i.e., in the downstream direction. Clearly, in order to predict accurately the overall sheet breakup process, a three-dimensional theory is needed.

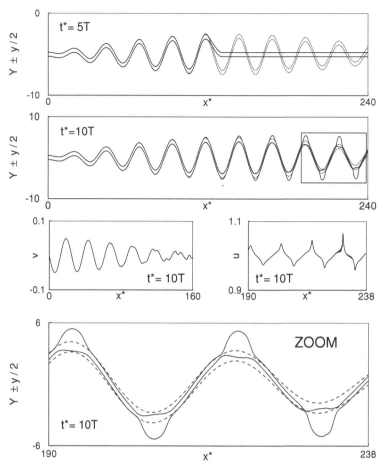

Figure 12.11. Sinuous distortion of semi-infinite planar liquid sheet harmonically forced at $x = 0$ with forcing frequency ω; We = 1000; $T = 2\pi/\omega = 25$; - - - linear limit-cycle solution at $\tau = nT$, where n is an integer; — nonlinear transient solution (Mehring and Sirignano, 1999).

NONLINEAR AERODYNAMIC EFFECTS. The nonlinear evolution of a thin planar liquid sheet under the influence of capillary and aerodynamic effects and with disturbances in the main flow direction (x direction) only has been modeled (Mehring and Sirignano, 2003) by use of their reduced-dimension approach to describe the thin planar sheet and a BEM formulation (Hilbing et al., 1995) for the inviscid incompressible gas phase. The BEM in combination with the unsteady Bernoulli equation is used to determine the instantaneous gas pressure at the liquid–gas interfaces, which is needed within the reduced-dimension equations governing the liquid phase. The use of a discrete BEM for the gas phase allows, in principle, the consideration of more practical applications in which sheets of liquid are injected into a gaseous flow field with its own physical constraints or boundary conditions. The latter is of par-ticular importance with regard to the complicated flow fields within and/or around

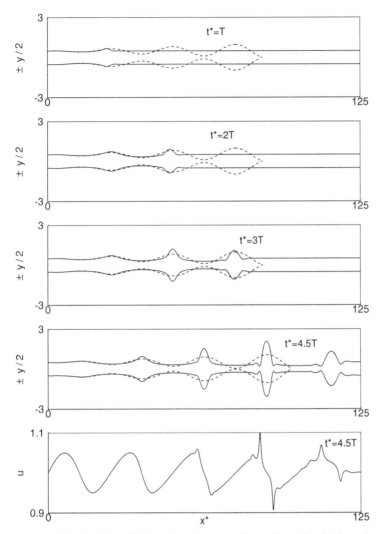

Figure 12.12. Dilational distortion of semi-infinite planar liquid sheet harmonically forced at $x = 0$ with forcing frequency ω; We = 500; $T = 2\pi/\omega = 25$; - - - linear limit-cycle solutions at $\tau = nT$, where n is an integer; — nonlinear transient solution (Mehring and Sirignano, 1999).

fuel-injection elements or atomizers used for spray combustion purposes, e.g., prefilming airblast atomizers.

The model is applicable for both dilational and sinuous modes. However, only dilational sheet distortions have been considered by Mehring and Sirignano. The solution procedure for the liquid-phase assumes that the sheet is disturbed only locally and that the disturbance is prescribed by the initial conditions. There is no time-dependent forcing imposed onto the sheet, and the sheet remains undisturbed at the boundaries of the computational domain.

Figure 12.13 illustrates two typical results generated with the new model. Both solutions are for the case $We_g = 2$ with zero liquid-stream velocity and a local

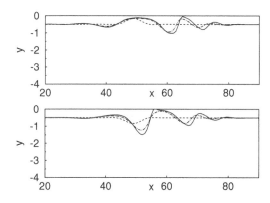

Figure 12.13. Dilational sheet distortion that is due to parallel gas flow and local sheet thickness disturbance: (a) local sheet depression, (b) local sheet thickness increase. Symmetry axis at $y = 0$; $\mathrm{We}_g = 2$; $\rho_g/\rho_l = 1.2 \times 10^{-3}$; - - - $\tau = 0$; ···· $\tau = 24$; — $\tau = 28$. [$\tau = t/t_{\mathrm{ref}}$, $t_{\mathrm{ref}} = h\sqrt{\sigma/(2\rho_l h)}$] (Mehring and Sirignano, 2003).

thinning or thickening of the sheet initially, i.e.,

$$y_\pm (x, t = 0) = \pm 0.5 \text{ for } \left[x > x_0 - \frac{\Delta x}{2} \right],$$

$$y_\pm (x, t = 0) = \pm 0.5 + 0.35 \theta \left\{ 1 - \cos \left[\frac{2\pi}{\Delta x} (x - \Delta x) \right] \right\}$$

$$\text{for } \left[x_0 - \frac{\Delta x}{2} < x < x_0 + \frac{\Delta x}{2} \right],$$

$$y_\pm (x, t = 0) = \pm 0.5 \text{ for } \left[x < x_0 + \frac{\Delta x}{2} \right],$$

and $u(x, t = 0) = 0$. Here $x_0 = 50$ and $\theta = \pm 1$ for local increasing of the sheet thickness and $\theta = \mp 1$ for local thinning of the sheet. As seen from Fig. 12.13, the gas stream flowing along the disturbed liquid–gas interface causes gas-pressure fluctuations that result in significant sheet distortion and that eventually cause the liquid stream to break. Steepening of the originally symmetric disturbances is observed on the "valley" or "hillside" opposite to the direction from which the gas flow impacts the disturbances. For the case with locally increased sheet thickness, the breakup time is shorter than predicted for the similar case but with locally reduced sheet thickness. In the latter case, the sheet tends to break "simultaneously" at two locations (creating a free liquid column), whereas in the former case the sheet breaks only once.

Note that for large values of We_g destabilizing aerodynamic effects dominate over stabilizing capillary effects, and the numerical error associated with the solution of Laplace's equation and the unsteady Bernoulli equation in the gas phase becomes more important. More accurate predictions of the sheet-disintegration process at larger values of We_g require a larger number of nodes for the BEM solution of Laplace's equation in the gas phase as well as a more accurate time-integration scheme.

THREE-DIMENSIONAL THEORY. Although liquid sheet breakup is an inherently three-dimensional problem, only a few three-dimensional or quasi-three-dimensional analyses on planar liquid sheets have been reported so far. Ibrahim and Akpan

(1996) presented a fully three-dimensional linear analysis of a plane viscous liquid sheet in an inviscid gas medium. At low Weber numbers, two-dimensional disturbances were found to always dominate the instability of symmetric and antisymmetric waves. Furthermore, when the Weber number is high, long-wave three-dimensional symmetric disturbances were found to have larger growth rates than their two-dimensional counterparts, whereas the opposite was true for antisymmetric disturbances. For short waves, both two- and three-dimensional disturbances grow at approximately the same rate. Increasing the gas-to-liquid density ratio or decreasing the Ohnesorge number was found to enhance the departure in the growth rates of two- and three-dimensional symmetric disturbances of long wavelength. Both the maximum growth rate and the dominant wave number were predicted to increase with Weber number and density ratio but decrease with Ohnesorge number. Using the reduced-dimension approach previously used for capillary waves on a two-dimensional planar sheet (Ponstein, 1959), a linear temporal analysis provides the following dispersion relations in a coordinate system moving with undisturbed (uniform) sheet velocity:

$$\omega = \frac{1}{2}\frac{\sigma}{\rho_l a}\left(k^2 + n^2\right)$$

for the dilational mode and

$$\omega = \frac{\sigma}{\rho_l a}\sqrt{l^2 + m^2}$$

for the sinuous mode, where k and l and n and m denote dilational and sinuous mode wave numbers in the axial and transverse directions, respectively. This illustrates that, in contrast to planar sheets, both sinuous and dilational capillary waves are dispersive in the presence of disturbances parallel and transverse to the main flow direction. Nonlinear temporal and spatially reduced-dimension analyses of three-dimensional distorting planar sheets are able to predict sheet breakup into droplets without further simplifications. They also provide information on the nonlinear interaction between transverse and longitudinal waves, both of which are observed experimentally.

Three-dimensional nonlinear sinuous and dilational oscillations on a thin planar sheet were considered by Kim and Sirignano (2000) who extended the Mehring and Sirignano (1999) analysis by using the reduced-dimension approach to yield an approximate system of two-dimensional, unsteady equations. They examined both temporal spatially periodic waves with prescribed initial conditions and spatially developing instabilities on sheets injected from slot orifices with exit-velocity modulation. Standing waves in the lateral direction were assumed while waves travelled in the mainstream direction. A linear analysis was also performed in order to have a basis for comparison with the numerical nonlinear results. This linear analysis was also useful in setting the boundary conditions for the spatially developing instability by determining the sign of the group velocities. In both the linear analysis and the numerical analysis, the viscous effects and the inertia of the surrounding gas were neglected.

Interesting effects occur when the streamwise and lateral wave numbers are close to each other. For the dilational temporal mode, the kinetic energy and the surface energy oscillate on both a long time scale and a short time scale. The surface energy and the kinetic energy can each be divided into two parts, one part for lateral wave motion and the other for streamwise wave motion. On the long time scale, there is no oscillation of the total kinetic energy or the total surface energy. However, energy is continually transferred from lateral motion to streamwise motion and back to lateral motion. Over the long time scale, lateral waviness in the thickness disappears whereas higher harmonics appear in the streamwise thickness wave. Then the higher harmonics disappear and the lateral waviness reappears. This continues in a cyclic manner on the long period. Similar results occur for spatially developing instabilities at low Weber numbers with small differences between the two wave numbers. At high Weber numbers, with similar wave numbers for each direction, fluid lumps are formed in the dilational mode at wavelength intervals in each direction. These fluid lumps will be the first ligaments in a capillary disintegration. Sinuous three-dimensional oscillations are found to be dispersive unlike two-dimensional oscillations. In similar fashion to the two-dimension case, the three-dimension sinuous waves involve the appearance of dilational waves through nonlinear effects. When the lateral and streamwise wave numbers are close in value, fluid lumps are formed at half-wavelength intervals in both directions for the temporal oscillation. This same phenomenon occurs at high Weber numbers in the spatially developing case.

For the sinuous case with lower Weber numbers, two waves of very different speeds and wavelengths will propagate differently. Because one wave propagates much faster than the other, there is a domain where only one wave exists.

12.5 Annular Free Films

A free film (sheet) is annular when, in the undisturbed state, its inner and outer surfaces are cylindrical and concentric. A "conical" film (sheet), as noted earlier, does not necessarily have a conical surface but rather a curved axisymmetric surface that can have an approximate cone or bell shape. For both annular and conical films, the curvatures of the outer surface and of the inner surface usually result in positive pressure jumps across the surfaces in the direction of decreasing radial position in cylindrical coordinates aligned with the axis of the flow. If the gas pressure outside of the annular or conical film equals the pressure in the gaseous core surrounded by the film, the capillary pressure causes a radial pressure gradient in the liquid that will tend to collapse the sheet toward the axis of symmetry. This can be stabilized in one of two ways: pressurization of the gaseous core or swirl of the liquid. If the swirl, pressurization, or both balance the mean capillary pressure, an annular cylindrical sheet forms. If the swirl, pressurization, or both are more than sufficient to balance the capillary pressure, the sheet radius will increase as it flows downstream, causing a "conical" film to form.

Both the annular and conical films have a surface curvature in the undisturbed state. There is one radius of curvature for a surface in the annular case in the undisturbed state. That radius of curvature is uniform over the surface. On the contrary, the conical film surface has two radii of curvature, each of which can vary with position on the surface. The conical film also differs because of the stream divergence as the flow proceeds downstream. As a consequence, the mean local thickness of the conical film decreases with downstream distance while the thickness of the undisturbed annular film is uniform. From a practical point of view, annular liquid films or conical films are of greater interest than the planar film. From a theoretical point of view, an annular film is a general two-dimensional geometry and other common geometries, such as the cylindrical jet and the planar film can be treated as special cases (Ponstein, 1959; Meyer and Weihs, 1987; Lee and Chen, 1991; Panchagnula et al., 1995; Shen and Li, 1996). For annular films or jets, a pure dilational or sinuous mode does not exist, and dominantly sinuous or dilational film disturbances are often also referred to as "parasinuous" or "paradilational" disturbances (Shen and Li, 1996).

12.5.1 Linear Theory

A linear analysis of an inviscid annular liquid film subject to aerodynamic forces and without the thin-film assumption was studied by Crapper et al., (1975), recovering Squire's result (Squire, 1953) for the limiting planar case. Crapper et al. presented the general dispersion relation for waves of wave number k on an inviscid thin annular film moving at constant velocity through a quiescent gas. Numerical solutions for the growth rate ω and the wave velocity ω/k are obtained from the dispersion relation (which is of fourth order in ω) for a range of values for the film velocity, film thickness, and cylinder radius. The authors do not address the simultaneous appearance of both modes, but merely note that, in the limit of planar films, the dispersion relation factors into two quadratic equations in ω. The growth rate and range of unstable symmetric (dilational) and antisymmetric (sinuous) modes increase with decreasing annular radius. Also, for symmetric (aerodynamically unstable) waves, the optimum frequency is independent of the film thickness, but that, for antisymmetric waves, the optimum value (i.e., wave number of maximum growth) is affected by both the radius and film thickness. The velocities of symmetric waves relative to the liquid are the same as the film velocity for all frequencies and are unaffected by the film curvature and thickness. On the other hand, antisymmetric wave velocities are always less than the film velocity and increase with increasing radius and diminishing thickness, whereby the differences between wave and film velocity decrease with increasing frequency. The increase in growth rates with increasing film curvature might be a reason for the observed shortness of conical films produced from swirl spray nozzles, compared with flat films. Note that the waves analyzed by Crapper et al. are dominated by aerodynamic effects.

A linear temporal analysis of the capillary instability for thin axisymmetric annular films (Mehring and Sirignano, 2000a, 2000b), stabilized in their undisturbed configuration by a constant pressure difference between the inner and outer fluids surrounding the film (i.e., pressure stabilization), yields

$$\frac{\partial^2 h}{\partial \tau^2} + \epsilon^2 \left\{ \frac{\partial^4 h}{\partial \xi^4} - C \left[2R \frac{\partial^2 r'}{\partial \xi^2} - \left(R^2 + 1/4\right) \frac{\partial^2 h}{\partial \xi^2} \right] \right\}$$
$$= -\frac{1}{R} \epsilon^2 \left\{ 4 \frac{\partial^2 r'}{\partial \xi^2} + 2C \left[\left(2R^2 + 1/2\right) r' - Rh \right] \right\}, \tag{12.37}$$

$$\frac{\partial^2 r'}{\partial \tau^2} = \epsilon^2 \left\{ 4 \frac{\partial^2 r'}{\partial \xi^2} + 2C \left[\left(2R^2 + 1/2\right) r' - Rh \right] \right\}, \tag{12.38}$$

where h denotes the nondimensional disturbance of the film thickness, r' is the nondimensional disturbance of the annular radius, and R denotes its nondimensional undisturbed value. Here, pressure stabilization of the annulus refers to the stabilization of the film in its undisturbed configuration by a constant pressure difference between the inside and outside of the annular film. The undisturbed film thickness $2a$ is used as characteristic length in the nondimensionalization process, and τ and ξ denote the nondimensional time and space variables in a coordinate system moving with the undisturbed film velocity. Also, $C = \left(R^2 - 1/4\right)^{-2}$ in the preceding equations, and $\epsilon^2 = 1/(2\mathrm{We}) = 1$, if $0.5\sqrt{\sigma/(\rho_l a)}$ is chosen as the characteristic velocity in the nondimensionalization process. Neglecting terms of the order of R^{-3}, the previous equations reduce to the same equations governing linear waves along a thin rod under prestress or along a string on an elastic foundation; however, the "spring constant" in Eq. (12.38) is negative. Assuming solutions of the form $e^{i(\omega\tau + k\xi)}$, the corresponding nondimensional growth rates of dilational and sinuous disturbances are decoupled and given by

$$\omega_d = \pm \epsilon\, k\sqrt{k^2 - R^{-2}}; \quad \omega_s = \pm 2\,\epsilon\,\sqrt{k^2 - R^{-2}}. \tag{12.39}$$

In other words, if terms of the order of R^{-3} can be neglected, then a dilational distortion does not couple with a sinuous deformation, and the sinuous mode leads to variations in the film thickness only through the conservation of mass equation. Also note that the previous analysis assumes that the gas-core pressure is constant and unaffected by perturbations of the liquid interfaces. Eqations (12.39) indicate that instability occurs if the disturbance wave number is smaller than the reciprocal of the undisturbed annular radius, i.e., $k < 1/R$.

In contrast to the planar geometry, not only dilational but both sinuous and dilational waves are dispersive. Linear theory also shows that, in the absence of swirl, disturbances that are not cylindrically symmetric, i.e., circumferential or azimuthal disturbances, are always stable (Dumbleton and Hermans, 1970). For the annular geometry, linear coupling between sinuous and dilational waves occurs if terms of the order of R^{-3} are considered. Note that Rcorresponds to the nondimensional ratio between annular radius and thickness of the undisturbed film. Similarly to thin planar films (Matsuuchi, 1974, 1976), a nonlinear reduced-dimension analysis for thin annular films also demonstrates the presence of a nonlinear capillary instability for the dilational (and sinuous) mode in the presence of subharmonic dilational film disturbances.

Linear reduced-dimension spatial analysis shows that, for annular films modulated at the nozzle exit, capillary dilational wave propagation is similar to that observed on thin planar films.

Only two of the four dilational waves obtained from the dispersion relation, i.e., $k_{2,3}$, will appear on the film for any forcing frequency. For We > 2, both sinuous mode waves that are solutions to the dispersion relation will appear even though one of them has negative group velocity in some range of the forcing frequency; for We > 2, linear theory predicts that only one of the sinuous mode waves appears downstream from the nozzle. For sufficiently large Weber numbers, i.e., We < 2, and stable films; the film distortion with a beat in the envelope of the film-centerline location or the film thickness (depending on whether sinuous or dilational modulations are forced onto the film) is analogous to the one observed on modulated planar films. Sheet breakup or collapse might occur if the beat amplitude in the variation of film thickness or annular centerline location is sufficiently large. However, in contrast to planar films and because of the nonzero second radius of curvature, there are now ranges of forcing frequencies for which dilational and sinuous film modulations lead to exponential growth in the film thickness disturbance or the film-centerline disturbance downstream from the nozzle.

12.5.2 Nonlinear Theory

Nonlinear analyses of annular liquid films were presented by Lee and Wang (1986, 1989, 1990) and Ramos (1992), who considered the formation of closed cylindrical liquid shells through the collapse of annular liquid films. The analyses assumed the liquid layer to be a structureless sheet and neglected axial pressure gradients, omitting the possibility for dilational waves. Time-independent boundary conditions were specified at the nozzle exit. Boundary conditions were specified for the film thickness, the annular radius, and the axial- and transverse-velocity components at the nozzle exit. Wave-propagation properties at this boundary were not taken into consideration. Ramos concluded that the convergence length and the volume enclosed by annular films increase as the Froude and Weber numbers, the pressure coefficient (i.e., the nondimensional pressure difference across the film), and the nozzle exit angle are increased, but decrease as the thickness-to-radius ratio at the nozzle is increased. Depending on the balance between pressure forces and surface-tension forces, the annular liquid jets remain cylindrical (if the nozzle exit angle is zero) and can converge or not converge.

The thin-sheet assumption leads to a reduced-dimension subsequently approach for axisymmetric annular and conical (swirling or nonswirling) sheets, as shown (Mehring and Sirignano, 2000a, 2000b, 2001). The problem is cast in cylindrical coordinates. We can assume an analytical behavior of the governing equations away from $r = 0$, as a function of r. Also, assuming that the sheet thickness is small compared with the streamwise disturbance wavelength, it is consistent to consider v_z and v_θ to be nearly constant and v_r and p to be linearly varying with r. It is convenient, therefore, to reduce the problem to a one-dimensional, unsteady formulation by integrating Eqs (12.1)–(12.4) over the sheet thickness.

Mehring and Sirignano (1999) showed that, for the case of a planar sheet, the assumptions used for the velocity and pressure profiles across the sheet agree with the lowest-order expansion of the full two-dimensional problem in terms of $(y - \bar{y})/\lambda$, where y denotes the direction perpendicular to the undisturbed sheet, $\bar{y}(x, t)$ is the instantaneous location of the sheet centerline in the y direction, and λ is the wavelength of a disturbance in the x direction. In other words, for the planar case, the reduced-dimension equations are exact in the limit where the ratio between sheet thickness and λ equals zero.

For the steady-state annular liquid membranes with sinuous distortions, the analogy between integral (control-volume) formulation and rigorous Taylor series expansions was demonstrated by Ramos (1996). The integral representation used

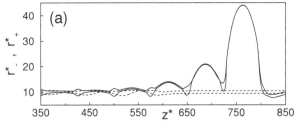

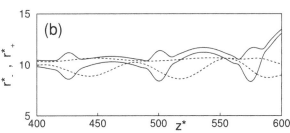

Figure 12.14. Linear (- - -) and nonlinear (—) distortions of dilationally modulated and linearly stable film at breakup: (a) interface location, (b) nondimensional axial-velocity component [We = 150, $R = 20$, $T = 2\pi/\omega = 100$]. Nozzle exit at $z^* = 0$.

by Mehring and Sirignano (2000a, 2000b) for nonswirling annular sheets follows the Ramos (1992) analysis for the steady version of the configuration but extends it to the unsteady cases of both sinuous and dilational sheet distortions.

The limitations of the Mehring and Sirignano formulation, in particular during pinch-off (when short-wavelength disturbances cannot be neglected), were addressed. It is noted, however, that, despite these constraints, the usefulness of the approach has been demonstrated conclusively for planar sheets (Mehring and Sirignano, 1999), by comparison with accurate two-dimensional vortex-dynamics simulations (Rangel and Sirignano, 1991b). Using this reduced-dimension analysis, Mehring and Sirignano provided time-dependent solutions for nonlinearly distorting, initially undisturbed axisymmetric annular liquid films modulated at the nozzle exit. Boundary conditions were specified according to wave-propagation information obtained from the corresponding linear boundary-value problem. Linearly stable or unstable dilationally and sinusoidally modulated films were also found to be nonlinearly stable or unstable, respectively. For stable dilationally modulated films, nonlinear effects lead to an increase in breakup time and breakup length (Figure 12.14).

The same observation is made for linearly stable dilationally modulated films for which large beat amplitudes result in film breakup. Strong nonlinear coupling between the dilational and sinuous modes might also lead to the formation of large gas bubbles between the fluid rings if the gas core is assumed to have constant pressure.

For linearly unstable sinuous modulated annular films, nonlinear effects always decrease film-collapse lengths and time. For linearly stable sinuous modulations, they result in the agglomeration of fluid into liquid rings that for small values of R might have considerably different cross-sectional areas and circumferential lengths along the film. Depending on the breakup length and the wavelength of the beat envelope in the variation of the annular radius, the disintegrating annulus will show the presence of satellite rings between the larger liquid rings. The variation of film breakup or collapse length and time with a variation of Weber number We, annular radius R, forcing period T, or disturbance amplitude A depends on the location in parameter space (We, R, T, A).

12.5.3 Effect of Swirl

A linear temporal analysis of three-dimensional swirling inviscid annular films with inner and outer gas flows of different velocities was conducted by Panchagnula et al. (1996). For small Weber numbers inner gas flows lead to significantly faster growing instability modes than outer gas flows at the corresponding Weber number. For large Weber numbers, inner and outer gas flows have the same effects. In the absence of swirl, the disturbance mode with maximum growth rate has always a circumferential mode number of zero. For nonzero axial Weber numbers, an increase in swirl increases the range of unstable axial and circumferential modes as well as their growth rates. It also increases the axial wave number with maximum growth. For large-enough swirl Weber numbers, the fastest-growing disturbance was predicted to have nonzero circumferential but zero axial wave number.

An extension of the prescribed reduced-dimension analysis for annular non-swirling films to swirling annular films allows for the study of the influence of the previously mentioned constant-gas-core-pressure assumption on the overall film distortion; it is assumed that the undisturbed annular film is now stabilized by a constant swirl. It also allows for the study of film divergence if the angular momentum at the nozzle exit exceeds the amount required to stabilize the undisturbed annular configuration.

Neglecting terms of the order of R^{-3}, the linear spatial stability analysis for capillary waves on swirling annular films shows that the stability limits for the forcing frequency ω are larger than in the corresponding pressure-stabilized cases. Also, dispersion relations for dilational and sinuous waves cannot be obtained separately and sheet distortion is characterized by predominantly dilational or predominantly sinuous waves. For the predominantly sinuous mode, no spatial instability is observed if the film is swirl stabilized rather than pressure stabilized. The argument here is that the stabilizing effect of the centrifugal forces due to swirl exceeds the destabilizing capillary forces if variations of the annular radius occur. This is in contrast to the pressure-stabilized annular film in which both capillary forces and pressure forces, exerted on the film by the gas core, destabilize the thin sinusoidally deformed annular film. Spatial stability analysis shows that, for sinuous modulated swirl-stabilized films, exponential solutions exist only for We < 2. However, the group velocities of the corresponding waves are negative, so that these waves are not expected to appear downstream from the nozzle. Pressure-stabilized films with modulated sinuous mode disturbances might be unstable only if We > 2.

For swirling annular films with inertia and surface tension only, energy is conserved with energy exchange between potential (surface) energy and kinetic energy (because of angular and radial momentum). However, with an imposed pressure difference, energy exchange with the ambient gas takes place, which might render the film unstable.

12.6 "Conical" Free Films

Similar to the prescribed oscillatory behavior of cylindrical liquid layers with excess swirl in a passive ambient gas, the spatial development of annular films exiting from a nozzle or atomizer, with excess swirl, into a void, is characterized by film divergence close to the nozzle exit and subsequent oscillatory variations of the annular radius in the downstream direction. Referring to the diverging part of the film directly at the nozzle exit, a spatial analysis of swirling "conical" films can be provided.

Mehring and Sirignano (2001) obtained steady-state nonlinear solutions for the pure initial-value problem of a spatially developing swirling axisymmetric annular film exiting from a nozzle or atomizer (into a passive ambient gas) for various sets of boundary conditions at the nozzle exit. Similar steady-state analyses were also presented by Yarin (1993) for the case of swirling liquid membranes and by Ramos (1992) for nonswirling thin-liquid films. Within those analyses, the authors specified boundary conditions only at the location where the film exits the nozzle or atomizer.

For the steady-state problem, the specification of the first- and second-order derivatives of the film thickness and the film-centerline location at the nozzle exit is a delicate matter, because it is not possible to determine a priori the fate of the film downstream, i.e., with respect to the encounter of singular points that render the existence of a steady-state solution impossible for the specified combination of nozzle exit conditions.

For the unsteady transient or time-periodic problem, in which capillary waves are generated on the film because of some forcing conditions imposed along the film, constraints that apply to the propagation of these waves and their energy facilitate the specification of boundary conditions. For example, for the time-periodic linear problem, the number of boundary conditions to be prescribed at the inflow and outflow planes of the considered spatial domain will depend, as already discussed for planar films, on the group velocities or energy-propagation characteristics associated with the (capillary) waves travelling along the film. For example, if energy is imposed onto the semi-infinite film only at the nozzle exit, then capillary waves associated with energy transport upstream, i.e., with group velocities directed in the upstream direction, have to be excluded from the solution of the linear problem. The rejection of (linear) solutions associated with such wave numbers effectively results from the use of the Sommerfeld radiation (boundary) condition at infinity, i.e., the requirement that energy has to be propagated downstream (or outwards) at infinity.

Richer in content and of more practical interest than time-periodic solutions are transient solutions to the initial- and boundary-value problems. Transient linear solutions can be obtained numerically or analytically by use of Fourier–Laplace transforms to obtain the dispersion relation in terms of complex wave number and frequencies. A numerical solution also allows ready access to the solution of the nonlinear problem. Here, the specification of boundary conditions is even more problematic as wave reflection at the nozzle exit might have to be considered. The latter occurs, for example, in the development of an absolute film instability triggered by some downstream disturbance on the liquid film at low Weber numbers. It remains a challenge to specify boundary conditions (in number and type) that result in an accurate representation of wave reflection within a simplified mathematical model for the overall film dynamics and provide reasonable agreement with solutions obtained by more elaborate models or with experimental observation.

Steady-state or time-periodic solutions can be obtained by transient simulations starting from some reasonable initial conditions (Lighthill, 1965). Accordingly, the selection of boundary conditions for the linear time-periodic problem, based on the group velocity of linear capillary waves (as discussed earlier) can, in many cases, be validated by linear transient numerical solutions using these boundary conditions. Similarly, nonlinear steady-state solutions can be obtained by nonlinear transient numerical simulations. If disturbances in the flow quantities are not too large at the nozzle exit, and nonlinear effects remain contained further downstream, then it is reasonable to consider the same nozzle exit conditions for both linear and nonlinear transient simulations. If the angular momentum at the nozzle exit is smaller than the amount needed to stabilize the undisturbed annular film with radius $\bar{r}_0$ and thickness Δr_0, the film collapses onto the symmetry axis. In this case, the centrifugal forces cannot compensate for the capillary pressure that promotes a minimization of the surface area of the swirling film. If the angular velocity at the nozzle exit exceeds the critical value that stabilizes the annular configuration, i.e.,

$$\bar{v}_{\theta,0}^* = \sqrt{2/\left\{\mathrm{We}\left[1 - \Delta r_0^2/\left(4\bar{r}_0^2\right)\right]\right\}},$$

the radial centerline position of the film oscillates in the downstream direction of the nozzle according to the energy transfer between surface energy and energy stored within the swirling motion of the film. These films are often described as "conical" in the engineering practice, although they can deviate substantially from a true conical shape. For the large Weber number cases, steady-state solutions were readily obtained. However, for low Weber numbers, the steady-state equations might become singular at some location downstream from the nozzle. See the equation for $ds/dz^* = d^2\bar{r}^*/dz^{*2}$ in Mehring and Sirignano (2001). The singularity is removable by appropriate choices for the amount of swirl imposed at the nozzle exit and/or by changing the boundary conditions for $d\bar{r}^*/dz^*$ and $d^2\bar{r}^*/dz^{*2}$ at the nozzle exit.

The nonlinear analyses of Mehring and Sirignano (2001, 2004) on "conical" films with modulations of the axial- or transverse-velocity component at the nozzle exit show that film thinning because of film divergence does not fundamentally change the characteristics of the nonlinear capillary breakup observed for swirling but nondiverging annular films modulated at the nozzle exit. See, for example, Fig. 12.15. However, an increase in the thinning rate or cone angle of diverging annular films, resulting from an increase in the angular momentum or swirl at the nozzle exit, leads to smaller disturbance amplitudes and longer breakup lengths for the same modulation at the nozzle exit. These breakup lengths and times might be significantly larger than those observed for the corresponding modulated planar films. For dilational film modulations, the effect of Weber number changes on film-breakup length l_b and breakup time t_b is determined by the relative importance of the radius of curvature in the main flow direction R_1, the radius of curvature in the corresponding perpendicular direction R_2, and the relative changes of the thinning rate with changes in We at constant swirl number. Here, the term "swirl number"

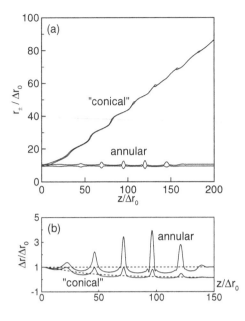

Figure 12.15. Annular and "conical" swirl-stabilized liquid film dilationally modulated at the nozzle exit $z = 0$, $r_\pm$ = radius of the outer ($+$) and inner ($-$) interfaces, Δr = local film thickness (subscript "0" denotes its value at $z = 0$) (Mehring and Sirignano, 2001).

refers to the ratio k of swirl c_2 imposed at the nozzle exit to the amount of swirl $c_{2,0}$ necessary to stabilize the film in its annular configuration, i.e., $c^2 = kc_{2,0}$. For sinuous film modulations, film breakup occurs because of nonlinear coupling with the dilational mode. Here, the dependency of nonlinear breakup length and time on Weber number is greatly influenced by the nonlinear mode coupling and the linear and nonlinear dilational mode behavior. The former decreases with increasing Weber number (and annular radius). A description of the dependencies for both modes is given in Mehring and Sirignano (2001).

Comparison between thin annular swirl- or pressure-stabilized films shows that swirl stabilization for annular liquid films can significantly reduce nonlinear and linear breakup lengths for thin dilationally modulated films. The comparison also shows that, because of the constant gas-core pressure, bubble formation between fluid rings observed for pressure-stabilized films is not found on swirl-stabilized films.

The analysis of axisymmetric "conical" films that are modulated at the nozzle exit shows that an increase in the thinning rate results in smaller absolute amplitude disturbances downstream for the same amplitude at the nozzle. Consequently, the appearance of nonlinear effects (i.e., higher harmonic disturbances) is typically delayed, thus resulting in longer breakup times and larger breakup lengths. This is particularly true for sinuous modulated films (i.e., modulation of the transverse velocity at the nozzle exit). Also, for large-enough thinning rates (i.e., swirl numbers), film breakup might not occur, even though breakup was predicted at lower thinning rates under the same forcing conditions. The results show that the breakup lengths and times for diverging films (with excess swirl at the nozzle exit) are generally significantly greater than for the similar planar case. This is found to be particularly true for larger swirl numbers k and sinuous mode simulations. On the other hand, breakup lengths and times for swirl-stabilized modulated annular films are in

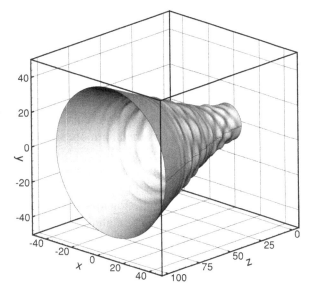

Figure 12.16. Instantaneous location of the outer interface for swirling conical liquid film sinusoidally modulated at the nozzle $z = 0$ (Mehring and Sirignano, 2004).

general smaller than those predicted for the corresponding planar cases. Planar films have been approximated by simulating swirl-stabilized annular films with an annular radius of $R = \bar{r}^* = 10^4$ at the nozzle exit.

Following the reduced-dimension approach, Mehring and Sirignano (2004) extended their analysis of swirling conical films to three-dimensional disturbances. Initial conditions for the nonlinear numerical simulations were obtained from the solution to the corresponding nonlinear steady-state problem without modulation imposed at the nozzle exit. The similar boundary conditions as previously used for axisymmetric swirling conical films were utilized, with the exception that now modulations were imposed onto the film thickness and film-centerline location rather than onto the axial and transverse film velocities. Also now, the modulations varied spatially in the circumferential direction at the nozzle exit.

Figure 12.16 shows the outer film surface location for the case of a swirling conical sheet generated by liquid that exits an annular slit nozzle with 10 times the amount of swirl needed to stabilize the film in its annular configuration. Initial conditions for this case were obtained by solving the nonlinear steady-state axisymmetric equations. The liquid film is modulated sinusoidally at the nozzle ($z = 0$) according to

$$v_0 = C \exp(-t^* / T_{\text{ex}}) \sin(2\pi \tau^* / T_p) \cos(n\theta),$$

with $C = 0.1$ and $n = 5$ (i.e., five standing waves imposed at the nozzle in the circumferential direction). The Weber number for this case is $\text{We} = 1000$, $R = 10$, and $T_{\text{ex}} = T_p = 10$. Areas of the liquid film, with a local maximum of the film thickness at the nozzle exit, develop into areas with two local maxima in the film thickness. The latter is not found for the similar swirling annular sheet and results from sheet divergence in the conical case. Sheet divergence that is due to excess swirl also causes the regular cellular structure observed at the nozzle to stretch, forming a netlike structure of thicker fluid ligaments imposed onto a thinner layer of liquid (see Figure 12.16).

For "conical" films, increased film divergence leads to increased film thinning and smaller droplets once the thin ligaments detached from the film do disintegrate further (e.g., according to Rayleigh's mechanism for jet breakup). However, for purely capillary film breakup and diverging films, the decreasing effect of the increasing second radius of curvature and the prescribed decrease in disturbance amplitude renders the film more stable. This contrasts to the case in which aerodynamic effects dominate the film breakup. Here, film divergence leads to an increase in surface area onto which the surrounding gas can impose forces or transfer energy. Within the present analysis of capillary waves on diverging annular films, energy is only transferred onto the film at the nozzle exit. Nonlinear analyses considering the aerodynamic stability of annular and conical films have not been published so far.

The reduced one-dimensional models, discussed here and used to analyze the nonlinear distortion and disintegration of thin films subject to long-wavelength disturbances on infinite periodic or semi-infinite modulated annular or conical liquid films, provide important qualitative and quantitative information on the stability and on the breakup of liquid films used in practical applications. As already mentioned, they allow rough predictions of film breakup length and resulting mean drop size.

A nonlinear numerical analysis of free liquid films by using a general-purpose multifluid method was reported by Poo and Ashgriz (1998). The accuracy of these predominantly Eulerian techniques is difficult to maintain at an acceptable level when thin films with considerable nonlinear distortion are considered. More accurate results can be expected by Lagrangian methods, e.g., boundary-element techniques or discrete-vortex methods as well as the prescribed Eulerian methods with dynamic regridding at the location of the fluid interfaces. However, such simulations are computationally very intensive and do not typically yield the same insights as the prescribed simplified analyses, e.g., with respect to the propagation properties of capillary waves and their interaction among each other.

12.7 Concluding Remarks

Some linear and nonlinear analyses were performed for round jets and planar sheets and for the annular and conical domains with and without swirl. Aerodynamic effects have been considered for the round jets and planar sheet, but remain to be studied for nonlinear annular and linear and nonlinear conical films. There are clear indications that nonlinear effects are important, especially in the latter stages of deformation. Substantial deviations from sinusoidal-shaped waveforms of linear theory have been demonstrated in various configurations, because of nonlinear effects. Viscous effects are critical in the pinch-off process. It remains to determine the role of intermolecular forces in the final stages of pinch-off.

The setting of the boundary conditions in the spatially developing wave case must be done with care. The required number of upstream boundary conditions can change from one regime to another. Group-velocity analysis is very useful in setting the proper boundary conditions.

The fan sheet stability should be examined for high-pressure effects in both the dilational and sinuous modes. Nonlinear effects for fan sheets also remain to be determined.

Little theoretical analysis exists on the coupling of the wave phenomenon with behavior upstream in the orifice. The study of the effects of generated turbulence, nonuniform velocity profiles, cavitation, and upstream pressure disturbances should be studied in the future. The impact of gas streams colliding with (rather than flowing parallel to) the liquid stream should be studied further. Similarly, acoustic interactions with the liquid stream should be analyzed. More theoretical work on the interaction of electric fields with charged liquid streams should be performed. More three-dimensional analyses are needed.

The interesting work of Eggers (1997) on the process of stream pinch-off must be extended. Molecular-dynamics methodology offers some hope for that evaluation (Moseler and Landman, 2000). The breakup of the ligaments formed from pinch-off into smaller ligaments and droplets is not well understood. Some simple theories exist that model the process of cascading from large ligaments to small droplets. There is substantial need for refinement and further understanding of the physical behavior.

The area of actively controlled liquid-stream behavior and spray formation has broad and vital technological applications. Here the issues of modulation upstream in the orifice flow, modulated air-flow–liquid-stream interactions, acoustic interactions, and electromagnetic-field–liquid-stream interactions should be studied.

APPENDIX A

The Field Equations

The primitive equations governing the gas field surrounding droplets and the internal liquid field are extensions of the Navier–Stokes equation. The extensions account for diffusion within a multicomponent mixture, chemical reaction, and radiative heat transfer. First, they are presented for the surrounding gas in a three-dimensional, unsteady dimensional form by use of tensor notation. A complete derivation from the fundamental kinetic theory can be found in Hirschfelder et al. (1954).

The continuity equation is written as

$$\frac{\partial \rho}{\partial t} + \frac{\partial (\rho u_j)}{\partial x_j} = 0. \tag{A.1}$$

The momentum equation states that

$$\frac{\partial (\rho u_i)}{\partial t} + \frac{\partial (\rho u_i u_j)}{\partial x_j} + \frac{\partial p}{\partial x_i} = \frac{\partial \tau_{ij}}{\partial x_j} + \rho g_i, \tag{A.2}$$

where, for a Newtonian fluid under the Stokes hypothesis,

$$\tau_{ij} = \mu \left(\frac{\partial u_i}{\partial x_j} + \frac{\partial u_j}{\partial x_i} \right) - \frac{2}{3} \frac{\partial}{\partial x_i} \left(\mu \frac{\partial u_k}{\partial x_k} \right) \delta_{ij}.$$

Fourier heat conduction is assumed. The chemical-reaction rates $\dot{w}_m$ for each species, the diffusion velocity V_i^m for each species, and the radiative heat flux $q_{\mathrm{rad},i}$ will be left without further specification.

$$\frac{\partial (\rho h)}{\partial t} + \frac{\partial}{\partial x_j} (\rho u_j h) - \frac{\partial}{\partial x_j} \left(\lambda \frac{\partial T}{\partial x_j} \right) + \frac{\partial q_{\mathrm{rad}.j}}{\partial x_j} + \frac{\partial}{\partial x_j} \left(\sum_m \rho V_j^m h_m Y_m \right)$$

$$= \frac{\partial p}{\partial t} + u_j \frac{\partial p}{\partial x_j} + \Phi + \sum_m \rho \dot{w}_m Q_m, \tag{A.3}$$

where Q_m is the negative of the heat-of-formation value for species m,

$$h = \int_{T_{\text{ref}}}^{T} c_p \, \mathrm{d}T' = \int_{T_{\text{ref}}}^{T} \left[\sum_m Y_m c_{pm}(T') \right] \mathrm{d}T' = \sum_m Y_m h_m.$$

$$\Phi = \tau_{ij} \frac{\partial u_i}{\partial x_j}.$$

The mass-conservation equation for each species equation can be written as

$$\frac{\partial(\rho Y_m)}{\partial t} + \frac{\partial(\rho Y_m u_j)}{\partial x_j} + \frac{\partial \left(\rho Y_m V_j^m \right)}{\partial x_j} = \rho \dot{w}_m, \quad m = 1, 2, \ldots, N, \tag{A.4}$$

where the diffusion velocity for each species in an N-component gas mixture is governed by

$$\frac{\partial X_m}{\partial x_i} = \sum_{k=1}^{N} \frac{X_m X_k}{D_{mk}} \left(V_i^k - V_i^m \right), \quad m = 1, 2, \ldots, N. \tag{A.5a}$$

The preceding linear system of N equations yields $N - 1$ independent differences of the V_i^m values in terms of the mole fractions and their gradients. Then the mass fractions can be related to mole fractions by the linear system

$$X_m = \frac{Y_m / W_m}{\sum_{k=1}^{N} Y_k / W_k}, \quad m = 1, 2, \ldots, N, \tag{A.5b}$$

together with the condition $\sum_{k=1}^{N} Y_k = 1$. Finally, the diffusion velocities are related to the mass fractions and their gradients. In the case of equal binary diffusion coefficients $D_{mi} = D$ and equal molecular weights for all species, the diffusion velocities are most conveniently given by the Fickian diffusion relations

$$Y_m V_i^m = -D \frac{\partial Y_m}{\partial x_i}, \quad m = 1, 2, \ldots, N. \tag{A.5c}$$

Unless otherwise noted, we generally make the approximation of Fickian diffusion even if molecular weights and binary diffusion coefficients differ with species. Megaridis and Sirignano (1991) showed that the non-Fickian corrective terms added to the right-hand side of Eq. (A.5) do have some local effects but no major global effect on the droplet heating, vaporization, or acceleration. So, in this approximation, Eq. (A.5c) is used in place of the more exact combination of Eqs. (A.5a) and (A.5b). A less severe approximation can be used when there exists a dominant species (e.g., nitrogen) in terms of its mass or mole fraction. Then the diffusivity D in Eq. (A.5c) should be replaced with D_m, where that diffusivity becomes the binary diffusion coefficient for species m and the dominant species. Of course, then, Eqs. (A.4) and (A.5c) may not be used for the dominant species. Rather, the condition $\sum Y_m = 1$ will determine the mass fraction for the dominant species. Also, the constraint that $\sum Y_m V_j^m = 0$ will determine the diffusion velocity of the dominant species to use in energy equation (A.3). Mass diffusion that is due to thermal gradients or pressure gradients is typically negligible.

We consider the gas phase to be thermodynamically perfect so that

$$p = \rho R T, \tag{A.6}$$

where R is the specific gas "constant" that actually can vary in a multicomponent flow because of its molecular weight and mass fraction dependencies. That is, $R = \sum_{m=1}^{N} (Y_m \mathcal{R})/W_m$, where $\mathcal{R}$ and W_m are the universal gas constant and the molecular weight of species m, respectively.

The equations for flow and heating of the internal multicomponent liquid in a droplet can readily be extracted from Eqs. (A.1)–(A.5). The liquid-phase-dependent variables carry a special subscript to distinguish them from the gas-phase variables. We have, for example, $u_{l,i}$, ρ_l, $Y_{l,m}$, $V_{l,m,i}$, p_l, $\tau_{l,ij}$, h_l, $q_{l,i}$, Φ_l, T_l, and $h_{l,m}$. The equations can be written for low Mach number flows; typically, constant density is assumed, which simplifies Eqs. (A.1) and (A.2) to yield the incompressible equations

$$\frac{\partial u_{l,j}}{\partial x_j} = 0, \tag{A.7}$$

$$\frac{\partial u_{l,i}}{\partial t} + \frac{\partial (u_{l,i} u_{l,j})}{\partial x_j} + \frac{1}{\rho_l}\frac{\partial p_l}{\partial x_i} = \frac{\mu_l}{\rho_l}\frac{\partial^2 u_{l,i}}{\partial x_j^2} + \frac{1}{\rho_l}\frac{\partial \mu_l}{\partial x_j}\left(\frac{\partial u_{l,i}}{\partial x_j} + \frac{\partial u_{l,j}}{\partial x_i}\right) + g_i. \tag{A.8}$$

Radiation and chemical reaction are typically neglected in the liquid. Kinetic energy and viscous dissipation can be neglected in the energy equation. Also, rates of mass diffusion are usually much less than heat diffusion rates in interesting liquids, so that transport of energy that is due to mass diffusion is neglected, yielding

$$\rho_l \frac{\partial e_l}{\partial t} + \rho_l u_{l,j}\frac{\partial e_l}{\partial x_j} - \frac{\partial}{\partial x_j}\left(\lambda_l \frac{\partial T_l}{\partial x_j}\right) = 0. \tag{A.9}$$

With Fickian diffusion, the liquid-species equation becomes

$$\frac{\partial Y_{l,m}}{\partial t} + u_{l,j}\frac{\partial Y_{l,m}}{\partial x_j} = \frac{\partial}{\partial x_j}\left(D_l \frac{\partial Y_{l,m}}{\partial x_j}\right), \quad m = 1, 2, \ldots, N. \tag{A.10}$$

Note that, although Eq. (A.10) is given for all species, the mass fraction of some species (present in the gas phase) will be zero in the liquid phase. The set of N species will include any species that appears in at least of one of the two phases.

The combined First and Second Laws of Thermodynamics lead to a differential relationship for an element of mass (Williams, 1985); $T \mathrm{d}s = \mathrm{d}h - (1/\rho)\mathrm{d}p - \sum_{m=1}^{N} e_m \mathrm{d}Y_m$, where s is the specific entropy and e_m is the internal energy (without the energy of formation) of the mth species. Note that the last term on the right-hand side goes to zero when the specific heats of all species have the same value at any given temperature. This equation can be recast in terms of Lagrangian time derivatives following the mass element and combined with Eqs. (A.1) and (A.3) to yield

$$\rho\left\{\frac{\partial s}{\partial t} + u_j\frac{\partial s}{\partial x_j}\right\} = \frac{\partial(\rho s)}{\partial t} + \frac{\partial(\rho u_j s)}{\partial x_j} = \frac{p}{T}\frac{\partial u_j}{\partial x_j} + \frac{\rho c_v}{T}\left\{\frac{\partial T}{\partial t} + u_j\frac{\partial T}{\partial x_j}\right\}$$

$$= \frac{1}{T}\left\{-\frac{\partial q_j}{\partial x_j} + \Phi + \sum_{m=1}^{N}\rho\omega_m[Q_m - e_m] + \sum_{m=1}^{N}e_m\frac{\partial(\rho V_{m,j}Y_m)}{\partial x_j}\right\}$$

$$\stackrel{\mathrm{def}}{=} \frac{R_1}{T}. \tag{A.11}$$

In the special case in which the specific heats of all species have the same value, the simplification results that

$$\frac{R_1}{T} = \frac{1}{T}\left\{-\frac{\partial q_j}{\partial x_j} + \Phi + \sum_{m=1}^{N}\rho\omega_m Q_m\right\}. \tag{A.12}$$

Axisymmetric Formulation

In many of our droplet applications, we consider axisymmetric flow around and within the droplet. For that reason, it is convenient to recast the equations in cylindrical coordinates. Equations (A.1) and (A.2) for the gas phase become

$$\frac{\partial(\rho r)}{\partial t} + \frac{\partial}{\partial z}(\rho r u_z) + \frac{\partial}{\partial r}(\rho r u_r) = 0, \tag{A.13}$$

$$\frac{\partial}{\partial t}(\rho r u_z) + \frac{\partial}{\partial z}(\rho r u_z^2 + pr) + \frac{\partial}{\partial r}(\rho r u_r u_z)$$

$$= \frac{2}{3}\frac{\partial}{\partial z}\left\{\mu\left[2\frac{\partial(r u_z)}{\partial z} - \frac{\partial(r u_r)}{\partial r}\right]\right\} + \frac{\partial}{\partial r}\left[r\mu\left(\frac{\partial u_r}{\partial z} + \frac{\partial u_z}{\partial r}\right)\right] + \rho g_z, \tag{A.14}$$

$$\frac{\partial}{\partial t}(\rho r u_r) + \frac{\partial}{\partial z}(\rho r u_z u_r) + \frac{\partial}{\partial r}(\rho r u_r^2 + p)$$

$$= \frac{\partial}{\partial z}\left[\mu r\left(\frac{\partial u_r}{\partial z} + \frac{\partial u_z}{\partial r}\right)\right] + \frac{2}{3}\frac{\partial}{\partial r}\left[r\mu\left(2\frac{\partial u_r}{\partial r} - \frac{u_r}{r} - \frac{\partial u_z}{\partial z}\right)\right]$$

$$+ \frac{2}{3}\mu\left(\frac{\partial u_z}{\partial z} + \frac{\partial u_r}{\partial r} - 2\frac{u_r}{r}\right). \tag{A.15}$$

Note that, for axisymmetry, we consider the body force to act only in the axial direction. Also, the momentum equation is separated into two scalar equations.

Energy and species equations (A.3) and (A.4) become

$$\frac{\partial}{\partial t}(\rho r h) + \frac{\partial}{\partial z}(\rho r u_z h) + \frac{\partial}{\partial r}(\rho r u_r h) - \frac{\partial}{\partial z}\left(\lambda r\frac{\partial T}{\partial z}\right) - \frac{\partial}{\partial r}\left(\lambda r\frac{\partial T}{\partial r}\right)$$

$$+ r\frac{\partial q_{rad,z}}{\partial z} + \frac{\partial}{\partial r}(r q_{rad,r}) + \frac{\partial}{\partial z}\left(\sum_m \rho r V_z^m h_m Y_m\right) + \frac{\partial}{\partial r}\left(\sum_m \rho r V_r^m h_m Y_m\right)$$

$$= r\left(\frac{\partial p}{\partial t} + u_z\frac{\partial p}{\partial z} + u_r\frac{\partial p}{\partial r}\right) + r\Phi + \sum_m \rho r \dot{w}_m Q_m, \tag{A.16}$$

$$\frac{\partial}{\partial t}(\rho r Y_m) + \frac{\partial}{\partial z}(\rho r u_z Y_m) + \frac{\partial}{\partial r}(\rho r u_r Y_m)$$

$$= \frac{\partial}{\partial z}\left(\rho r D\frac{\partial Y_m}{\partial z}\right) + \frac{\partial}{\partial r}\left(\rho r D\frac{\partial Y_m}{\partial r}\right) + \rho r \dot{w}_m, \quad m = 1, 2, \ldots, N. \tag{A.17}$$

For axisymmetric flow inside the liquid droplet, Eqs. (A.7)–(A.10) can be readily recast in cylindrical coordinates. However, it is convenient to proceed further with stream function and vorticity as the dependent variables in place of the two velocity components. Then, with the definitions

$$u_{l,r} = -\frac{1}{r}\frac{\partial \psi}{\partial z}, \qquad u_{l,z} = \frac{1}{r}\frac{\partial \psi}{\partial r},$$

$$\omega = \frac{\partial u_{l,r}}{\partial z} - \frac{\partial u_{l,z}}{\partial r}, \tag{A.18}$$

we obtain from Eqs. (A.7) and (A.8), from these definitions, and from the assumption of constant viscosity

$$\frac{\partial^2 \psi}{\partial z^2} + r\frac{\partial}{\partial r}\left(\frac{1}{r}\frac{\partial \psi}{\partial r}\right) = -\omega r, \tag{A.19}$$

$$\frac{\partial \omega}{\partial t} + \frac{1}{r}\frac{\partial \psi}{\partial r}\frac{\partial \omega}{\partial z} - \frac{1}{r}\frac{\partial \psi}{\partial z}\frac{\partial \omega}{\partial r} + \frac{1}{r^2}\frac{\partial \psi}{\partial z}\omega = \frac{\partial}{\partial z}\left(\nu_l\frac{\partial \omega}{\partial z}\right) + \frac{\partial}{\partial r}\left[\frac{\nu_l}{r}\frac{\partial (r\omega)}{\partial r}\right]. \tag{A.20}$$

Note that Eq. (A.20) has advective and diffusive terms so that it can be solved by the same numerical methods used for the momentum equation. Poisson equation (A.19) for the stream function is of the same character as the equation for pressure obtained from momentum and continuity. So the same techniques that would be used for the pressure solver can be used for the stream-function solver. The vorticity–stream-function formulation has the advantage of two dependent variables versus three variables for the primitive formulation.

Energy and species equations (A.9) and (A.10) can be recast as

$$\rho_l\frac{\partial (re_l)}{\partial t} + \rho_l\frac{\partial \psi}{\partial r}\frac{\partial e_l}{\partial z} - \rho_l\frac{\partial \psi}{\partial z}\frac{\partial e_l}{\partial r} = \frac{\partial}{\partial z}\left(\lambda_l r\frac{\partial T_l}{\partial z}\right) + \frac{\partial}{\partial r}\left(\lambda_l r\frac{\partial T_l}{\partial r}\right), \tag{A.21}$$

$$\frac{\partial (rY_{l,m})}{\partial t} + \frac{\partial \psi}{\partial r}\frac{\partial Y_{l,m}}{\partial z} - \frac{\partial \psi}{\partial z}\frac{\partial Y_{l,m}}{\partial r} = \frac{\partial}{\partial z}\left(D_l r\frac{\partial Y_{l,m}}{\partial z}\right) + \frac{\partial}{\partial r}\left(D_l r\frac{\partial Y_{l,m}}{\partial r}\right). \tag{A.22}$$

Spherically Symmetric Formulation

Often, we consider a droplet in a stagnant field (no relative motion). Then the assumption of spherical symmetry might be reasonable. The gas-field continuity and momentum equations become

$$\frac{\partial (\rho r^2)}{\partial t} + \frac{\partial}{\partial r}(\rho r^2 u) = 0, \tag{A.23}$$

$$\frac{\partial (\rho r^2 u)}{\partial t} + \frac{\partial}{\partial r}(\rho r^2 u^2) + r^2\frac{\partial p}{\partial r} = \frac{4r^2}{3}\frac{\partial}{\partial r}\left(\mu\frac{\partial u}{\partial r} - \mu\frac{u}{r}\right) + 4r^2\mu\frac{\partial}{\partial r}\left(\frac{u}{r}\right). \tag{A.24}$$

The body force has been eliminated in Eq. (A.24) because gravity would not act in a spherically symmetric manner. (Of course, another force such as an electrostatic force on a charged droplet might produce such symmetry.)

The energy and species equations in the gas become

$$\frac{\partial}{\partial t}(\rho r^2 h) + \frac{\partial}{\partial r}(\rho r^2 u h) - \frac{\partial}{\partial r}\left(\lambda r^2 \frac{\partial T}{\partial r}\right) + \frac{\partial}{\partial r}(r^2 q_{rad}) + \frac{\partial}{\partial r}\left(\sum_m \rho r^2 V^m h_m Y_m\right)$$

$$= r^2 \frac{\partial p}{\partial t} + r^2 u \frac{\partial p}{\partial r} + r^2 \Phi + \sum_m \rho r^2 \dot{w}_m Q_m, \qquad (A.25)$$

$$\frac{\partial}{\partial t}(\rho r^2 Y_m) + \frac{\partial}{\partial r}(\rho r^2 u Y_m) + \frac{\partial}{\partial r}(\rho r^2 V^m Y_m) = \rho r^2 \dot{w}_m, \quad m = 1, 2, \ldots, N, \quad (A.26)$$

where

$$V^m = -(D/Y_m)\frac{\partial Y_m}{\partial r}.$$

For the liquid interior in the spherically symmetric case, we have a stagnant interior with zero velocity and uniform pressure. Therefore we need not solve the continuity and momentum equations; we have their trivial solutions. The liquid-phase energy and species equations become

$$\rho_l \frac{\partial}{\partial t}(r^2 e_l) - \frac{\partial}{\partial r}\left(\lambda_l r^2 \frac{\partial T_l}{\partial r}\right) = 0, \qquad (A.27)$$

$$\frac{\partial}{\partial t}(r^2 Y_{l,m}) - \frac{\partial}{\partial r}\left(r^2 D_l \frac{\partial Y_{l,m}}{\partial r}\right) = 0, \quad m = 1, 2, \ldots, N. \qquad (A.28)$$

Conserved Scalars

Consider the gas phase that surrounds or adjoins liquid droplets, films, pools, and/or streams. Figure B.1 presents a diagram that portrays a wide variety of two-phase reacting flow situations included here. Laminar multicomponent flow with viscosity, Fourier heat conduction, and Fickian mass diffusion will be analyzed. Diffusivities for all species will be assumed identical; the Lewis number value will be unity; radiation will be neglected; and kinetic energy will be neglected in comparison with thermal energy so that, with regard to the energetics, pressure will be considered uniform over the space although the pressure gradient can be significant in the momentum balance (small Mach number). We will analyze, as a general case, the situation in which a one-step exothermic reaction occurs between the ambient gas and the vapor from the bulk liquid. We will refer to the liquid as the fuel and the ambient gas as the oxidizer, although the opposite situation, e.g., oxygen droplet in hydrogen gas, can be analyzed in exactly the same fashion. The nonreacting case can be the special subcase in which the reaction rate becomes zero.

Both steady and unsteady scenarios will be discussed. The roles of the gas-phase energy equation and the species equations will be emphasized. Of course, they must be coupled with the continuity and momentum equations and with a gas equation of state. In addition, the solution of conservation equations for the liquid phase might be required.

Because the Lewis number $Le = 1$, the thermal and mass diffusivities are equal so that

$$\rho D = \lambda/c_p \tag{B.1}$$

With the one-step reaction, the various reaction rates are proportioned by the stoichiometric ratios:

$$\dot{w}_F = v\dot{w}_O = -\frac{v}{v+1}\dot{w}_P \tag{B.2}$$

where v is the stoichiometric fuel-to-oxidizer mass ratio. The species equations become

$$L(Y_i) \equiv \rho\frac{\partial Y_i}{\partial t} + \rho\vec{u}\cdot\nabla Y_i - \nabla\cdot(\rho D\nabla Y_i) = \rho\dot{w}_i; \ i = 1, 2, \ldots, N \tag{B.3}$$

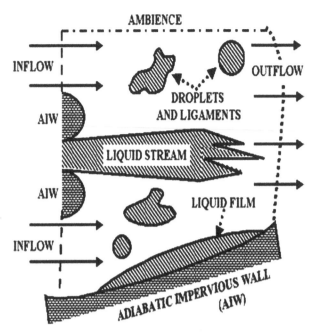

Figure B.1. Sketch of types of liquid elements in relevant two-phase flows: droplets, ligaments, wall films, and liquid streams, e.g., jets. Types of boundaries and interfaces are also displayed: inflow boundaries, outflow boundaries, adiabatic, impervious walls, and vaporizing (or condensing) interfaces. (Sirignano, 2002, with permission of The Combustion Institute.)

The energy equation may be written as

$$L(h) \equiv \rho \frac{\partial h}{\partial t} + \rho \vec{u} \cdot \nabla h \nabla \cdot (\rho D \nabla h) = \rho \dot{w}_F Q + \frac{\partial p}{\partial t}, \tag{B.4}$$

where the sensible enthalpy h is given as

$$h = \sum_{i=1}^{N} \int_{T_{\text{ref}}}^{T} c_{pi}(T') dT' = \sum_{i=1}^{N} Y_i h_i = \int_{T_{\text{ref}}}^{T} c_p(T') dT'. \tag{B.5}$$

It follows that

$$c_p \nabla T = \nabla h - \sum_{i=1}^{N} h_i \nabla Y_i. \tag{B.6}$$

We define four Shvab–Zel'dovich (or coupling) variables for the gas phase from linear combinations of the variables h, Y_F, Y_O, Y_P, and Y_N:

$$\beta_1 = Y_F - \nu Y_O; \quad \beta_2 = h + \nu Q Y_O; \quad \beta_3 = Y_O + Y_P/(1+\nu); \quad \beta_4 = Y_N. \tag{B.7}$$

In the nonreacting case, we would have $\beta_1 = Y_F$ and $\beta_2 = h$. Consider now that pressure is steady although other variables can vary with time. Combinations of Eqs. (B.3) and (B.4) will yield

$$L(\beta_i) = 0; \quad i = 1, 2, \ldots, N. \tag{B.8}$$

Each portion of the boundary on the gas domain will be one of four types, as indicated in Fig. B.1: adiabatic, impervious walls; inflow or ambient boundaries; outflow boundaries; and vaporizing (or condensing) liquid surfaces. At the adiabatic and impervious walls, the normal gradients of temperature and mass fraction are zero so that

$$\frac{\partial \beta_i}{\partial n} = 0; \quad i = 1, 2, \ldots, N, \tag{B.9}$$

where n is the coordinate measuring normal distance form the boundary.

At all inflow boundaries, the β_i will each maintain uniform but possibly time-varying values. Ambient conditions for stagnant (Stefan convection but no forced or natural convection) gas fields are considered equivalent to inflow conditions. So at those boundaries,

$$\beta_i = \beta_{i\infty}(t). \tag{B.10}$$

At the outflow boundaries, the spatial normal derivatives of the primitive variables can be set to zero in standard fashion. This assumes that mixing and combustion are completed prior to outflow. It follows that, at those outflow boundaries,

$$\frac{\partial \beta_i}{\partial n} = 0; \quad i = 1, 2, \ldots, N. \tag{B.11}$$

The superscalar for gaseous diffusion flames can readily be found. Consider the same restrictions on Lewis number, chemical kinetics, ambient pressure, and ambient, inflow and outflow, and wall boundary conditions. So Eqs. (B.7), (B.8), (B.9), and (B.11) apply. Equation (B.10) is replaced with two sets of equations, one for ambient or inflow conditions on the "fuel" side of the flame (designated by subscript minus sign) and the other for ambient or inflow conditions on the "oxidizer" side of the flame (designated by subscript plus sign). The analysis actually allows for fuel and oxidizer to be present on more than one side of the flame zone; so a combination of diffusion flame and premixed (lean or rich) flame or a pure premixed flame is allowed. It follows that, with the indices i and j unequal,

$$S = \beta_i - \frac{\beta_{i+} - \beta_{i-}}{\beta_{j+} - \beta_{j-}} \beta_j \tag{B.12}$$

will be uniform over the domain. The subscripts $-$ and $+$ imply ambient or inflow conditions. The superscalar is instantaneously uniform for steady-state conditions, quasi-steady conditions, and also for unsteady cases with steady ambient and/or inflow conditions. The limiting temperature at the diffusion flame depends on only the scalar boundary values and the fuel properties. In particular, because fuel vapor and oxidizer exist only on one side of the flame, the enthalpy at the diffusion flame is

$$h_f = h_- + \frac{h_+ + \nu Q Y_{O+} - h_-}{\nu Y_{O+} + Y_{F-}} Y_{F-}. \tag{B.13}$$

So the limiting flame enthalpy and temperature can be calculated without attention to the configuration or the velocity-field details.

Now, consider the case in which vaporizing liquid is present. The fourth type of boundary condition placed at the vaporizing (or condensing) liquid surface is the most complicated and the most interesting, because it couples the two Shvab–Zel'dovich variables and prevents a solution whereby β_i is uniform in value over the gas field. In its most primitive form, it is stated as an energy boundary condition plus a boundary condition for each species. For the oxidizer gas, we assume that it is consumed in a surrounding flame before it reaches the liquid surface so the oxidizer mass fraction is zero at and in the vicinity of the surface. The combustion product gas does not penetrate the surface so that its Stefan advection and diffusion are in balance there. For the fuel vapor species, we have

$$\dot{m}(1 - Y_{Fs}) = -\rho D \frac{\partial Y_F}{\partial n}\bigg)_s, \tag{B.14}$$

where $\dot{m}$ is the vaporization rate per unit surface area and n is the normal coordinate (positive into the gas). The subscript s implies conditions at the liquid surface. For the energy boundary condition, the energy flux from the gas to the surface (because of combined heat conduction and energy transfer by mass diffusion) is balanced by the combination of the heat required for vaporization and the heat conducted to the liquid interior. Using Eqs. (B.6) and (B.14) with the unitary Lewis number condition, we can show that, at the surface,

$$\frac{\partial h}{\partial n}\bigg)_s = -\frac{L_{\text{eff}}}{1 - Y_{Fs}} \frac{\partial Y_F}{\partial n}\bigg)_s. \tag{B.15}$$

The effective heat of vaporization is explained in Chapter 2; it combines the latent heat of vaporization with the heat conducted into the the liquid interior per unit of vaporized mass.

Now we can create by inspection the following supervariable as a certain linear combination of the of primitive conserved scalars.

$$S = \beta_1 + \frac{L_{\text{eff}}}{1 - Y_{Fs}}\beta_2 = h + \nu\left(Q - \frac{L_{\text{eff}}}{1 - Y_{Fs}}\right)Y_O + \frac{L_{\text{eff}}}{1 - Y_{Fs}}Y_F. \tag{B.16}$$

Here L_{eff} and Y_{Fs} each have the same instantaneous values uniformly across all of the liquid surfaces in the particular configuration. Y_{Fs} is a function of the surface temperature T_s through the phase-equilibrium law. We consider first two cases: steady state and quasi-steady state. In the steady state, the time derivatives in Eqs. (B.3) and (B.4) are zero and the parameters in boundary conditions (B.10) and (B.15) are steady. In the quasi-steady state, the time derivatives in Eqs. (B.3) and (B.4) are negligible compared with other terms in the equations although parameters in boundary conditions (B.10) and (B.15) can be time varying. Note that the thermal behavior of the liquid can be unsteady provided that the surface temperature is steady or quasi-steady. It can now easily be shown that Eq. (B.16), when subjected to Eq. (B.8) and boundary conditions (B.9)–(B.11), and (15) produces the spatially uniform value

$$S = S_\infty(t) = h_\infty + \nu\left(Q - \frac{L_{\text{eff}}}{1 - Y_{Fs}}\right)Y_{O\infty} + \frac{L_{\text{eff}}Y_{F\infty}}{1 - Y_{Fs}}, \tag{B.17}$$

where the infinity subscript implies ambient or inflow conditions. This result is valid if the primitive variables h_∞, $Y_{O\infty}$, and $Y_{F\infty}$ vary over the boundary, provided that the linear combination S_∞ is instantaneously uniform. Equation (B.17) can also be considered as a valid solution for the unsteady case if the pressure p and the parameter $\frac{L_{\text{eff}}}{1 - Y_{Fs}}$ are either steady or vary sufficiently slowly compared with other variables that the time derivatives of pressure and of the parameter are negligible. Of course,

Eq. (B.17) can provide the solution to an unsteady-state case only asymptotically after some time because no regard is paid here to initial conditions.

We are allowing for two possibilities: (i) no fuel vapor exists on the oxygen side of the flame so that $Y_{F\infty}$ and a pure diffusion flame occurs and (ii) fuel vapor exists on both sides of each flame in the configuration so that both a premixed character and a diffusion character exists for the flame(s). Consequently, Eq. (B.17) implies that

$$h - h_\infty + v\left(Q - \frac{L_{\text{eff}}}{1 - Y_{Fs}}\right)(Y_O - Y_{O\infty}) + \frac{L_{\text{eff}}}{1 - Y_{Fs}}Y_F = 0, \tag{B.18}$$

$$h - h_s + v\left(Q - \frac{L_{\text{eff}}}{1 - Y_{Fs}}\right)Y_O + \frac{L_{\text{eff}}}{1 - Y_{Fs}}(Y_F - Y_{Fs}) = 0, \tag{B.19}$$

$$h_\infty - h_s + v\left(Q - \frac{L_{\text{eff}}}{1 - Y_{Fs}}\right)Y_{O\infty} - \frac{L_{\text{eff}}}{1 - Y_{Fs}}Y_{Fs} = 0. \tag{B.20}$$

Equation (B.20) may be rewritten as

$$\frac{h_\infty - h_s + vQY_{O\infty}}{L_{\text{eff}}} = \frac{Y_{Fs} + vY_{O\infty}}{1 - Y_{Fs}} = B. \tag{B.21}$$

So Eq. (B.21) defines the nondimensional transfer number B. More important, the equation together with the enthalpy-temperature relationship $h(T)$ and the phase-equilibrium law $Y_{Fs} = Y_{Fs}(T_s)$ determine the surface temperature T_s and the surface mass fraction as functions of fuel properties, ambient or inflow conditions, and the effective latent heat of vaporization L_{eff}. These surface values are not affected by the details of the configuration or of the hydrodynamics and transport phenomena. Given a characteristic length d and a local Nusselt number Nu, Eqs. (B.15) and (B.21) yield that the ratio of the local nondimensional vaporization rate (per unit area) to Nu is B and is uniform over the liquid surface:

$$\text{Nu} = \frac{d\left(\frac{\partial h}{\partial n}\right)_s}{h_\infty - h_s + vQY_{O\infty}} = \frac{\frac{\dot{m}d}{\rho D}}{B}. \tag{B.22}$$

In the limit of infinite reaction rate, the flame becomes a sheet of zero thickness at which $Y_O = Y_F = 0$. Equation (B.18) or (B.19) then yields a formula for the limiting flame temperature T_f:

$$h(T_f) = h_f = h_\infty + v\left(Q - \frac{L_{\text{eff}}}{1 - Y_{Fs}}\right)Y_{O\infty} \tag{B.23}$$

or

$$h(T_f) = h_f = h_s + \frac{L_{\text{eff}}}{1 - Y_{Fs}}Y_{Fs}. \tag{B.24}$$

Flame temperature as well as liquid-surface temperature does not depend on the dynamics or geometrical configuration. Rather, it depends on the ambient or inflow conditions, the fuel properties, and the heat per unit mass transferred to the liquid.

The results indicated by Eqs. (B.17)–(B.24) are remarkable in that they apply independently of the details concerning the geometrical configuration, velocity field, transport properties (other than unitary Lewis number) chemical-kinetic constants, and degree of unsteadiness for the liquid. The liquid surface where Eqs. (B.14) and (B.15) are applied is allowed to regress, advance, or move in any fashion; the vaporization rate per unit area equals the product of density and normal velocity only when the velocity is measured in a frame of reference attached to the surface. Although it is required that L_{eff}, T_s, and Y_{Fs} be uniform instantaneously over all liquid surfaces, the vaporization rate per unit area $\dot{m}$ may vary spatially as well as temporally. The superscalar S will be uniform even if $\dot{m}$ is nonuniform.

The exit condition given by Eq. (B.11) deserves further discussion. It indicates that the flow is well mixed by the time that it reaches the exit. We can relax the exit condition somewhat by insisting that only some higher derivative of α_i in the normal direction to the outflow boundary goes to zero. Still, the solutions given by Eqs. (B.17)–(B.24) would apply. However, if exit conditions (B.11) were that the Lagrangian time-derivative of α_i at the exit goes to zero, there would be a different conclusion in the unsteady case. If the superscalar is time varying but spatially uniform, the zero Lagrangian time derivative implies that the local time derivative is zero. However, that contradicts the hypothesis so that Eq. (B.17) would not be a solution in the unsteady case. There is no problem but just a subtle difference in the steady or quasi-steady case. There, the zero Lagrangian time derivative means that the first spatial derivative in the flow direction at the boundary is zero. Note that the flow direction need not be normal to the surface. Still, in the steady or quasi-steady case, Eqs. (B.17)–(B.24) will be valid solutions when the Lagrangian derivative exit condition is used.

The uniformity of the superscalar S over the gas phase does require that L_{eff} be instantaneously uniform over a liquid surface. Note that $L_{\text{eff}} = L + \dot{q}/\dot{m}$ and the latent heat of vaporization L is a fuel property and therefore will be uniform and constant. The uniformity of the heat flux to the interior per unit mass flux $\dot{q}/\dot{m}$ is a more demanding requirement. It can be satisfied in various situations. For example, if the wet-bulb temperature has been reached, $\dot{q} = 0$. When a quasi-steady thermal layer exists in the liquid near the surface, $\dot{q}/\dot{m} = c_l(T - T_l)$, where c_l is the liquid specific heat and T_o is the liquid interior core temperature. So, there L_{eff} will have a uniform value. In various cases with certain symmetries $\dot{q}/\dot{m}$ can be uniform. In cases in which the liquid thermal layer is unsteady, $\dot{q}/\dot{m}$ can still be approximately uniform if the liquid thermal layer thickness is small compared with liquid dimensions (e.g., surface radius of curvature, droplet diameter, pool depth, or film thickness).

In the case without chemical reaction, the reaction rates in Eqs. (B.2)–(B.4) are zero, the oxidizer is merely another inert so that $Y_O = Y_{O\infty} = 0$, the flame temperature has no meaning, and α_1 and α_2 become merely h and Y_F, respectively. Then the supervariable for the scalar properties is

$$S_v = h + \frac{L_{\text{eff}}}{1 - Y_{Fs}} Y_{Fs}, \tag{B.25}$$

where the subscript v is used for this nonburning case. The transfer number becomes

$$\frac{h_\infty - h_s}{L_{\text{eff}}} = \frac{Y_{Fs} - Y_{F\infty}}{1 - Y_{Fs}} = B_v. \tag{B.26}$$

The phase-equilibrium law for $Y_{Fs}(T_s)$ and the equation of state for $h(T)$ together with Eq. (B.26) allow the determination of the surface temperature T_s. Again, the results of Eqs. (B.25) and (B.26) are very robust.

The superscalar can be developed for case of vaporization of a multicomponent liquid if all species have the same gas-phase mass diffusivity, Lewis number has unity value, and surface temperature is uniform. The boundary conditions at the surface by Eqs. (2.6) and (4.1) are used. The superscalar

$$S_{mv} = h + \frac{L_{\text{eff}}(Y_m - Y_{m,\infty})}{\varepsilon_m - Y_{ms}} \tag{B.27}$$

will have the same value regardless of the value of the index m, denoting the particular vaporizing species; that value will be uniform through the gas phase. We have defined here that

$$L_{\text{eff}} = \sum \varepsilon_m \left[L_m + \frac{4\pi R^2}{\dot{m}} \lambda \frac{\partial T}{\partial n} \right] = \left(\sum \varepsilon_m L_m \right) + \frac{4\pi R^2}{\dot{m}} \lambda \frac{\partial T}{\partial n}, \tag{B.28}$$

where L_m, ε_m, and n are respectively the latent heat for species m, the mass-flux fraction for species m, and the direction locally normal to the liquid–gas interface, which is positive outward from the liquid. Clearly, $\sum \varepsilon_m = 1$. Now it can be deduced from Eq. (B.27) that Eq. (4.4) holds and

$$B_H = B_M = B_m \tag{B.29}$$

or, equivalently,

$$\frac{h_\infty - h_s}{L_{\text{eff}}} = \frac{Y_{Fs} - Y_\infty}{1 - Y_{Fs}} = \frac{Y_{ms} - Y_\infty}{\varepsilon_m - Y_{ms}}, \tag{B.30}$$

where $Y_F = \sum Y_m$ with the sum over only the components present in the liquid.

APPENDIX C

Droplet-Model Summary

The purpose of this appendix is to summarize the algorithms available to predict in an approximate but reasonable manner droplet heating, vaporization, and trajectory in a user-friendly manner. We consider single-component and multicomponent droplets, isolated droplets and droplets interacting with droplets, and stagnant and moving droplets. We do not consider turbulence or critical thermodynamic conditions in this section. The underlying theories are detailed in Chapters 2, 3, and 4.

The description of the quasi-steady gas film can be summarized by use of drag coefficient C_D, lift coefficient C_L, Nusselt number Nu, Sherwood number Sh, and a nondimensional vaporization rate $\dot{m}/4\pi\rho DR$. These quantities can be prescribed as functions of the transfer number B, the droplet Reynolds number Re, and a nondimensional spacing between neighboring droplets.

The vaporization rate and the other parameters will depend (through the parameters B and Re) on the ambient conditions outside the gas film and on the droplet-surface conditions. To determine the vaporization rate, Nusselt number, and Sherwood number quantitatively, we must simultaneously solve for the temperature and composition at the droplet surface. Typically the solution of diffusion equations for heat and mass in the droplet interior must be solved. These solutions of the diffusion equations require some determination of the internal droplet-velocity field.

For the single-component, isolated droplet, there are three approximate models that have been considered in Chapters 2 and 3. Table C.1 presents the summary for the spherically symmetric case; Tables C.2 and C.3 present the vortex model and the effective-conductivity model, respectively.

The tables are intended to give samples of logical combinations of gas-film and liquid-core models. Other combinations could be useful in special circumstances. For example, the type (iv) and type (v) models in Tables C.2 and C.3 could be interchanged. Or a type (ii) model might be substituted into Table C.2 or C.3. Each model has its advantages and limitations. Note that the outer boundary condition on the liquid-core heat diffusion equation has been constructed in each table with the gas-film model in that table. If a change in the liquid-heating model is made, the boundary condition must be reformulated.

Table C.1. *Spherically symmetric droplet model. Primary equations and sources*

Vaporization rate

$$\frac{\dot{m}}{4\pi\rho DR} = \ln(1 + B). \tag{2.10}$$

Nusselt number

$$\mathrm{Nu} = 2\frac{\ln(1 + B)}{B}. \tag{2.15a}$$

Sherwood number

$$\mathrm{Sh} = 2\frac{\ln(1 + B)}{B}. \tag{2.15b}$$

Liquid heat diffusion: type (iii) model

$$\frac{\partial T_l}{\partial r^2} + \frac{2}{r}\frac{\partial T_l}{\partial r}\Bigg), \tag{2.16a}$$

$$\left.\frac{\partial T_l}{\partial r}\right)_s = \frac{\ln(1 + B)}{R}\frac{\lambda}{\lambda_l}\left[\frac{(T_\infty - T_s + \nu Q Y_{O\infty}/c_p)}{B} - \frac{L}{c_p}\right], \tag{2.16b}$$

$$\left.\frac{\partial T_l}{\partial r}\right)_{r=0} = 0. \tag{2.16c}$$

Comments: See Eq. (2.11) for the definition of B. Le $= 1$ has been assumed. $\rho D =$ constant. No radiation. Uniform initial droplet temperature is typically chosen. Phase equilibrium is given by Eq. (2.4a).

With multicomponent droplets, additional analysis of the liquid-phase mass diffusion is necessary for the prediction of the vaporization rates of each component. Several model choices are identified in Chapter 4. They are summarized in Tables C.4–C.6.

Table C.2. Re $\gg 1$, Pe$_l \gg 1$ *Droplet-vortex model. Primary equations and sources*

Vaporization rate

$$\frac{\dot{m}}{4\pi\rho DR} = \frac{k}{2}[-f(0)]\mathrm{Re}^{1/2}. \tag{3.30}$$

Nusselt number

$$\mathrm{Nu} = k[-f(0)]\mathrm{Re}^{1/2}/B. \tag{3.29}$$

Sherwood number

$$\mathrm{Sh} = k[-f(0)]\mathrm{Re}^{1/2}/B. \tag{3.29}$$

Liquid heat diffusion: Type (v) model

$$\frac{\partial T_l}{\partial T_l}\partial\phi^2 + b(\phi, \tau)\frac{\partial T_l}{\partial\phi}, \tag{3.48a}$$

$$\left.\frac{\partial T_l}{\partial\phi}\right)_s = k[-f(0)]\frac{\mathrm{Re}^{1/2}}{8}\frac{\lambda}{\lambda_l}\left[\frac{(T_\infty - T_s + \nu Q Y_{O\infty}/c_p)}{B} - \frac{L}{c_p}\right], \tag{3.48b}$$

$$\left.\frac{\partial T_l}{\partial\tau}\right)_{\phi=0} = b(0, \tau)\frac{\partial T_l}{\partial\phi}. \tag{3.48c}$$

Comments: See Eq. (2.8) for the definition of B. Le $= 1$ and $\rho D =$ constant have been assumed. No radiation is considered. Uniform initial droplet temperature is typically chosen. $k = 0.552\sqrt{2}$ to $0.6\sqrt{2}$; $a(\phi, \tau)$ and $b(\phi, \tau)$ are prescribed in Chapter 3. Phase equilibrium is given by Eq. (2.4a).

Table C.3. *Robust droplet model. Primary equations and sources*

Vaporization rate

$$\frac{\dot{m} c_p}{4\pi \lambda R} = \ln(1 + B_H) \left[1 + \frac{k}{2} \frac{\mathrm{Pr}^{1/3} \mathrm{Re}^{1/2}}{F(B_H)} \right], \tag{3.34b}$$

$$\frac{\dot{m}}{4\pi \rho D R} = \ln(1 + B_M) \left[1 + \frac{k}{2} \frac{\mathrm{Sc}^{1/3} \mathrm{Re}^{1/2}}{F(B_M)} \right]. \tag{3.34b}$$

Nusselt number

$$\mathrm{Nu} = \frac{2 \ln(1 + B_H)}{B_H} \left[1 + \frac{k}{2} \frac{\mathrm{Pr}^{1/3} \mathrm{Re}^{1/2}}{F(B_H)} \right]. \tag{3.34a}$$

Sherwood number

$$\mathrm{Sh} = \frac{2 \ln(1 + B_M)}{B_M} \left[1 + \frac{k}{2} \frac{\mathrm{Sc}^{1/3} \mathrm{Re}^{1/2}}{F(B_M)} \right]. \tag{3.34b}$$

Liquid heat diffusion: Type (iv) model

$$r_s^2 \frac{\partial z}{\partial \tau} - \beta \zeta \frac{\partial z}{\partial \zeta} = \frac{\chi}{\zeta^2} \frac{\partial}{\partial \zeta} \left(\zeta^2 \frac{\partial z}{\partial \zeta} \right), \tag{3.51}$$

$$\left. \frac{\partial z}{\partial \zeta} \right)_s = \ln(1 + B_H) \left[1 + \frac{k}{2} \frac{\mathrm{Pr}^{1/3} \mathrm{Re}^{1/2}}{F(B_H)} \right] \frac{\lambda}{\lambda_l} \left[\frac{T_\infty - T_s}{T_0 B_H} - \frac{L}{c_p T_0} \right], \tag{3.52a}$$

$$\left. \frac{\partial z}{\partial \zeta} \right)_{\zeta=0} = 0. \tag{3.52b}$$

Comments: Note that Eqs. (3.62a) and (3.62b) can replace Eqs. (3.34a) and (3.34b) to give a potentially more robust model. See Eqs. (3.33), (3.36), (3.51), (3.52c), and (3.52d) for the definitions of B_H, B_M, $F(B)$, β, ζ, τ, z, α_{leff}, and Pe_l. These equations apply for a wide range of Re, Pe_l, Pr, and Sc. No radiation is considered. Uniform initial droplet temperature is typically chosen. $k = 0.552\sqrt{2}$ to $0.6\sqrt{2}$. Phase equilibrium is given by Eq. (2.4a).

Table C.4. *Spherically symmetric multicomponent-droplet model. Primary equations and sources*

Fractional vaporization rate

$$B_m = \frac{\left(\sum_{i,F} Y_i \right)_s - \left(\sum_{i,F} Y_i \right)_\infty}{1 - \left(\sum_{i,F} Y_i \right)_s} = \frac{Y_{m,F,s} - Y_{m,F,\infty}}{\varepsilon_{m,F} - Y_{m,F,s}}. \tag{4.4}$$

Species-vaporization rate

$$\dot{m}_m = \varepsilon_m \dot{m} = 4\pi \rho D R \varepsilon_m \ln(1 + B_m). \tag{2.10, 4.1b}$$

Liquid mass diffusion: Type (iii) model

$$\frac{\partial Y_{l,m}}{\partial t} = D_l \left(\frac{\partial^2 Y_{l,m}}{\partial r^2} + \frac{2}{r} \frac{\partial Y_{l,m}}{\partial r} \right), \tag{4.5a}$$

$$\left. \frac{\partial Y_{l,m}}{\partial r} \right)_s = \frac{\rho D}{\rho_l D_l} \frac{\ln(1 + B)}{R} (Y_{l,ms} - \varepsilon_m), \tag{4.5b}$$

$$\left. \frac{\partial Y_{l,m}}{\partial r} \right)_{r=0} = 0. \tag{4.5c}$$

Comments: All of the equations in Table C.1 apply simultaneously. Phase equilibrium is now provided by Eq. (4.3). Uniform initial liquid-phase concentration is typically chosen. All binary mass diffusion coefficients in the liquid phase are identical. All gas-phase binary diffusion coefficients are equal.

Table C.5. Re $\gg$ 1, Pe$_l$ $\gg$ 1, *Multicomponent-droplet-vortex model. Primary equations and sources*

Fractional vaporization rate

$$B_m = \frac{\left(\sum_{i,F} Y_i\right)_s - \left(\sum_{i,F} Y_i\right)_\infty}{1 - \left(\sum_{i,F} Y_i\right)_s} = \frac{Y_{m,F,s} - Y_{m,F,\infty}}{\varepsilon_{m,F} - Y_{m,F,s}}. \quad (4.4)$$

Species-vaporization rate

$$\dot{m}_m = \varepsilon_m \dot{m} = 2\pi\rho D R \varepsilon_m k[-f(0)]\mathrm{Re}^{1/2}. \quad (3.30, 4.1b)$$

Liquid mass diffusion: Type (v) model

$$\frac{1}{\gamma}\frac{\partial Y_{l,m}}{\partial \tau} = a(\phi, \tau)\frac{\partial^2 Y_{l,m}}{\partial \phi^2} + b(\phi, \tau)\frac{\partial Y_{l,m}}{\partial \phi}, \quad (4.6a)$$

$$\left.\frac{\partial Y_{l,m}}{\partial \phi}\right) = \frac{k[-f(0)]\mathrm{Re}^{1/2}}{8}\frac{\rho D}{\rho_l D_l}(Y_{l,ms} - \varepsilon_m), \quad (4.6b)$$

$$\left.\frac{\partial Y_{l,m}}{\partial \tau}\right)_{\phi=0} = \gamma b(0, \tau)\left.\frac{\partial Y_{l,m}}{\partial \phi}\right)_{\phi=0}. \quad (4.6c)$$

Comments: All of the equations in Table C.2 apply simultaneously. Phase equilibrium is now provided by Eq. (4.3). Uniform initial liquid-phase concentration is typically chosen. All binary diffusion coefficients in the liquid phase are identical. All gas-phase binary diffusion coefficients are equal.

Table C.6. *Robust multicomponent-droplet model. Primary equations and sources*

Fractional vaporization rate

$$B_M = \frac{\left(\sum_{i,F} Y_i\right)_s - \left(\sum_{i,F} Y_i\right)_\infty}{1 - \left(\sum_{i,F} Y_i\right)_s} = \frac{Y_{m,F,s} - Y_{m,F,\infty}}{\varepsilon_{m,F} - Y_{m,F,s}}. \quad (4.4)$$

Species-vaporization rate

$$\dot{m}_m = \varepsilon_m \dot{m} = 4\pi\rho D R \varepsilon_m \ln(1 + B_M)\left[1 + \frac{k}{2}\frac{\mathrm{Sc}^{1/3}\mathrm{Re}^{1/2}}{F(B_M)}\right]. \quad (3.34c, 4.1b)$$

Liquid mass diffusion: Type (iv) model

$$\frac{1}{\gamma}\left(r_s^2\frac{\partial Y_{l,m}}{\partial \tau} - \beta\xi\frac{\partial Y_{l,m}}{\partial \zeta}\right) = \frac{\chi_d}{\zeta^2}\frac{\partial}{\partial \zeta}\left(\zeta^2\frac{\partial Y_{l,m}}{\partial \zeta}\right), \quad (4.8a)$$

$$\left.\frac{\partial Y_{l,m}}{\partial \zeta}\right)_{\zeta=0} = 0. \quad (4.8c)$$

Comments: All of the equations in Table C.3 apply simultaneously. Phase equilibrium is now provided by Eq. (4.3). Uniform initial liquid-phase concentration is typically chosen. All binary diffusion coefficients in the liquid phase are identical. All gas-phase binary diffusion coefficients are equal.

Bibliography

Abdel-Hameed, H., and Bellan, J., 2002, "Direct Numerical Simulations of Two-Phase Laminar Jet Flows with Different Cross-Section Injection Geometries," *Phys. Fluids*, **14**, 3655–74.

Abdulagatov, I.M. and Rasulov, S.M., 1997, "Transport Properties of *n*-Pentane in the Vicinity of the Critical Point," *High Temp.* **35**, 220–25.

Abramzon, B. and Sirignano, W.A., 1989, "Droplet Vaporization Model for Spray Combustion Calculations," *Int. J. Heat Mass Transfer* **32**, 1605–18.

Ackerman, M. and Williams, F.A., 2005, "Simplified Model for Droplet Combustion in a Slow Convective Flow," *Combust. Flame* **143**, 599–612.

Acrivos, A. and Goddard, J., 1965, "Asymptotic Expansion for Laminar Forced Convection Heat and Mass Transfer," *J. Fluid Mech.* **23**, 273–91.

Acrivos, A. and Taylor, T.D., 1962, "Heat and Mass Transfer from Single Spheres in Stokes Flow," *Phys. Fluids* **5**, 387–94.

Aggarwal, S.K., 1987, "Modeling of Multicomponent Fuel Spray Vaporization," *Int. J. Heat Mass Transfer* **30**, 1949–61.

Aggarwal, S.K., 1989, "Ignition Behavior of a Dilute Multicomponent Fuel Spray," *Combust. Flame* **76**, 5–15.

Aggarwal, S.K., 1998, "A Review of Spray Ignition Phenomenon: Present Status and Future Research," *Prog. Energy Combust. Sci.* **24/6**, 565–600.

Aggarwal, S.K., Chen, G., Yapo, J.B., Grinstein, F.F., and Kailasnath, K., 1996, "Numerical Simulations of Particle Dynamics in Planar Shear Layer," *Comput. Fluids* **25**, 39–59.

Aggarwal, S.K., Fix, G.J., Lee, D.N., and Sirignano, W.A., 1983, "Numerical Optimization Studies of Axisymmetric Unsteady Sprays," *J. Comput. Phys.* **50**, 101–15.

Aggarwal, S.K., Fix, G.J., and Sirignano, W.A., 1985, "Two-Phase Laminar Axisymmetric Jet Flow: Explicit, Implicit and Split-Operator Approximations," *Numer. Methods Partial Differential Equations* **1**, 279–94.

Aggarwal, S.K., Lee, D.N., Fix, G.J., and Sirignano, W.A., 1981, "Numerical Computation of Fuel Air Mixing in a Two-Phase Axisymmetric Coaxial Free Jet Flow," in *Proceedings of the Fourth IMACS International Symposium on Computer Methods for Partial Differential Equations.* IMACS.

Aggarwal, S.K. and Mongia, H., 2002, "Multicomponent and High-Pressure Effects on Droplet Vaporization," *J. Eng. Gas Turbines Power* **124**, 248–57.

Aggarwal, S.K., Park, T.W., and Katta, V.R., 1996, "Unsteady Spray Behavior in a Heated Jet Shear Layer: Droplet–Vortex Interactions," *Combust. Sci. Technol.* **113–114**, 429–49.

Aggarwal, S.K. and Sirignano, W.A., 1980, "One-Dimensional Turbulent Flame Propagation in an Air Fuel Droplet Mixture," *ASME Preprint* 80-WA/HT-37, American Society of Mechanical Engineers: New York.

Aggarwal, S.K. and Sirignano, W.A., 1982, "Computation of Laminar Spray Flames; Hybrid-Eulerian-Lagrangian Scheme," Eastern Section Combustion Institute Fall Technical Meeting, Atlantic City, New Jersey.

Aggarwal, S.K. and Sirignano, W.A., 1984, "Numerical Modeling of One-Dimensional Enclosed Homogeneous and Heterogeneous Deflagrations," *Comput. Fluids* **12**, 145–58.

Aggarwal, S.K. and Sirignano, W.A., 1985a, "Ignition of Fuel Sprays: Deterministic Calculations for Idealized Droplet Arrays," in *Proceedings of the Twentieth Symposium (International) on Combustion*, 1773–80. The Combustion Institute: Pittsburgh, PA.

Aggarwal, S.K. and Sirignano, W.A., 1985b, "Unsteady Spray Flame Propagation in a Closed Volume," *Combust. Flame* **62**, 69–84.

Aggarwal, S.K. and Sirignano, W.A., 1986, "An Ignition Study of Polydisperse Sprays," *Combust. Sci. Technol.* **46**, 289–300.

Aggarwal, S.K., Tong, A.Y., and Sirignano, W.A., 1984, "A Comparison of Vaporization Models in Spray Calculations," *AIAA J.* **22**, 1448–57.

Aggarwal, S.K., Yan, C., and Zhu, G., 2002, "On the Transcritical Vaporization of a Liquid Fuel Droplet in Supercritical Ambient," *Combust. Sci. Technol.* **174** (9), 103–30.

Ahmed, A.M. and Elghobashi, S.E., 2000, "On the Mechanisms of Modifying the Structure of Turbulent Homogeneous Shear Flows by Dispersed Particles," *Phys. of Fluids* **12**, 2906–30.

Ahmed, A.M. and Elghobashi, S.E., 2001, "Direct Numerical Simulation of Particle Dispersion in Turbulent Homogeneous Shear Flows," *Phy. Fluids* **13**, 3346–64.

Alder, B.J. and Wainwright, T.E., 1959, "Studies in Molecular Dynamics. 1. General Method," *J. Chem. Phys.* **31**, 459–66.

Alymore, D.W., Gregg, S.J., and Jepson, W.B., 1960, "The Oxidation of Aluminum in Dry Oxygen in the Temperature Range 400–600°C," *J. Inst. Met.* **88**, 205.

Ames, W.R., 1977, *Numerical Methods for Partial Differential Equations*, 2nd ed. Academic: New York.

Annamalai, K. and Ryan, W., 1992, "Interactive Processes in Gasification and Combustion. Part I: Liquid Drop Arrays and Clouds," *Prog. Energy Combust. Sci.* **18**, 221–95.

Antaki, P.J., 1986 "Transient Process in a Rigid Slurry Droplet During Liquid Vaporization and Combustion," *Combust. Sci. Technol.* **46**, 113–35.

Antaki, P.J. and Williams, F.A., 1986, "Transient Process in a Non-Rigid Slurry Droplet During Liquid Vaporization and Combustion," *Combust. Sci.* **49**, 289–96.

Antaki, P.J. and Williams, F.A., 1987, "Observations on the Combustion of Boron Slurry Droplets in Air," *Combust. Flame* **67**, 1–8.

Archambault, M.R., Edwards, C.F., and MacCormack, R.W., 2003a, "Computation of Spray Dynamics by Moment Transport Equations I: Theory and Development," *Atomization Sprays* **13**, 63–87.

Archambault, M.R., Edwards, C.F., and MacCormack, R.W., 2003b, "Computation of Spray Dynamics by Moment Transport Equations II: Application to Calculation of a Quasi-One-Dimensional Spray," *Atomization Sprays* **13**, 89–115.

Archambault, M., MacCormack, R.W., and Edwards, C.F., 1998, "A Maximum Entropy Moment Closure Approach to Describing Spray Flows," in *Proceedings of the ILASS Americas Eleventh Annual Conference on Liquid Atomization and Spray Systems*, 381–85. Institute of Liquid Atomization and Spray Systems: Irvine, CA.

Archer, L.A., Ternet, D., and Larson, R.G., 1997, "Fracture Phenomena in Shearing Flow of Viscous Liquids," *Rheol. Acta* **36**, 579.

Asano, K., Taniguchi, I., and Kawahara, T., 1988, "Numerical and Experimental Approaches to Simultaneous Vaporization of Two Adjacent Volatile Drops," in *Proceedings of the Fourth International Conference on Liquid Atomization and Sprays*, 411–18. Institute of Liquid Atomization and Spray Systems: Irvine, CA.

Ashgriz, N. and Mashayek, F., 1995, "Temporal Analysis of Capillary Jet Breakup," *J. Fluid Mech.* **219**, 163–90.

Ashgriz, N. and Poo, J.Y., 1990, "Coalescence and Separation in Binary Collisions of Liquid Drops," *J. Fluid Mech.* **221**, 183–204.

Bankston, C.P., Bock, L.H., Kwock, E.Y., and Kelly, A.J., 1988, "Experimental Investigation of Electrostatic Dispersion and Combustion of Diesel Fuel Jets," *J. Eng. Gas Turbines Power* **110**, 361–68.

Basset, A.B., 1888, "On the Motion of a Sphere in a Viscous Liquid," *Philos. Trans. R. Soc. London Ser. A* **179**, 43–63. Also in, 1961, *A Treatise on Hydrodynamics* 2, 285. Dover: New York.

Batchelor, G.K., 1990, *An Introduction to Fluid Dynamics.* Cambridge University Press: Cambridge.

Batchelor, G.K. and Green, J.T., 1972, "The Hydrodynamic Interaction of Two Small Freely-Moving Spheres in a Linear Flow Field," *J. Fluid Mech.* **56**, 375–400.

Bayvel, L. and Orzechowski, Z., 1993, *Liquid Atomization.* Taylor & Francis: Washington, D.C.

Bear, J., 1972, *Dynamics of Fluids in Porous Media.* American Elsevier: New York.

Bechtel, S.E., Carlson, S.D., and Forest M.G., 1995, "Recovery of the Rayleigh Capillary Instability from Slender 1-D Inviscid and Viscous Models," *Phys. Fluids* **7**, 2956–71.

Bechtel, S.E., Forest, M.G., and Lin, K.J., 1992, "Closure to All Orders in 1-D Models for Slender Viscoelastic Jets: An Integrated Theory of Axisymmetric, Torsionless Flow," *Stability Appl. Anal. Continuous Media* **2**, 59.

Bellan, J., 2000a, "Supercritical (and Subcritical) Fluid Behavior and Modeling: Drops, Streams, Shear and Mixing Layers, Jets and Sprays," *Prog. Energy Combust. Sci.* **26**, 329–66.

Bellan, J., 2000b, "Perspectives on Large Eddy Simulations for Sprays: Issues and Solutions," *Atomization Sprays* **10**, 409–25.

Bellan, J., 2006, "Theory, Modeling and Analysis of Turbulent Supercritical Mixing," *Combust. Sci. Technol.* **178**, 253–81.

Bellan, J. and Cuffel, R., 1983, "A Theory of Non-Dilute Spray Evaporation Based Upon Multiple Drop Interactions," *Combust. Flame* **51**, 55–67.

Bellan, J. and Harstad, K., 1987, "Analysis of the Convective Evaporation of Non-Dilute Clusters of Drops," *Int. J. Heat Mass Transfer* **30**, 125–36.

Bellan, J. and Harstad, K., 1988, "Turbulence Effects During Evaporation of Drops in Clusters," *Int. J. Heat Mass Transfer* **31**, 1655–68.

Bellan, J. and Harstad, K., 1992, "Evaporation of Steadily Injected, Identical Clusters of Drops Embedded in Jet Vortices," in *Proceedings of the Twenty-Fourth Symposium (International) on Combustion*, Poster Session. The Combustion Institute: Pittsburgh, PA.

Bellan, J. and Summerfield, M., 1978a, "Theoretical Examination of Assumptions Commonly Used for the Gas Phase Surrounding a Burning Droplet," *Combust. Flame* **33**, 107–22.

Bellan, J. and Summerfield, M., 1978b, "A Preliminary Theoretical Study of Droplet Extinction by Depressurization," *Combust. Flame* **32**, 257–70.

Berlemont, A., Desjonqueres, P., and Gouesbet, G., 1990, "Particle Lagrangian Simulation in Turbulent Flows," *Int. J. Multiphase Flow* **16**(1), 19–34.

Berlemont, A., Grancher, M., and Gousebet, G., 1991, "On the Lagrangian Simulation of Turbulence Influence on Droplet Evaporation," *Int. J. Heat Mass Transfer* **34**, 2805–12.

Bhatia, R., 1993, "Vaporization and Combustion of Metal Slurry Droplets," Ph.D. Dissertation, Department of Mechanical and Aerospace Engineering, University of California, Irvine.

Bhatia, R. and Sirignano, W.A., 1991, "A One-Dimensional Analysis of Liquid-Fueled Combustion Instability," *J. Propul. Power* **7**, 953–61.

Bhatia, R. and Sirignano, W.A., 1992, "Transient Heating and Burning of Droplet Containing a Single Metal Particle," *Combust. Sci. Technol.* **84**, 953–61.

Bhatia, R. and Sirignano, W.A., 1993a, "Convective Burning of a Droplet Containing a Single Metal Particle," *Combust. Flame* **93**, 215–29.

Bhatia, R. and Sirignano, W.A., 1993b, "Liquid Vaporization from Fine-Metal Slurry Droplets," *Dynamics of Heterogeneous Combustion and Reacting Systems.* Prog. Astronaut. Aeronaut. **152**, 235–62.

Bhatia, R. and Sirignano, W.A., 1995, "Flame Propagation and Metal Burning in Metal Slurry Sprays," *Combust. Flame* **100**, 605–20.

Birouk, M. and Gökalp, I., 2002, "A New Correlation for Turbulent Mass Transfer from Liquid Droplets," *Int. J. Heat Mass Transfer* **45**, 37–45.

Birouk, M. and Gökalp, I., 2006, "Current Status of Droplet Evaporation in Turbulent Flows," *Prog. Energy Combust. Sci.* **32**, 408–23.

Bogy, D.B., 1978a, "Wave Propagation and Instability in a Circular Semi-Infinite Liquid Jet Harmonically Forced at the Nozzle," *J. Appl. Mech.* **45**, 469–74.

Bogy, D.B., 1978b, "Use of One-Dimensional Cosserat Theory to Study Instability in a Viscous Liquid Jet," *Phys. Fluids* **21**, 190–97.

Bogy, D.B., 1979, "Drop Formation in a Circular Jet," *Annu. Rev. Fluid Mech.* **11**, 207–28.

Boivin, M., Simonin, O., and Squires, K.D., 1998, "Direct Numerical Simulation of Turbulence Modulation by Particles in Isotropic Turbulence," *J. Fluid Mech.* **375**, 235–63.

Borghi, R. and Loison, S., 1992, "Studies of Dense Sprays Combustion by Numerical Simulation with a Cellular Automaton," in *Proceedings of the Twenty-Fourth Symposium (International) on Combustion*, 1541–47. The Combustion Institute: Pittsburgh, PA.

Botros, P., Law, C.K., and Sirignano, W.A., 1980, "A Droplet Combustion in a Reactive Environment," *Combust. Sci. Technol.* **21**, 123–30.

Boussinesq, J., 1903, *Theorie Analytique de la Chaleur*, Vol. 2, 224. L'Ecole Polytechnique: Paris.

Bowman, J.R. and Edmister, W.C., 1951, "Flash Distillation of an Indefinite Number of Components," *Ind. Chem. Eng.* **43**, 2625–28.

Bracco, F.V., 1973, "Nitric Oxide Formation in Droplet Diffusion Flames," in *Proceedings of the Fourteenth Symposium (International) on Combustion*, 831–42. The Combustion Institute: Pittsburgh, PA.

Bracco, F.V., 1974, "Unsteady Combustion of a Confined Spray," *AIChe Symp. Ser. Heat Transfer Res. Des.* **70** (138), 48–56.

Bracco, F.V., 1985, "Modeling of Engine Sprays," SAE Preprint 850394. Society of Automotive Engineers: Warrendale, PA.

Brenner, M.P., Shi, X.D., and Nagel, S.R., 1994, "Iterated Instability During Droplet Fission," *Phys. Rev. Lett.* **73**, 3391–94.

Briano, J.G. and Glandt, E.D., 1983, "Molecular Thermodynamics of Continuous Mixtures," *Fluid Phase Equilibria* **14**, 91–102.

Briley, W.R. and McDonald, H., 1977, "Solution of the Multidimensional Navier–Stokes Equations by a Generalized Implicit Method," *J. Comput. Phys.* **24**, 372.

Brunn, P.O., 1982, "Heat or Mass Transfer From Single Spheres in Low Reynolds Number Flow," *Int. J. Eng. Sci.* **20**, 817–22.

Buckholz, E.K. and Tapper, M.L., 1978, "Time to Extinction of Liquid Hydrocarbon Fuel Droplets Burning in a Transient Diesel-Like Environment," *Combust. Flame* **31**, 161–72.

Bulthuis, H.F., Prosperetti, A., and Sangani, A.S., 1995, "Particle Stress in Disperse Two-Phase Potential Flow," *J. Fluid Mech.* **294**, 1–16.

Cai, J., Liu, F., and Sirignano, W.A., 2002, "Three-Dimensional Flame Propagation Above Liquid Fuel Pools," *Combust. Sci. Technol.* **174**, 5–34.

Cai, J., Liu, F., and Sirignano, W.A., 2003, "Three-Dimensional Structures of Flames Over Liquid Fuel Pools," *Combust. Sci. Technol.* **175**, 2113–39.

Card, J.M. and Williams, F.A., 1992a, "Asymptotic Analysis of the Structure and Extinction of Spherically Symmetrical *n*-Heptane Diffusion Flames," *Combust. Sci. Technol.* **84**, 91–119.

Card, J.M. and Williams, F.A., 1992b, "Asymptotic Analysis with Reduced Chemistry for the Burning of *n*-Heptane Droplets," *Combust. Flame* **91**, 187–99.

Cetegen, B.M. and Sirignano, W.A., 1988, "Study of Molecular Mixing and Finite-Rate Chemical Reaction in a Mixing Layer," in *Proceedings of the International Symposium on Combustion*, Vol. 22, 489–94, The Combustion Institute: Pittsburgh, PA.

Cetegen, B.M. and Sirignano, W.A., 1990, "Study of Mixing and Reaction in the Field of a Vortex," *Combust. Sci. Technol.* **72**, 157–81.

Chakravarthy, V.K. and Menon, S., 2000, "Large-Eddy Simulations of Turbulent Premixed Flames in the Flamelet Regime," *Combust. Sci. Technol.* **162**, 1–48.

Chakravarthy, V.K. and Menon, S., 2001, "Linear-Eddy Simulations of Reynolds and Schmidt Number Dependencies in Turbulent Scalar Mixing," *Phys. Fluids* **13**, 488–99.

Chang, E.J. and Maxey, M.R., 1994, "Accelerated Motion of Rigid Spheres in Unsteady Flow at Low to Moderate Reynolds Numbers. Part 1. Oscillatory Motion," *J. Fluid Mech.* **277**, 347–79.

Chang, E.J. and Maxey, M.R., 1995, "Unsteady Flow About a Sphere at Low to Moderate Reynolds Numbers. Part 2. Accelerated Motion," *J. Fluid Mech.* **303**, 133–53.

Chao, B.T., 1962, "Motion of Spherical Gas Bubbles in a Viscous Liquid at Large Reynolds Numbers," *Phys. Fluids* **5**, 69–79.

Chao, B.T., 1964, "Turbulent Transport Behavior of Small Particles in Dilute Suspension," *Osterr. Ing. Arch.* **18**, 7.

Chao, Y.-C., Chen, G.-B., and Tsai, M.-H., 2001, "Investigation of the Performance Characteristics of a Miniature Catalytic Gas Turbine Combustor," 4th Pacific International Conference on Aerospace Science and Technology, Kaohsiung, Taiwan, May 21–23.

Charlesworth, D.H. and Marshall Jr., W.R., 1960, "Evaporation from Drops Containing Dissolved Solids," *AIChE J.* **6**, 9–23.

Chaudhary, K.C. and Maxworthy, T., 1980a, "The Nonlinear Capillary Instability of a Liquid Jet: Part 2. Experiments on Jet Behavior Before Droplet Formation," *J. Fluid Mech.* **96** 275–86.

Chaudhary, K.C. and Maxworthy, T., 1980b, "The Nonlinear Capillary Instability of a Liquid Jet: Part 3. Experiments on Satellite Drop Formation and Control," *J. Fluid Mech.* **96**, 287–97.

Chaudhary, K.C. and Redekopp, L., 1980, "The Nonlinear Capillary Instability of a Liquid Jet: Part 1.Theory," *J. Fluid Mech.* **96**, 257–74.

Chein, R. and Chung, J.N., 1987, "Effects of Vortex Pairing on Particle Dispersion in Turbulent Shear Flows," *Int. J. Multiphase Flow* **13**, 785–802.

Chein, R. and Chung, J.N., 1988, "Simulation of Particle Dispersion in a Two-Dimensional Mixing Layer," *AIChE J.* **34**, 946–54.

Chen, N.-H., Rogg, B., and Bray, K.N.C., 1992, "Modelling Laminar Two-Phase Counterflow Flames with Detailed Chemistry and Transport," in *Proceedings of the Twenty-Fourth Symposium (International) on Combustion*, 1513–21. The Combustion Institute: Pittsburgh, PA.

Chen, Z.H., Lin, T.H., and Sohrab, S.H., 1988, "Combustion of Liquid Fuel Sprays in Stagnation-Point Flow," *Combust. Sci. Technol.* **60**, 63–77.

Cheng, H.Y., Anisimov, M.A., and Sengers, J.V., 1997, "Prediction of Thermodynamic and Transport Properties in the One-Phase Region of Methane (*n*-Hexane Mixtures Near Their Critical End Points)," *Fluid Phase Equilibria* **128**, 67–96.

Chervinsky, A., 1969, "Transient Burning of Spherical Symmetric Fuel Droplets," *Isr. Ins. Technol.* **7**, 35–42.

Chiang, C.H., 1990, "Isolated and Interacting, Vaporizing Fuel Droplets: Field Calculation with Variable Properties," Ph.D. Dissertation, University of California, Irvine.

Chiang, C.H., Raju, M.S., and Sirignano, W.A., 1992, "Numerical Analysis of Convecting, Vaporizing Fuel Droplet with Variable Properties," *Int. J. Heat Mass Transfer* **35**, 1307–24.

Chiang, C.H. and Sirignano, W.A., 1991, "Axisymmetric Vaporizing Oxygen Droplet Computations," Preprint 91-0281, AIAA 29th Aerospace Sciences Meeting, Reno, NV.

Chiang, C.H. and Sirignano, W.A., 1993a, "Interacting, Convecting, Vaporizing Fuel Droplets With Variable Properties," *Int. J. Heat Mass Transfer* **36**, 875–86.

Chiang, C.H. and Sirignano, W.A., 1993b, "Axisymmetric Calculations of Three-Droplet Interactions," *Atomization Sprays* **3**, 91–107.

Chigier, N.A., 1976, "The Atomization and Burning of Liquid Fuel Sprays," *Prog. Energy Combust. Sci.* **2**, 97–114.

Chigier, N.A., 1981, *Energy, Combustion, and Environment.* McGraw-Hill: New York.

Chigier, N.A. and McCreath, C.G., 1974, "Combustion of Droplets in Sprays," *Acta Astronaut.* **1**, 687–710.

Chin, L.P., Hsing, P.C., Tankin, R.S., and Jackson, T., 1995, "Comparisons Between Experiments and Predictions Based on Maximum Entropy for the Breakup of a Cylindrical Jet," *Atomization Sprays* **5**, 603–20.

Chiu, H.H., Kim, H.Y., and Croke, E.J., 1983, "Internal Group Combustion of Liquid Droplets," in *Proceedings of the Nineteenth Symposium (International) on Combustion*, 971–80. The Combustion Institute: Pittsburgh, PA.

Chiu, H.H. and Liu, T.M., 1977, "Group Combustion of Liquid Droplets," *Combust. Sci. Technol.* **17**, 127–31.

Cho, S.Y., Choi, M.Y., and Dryer, F.L., 1991, "Extinction of a Free Methanol Droplet in Microgravity," in *Proceedings of the Twenty-Third Symposium (International) on Combustion*. The Combustion Institute: Pittsburgh, PA.

Cho, S.Y., Takahashi, F., and Dryer, F.L., 1989, "Some Theoretical Consideration on the Combustion and Disruption of Free Slurry Droplets," *Combust. Sci. Technol.* **67**, 37–57.

Chou, G.F. and Prausnitz, J.M., 1986, "Adiabatic Flash Calculations for Continuous or Semicontinuous Mixtures Using an Equation of State," *Fluid Phase Equilibria*, **30**, 75–82.

Chueh, P.L. and Prausnitz, J.M., 1967, "Vapor-Liquid Equilibrium at High Pressures. Vapor-Phase Fugacity Coefficients in Nonpolar and Quantum-Gas Mixtures," *I \& EC Fund.* **6**, 493–98.

Chung, J.N., 1982, "The Motion of Particles Inside a Droplet," *J. Heat Transfer* **104**, 438–45.

Chung, J.N., Ayyaswamy, P.S., and Sadhal, S.S., 1984a, "Laminar Condensation on a Moving Drop. Part I. Singular Perturbation Technique," *J. Fluid Mech.* **139**, 105–30.

Chung, J.N., Ayyaswamy, P.S., and Sadhal, S.S., 1984b, "Laminar Condensation on a Moving Drop. Part 2. Numerical Solutions," *J. Fluid Mech.* **139**, 131–44.

Chung, J.N. and Troutt, T.R., 1988, "Simulation of Particle Dispersion in an Axisymmetric Jet," *J. Fluid Mech.* **186**, 199–227.

Chung, M. and Rangel, R.H., 2000, "Simulation of Metal Droplet Deposition with Solidification Including Undercooling and Contact Resistance Effects," *Numer. Heat Transfer* **37**, 201–26.

Chung, M. and Rangel, R.H., 2001, "Parametric Study of Metal Droplet Deposition and Solidification Process Including Contact Resistance and Undercooling Effects," *Int. J. Heat Mass Transfer* **44**, 605–18.

Chung, P.M., 1965, "Chemically Reacting Nonequilibrium Boundary Layers," Vol. 2 of Advances in Heat Transfer Series, Hartnett, J.P. and Irvine, T.F., eds. Academic: New York.

Clark, C.J. and Dombrowski N., 1972, "Aerodynamic Instability and Disintegration of Inviscid Liquid Sheets," *Proc. R. Soc. London Ser. A, Math. Phys. Sci.* **329**, 467–78.

Clift, R., Grace, J.R., and Weber, M.E., 1978, *Bubbles, Drops, and Particles.* Academic: New York.

Coimbra, C.F.M., Edwards, D.K., and Rangel, R.H., 1998, "Heat Transfer in a Homogeneous Suspension Including Radiation and History Effects," *J. Thermophys. Heat Transfer* **12**, 304–12.

Coimbra, C.F.M. and Rangel, R.H., 1998, "General Solution of the Particle Momentum Equation in Unsteady Stokes Flows," *J. Fluid Mech.* **370**, 53–72.

Colby, J., Menon, S., and Jagoda, J., 2005, "Flow Field Measurements in a Counter-Swirling Spray Combustor," AIAA Preprint 2005-4143.

Conner, J.M., 1984, "Calculation of Mass and Momentum Transfer for Laminar Flow Around a Cylinder and a Sphere," M.S. Thesis, University of California, Irvine.

Conner, J.M. and Elghobashi, S.E., 1987, "Numerical Solution of Laminar Flow Past a Sphere with Surface Mass Transfer," *Numer. Heat Transfer* **122**, 57–82.

Connon, C.S. and Dunn-Rankin, D., 1996, "Droplet Stream Dynamics at High Ambient Pressure," *Atomization Sprays* **6**, 485–97.

Consilini, L., Aggarwal, S.K., and Murad, S., 2003, "A Molecular Dynamics Simulation of Droplet Evaporation," *Int. J. Heat Mass Transfer* **46**, 3179–88.

Continillo, G. and Sirignano, W.A., 1988, "Numerical Study of Multicomponent Fuel Spray Flame Propagation in a Spherical Closed Volume," in *Proceedings of the Twenty-Second Symposium (International) on Combustion*, 1941–49. The Combustion Institute: Pittsburgh, PA.

Continillo, G. and Sirignano, W.A., 1990, "Counterflow Spray Combustion Modelling," *Combust. Flame* **81**, 325–40.

Continillo, G. and Sirignano, W.A., 1991, "Unsteady, Spherically Symmetric Flame Propagation Through Multicomponent Fuel Spray Clouds," in *Modern Research Topics in Aerospace Propulsion*, G. Angelino, L. DeLuca, and W.A. Sirignano, eds. Springer-Verlag: New York.

Cook, A.W. and Riley, J.J., 1994, "A Subgrid Model for Equilibrium Chemistry in Turbulent Flows," *Phys. Fluids* **6**, 2868–70.

Cook, A.W. and Riley, J.J., 1998, "Subgrid Scale Modeling for Turbulent Reactive Flows," *Combust. Flame* **112**, 593–606.

Correa, S.M. and Sichel, M., 1983, "The Group Combustion of a Spherical Cloud of Monodisperse Fuel Droplets," in *Proceedings of the Nineteenth Symposium (International) on Combustion*, 981–91. The Combustion Institute: Pittsburgh, PA.

Cotterman, R.L., Bender, R., and Prausnitz, J.M., 1985, "Phase Equilibria for Mixtures Containing Very Many Components. Development and Application of Continuous Thermodynamics for Chemical Process Design," *Ind. Eng. Chem. Process Des. Dev.* **24**, 94–203.

Courant, R. and Hilbert, D., 1962, *Methods of Mathematical Physics*, Vol. 2, Wiley Interscience: New York.

Crapper, G.D. and Dombrowski, N., 1984, "A Note on the Effect of Forced Disturbances on the Stability of Thin Liquid Sheets and on the Resulting Drop Size," *Int. J. Multiphase Flow* **10**, 731–6.

Crapper, G.D., Dombrowski, N., Jepson, W.P., Pyott, G.A.D., 1973, "A Note on the Growth of Kelvin–Helmholtz Waves on Thin Liquid Sheets," *J. Fluid Mech.* **57**, 671–72.

Crapper, G.D., Dombrowski, N. and Pyott, G.A.D, 1975, "Large Amplitude Kelvin–Helmholtz Waves on Thin Liquid Sheets," *Proc. R. Soc. London Ser. A, Phys. Sci.* **342**, 209–24.

Crespo, A. and Linan. A., 1975, "Unsteady Effects in Droplet Evaporation and Combustion," *Combust. Sci. Technol.* **1**, 9–18.

Crocco, L. and Cheng, S.I., 1956, *Theory of Combustion Instability in Liquid Propellant Rocket Motors*, AGARDograph 8. Buttersworth: London.

Crocco, L., Harrje, D.T., and Reardon, F.H., 1962, "Transverse Combustion Instability in Liquid Propellant Rocket Motors," *ARS J.* **32**, 366–73.

Crocco, L. and Sirignano, W.A., 1966, "Effect of the Transverse Velocity Component on the Nonlinear Behavior of Short Nozzles," *AIAA J.* **4**, 1428–30.

Crowe, C.T., 1978, "On the Vapor Droplet Flows Including Boundary Droplet Effects," *Two-Phase Transp. Reactor Saf.* **1**, 385–405.

Crowe, C.T., 1982, "Numerical Models for Dilute Gas-Particle Flows," *J. Fluids Eng.* **104**, 297–301.

Crowe, C.T., Chung, J.N., and Troutt, T.R., 1988, "Particle Mixing in Free Shear Flows," *Prog. Energy Combust. Sci.* **14**, 171–99.

Crowe, C.T., Sharma, M.P., and Stock, D.E., 1977, "The Particle-Source-in Cell (PSI-CELL) Model for Gas-Droplet Flows," *J. Fluids Eng.* **99**, 325.

Crowe, C.T., Sommerfeld, M., and Tsuji, Y., 1998, *Multiphase Flows with Droplets and Particles*. CRC Press: Boca Raton, FL.

Crowe, C.T., Troutt, T.R., and Chung, J.N., 1996, "Numerical Models for Two-Phase Turbulent Flows," *Annu. Rev. Fluid Mech.* **28**, 11–43.

Curtis, E.W. and Farrell, P.V., 1988, "Droplet Vaporization in a Supercritical Microgravity Environment," *Astronaut. Acta* **17**, 1189–93.

Dabiri, S., Sirignano, W.A., and Joseph, D.D., 2007, "Cavitation in an Orifice Flow," *Phys. Fluids* **19**, paper 072112.

Dabiri, S., Sirignano, W.A., and Joseph, D.D., 2008, "Two-Dimensional and Axisymmetric Viscous Flow in Apertures," *J. Fluid Mech.* **605**, 1–18.

Dakka, S.M. and Shaw, B.D., 2006, "Influences of Pressure on Reduced-Gravity Combustion of Propanol Droplets," *Microgravity Sci. Technol.* **18**, 5–13.

Dandy, D.S. and Dwyer, H.H., 1990, "A Sphere in Shear Flow at Finite Reynolds Number: Effect of Shear on Particle Lift, Drag, and Heat Transfer," *J. Fluid Mech.* **216**, 381–410.

Daou, J., Haldenwang, P., and Nicoli, C., 1995, "Supercritical Burning of Liquid Oxygen (LOX) Droplet with Detailed Chemistry," *Combust. Flame* **101**, 153–69.

Dee, V. and Shaw, B.D., 2004, "Combustion of Propanol-Glycerol Mixture Droplets in Reduced Gravity," *Int. J. Heat Mass Transfer* **47**, 4857–67.

Del Alamo, G. and Williams, F.A., 2007, "Theory of Vaporization of a Rigid Spherical Droplet in Slowly Varying Rectilinear Flow at Low Reynolds Number," *J. Fluid Mech.* **580**, 219–49.

Delplanque, J.-P., Lavernia, E.J., and Rangel, R.H., 1996, "Multidirectional Solidification Model for the Description of Micropore Formation in Spray Deposition Processes," *Numer. Heat Transfer A* **30**, 1–18.

Delplanque, J.-P., Lavernia, E.J. and Rangel, R.H., 2000, "Analysis of In-Flight Oxidation During Reactive Spray Atomization and Deposition Processing of Aluminum," *J. Heat Transfer* **122**, 126–32.

Delplanque, J.-P. and Potier, B., 1995, "Rocket Engine Supercritical Combustion: A Review of the Issues and Current Research," *Rech. Aerosp.* **5**, 299–309.

Delplanque, J.-P. and Rangel, R.H., 1991, "Droplet Stream Combustion in the Steady Boundary Layer Near a Wall," *Combust. Sci. Technol.* **78**, 97–115.

Delplanque, J.-P. and Rangel, R.H., 1998, "A Comparison of Models, Numerical Simulation and Experimental Results in Droplet Deposition Processes," *Acta Mater.* **14**, 4925–33.

Delplanque, J.-P. and Rangel, R.H., 1999, "Simulation of Liquid Jet Overflow in Droplet Deposition Processes," *Acta Mater.* **47**, 2207–13.

Delplanque, J.-P., Rangel, R.H., and Sirignano, W.A., 1990, "Liquid-Waste Incineration in a Parallel-Stream Configuration: Parametric Study," in *Proceedings of the Twenty-Third Symposium (International) on Combustion*, 887–94. The Combustion Institute: Pittsburgh, PA.

Delplanque, J.-P., Rangel, R.H., and Sirignano, W.A., 1991, "Liquid-Waste Incineration in a Parallel-Stream Configuration: Effect of Auxiliary Fuel," in *Dynamics of Deflagrations and Reactive Systems: Heterogeneous Combustion*, Vol. 132 of Progress in Astronautics and Aeronautics Series, 164–86. AIAA: Washington, DC.

Delplanque, J.-P. and Sirignano, W.A., 1992, "Oscillatory and Transcritical Vaporization of an Oxygen Droplet," AIAA-92-0104, 30th AIAA Aerospace Sciences Meeting and Exhibit, Reno, NV.

Delplanque, J.-P. and Sirignano, W.A., 1993, "Numerical Study of the Transient Vaporization of an Oxygen Droplet at Sub- and Super-critical Conditions," *Int. J. Heat Mass Transfer* **36**, 303–14.

Delplanque, J.-P. and Sirignano, W.A., 1994, "Boundary Layer Stripping Effects on Droplet Transcritical Convective Vaporization," *Atomization Sprays* **4**, 325–49.

Delplanque, J.-P. and Sirignano, W.A., 1995, "Transcritical Vaporization and Combustion of LOX Droplet Arrays in a Convective Environment," *Combust. Sci. Technol.* **105**, 327–44.

Delplanque, J.-P. and Sirignano, W.A., 1996, "Transcritical Liquid Oxygen Droplet Vaporization: Effect on Rocket Combustion Instability," *J. Propul. Power* **12**, 349–57.

Delplanque, J.-P. and Sirignano, W.A., 1999, "Transcritical Vaporization of Liquid Fuels and Propellants", *J. Propul. Power* **15**, 896–902.

DesJardin, P.E. and Frankel, S.H., 1998, "Large-Eddy Simulation of a Turbulent Non-premixed Reacting Jet: Application and Assessment of Subgrid-Scale Combustion Models," *Phys. Fluids* **10**, 2298–314.

Dombrowski, N. and Fraser, R.P., 1954, "A Photographic Investigation Into the Disintegration of Liquid Sheets," *Philos. Trans. R. Soc. London Ser. A, Math. Phys. Sci.* **247**, 101–30.

Dombrowski, N., Hasson, D., and Ward, D.E., 1960, "Some Aspects on Liquid Flow Through Fan Spray Nozzles," *Chem. Eng. Sci.* **12**, 35–50.

Dombrowski, N. and Johns, W.R., 1963, "The Aerodynamic Instability and Disintegration of Viscous Liquid Sheets," *Chem. Eng. Sci.* **18**, 203–14.

Dorman, R.G., 1952, "The Atomization of Liquid in a Flat Spray," *Brit. J. Appl. Phys.* **3**, 189–192.

Drazin, P.G. and Reid, W.H., 1981, *Hydrodynamic Stability.* Cambridge University Press: New York.

Drew, D.A., 1983, "Mathematical Modeling of Two-Phase Flow," *Annu. Rev. Fluid Mech.* **15**, 261–91.

Drew, D.A. and Passman, S.L., 1999, *Theory of Multicomponent Fluids.* Springer-Verlag: New York.

Drew, T.B., 1931, "Mathematical Attacks on Forced Convection Problems: A Review," *Transactions of the Amer. Inst. Chem. Eng.,* 2626–80.

Druzhinin, O.A. and Elghobashi, S., 1998, "Direct Numerical Simulations of Bubble-laden Turbulent Flows Using the Two-Fluid Formulation," *Phys. Fluids* **10**, 685–97.

Druzhinin, O.A. and Elghobashi, S.E., 2001, "Direct Numerical Simulation of a Three-Dimensional Spatially Developing Droplet-Laden Mixing Layer with Two-Way Coupling," *J. Fluid Mech.* **429**, 23–61.

Dryer, F.L., 1976, "Water Addition to Practical Combustion Systems-Concepts and Applications," in *Proceedings of the Sixteenth Symposium (International) on Combustion,* 279–95. The Combustion Institute: Pittsburgh, PA.

Dukowicz, J.K., 1980, "A Particle-Fluid Numerical Model for Liquid Sprays," *J. Comput. Phys.* **35**, 229–53.

Dumbleton, J.H., and Hermans, J.J., 1970, "Capillary Stability of a Hollow Inviscid Cylinder," *Phys. Fluids* **13**, 12–17.

Dunn-Rankin, D., Leal, E., and Walther, D., 2006, "Personal Power Systems," *Prog. Energy Combust. Sci.* **31**, 422–65.

Duvvur, A., Chiang, C.H., and Sirignano, W.A., 1996, "Oscillatory Fuel Droplet Vaporization: Driving Mechanism for Combustion Instability," *J. Propul. Power* **12**, 358–65.

Düwel, I., Ge , H.-W., Kronemayer, H., Dibble, R., Gutheil, E., Schulz , C., and Wolfrum, J., 2007, Experimental and Numerical Characterization of a Turbulent Spray Flame," *Proc. Combust. Inst.* **31**, 2247–55.

Dwyer, H.A., Aharon, I., Shaw, B.D., and Niazmand, H., 1996, "Surface Tension Influences on Methanol Droplet Vaporization in the Presence of Water," in *Proceedings of the Twenty-Sixth Symposium (International) on Combustion,* 1613–19. The Combustion Institute: Pittsburgh, PA.

Dwyer, H.A. and Sanders, B.R., 1984a, "Detailed Computation of Unsteady Droplet Dynamics," in *Proceedings of the Twentieth Symposium (International) on Combustion,* 1743–49. The Combustion Institute: Pittsburgh, PA.

Dwyer, H.A. and Sanders, B.R., 1984b, "Comparative Study of Droplet Heating and Vaporization at High Reynolds and Peclet Numbers," in *Dynamics of Flames and Reactive Systems,* Vol. 95 of AIAA Progress in Astronautics and Aeronautics Series, 464–83. AIAA: Washington, D.C.

Dwyer, H.A. and Sanders, B.R., 1984c, "Droplet Dynamics and Vaporization with Pressure as a Parameter," ASME Winter Annual Meeting, paper 84-WA/HT-20.

Dwyer, H.A. and Shaw, B.D., 2001, "Marangoni and Stability Studies on Fiber-Supported Droplets Evaporating in Reduced Gravity," *Combust. Sci. Technol.* **162**, 331–46.

Dwyer, H.A., Shaw, B.D., Niazmand, H., 1998, "Droplet Flame Interactions Including Surface Tension Influences," *Proc. Combust. Inst.* **27**, 1951.

Edmister, W.C. and Bowman, J.R., 1952, "Equilibrium Conditions of Flash Vaporization of Petroleum Fractions," *Chem. Eng. Prog. Symp. Ser.* **48**, 46–51.

Eggers, J., 1995, "Theory of Drop Formation," *Phys. Fluids* **7**, 941–53.

Eggers, J., 1997, "Nonlinear Dynamics and Breakup of Free-Surface Flows," *Rev. Mod. Phys.* **69**, 865–929.

Eggers, J. and Dupont, T.F., 1994, "Drop Formation in a One-dimensional Approximation of the Navier–Stokes Equation," *Journal of Fluid Mechanics* **262**, 205–21.

Elghobashi, S., 1991, "Particle-Laden Turbulent Flows: Direct Simulation and Closure Models," *Appl. Sci. Res.* **48**, 301–14.

Elghobashi, S., 1994, "On Predicting Particle-Laden Turbulent Flows," *Appl. Sci. Res.* **52**, 309–29.

Elghobashi, S., 2006, "An Updated Classification Map of Particle-Laden Turbulent Flows," in *Proceedings of the IUTAM Symposium on Computational Multiphase Flow*, S. Balachandar and A. Prosperetti, eds., 3–10. Springer-Verlag: Dordrecht, The Netherlands.

Elghobashi, S.E. and Abou-Arab, T.W., 1983, "A Two-Equation Turbulence Model for Two-Phase Flows," *Phys. Fluids* **26**, 931–38.

Elghobashi, S., Abou-Arab, T., Rizk, M., and Mostafa, A., 1984, "Prediction of the Particle-Laden Jet with a Two-Equation Turbulence Model," *Int. J. Multiphase Flow* **10**, 697–710.

Elghobashi, S.E. and Truesdell, G.C., 1991, "On the Interaction Between Solid Particles and Decaying Turbulence," Eighth Symposium on Turbulent Shear Flows, Munich.

Elghobashi, S. and Truesdell, G.C., 1992, "Direct Simulation of Particle Dispersion in a Decaying Isotropic Turbulence," *J. Fluid Mech.* **242**, 655–706.

Elghobashi, S. and Truesdell, G.C., 1993, "On the Two-Way Interaction Between Homogeneous Turbulence and Dispersed Solid Particles: Part I: Turbulence Modification," *Phys. Fluids A*, **5**, 1790–1801.

El-Wakil, M.M., Priem, R.J., Brikowski, H.J., Meyer, P.S., and Uyehara, O.A., 1956, "Experimental and Calculated Temperature and Mass Histories of Vaporizing Fuel Drop," NACA TN 2490.

Emmons, H.W., 1956, "The Film Combustion of Liquid Fuel," *Z. Angew. Math. Mech.* **36**, 60–71.

Erlebacher, G., Hussaini, M.Y., Speziale, C.G., and Zang, T.A., 1992, "Toward the Large-Eddy Simulation of Compressible Turbulent Flows," *J. Fluid Mech.* **238**, 155–85.

Fachini, F.F., Liñán, A., And Williams, F.A., 1999, "Theory of Flame Histories in Droplet Combustion at Small Stoichiometric Fuel-Air Ratios," *AIAA J.* **37**, 1426–35.

Faeth, G.M., 1977, "Current Status of Droplet and Liquid Combustion," *Prog. Energy Combust. Sci.* **3**, 191–224.

Faeth, G.M., 1983, "Evaporation and Combustion of Sprays," *Prog. Energy Combust. Sci.* **9**, 1–76.

Faeth, G.M., 1987, "Mixing, Transport, and Combustion in Sprays," *Prog. Energy Combust. Sci.* **13**, 293–345.

Faxen, H., 1922, "The Resistance to the Motion of a Solid Sphere in a Viscous Liquid Enclosed Between Parallel Walls," *Ann. Phys.* **68**, 89–119.

Fendell, F.E., 1968, "Decompositional Burning of a Droplet in a Small Peclet Number Flow," *AIAA J.* **6**, 1946–53.

Fendell, F.E., Coats, D.E., and Smith, E.B., 1968, "Compressible Slow Viscous Flow Past a Vaporizing Droplet," *AIAA J.* **6**, 1953–60.

Fendell, F.E., Sprankle, M.L., and Dodson, D.S., 1966, "Thin-Flame Theory for a Fuel Droplet in a Slow Viscous Flow," *J. Fluid Mech.* **26**, 267–80.

Fernandez-Pello, A.C., 1983, "Theory of the Mixed Convective Combustion of a Spherical Fuel Particle," *Combust. Flame* **53**, 23–32.

Fernandez-Pello, A.C., 2002, "Micropower Generation Using Combustion Issues and Approaches," *Proc. Combust. Inst.* **29**, 883–99.

Fernandez-Pello, A.C. and Law, C.K., 1982a, "A Theory for the Free-Convective Burning of a Condensed Fuel Particle," *Combust. Flame* **44**, 97–112.

Fernandez-Pello, A.C. and Law, C.K., 1982b, "On the Mixed-Convective Flame Structure in the Stagnation Point of a Fuel Particle," in *Proceedings of the Nineteenth Symposium (International) on Combustion*, 1037–44. The Combustion Institute: Pittsburgh, PA.

Ferrante, A. and Elghobashi, S.E., 2003, "On the Physical Mechanism of Two-Way Coupling in Particle-Laden Isotropic Turbulence," *Phys. Fluids* **15**, 315–29.

Ferrante, A. and Elghobashi, S.E., 2004a, "A Robust Method for Generating Inflow Conditions for Direct Simulation of Spatially Developing Turbulent Boundary Layers," *J. Comput. Phys.* **198**, 372–87.

Ferrante, A. and Elghobashi, S.E., 2004b, "On the Physical Mechanism of Drag Reduction in a Spatially Developing Turbulent Boundary Layer Laden with Microbubbles," *J. Fluid Mech.* **503**, 345–355.

Ferrante, A. and Elghobashi, S.E., 2005, "Reynolds Number Effect on Drag Reduction in a Microbubble-Laden Spatially Developing Turbulent Boundary Layer," *J. Fluid Mech.* **543**, 93–106.

Ferrante, A. and Elghobashi, S.E., 2007a, "On the Effects of Microbubbles on Taylor–Green Vortex Flow," *J. Fluid Mech.* **572**, 145–77.

Ferrante, A. and Elghobashi, S.E., 2007b, "On the Accuracy of the Two-Fluid Formulation in Direct Numerical Simulation of Bubble-Laden Turbulent Boundary layers," *Phys. Fluids* **19**, Paper 045105.

Ferrante, A. and Elghobashi, S.E., 2007c, "On the Effects of Finite-Size Particles on Decaying Isotropic Turbulence," *International Conference on Multiphase Flow ICMF 2007*, Leipzig, July.

Ferrenberg, A., Hunt, K., and Duesberg, J., 1985, *Atomization and Mixing Study*, Tech. Rep. RI/RD85-312, NAS8-34504, Rocketdyne, Canoga Park, CA.

Fix, G. and Gunzburger, M., 1977, "Downstream Boundary Conditions for Viscous Flow Problems," *Comput. Math. Appl.* **3**, 53.

Fletcher, C.A.J., 1991, *Computational Techniques for Fluid Dynamics*, Vol. II. Springer-Verlag: New York.

Forest, M.G. and Wang, Q., 1990, "Change-of-Type Behavior in Viscoelastic Slender Jet Models," *J. Theor. Comput. Fluid Dyn.* **2**, 1–25.

Fornes, V.G., 1968, *A Literature Review and Discussion of Liquid Particle Breakup in Gas Streams*, Tech. Rep. NWC-TP-4589, Naval Weapons Center, China Lake, CA.

Fraser, R.P. and Eisenklam, P., 1953, "Research into the Performance of Atomizers of Liquids," *Imp. Coll. Chem. Eng. Soc. J.* **7**, 52–68.

Friedman, R. and Macek, A., 1962, "Ignition and Combustion of Aluminum Particles in Hot Ambient Gases," *Combust. Flame* **6**, 9.

Friedman, R. and Macek, A., 1963, "Combustion Studies of Single Aluminum Particles," in *Proceedings of the Ninth Symposium (International) on Combustion*, p. 703. The Combustion Institute: Pittsburgh, PA.

Frohn, A. and Roth, N., 2000, *Dynamics of Droplets*, Springer-Verlag: Berlin.

Frossling, N., 1938, "Evaporation of Falling Drops," *Gerlands Beitr. Geophys.* **52**, 170.

Fu, K., Knobloch, A., Martinez, F., Walther, D.C., Fernandez-Pello, A.C., Pisano, A., Liepmann, D., Miyasaka, K., and Maruta, K., 2001, "Design and Experimental Results of Small-scale Rotary Engines," in *Proceedings of the 2001 International Mechanical Engineering Congress and Exposition (ICECE)*. American Society of Mechanical Engineers: New York.

Fukai, J., Shiba, Y., Yamamoto, T., Miyatake, O., Poulikakos, D., Megaridis, C.M., and Zhao, Z., 1995, "Wetting Effects on the Spreading of a Liquid Droplet Colliding with a Flat Surface: Experiment and Modeling," *Phys. Fluids* **7**, 236–47.

Fukai, J., Zhao, Z., Poulikakos, D., Megaridis, C.M., and Miyatake, O., 1993, "Modeling of the Deformation of a Liquid Droplet Impinging Upon a Flat Surface," *Phys. Fluids A* **5**, 2588–99.

Fureby, C., 2000, "Large-Eddy Simulation of Combustion Instabilities in a Jet Engine After-burner Model," *Combust. Sci. Technol.* **161**, 213–43.

Gal-Or, B., Cullinan, Jr., H.T., and Galli, R., 1975, "New Thermodynamic-Transport Theory for Systems with Continuous Component Density Distributions," *Chem. Eng. Sci.* **30**, 1085–92.

Ganan-Calvo, A.M., Lasheras, J.C., Davila, J., and Barrero, A., 1994, "The Electrostatic Spray Emitted From An Electrified Conical Meniscus," *J. Aerosol Sci.* **25**, 1121–42.

Gao, L.P., D'Angelo, Y., Silverman, I., Gomez, A., and Smooke, M.D., 1996, "Quantitative Comparison of Detailed Numerical Computations and Experiments in Counterflow Spray Diffusion Flames," in *Proceedings of the Twenty-Fourth Symposium (International) on Combustion*, 1739–46. The Combustion Institute: Pittsburgh, PA.

Garcia, F.J., and Castellanos, A., 1994, "One-Dimensional Models for Slender Axisymmetric Viscous Liquid Jets," *Phys. Fluids* **6**, 2676–89.

Gardiner, W.C., 1984, *Combustion Chemistry*, Springer Verlag, New York.

Ge, H.-W., Düwel, I., Kronemayer, H., Dibble, R., Gutheil, E., Schulz , C., and Wolfrum, J., 2008, "Laser-Based Experimental and Monte Carlo PDF Numerical Investigation of an Ethanol/Air Spray Flame," *Combust. Sci. Technol.* **180**, 1529–47.

Ge, H.-W. and Gutheil, E., 2006, "Probability Density Function (PDF) Simulation of Turbulent Spray Flows," *Atomization Sprays* **16**, 531–42,

Ge, H.-W. and Gutheil, E., 2008, "Simulation of a Turbulent Spray Flame Using Coupled PDF Gas Phase and Spray Flamelet Modeling," *Combust. Flame* **153**, 173–85.

Ge, H.-W., Urzica, D., Vogelgesang, M. and Gutheil, E., 2006, "Modeling and Simulation of Turbulent Non-reacting and Reacting Spray Flows," in *Reactive Flows, Diffusion and Transport*, W. Jäger, R. Rannacher, and J. Warnatz, eds. Springer-Verlag: Berlin.

Gear, C.W., 1971, *Numerical Initial Value Problems in Ordinary Differential Equations*. Prentice-Hall: Englewood Cliffs, NJ.

Ghosal, S. and Moin, P., 1995, "The Basic Equations for the Large Eddy Simulations of Turbulent Flows in Complex Geometry," *J. Comput. Phys.* **118**, 24–37.

Givi, P., 2003, "Subgrid Scale Modeling in Turbulent Combustion: A Review," Preprint 2003-5081, AIAA.

Givler, S.D. and Abraham, J., 1996, "Supercritical Droplet Vaporization and Combustion Studies," *Prog. Energy Combust. Sci.* **22**, 1–28.

Glassman, I., 1960, "Combustion of Metals Physical Considerations," in *Solid Propellant Rocket Research*, Vol. 1 of Progress in Astronautics and Rocketry Series, M. Summerfield, ed., 253–58. Academic: New York.

Glassman, I., 1987, *Combustion*, 2nd ed. Academic: New York.

Godsave, G.A.E., 1953, "Burning of Fuel Droplets," in *Proceedings of the Fourth Symposium (International) on Combustion*, 818–30. The Combustion Institute: Pittsburgh, PA.

Gogos, G. and Ayyaswamy, P.S., 1988, "A Model for the Evaporation of a Slowly Moving Droplet," *Combust. Flame* **74**, 111–29.

Gogos, G., Sadhal, S.S., Ayyaswamy, P.S., and Sundararajan, T., 1986, "Thin-Flame Theory for the Combustion of a Moving Liquid Droplet: Effects Due to Variable Density," *J. Fluid Mech.* **171**, 121–44.

Gollahalli, S.R., 1977, "Buoyancy Effects on the Flame Structure in the Wakes of Burning Liquid Drops," *Combust. Flame* **29**, 21–31.

Gomez, A. and Chen, G., 1997, "Monodisperse Electrosprays: Combustion, Scale-Up and Implications for Pollutant Formation," in *A Tribute to Irvin Glassman*, Vol. 4 of Combustion Science and Technology Book Series, F.L. Dryer and R.F. Sawyer, eds., 461–505. Gordon & Breach: New York.

Gosman, A.D. and Ioannides, E., 1981, "Aspects of Computer Simulation of Liquid-Fueled Combustors," AIAA Preprint 91-0323. AIAA: New York.

Gosman, A.D. and Johns, R.J.R., 1980, "Computer Analysis of Fuel Air Mixing in Direct Injection Engines," SAE Preprint, 80–91. Society of Automotive Engineers: Warrendale, PA.

Graff, K.F., 1975, *Wave Motion in Elastic Solids*. Dover: New York.

Gray, W.G. and Lee, P.C.Y., 1977, "On the Theorems for Local Volume Averaging of Multiphase Systems," *Int. J. Multiphase Flow* **3**, 333–400.

Greenberg, J.B. and Sarig, N., 1996, "An Analysis of Multiple Flames in Counterflow Spray Combustion," *Combust. Flame* **104**, 431–59.

Greenberg, J.B., Silverman, I., and Tambour, Y., 1993, "On the Origins of Spray Sectional Conservation Equations," *Combust. Flame* **93**, 90–96.

Gupta, H.C. and Bracco, F.V., 1978, "Numerical Computations of the Two Dimensional Unsteady Sprays for Application to Engines," *AIAA J.* **16**, 1053–61.

Gutheil, E., 2001, "Structure and Extinction of Laminar Ethanol–Air Spray Flames," *Combust. Theory Model.* **5**, 1–15.

Gutheil, E. and Sirignano, W.A., 1998, "Counterflow Spray Combustion Modeling with Detailed Transport and Detailed Chemistry," *Combust. Flame* **113**, 92–105.

Hadamard, J.S., 1911, "Mouvement Permanent Lent d'une Sphere Liquide et Visqueuse dans Une Liquid Visquese," *C.R. Acad. Sci.* **152**, 1735–38.

Hagerty, W.W., Shea, J.F., 1955, "A Study of the Stability of Plane Fluid Sheets," *J. Appl. Mech.* **22**, 509–14.

Haldenwang, P., Nicoli, C., and Daou, J., 1996, "High Pressure Vaporization of LOX Droplet Crossing the Critical Conditions," *Int. J. Heat Mass Transfer* **39**, 3453–64.

Hallett, W.L.H., 2000, "A Simple Model for the Vaporization of Droplets With Large Numbers of Components," *Combust. Flame* **121**, 334–44.

Harper, E.Y., Grube, G.W., and Chang, I-D., 1972, "On the Breakup of Accelerating Liquid Drops," *J. Fluid Mech.* **52**, 565–91.

Harper, J.F., 1970, "Viscous Drag in Steady Potential Flow Past a Bubble," *Chem. Eng. Sci.* **25**, 342–43.

Harper, J.F. and Moore, D.W., 1968, "The Motion of a Spherical Liquid Drop at High Reynolds Number," *J. Fluid Mech.*, 367–91.

Harrje, D.T. and Reardon, F.H., eds., 1972, *Liquid Propellant Rocket Combustion Instability*, NASA SP194. Government Printing Office: Washington, D.C.

Harstad, K. and Bellan, J., 1997, "Behavior of a Polydisperse Cluster of Interacting Drops in an Inviscid Vortex," *Int. J. Multiphase Flow* **23**, 899–925.

Harstad, K.G. and Bellan, J., 1998a, "Isolated Liquid Oxygen Drop Behavior in Fluid Hydrogen at Rocket Chamber Pressures," *Int. J. Heat Mass Transfer* **41**, 3537–50.

Harstad, K.G. and Bellan, J., 1998b, "Interactions of Liquid Oxygen Drops in Fluid Hydrogen at Rocket Chamber Pressures," *Int. J. Heat Mass Transfer* **41**, 3551–58.

Harstad, K.G. and Bellan, J., 1999, "The Lewis Number Under Supercritical Conditions," *Int. J. Heat Mass Transfer* 42, 961–70.

Harstad, K.G. and Bellan, J., 2000, "An All-Pressure Fluid Drop Model Applied to a Binary Mixture: Heptane in Nitrogen," *Int. J. Multiphase Flow* **26**, 1675–1706.

Harstad, K.G. and Bellan, J., 2001a, "The d^2 Variation for Isolated LOX Drops and Polydisperse Clusters in Hydrogen at High Temperature and Pressures," *Combust. Flame* **124**, 535–50.

Harstad, K.G. and Bellan, J., 2001b, "Evaluation of Commonly Used Assumptions for Isolated and Cluster Heptane Drops in Nitrogen at All Pressures," *Combust. Flame* **127**, 1861–79.

Harstad, K.G. and Bellan, J., 2004a, "Modeling Evaporation of Jet A, JP-7 and RP-1 drops at 1 to 15 bars," *Combust. Flame* **137**, 163–77.

Harstad, K.G. and Bellan, J., 2004b, "Modeling of Multicomponent Homogeneous Nucleation Utilizing Continuous Thermodynamics," *Combust. Flame* **139**, 252–62.

Harstad, K.G., Le Clercq, P.C., and Bellan, J., 2003, "A Statistical Model of Multicomponent-Fuel Drop Evaporation for Many-Droplet Gas-Liquid Flow Simulations," *AIAA J.* **41**, 1858–74.

Hashimoto, H. and Suzuki, T., 1991, "Experimental and Theoretical Study of Fine Interfacial Waves on Thin Liquid Sheet," *JSME Int. J., Ser. II* **34**, 277–83.

Hatch, J.E., ed., 1984, *Aluminum Properties and Physical Metallurgy*, Chap. 1. American Society for Metals: Metals Park, OH.

Haywood, R.J., Nafziger, N., and Renksizbulut, M., 1989, "A Detailed Examination of Gas and Liquid Phase Transient Processes in Convective Droplet Evaporation," *J. Heat Transfer* **111**, 495–502.

Haywood, R.J. and Renksizbulut, M., 1986, "On Variable Property, Blowing and Transient Effects in Convective Droplet Evaporation with Internal Circulation," in *Proceedings of the Eighth International Heat Transfer Conference*, 1861–66. Hemisphere: New York.

Heidmann, M.F. and Wieber, P.R., 1965, "Analysis of *n*-heptane Vaporization in Unstable Combustor with Travelling Transverse Oscillations," NASA TN-3424.

Heidmann, M.F. and Wieber, P.R., 1966, "Analysis of Frequency Response Characteristics of Propellant Vaporization," NASA TM-X-52195.

Hilbing, J.H. and Heister, S.D., 1996, "Droplet Size Control in Liquid Jet Breakup," *Phys. Fluids* **8**, 1574–81.

Hilbing, J.H. and Heister, S.D., 1997, "Nonlinear Simulation of a High-Speed, Viscous Liquid Jet," *Atomization Sprays* **8**, 155–78.

Hilbing, J.H., Heister, S.D. and Spangler, C.A., 1995, "A Boundary-Element Method for Atomization of a Finite Liquid Jet," *Atomization Sprays* **5**, 621–38.

Hinze, J.O., 1972, "Turbulent and Fluid Particle Interaction," *Prog. Heat Mass Transfer* **6**, 433–52.

Hinze, J.O., 1975, *Turbulence.* McGraw-Hill: New York.

Hiroyasu, H., 2000, "Spray Breakup Mechanism From the Hole-Type Nozzle and Its Applications," *Atomization Sprays* **10**, 511–27.

Hirschfelder, J.O., Curtiss, C.F., and Bird, R.B., 1954, *Molecular Theory of Gases and Liquids.* Wiley: New York.

Hirt, C.W. and Nichols, B.D., 1981, "Volume of Fluid (VOF) Method for the Dynamics of Free Boundaries," *J. Comput. Phys.* **39**, 201–25.

Hishida, K., Ando, A. and Maeda, M., 1992, "Experiments on Particle Dispersion in a Turbulent Mixing Layer," *Int. J. Multiphase Flow* **18**, 181–94.

Hollmann, C. and Gutheil, E., 1996, "Modeling of Turbulent Spray Diffusion Flames Including Detailed Chemistry," in *Proceedings of the Twenty-Sixth Symposium (International) on Combustion*, 1731–38. The Combustion Institute: Pittsburgh, PA.

Hollmann, C. and Gutheil, E., 1998, "Flamelet-Modeling of Turbulent Spray Diffusion Flames Based on a Laminar Spray Flame Library, "*Combust. Sci. Technol.* **135**, 175–92.

Hornbeck, R.W., 1973, "Numerical Marching Techniques for Fluid Flows with Heat Transfer," NASA-SP-297.

Howarth, L., ed., 1964, *Modern Developments in Fluid Dynamics-High Speed Flow*, Vol. 1. Clarendon: Oxford.

Hsieh, K.C., Shuen, J.S., and Yang, V., 1991, "Droplet Vaporization in High-Pressure Environments. 1. Near Critical Conditions," *Combust. Sci. Technol.* **76**, 111–32.

Hubbard, G.L., Denny, V.E., and Mills, A.F., 1975, "Droplet Evaporation: Effects of Transients and Variable Properties," *Int. J. Heat Mass Transfer* **18**, 1003–1008.

Ibrahim, E.A. and Akpan E.R., 1996, "Three-Dimensional Instability of Viscous Liquid Sheets," *Atomization Sprays* **6**, 649–65.

Imaoka, R. and Sirignano, W.A., 2005a, "Vaporization and Combustion in Three-Dimensional Droplet Arrays," *Proc. Combust. Inst.* **30**, 1981–89.

Imaoka, R. and Sirignano, W.A., 2005b, "A Generalized Analysis for Liquid-Fuel Vaporization and Burning," *Int. J. Heat Mass Transfer* **48**, 4342–53.

Imaoka, R. and Sirignano, W.A., 2005c, "Transient Vaporization and Burning in Dense Droplet Sprays," *Int. J. Heat Mass Transfer* **48**, 4354–66.

Ingebo, R.D., 1956, "Drag Coefficients for Droplets and Solid Spheres in Clouds Accelerating in Airstreams," NACA Tech. Note 3762.

Ingebo, R.D., 1967, "Heat Transfer Rates and Drag Coefficients for Ethanol Drops in a Rocket Motor," in *Proceedings of the Eighth Symposium (International) on Combustion*, 104–13. The Combustion Institute: Pittsburgh, PA.

Jackson, G.S. and Avedisian, C.T., 1996, "Modeling of Spherically Symmetric Droplet Flames Including Complex Chemistry: Effect of Water Addition on *n*-Heptane Droplet Combustion," *Combust. Sci. Technol.* **115**, 125–49.

Jackson, G.S. and Avedisian, C.T., 1998, "Combustion of Unsupported Water-in-*n*-Heptane Emulsion Droplets in a Convection-Free Environment," *Int. J. Heat Mass Transfer* **41**, 2503–15.

Jazayeri, S.A. and Li, X., 1996, "Nonlinear Breakup of Liquid Sheets," in *Proceedings of the Ninth Annual Conference on Liquid Atomization and Spray Systems, ILASS, North and South America*, 114–19. Institute of Liquid Atomization and Spray Systems: Irvine CA.

Jeffrey, D.J. and Onishi, Y., 1984, "Calculation of the Resistance and Mobility Functions for Two Unequal Rigid Spheres in Low Reynolds Flow," *J. Fluid Mech.* **139**, 261–90.

Jiang, T.L. and Chiang, W.-T., 1994a, "Effects of Multiple Droplet Interactions on Droplet Vaporization in Subcritical and Supercritical Pressure Environments," *Combust. Flame* **97**, 17–34.

Jiang, T.L. and Chiang, W.-T., 1994b, "Vaporization of a Dense Spherical Cloud of Droplets at Subcritical and Supercritical Conditions," *Combust. Flame* **99**, 355–62.

Jiang, T.L. and Chiang, W.-T., 1996, "Transient Heating and Vaporization of a Cool Dense Cloud of Droplets in Hot Supercritical Surroundings," *Int. J. Heat Mass Transfer* **39**, 1023–31.

Jiang, Y.J., Umemura, A., and Law, C.K., 1992, "An Experimental Investigation on the Collision Behaviour of Hydrocarbon Droplets," *J. Fluid Mech.* **234**, 171–90.

Jog, M.A., Ayyaswamy, P.S., and Cohen, I.M., 1996, "Evaporation and Combustion of Slowly Moving Fuel Droplets," *J. Fluid Mech.* **307**, 135–65.

Johns, L.E. and Beckman, R.B., 1966, "Mechanism of Dispersed-Phase Mass Transfer in Viscous, Single-Drop Extraction System," *AIChE J.* **12**, 10–16.

Joseph, D.D., 1998, "Cavitation and the State of Stress in a Flowing Liquid," *J. Fluid Mech.* **366**, 367–78.

Joseph, D.D., 2003, "Viscous Potential Flow," *J. Fluid Mech.* **479**, 191–97.

Joseph, D.D., Funada, T., and Wang, J., 2007, *Potential Flows of Viscous and Viscoelastic Fluids*. Cambridge University Press: New York.

Joseph, D.D. and Liao, T.Y., 1994, "Potential Flow of Viscous and Viscoelastic Fluids," *J. Fluid Mech.* **265**, 1–23.

Kaltz, T.L., Long, L.N., and Micci, M.M., 1998, "Supercritical Vaporization of Liquid Oxygen Droplets Using Molecular Dynamics," *Combust. Sci. Technol.* **136**, 279.

Kanury, A.M., 1975, *Introduction to Combustion Phenomena*. Gordon & Breach: New York.

Karagozian, A.K. and Marble, F.E., 1986, "Study of a Diffusion Flame in a Stretched Vortex," *Combust. Sci. Technol.* **45**, 65.

Kassoy, D.R., Lui, M.K., and Williams, F.A., 1969, "Comments on 'Effects of Chemical Kinetics on Near Equilibrium Combustion in Nonpremixed Systems,'" *Phys. Fluids* **12**, 265–67.

Kassoy, D.R. and Williams, F.A., 1968, "Effects of Chemical Kinetics on Near Equilibrium Combustion in Nonpremixed Systems," *Phys. Fluids* **11**, 1343–52.

Kee, R.J. and Miller, J.A., 1978, "A Split-Operator Finite-Difference Solution for Axisymmetric Laminar Jet Diffusion Flames," *AIAA J.* **16**, 169.

Keller, J.B., Rubinow, S.I. and Tu, Y.O., 1973, "Spatial Instability of a Jet," *Phys. Fluids* **16**, 2052–55.

Kelly, A.J., 1984, "The Electrostatic Atomization of Hydrocarbons," *J. Inst. Energy*, June, 312–20.

Kerstein, A.R., 1984, "Prediction of the Concentration PDF for Evaporating Sprays," *Int. J. Heat Mass Transfer* **27**, 1291–1309.

Kerstein, A.R., 1989, "Linear-Eddy Model of Turbulent Transport II," *Combust. Flame* **75**, 397–413.

Kerstein, A.R. and Law, C.K., 1983, "Percolation in Combustion Sprays. I. Transition from Cluster Combustion to Percolate Combustion in Non-Premixed Sprays," in *Proceedings of the Nineteenth Symposium (International) on Combustion*, 961–70. The Combustion Institute: Pittsburgh, PA.

Khau, L.H., Dwyer, H.A., and Shaw, B.D., 2007, "Studies on Combustion of Double Streams of Methanol/Dodecanol Fuel Droplets Next to a Cooling Wall," *Combust. Sci. Technol.* **179**, 601–16.

Kim, I., Elghobashi, S.E., and Sirignano, W.A., 1992, "Three-Dimensional Flow Computation for Two Interacting, Moving Droplets," Preprint 92-0343, AIAA 30th Aerospace Sciences Meeting, Reno, NV.

Kim, I., Elghobashi, S.E., and Sirignano, W.A., 1993, "Three-Dimensional Flow Over Two Spheres Placed Side-by-Side," *J. Fluid Mech.* **246**, 465–88.

Kim, I., Elghobashi, S.E., and Sirignano, W.A., 1995, "Unsteady Flow Interactions Between an Advected Cylindrical Vortex Tube and a Spherical Particle," *J. Fluid Mech.* **288**, 123–55.

Kim, I., Elghobashi, S.E., and Sirignano, W.A., 1997, "Unsteady Flow Interactions Between a Pair of Advected Vortex Tubes and a Rigid Sphere," *Int. J. Multiphase Flow* **23**, 1–23.

Kim, I., Elghobashi, S.E., and Sirignano, W.A., 1998, "On the Equation for Spherical Particle Motion: Effects of Reynolds and Acceleration Numbers," *J. Fluid Mech.* **367**, 221–53.

Kim, I., Schiller, D.N., and Sirignano, W.A., 1998, "Axisymmetric Flame Spread Across Propanol Pools in Normal and Zero Gravities," *Combust. Sci. Technol.* **139**, 249.

Kim, I. and Sirignano, W.A., 2000, "Three-Dimensional Wave Distortion and Disintegration of Thin Planar Liquid Sheets," *J. Fluid Mech.* **410**, 147–83.

Kim, I. and Sirignano, W.A., 2003, "Computational Study of Opposed-Forced-Flow Flame Spread Across Propanol Pools," *Combust. Flame* **132**, 611–27.

Kim, W.-W. and Menon, S., 1999, "A New Incompressible Solver for Large-Eddy Simulations," *Int. J. Numer. Fluid Mech.* **31**, 983–1017.

Kim, W.-W. and Menon, S., 2000, "Numerical Simulations of Turbulent Premixed Flames in the Thin-Reaction-Zones Regime," *Combust. Sci. Technol.* **160**, 113–150.

Kim, W.-W., Menon, S. and Mongia, H.C., 1999, "Large-Eddy Simulation of a Gas Turbine Combustor Flow," *Combust. Sci. Technol.* **143**, 25–62.

Kleinstreuer, C., Chiang, H., and Wang., Y.Y., 1989, "Mathematical Modelling of Interacting Vaporizing Fuel Droplets," in *Heat Transfer Phenomena in Radiation, Combustion and Fire* 106, ASME Heat Transfer Division, 469–77.

Kotake, S. and Okuzaki, T., 1969, "Evaporation and Combustion of a Fuel Droplet," *Intl. J. Heat Mass Transfer* **12**, 595–609.

Kottke, P.A., Bair, S.S. and Winer, W.O., 2005, "Cavitation in Creeping Shear Flows," *AICHE J.* **51**, 2150.

Kuo, K.K., 1986, *Principles of Combustion*. Wiley: New York.

Kyritsis, D.C., Coriton, B., Faure, F., Roychoudhury, S., and Gomez, A., 2004a, "Optimization of a Catalytic Combustor Using Electrosprayed Liquid Hydrocarbons for Mesoscale Power Generation," *Combust. Flame* **139**, 1–2, 77–89.

Kyritsis, D.C., Guerrero-Arias, I., Roychoudhury, S., and Gomez, A., 2002, "Mesoscale Power Generation by a Catalytic Combustor Using Electrosprayed Liquid Hydrocarbons," *Proc. Combust. Inst.* **29**, 965–72.

Kyritsis, D.C., Roychoudhury, S., McEnally, C.S., Pfefferle, L.D., and Gomez, A., 2004b, "Mesoscale Combustion: A First Step Towards Liquid Fueled Batteries," *Exp. Thermal Fluid Sci.* **28**, 763–70.

Labowsky, M., 1976, "The Effects of Nearest Neighbor Interactions on the Evaporation Rate of Cloud Particles," *Chem. Eng. Sci.* **31**, 803–13.

Labowsky, M., 1978, "A Formalism for Calculating the Evaporation Rates of Rapidly Evaporating Interacting Particles," *Combust. Sci. Technol.* **18**, 145–51.

Labowsky, M., 1980, "Calculation of Burning Rates of Interacting Fuel Droplets," *Combust. Sci. Technol.* **22**, 217–26.

Labowsky, M. and Rosner, D.C., 1978, "Group Combustion of Droplets in Fuel Clouds, I. Quasi-Steady Predictions," in *Evaporation-Combustion of Fuels*, Vol. 166 of Advances in Chemistry Series, J.T. Zung, ed., 63–79. American Chemical Society: Washington, D.C.

Lacas, F., Darabiha, N., Versaevel, P., Rolon, J.C., and Candel, S., 1992, "Influence of Droplet Number Density on the Structure of Strained Laminar Spray Flame," in *Proceedings of the Twenty-Fourth Symposium (International) on Combustion*, 1523–29. The Combustion Institute: Pittsburgh, PA.

Lage, P.L.C. and Rangel, R.H., 1993a, "Total Thermal Radiation Absorption by a Single Spherical Droplet," *J. Thermophys. Heat Transfer* **7**, 101–109.

Lage, P.L.C. and Rangel, R.H., 1993b, "Single Droplet Vaporization Including Thermal Radiation Absorption," *J. Thermophys. Heat Transfer* **7**, 502–509.

Lage, P.L.C., Hackenberg, C.M., and Rangel, R.H., 1995, "Nonideal Vaporization of Dilating Binary Droplets with Radiation Absorption," *Combust. Flame* **101**, 36–44.

Lamb, H., 1945, *Hydrodynamics.* Dover: New York.

Landau, L.D. and Lifshitz, E.M., 1987, *Fluid Mechanics*, 2nd ed. Pergamon: New York.

Landis, R.B. and Mills, A.F., 1974, "Effects of Internal Diffusional Resistance on the Vaporization of Binary Droplets," Paper B7.9, Fifth International Heat Transfer Conference, Tokyo, Japan.

Lara-Urbaneja, P. and Sirignano, W.A., 1981, "Theory of Transient Multicomponent Droplet Vaporization in a Convective Field," in *Proceedings of the Eighteenth Symposium (International) on Combustion*, 1365–74. The Combustion Institute: Pittsburgh, PA.

Launder, B.E., Reece, G.J., and Rodi, W., 1975, "Progress in the Development of a Reynolds Stress Turbulence Closure," *J. Fluid Mech.* **68**, 537.

Lavernia, E.J. and Wu, Y., 1996, *Spray Atomization and Deposition.* Wiley: New York.

Law, C.K., 1973, "A Simplified Theoretical Model for the Vapor-Phase Combustion of Metal Particles," *Combust. Sci. Technol.* **7**, 197–212.

Law, C.K., 1976a, "Multicomponent Droplet Vaporization with Rapid Internal Mixing," *Combust. Flame* **26**, 219–33.

Law, C.K., 1976b, "Unsteady Droplet Combustion With Droplet Heating," *Combust. Flame* **26**, 17–22.

Law, C.K., 1977, "A Model for the Combustion of Oil/Water Emulsion Droplets," *Combust. Sci. Technol.* **17**, 29–38.

Law, C.K., 1982, "Recent Advances in Droplet Vaporization and Combustion," *Prog. Energy Combust. Sci.* **8**, 171–201.

Law, C.K., 1998, "Droplet Combustion of Energetic Liquid Fuels," in *Propulsion Combustion: Fuels to Emission*, G.D. Roy, ed., 63–92. Taylor & Francis: New York.

Law, C.K., Chung, S.H., and Srinivasan, N., 1980a, "Gas-Phase Quasi-Steadiness and Fuel Vapor Accumulation Effects in Droplet Burning," *Combust. Flame* **38**, 173–98.

Law, C.K. and Law, H.K., 1976, "Quasi-Steady Diffusion Flame Theory with Variable Specific Heats and Transport Coefficients," *Combust. Sci. Technol.* **12**, 207–16.

Law, C.K. and Law, H.K., 1977, "Theory of Quasi-Steady One-Dimensional Diffusional Combustion With Variable Properties Including Distinct Binary Diffusion Coefficients," *Combust. Flame* **29**, 269–75.

Law, C.K., Lee, C.H., and Srinivasan, N., 1980b, "Combustion Characteristics of Water-in-Oil Emulsion Droplets," *Combust. Flame* **37**, 125–44.

Law, C.K., Prakash, S., and Sirignano, W.A., 1977, "Theory of Convective, Transient, Multicomponent Droplet Vaporization," in *Proceedings of the Sixteenth Symposium (International) on Combustion*, 605–17. The Combustion Institute: Pittsburgh, PA.

Law, C.K. and Sirignano, W.A., 1977, "Unsteady Droplet Combustion With Droplet Heating II: Conduction Limit," *Combust. Flame* **28**, 175–86.

Lazaro, B.J. and Lasheras, J.C., 1989, "Particle Dispersion in a Turbulent, Plane, Free Shear Layer," *Phys. Fluids A* **1**, 1035–44.

Lazaro, B.J. and Lasheras, J.C., 1992a, "Particle Dispersion in the Developing Free Shear Layer. Part 1. Unforced Flow," *J. Fluid Mech.* **235**, 143–78.

Lazaro, B.J. and Lasheras, J.C., 1992b, "Particle Dispersion in the Developing Free Shear Layer. Part 2. Forced Flow," *J. Fluid Mech.* **235**, 179–221.

Leboissetier, A., Okong'o, N., and Bellan, J., 2005, "Consistent Large Eddy Simulation of a Temporal Mixing Layer Laden with Evaporating Drops. Part 2: *A Posteriori* Modeling," *J. Fluid Mech.* **523**, 37–78.

Le Clercq, P.C. and Bellan, J., 2004, "Direct Numerical Simulation of a Transitional Temporal Mixing Layer Laden with Multicomponent-Fuel Evaporating Drops Using Continuous Thermodynamics," *Phys. Fluids* **16**, 1884–1907.

Le Clercq, P.C. and Bellan, J., 2005a, "Modeling of Multicomponent-Fuel Drop-laden Mixing Layers Having a Multitude of Species," *Proc. Combust. Inst.* **30**, 2011–19.

Le Clercq, P.C. and Bellan, J., 2005b, "Direct Numerical Simulation of Gaseous Mixing Layers Laden With Multicomponent-Liquid Drops: Liquid-Specific Effects," *J. Fluid Mech.* **533**, 57–94; **540**, 439.

Lee, A. and Law, C.K., 1991, "Gasification and Shell Characteristics in Slurry Droplet Burning," *Combust. Flame* **85**, 77–93.

Lee, C.P. and Wang, T.G., 1986, "A Theoretical Model for the Annular Jet Instability – Revisted," *Phys. Fluids* **29**, 2076–85.

Lee, C.P., and Wang, T.G., 1989, "The Theoretical Model for the Annular Jet Instability – Revisited," *Phys. Fluids A* **1**, 967–74.

Lee, C.P. and Wang, T.G., 1990, "Dynamics of Thin Liquid Sheets," AIP Conference Proc., Vol. 197, 496–504, doi: 10.1063/1.38948.

Lee, H.C., 1974, "Drop Formation in a Liquid Jet," *IBM J. Res. Develop.* **18**, 364–69.

Lee, H.-S., Fernandez-Pello, A.C., Corcos, G.M., and Oppenheim, A.K., 1990, "A Mixing and Deformation Mechanism for a Supercritical Droplet," *Combust. Flame* **81**, 50–8.

Lee, J.G., and Chen, L.D., 1991, "Linear Stability Analysis of Gas-Liquid Interface," *AIAA J.* **29**, 1589–95.

Lees, L., 1956, "Laminar Heat Transfer Over Blunt-Nosed Bodies at Hypersonic Flight Speeds," *Jet Propul.* **26**, 259–69.

Lefebvre, A.H., 1989, *Atomization and Sprays*. Hemisphere: Washington, DC.

Leib, S.J. and Goldstein, M.E., 1986, "The Generation of Capillary Instabilities on a Liquid Jet," *J. Fluid Mech.* **168**, 479–500.

Leiroz, A.J.K. and Rangel, R.H., 1995, "Numerical Study of Droplet-Stream Vaporization at Zero Reynolds Number," *Numer. Heat Transfer Part A: Applications* **27**, 273–96.

Leiroz, A.J.K. and Rangel, R.H., 1997, "Flame and Droplet Interaction Effects During Droplet-Stream Combustion at Zero Reynolds Number," *Combust. Flame* **108**, 287–301.

Leiroz, A.J.K. and Rangel, R.H., 2007, "Transient Gas Phase Interaction Effects During Droplet-Stream Combustion," *J. Braz. Soc. Mech. Sci. Eng.* **23**, 329.

Lerner, S.L., Homan, H.S., and Sirignano, W.A., November 1980, "Multicomponent Droplet Vaporization at High Reynolds Numbers: Size, Composition, and Trajectory Histories," preprinted for AlChe Meeting, Chicago, IL.

Levich, V.G., 1962, *Physicochemical Hydrodynamics*. Prentice-Hall: Englewood Cliffs, NJ.

Li, S.C., 1997, "Spray Stagnation Flames," *Prog. Energy Combust. Sci.* **23**, 303–47.

Li, S.C., Libby, P.A., and Williams, F.A., 1992, "Experimental and Theoretical Studies of Counterflow Spray Diffusion Flames," in *Proceedings of the Twenty-Fourth Symposium (International) on Combustion*, 1503–12. The Combustion Institute: Pittsburgh, PA.

Li, X. 1993, "Spatial Instability of Plane Liquid Sheets," *Chem. Eng. Sci.* **48**, 2973–81.

Li, X. and Tankin, R.S., 1991, "On the Temporal Instability of a Two-Dimensional Viscous Liquid Sheet," *J. Fluid Mech.* **226**, 425–43.

Liang, P.Y., 1990, "Numerical Method for Calculation of Surface Tension Flows in Arbitrary Grids," *AIAA J.* **29**, 161–67.

Lighthill, J., 1978, *Waves in Fluids*. Cambridge University Press: Cambridge.

Lighthill, M.J., 1965, "Group Velocity," *J. Inst. Math. Appl.* **1**, 1–28.

Lin, S.P., 2003, *Breakup of Liquid Sheets and Jets*, Cambridge University Press, Cambridge, UK.

Lindsay, W., Teasdale, D., Milanovic, V., Pister, K., and Fernandez-Pello, A.C., 2001, "Thrust and Electrical Power from Solid Propellant Microrockets," in *Proceeding of MEMS 2001, 14th IEEE International Conference on Micro Electro Mechanical Systems*, IEEE Cat. No. 01CH37090, 606–10. IEEE: New York.

Ling, W., Chung, J.N., Trout, T.R., and Crowe, C.T., 1998, "Direct Numerical Simulation of a Three-Dimensional Temporal Mixing Layer with Particle Dispersion," *J. Fluid Mech.* **358**, 61–85.

Lippert, A.M. and Reitz, R.D., 1997, "Modeling of Multicomponent Fuels Using Continuous Distributions with Application to Droplet Evaporation and Sprays," SAE Paper 972882.

Litchford, R.J. and Jeng, S.-M., 1990, "LOX Vaporization in High-Pressure Hydrogen-Rich Gas," AIAA-90-2191, 26th Joint Propulsion Conference, Orlando, FL.

Litchford, R.J., Parigger, C., and Jeng, S.-M, 1992, "Supercritical Droplet Gasification Experiments with Forced Convection," AIAA 92-3118, 28th Joint Propulsion Conference, Nashville, TN.

Liu, H., Lavernia, E.J., and Rangel, R.H., 1994, "Numerical Investigation of Micropore Formation During Substrate Impact of Molten Droplets in Plasma Spray Processes," *Atomization Sprays* **4**, 369–84.

Long, L.N., Micci, M.M., and Wong, B.C., 1996, "Molecular Dynamics Simulations of Droplet Evaporation," *Comput. Phys. Commun.* **906**, 167–72.

Lozinski, D. and Matalon, M., 1992, "Vaporization of a Spinning Fuel Droplet," in *Proceedings of the Twenty-Fourth Symposium (International) on Combustion*, 1483–91. The Combustion Institute: Pittsburgh, PA.

Lucci, F., Ferrante, A., and Elghobashi, S., 2008, "DNS of Fully Resolved Particles Dispersed in Isotropic Turbulence," *American Physical Society Fluid Dynamics Meeting*, San Antonio, TX, November 2008.

Luettmerstrathmann, J. and Sengers, J.V., 1996, "The Transport Properties of Fluid Mixtures Near the Vapor-Liquid Critical Line," *J. Chem. Phys.* **104**, 3026–47.

Lundgren, T.S. and Mansour, N.N., 1988, "Computation of Drop Pinch-off and Oscillation," in *Proceedings of the 3rd International Colloquium on Drops and Bubbles*, T.G. Wang, ed., 208–15.

Luo, H., Ciccotti, G., Mareshal, M., Meyer, M., and Zappoli, B., 1995, "Thermal Relaxation of Supercritical Fluids by Equilibrium Molecular Dynamics," *Phys. Rev. E* **51**, 2013–21.

MacCormack, R.W., 1969, "The Effect of Viscosity in Hypervelocity Impact Cratering," AIAA Paper 69-354, AIAA Hypervelocity Impact Conference, Cincinnati, OH, April.

MacInnes, J.M. and Bracco, F.V., 1992a, "Stochastic Particle Dispersion Modelling and the Tracer-Particle Limit," *Phys. Fluids A* **4**, 2809–24.

MacInnes, J.M. and Bracco, F.V., 1992b, "A Stochastic Model of Turbulent Drop Dispersion in Dilute Sprays," ILASS-Europe Conference, Amsterdam.

Marble, F.E., 1970, "Dynamics of Dusty Gases," *Annu. Rev. Fluid Mech.* **2**, 397–446.

Marble, F.E., 1985, "Growth of a Diffusion Flame in the Field of a Vortex," in *Recent Advances in the Aerospace Sciences*, C. Casci, ed., 395–413. Plenum: New York.

Marchese, A.J. and Dryer, F.L., 1996, "Effect of Liquid Mass Transport on the Combustion and Extinction of Bi Component Liquid Droplets of Methanol and Water," *Combust. Flame* **105**, 104–22.

Marchese, A.J. and Dryer, F.L., 1997, "The Effect of Non-Luminous Thermal Radiation in Microgravity Droplet Combustion," *Combust. Sci. Technol.* **124**, 371–402.

Marchese, A.J., Dryer, F.L., and Colantonio, R., 1998, "Radiative Effects in Space-Based Methanol/Water Droplet Combustion Experiments," in *Symposium (International) on Combustion* **27**, 2627–34.

Marchese, A.J., Dryer, F.L., and Nayagam, V., 1999, "Numerical Modeling of Isolated *n*-Alkane Droplet Flames: Initial Comparison with Ground and Space-Based Microgravity Experiments," *Combust. Flame* **116**, 432–59.

Marchese, A.J., Dryer, F.L., Nayagam, M.V., and Colantonio, R., 1996, "Microgravity Combustion of Methanol and Methanol/Water Droplets: Drop Tower Experiments and Model Predictions," in *Proceedings of the Twenty-Sixth Symposium (International) on Combustion*, 1209–17. The Combustion Institute: Pittsburgh, PA.

Marcus, B.D., 1972, "Theory and Design of Variable Conductance Heat Pipes," NASA CR-2018.

Mashayek, F., 1998a, "Direct Numerical Simulation of Evaporating Droplet Dispersion in Forced Low-Mach-Number Turbulence," *Int. J. Heat Mass Transfer* **41**, 2601–17.

Mashayek, F., 1998b, "Droplet-Turbulence Interactions in a Low-Mach-Number Homogeneous Shear Two-Phase Flow," *J. Fluid Mech.* **367**, 163–203.

Mashayek, F. and Pandya, R.V.R., 2003, "Analytical Description of Particle/ Droplet-Laden Turbulent Flows," *Prog. Energy Combust. Sci.* **29**, 329–78.

Masoudi, M. and Sirignano, W.A., 1997, "The Influence of an Advecting Vortex on the Heat Transfer to a Liquid Droplet," *Int. J. Heat Mass Transfer* **40**, 3663–73.

Masoudi, M. and Sirignano, W.A., 1998, "Vortex Interaction With a Translating Sphere in a Stratified Temperature Field," *Int. J. Heat Mass Transfer* **41**, 2639–52.

Masoudi, M. and Sirignano, W.A., 2000, "Collision of a Vortex with a Vaporizing Droplet," *Int. J. Multiphase Flow* **26**, 1925–49.

Matsuuchi, K., 1974, "Modulational Instability of Nonlinear Capillary Waves on Thin Liquid Sheets," *J. Phys. Soc.* **37**, 1680–87.

Matsuuchi K., 1976, "Instability of Thin Liquid Sheet and Its Break-Up," *J. Phys. Soc.* **41**, 1410–16.

Mattioli, R., Pham, T.K., and Dunn-Rankin, D., 2009, "Secondary Air Injection in Miniature Liquid Fuel Film Combustors," *Proc. Combust. Inst.* **32**, 3091–98.

Maxey, M.R., 1993, "The Equation of Motion for a Small Rigid Sphere in a Nonuniform or Unsteady Flow," ASME/FED, *Gas-Solid Flows* **166**, 57–62.

Maxey, M.R. and Riley, J.J., 1983, "Equation of Motion for a Small Rigid Sphere in a Nonuniform Flow," *Phys. Fluids* **26**, 883–89.

Megaridis, C.M. and Sirignano, W.A., 1991, "Numerical Modelling of a Vaporizing Multicomponent Droplet," in *Proceedings of the Twenty-Third Symposium (International) on Combustion,* 1413–21. The Combustion Institute: Pittsburgh, PA.

Megaridis, C.M. and Sirignano, W.A., 1992a, "Multicomponent Droplet Vaporization in a Laminar Convecting Environment," *Combust. Sci. Technol.* **87**, 27–44.

Megaridis, C.M. and Sirignano, W.A., 1992b, "Numerical Modeling of a Slurry Droplet Containing a Spherical Particle," *J. Thermophys. Heat Transfer* **7**, 110–19.

Mehring, C. and Sirignano, W.A., 1999, "Nonlinear Capillary Wave Distortion and Disintegration of Thin Planar Liquid Sheets," *J. Fluid Mech.* **388**, 69–113.

Mehring, C., and Sirignano, W.A., 2000a, "Axisymmetric Capillary Waves on Thin Annular Liquid Sheets. Part I: Temporal Stability," *Phys. Fluids* **12**, 1417–39.

Mehring, C., and Sirignano, W.A., 2000b, "Axisymmetric Capillary Waves on Thin Annular Liquid Sheets. Part II: Spatial Development," *Phys. Fluids* **12**, 1440–60.

Mehring, C., and Sirignano, W.A., 2001, "Nonlinear Capillary Waves on Swirling, Axisymmetric Liquid Films," *Int. J. Multiphase Flow* **27**, 1707–34.

Mehring, C. and Sirignano, W.A., 2003, "Disintegration of Planar Liquid Film Impacted by Two-Dimensional Gas Jets," *Phys. Fluids* **15**, 1158–77.

Mehring, C. and Sirignano, W.A., 2004, "Capillary Stability of Modulated Swirling Liquid Sheets," *Atomization Sprays* **14**, 397–436.

Mei, R., 1994, "Flow Due To an Oscillating Sphere and an Expression for Unsteady Drag on the Sphere at Finite Reynolds Number," *J. Fluid Mech.* **270**, 133–74.

Mei, R. and Adrian, R.J., 1992, "Flow Past a Sphere With an Oscillation in the Free-Stream and Unsteady Drag at Finite Reynolds Number," *J. Fluid Mech.* **237**, 323–41.

Mei, R., Lawrence, C.J., and Adrian, R.J., 1991, "Unsteady Drag on a Sphere at Finite Reynolds Number with Small Fluctuations in the Free-Stream Velocity," *J. Fluid Mech.* **233**, 613–31.

Menon, S., 2004, "CO Emission and Combustion Dynamics Near Lean Blow-Out in Gas Turbine Engines," ASME Paper GT2004-53290.

Menon, S. and Calhoon, W., 1996, "Subgrid Mixing and Molecular Transport Modeling for Large-Eddy Simulations of Turbulent Reacting Flows," *Proc. Combust. Inst.* **26**, 59–66.

Menon, S. and Kim, W.-W., 1996, "High Reynolds Number Flow Simulations Using the Localized Dynamic Subgrid-Scale Model," AIAA Preprint 96–0425.

Menon, S. and Patel, N., 2005, "Perspective on Subgrid Spray Injection Modeling in Gas Turbine Engines," ISABE-2005–1105.

Menon, S. and Patel, N., 2006, "Subgrid Modeling for Simulation of Spray Combustion in Large-Scale Combustors," *AIAA J.* **44**, 709–23.

Menon, S., Stone, C., and Patel, N., 2004, "Multi-Scale Modeling for LES of Engineering Designs of Large-Scale Combustors," AIAA Preprint 04–0157.

Menon, S., Yeung, P.-K., and Kim, W.-W., 1996, "Effect of Subgrid Models on the Computed Interscale Energy Transfer in Isotropic Turbulence," *Comput. Fluids* **25**, 165–80.

Merzhanov, A.G., Grigorev, Yu.M., and Gal'chenko, Yu.A., 1977, "Aluminum Ignition," *Combust. Flame* **29**, 1.

Meyer, J. and Weihs, D., 1987, "Capillary Instability of an Annular Liquid Jet," *J. Fluid Mech.* **179**, 531–45.

Micci, M.M. and Ketsdever, A.D., eds., 2000, *Micropropulsion for Small Spacecraft*, Vol. 8187 of Progress in Astronautics and Aeronautics Series. AIAA: Washington, D.C.

Michaelides, E.E. and Feng, Z., 1994, "Heat Transfer From a Rigid Sphere in a Nonuniform Flow and Temperature Field," *Int. J. Heat Mass Transfer* **37**, 2069–76.

Miller, F.J., Ross, H.D., Kim, I., and Sirignano, W.A., 2000, "Parametric Investigations of Pulsating Flame Spread Across 1-Butanol Pools," *Proc. Combust. Inst.* **28**, 2827–34.

Miller, R.S. and Bellan, J., 1999, "Direct Numerical Simulation of a Confined Three-Dimensional Gas Mixing Layer with One Evaporating Hydrocarbon-Droplet Laden Stream," *J. Fluid Mech.* **384**, 293–338.

Miller, R.S. and Bellan, J., 2000, "Direct Numerical Simulation and Subgrid Analysis of a Transitional Droplet Laden Mixing Layer," *Phys. Fluids* **12**, 650–71.

Miller, R.S., Harstad, K.G., and Bellan, J., 2001, "Direct Numerical Simulations of Supercritical Fluid Mixing Layers Applied to Heptane – Nitrogen," *J. Fluid Mech.* **436**, 1–39.

Molodetsky, I.E., Dreizen, E.L., and Law, C.K., 1996, "Evolution of Particle Temperature and Internal Composition for Zirconium Burning in Air," in *Proceedings of the Twenty-Sixth Symposium (International) on Combustion*, 1919–23. The Combustion Institute: Pittsburgh, PA.

Molodetsky, I.E., Dreizen, E.L., Vincenzi, E.P., and Law, C.K., 1998, "Phases of Titanium Combustion in Air," *Combust. Flame* **112**, 522–32.

Mills, A.F., 1990, Personal communication.

Moin, P., Squires, W., Cabot, W.H., and Lee, S., 1991, "A Dynamic Subgrid-Scale Model for Compressible Turbulence and Scalar Transport," *Phys. Fluids A* **3**, 2746–57.

Molavi, K. and Sirignano, W.A., 1988, "Computational Analysis of Acoustic Instabilities in Dump Combustor Configurations," Preprint 88-2856, AIAA/ASME 24th Joint Propulsion Conference, Boston, MA.

Moseler, M., and Landman, U., 2000, "Formation, Stability, and Breakups of Nanojets," *Science* **289**, 1165–69.

Mostafa, A.A. and Elghobashi, S.E., 1985a, "A Two-Equation Turbulence Model for Jet Flows Laden with Vaporizing Droplets," *Int. J. Multiphase Flow* **11**, 515–33.

Mostafa, A.A. and Elghobashi, S.E., 1985b, "A Study of the Motion of Vaporizing Droplets in a Turbulent Flow," in *Dynamics of Flames and Reactive Systems*, Vol. 95 of Progress in Astronautics and Aeronautics Series, J.R. Bowen, N. Mason, A.K. Oppenheim, and R.I. Soloukhin, eds., 513–29. AIAA: Washington, D.C.

Mostafa, A.A. and Mongia, H.C., 1988, "On the Interaction of Particles and Turbulent Fluid Flow," *Int. J. Heat Mass Transfer* **31**, 2063–75.

Mostafa, A.A., Mongia, H.C., McDonell, V.G., and Samuelsen, G.S., 1989, "Evolution of Particle Laden Jet Flows: A Theoretical and Experimental Study," *AIAA J.* **27**, 167–83.

Naber, J.D. and Reitz, R.D., 1988, "Modelling Engine Spray/Wall Impingement," Preprint 88-0107, SAE Congress and Exposition, Detroit, Feb. 29–March 4.

Nayfeh, A.H., 1970, "Nonlinear Stability of a Liquid Jet," *Phys. Fluids* **13**, 841–7.

Nguyen, Q.V. and Dunn-Rankin, D., 1992, "Experiments Examining Drag in Linear Droplet Packets," *Exp. Fluids* **12**, 157–65.

Nguyen, Q.V., Rangel, R.H., and Dunn-Rankin, D., 1991, "Measurement and Prediction of Trajectories and Collision of Droplets," *Int. J. Multiphase Flow* **17**, 159–77.

Niazmand, H., Shaw, B.D., Dwyer, H.A., and Aharon, I., 1994, "Effects of Marangoni Convection on Transient Droplet Evaporation," *Combust. Sci. Technol.* **103**, 219–33.

Nichols, B.D. and Hirt, C.W., 1971, "Improved Free-Surface Boundary Conditions for Numerical Incompressible Flow Calculations," *J. Comput. Phys.* **8**, 434–48.

Nielsen, L.E., 1978, *Predicting the Properties of Mixtures: Mixture Rules in Science and Engineering*, Chap. 3, 49–72. Marcel Dekker: New York.

Nobari, M.R.H., Jan, Y.-J., and Tryggvason, G., 1996, "Head-On Collision of Drops – A Numerical Investigation," *Phys. Fluids* **8**, 29–42.

Nobari, M.R.H. and Tryggvason, G., 1996, "Numerical Simulations of Three-Dimensional Drop Collisions," *AIAA J.* **34**, 750–55.

Nwobi, O.C., Long, L.N., and Micci, M.M., 1998, "Molecular Dynamics Studies of Properties of Supercritical Fluids," *J. Thermophys. Heat Transfer*, **12**, 322–27.

Odar, F., 1966, "Verification of the Proposed Equation for Calculation of the Forces on a Sphere Accelerating in a Viscous Fluid," *J. Fluid Mech.* **25**, 591–92.

Odar, F. and Hamilton, W.S., 1964, "Forces on a Sphere Accelerating in a Viscous Fluid," *J. Fluid Mech.* **18**, 302–14.

Okai, K., Moriue, O., Araki, M., Tsue, M., Kono, M., Sato, J., Dietrich, D.L., and Williams, F.A., 2000, "Combustion of Single Droplets and Droplet Pairs in a Vibrating Field Under Microgravity," *Proc. Combust. Inst.* **28**, 977–83.

Okai, K., Tsue, M., Kono, M., Sato, J., Dietrich, D.L., and Williams, F.A., 2004, "Effects of DC Electric Fields on Combustion of Octane Droplet Pairs in Microgravity," *Combust. Flame* **136**, 390–3.

Okai, K., Ueda, T., Imamura, O., Tsue, M., Kono, M., Sato, J., Dietrich, D.L., and Williams, F.A., 2003, "An Experimental Study of Microgravity Combustion of a Droplet Near a Wall," *Combust. Flame* **133**, 169–72.

Okong'o, N. and Bellan, J., 2000a, "A Priori Subgrid Analysis of Temporal Mixing Layers with Evaporating Droplets," *Phys. Fluid* **12**, 1573–1591.

Okong'o, N. and Bellan, J., 2000b, "Entropy Production of Emerging Turbulent Scales in a Temporal Supercritical *n*-Heptane/Nitrogen Three-Dimensional Mixing Layer," *Proc. Combust. Inst.* **28**, 497–504.

Okong'o, N. and Bellan, J., 2002a, "Direct Numerical Simulation of a Transitional Supercritical Mixing Layer: Heptane and Nitrogen," *J. Fluid Mech.* **464**, 1–34.

Okong'o, N. and Bellan, J., 2002b, "Characteristics of Supercritical Transitional Temporal Mixing Layers," in *Proceedings of the IUTAM Symposium on Turbulent Mixing and Combustion*, Kingston, Pollard, A. and Candel S., eds., 59–71. Kluwer Academic: Dordrecht, The Netherlands.

Okong'o, N. and Bellan, J., 2004a, "Turbulence and Fluid-Front Area Production in Binary-species, Supercritical, Transitional Mixing Layers," *Phys. Fluids* **16**, 1467–92.

Okong'o, N. and Bellan, J., 2004b, "Consistent Large Eddy Simulation of a Temporal Mixing Layer Laden with Evaporating Drops. Part 1: Direct Numerical Simulation, Formulation and *A Priori* Analysis," *J. Fluid Mech.* **499**, 1–47.

Okong'o, N., Harstad, K., and Bellan, J., 2002, "Direct Numerical Simulations of O_2/H_2 Temporal Mixing Layers Under Supercritical Conditions," *AIAA J.* **40**, 914–926.

Okong'o, N., Leboissetier, A., and Bellan, J., 2008, "Detailed Characteristics of Drop-Laden Mixing Layers: LES Predictions Compared to DNS," *Phys. Fluids* **20**, paper 103305.

Orme, M., 1997, "Experiments on Droplet Collisions, Bounce, Coalescence and Disruption," *Prog. Energy Combust. Sci.* **23**, 65–79.

Orme, M. and Muntz, E.P., 1987, "New Technique for Producing Highly Uniform Droplet Streams Over an Extended Range of Disturbance Wavenumbers," Rev. Sci. Instrum. **58**, 279–84.

O'Rourke, P.J., and Bracco, F.V., 1980, "Modeling of Drop Interactions in Thick Sprays and a Comparison With Experiments," Stratified Charge Automotive Engines Conference, The Institution of Mechanical Engineers.

Oseen, C.W., 1927, *Hydrodynamk*. Akademische Verlagsgesellschaft: Leipzig.

Pan, K.L. and Law, C.K., 2007, "Dynamics of Droplet-film Collision," *J. Fluid Mech.* **587**, 1–22.

Panchagnula, M.V., Santangelo, P.J., and Sojka, P.E., 1995, "The Instability of an Inviscid Annular Liquid Sheet Subject to Two-Dimensional Disturbances," in *Proceedings of the ILASS Eighth Annual Conference on Liquid Atomization and Spray Systems*, 54–58. Institute of Liquid Atomization and Spray: Systems Irvine, CA.

Panchagnula, M.V., Sojka, P.E., and Santangelo, P.J., 1996, "On the Three-Dimensional Instability of a Swirling Annular, Inviscid Liquid Sheet Subject to Unequal Gas Velocities," *Phys. Fluids* **8**, 3300–12.

Park, T.W., Aggarwal, S.K., and Katta, V.R., 1994, "Effect of Gravity on the Structure of an Unsteady Spray Diffusion Flame," *Combust. Flame* **99**, 767–74.

Park, T.W., Aggarwal, S.K., and Katta, V.R., 1996, "A Numerical Study of Droplet-Vortex Interactions In an Evaporating Spray," *Int. J. Heat Mass Transfer* **39**, 2205–19.

Park, T.W., Katta, V.R., and Aggarwal, S.K., 1998, "On The Dynamics of a Two-Phase, Nonevaporating Swirling Jet," *Int. J. Multiphase Flow* **24**, 295–317.

Patankar, S.V., 1980, *Numerical Heat Transfer and Fluid Flow*. Hemisphere: New York.

Patel, N. and Menon, S., 2008, "Simulation of Spray – Turbulence – Flame Interactions in a Lean Direct Injection Combustor," *Combustion and Flame* **153**, 228–57.

Patel, N., Kirtas, M., Sankaran, M., and Menon, S., 2006, "Simulation of Spray Combustion in a Lean Direct Injection Combustor," *Proceedings of The Combustion Institute* doi:10.1016/j.proci.2006.07.232.

Patnaik, G. and Sirignano, W.A., 1986, "Axisymmetric, Transient Calculation for Two Vaporizing Fuel Droplets," Western States Section/Combustion Institute Technical Meeting, Banff, Alberta, Canada.

Patnaik, G., Sirignano, W.A., Dwyer, H.A., and Sanders, B.R., 1986, "A Numerical Technique for the Solution of a Vaporizing Fuel Droplet," in *Dynamics of Reactive Systems*, Vol. 105 of Progress in Astronautics and Aeronautics Series, 253–66. AIAA: Washington, D.C.

Payri, F., Bermudez, V., Payri, R., and Salvador, F. J., 2004, "The Influence of Cavitation on the Internal Flow and the Spray Characteristics in Diesel Injection Nozzles," *Fuel* **83**, 419.

Pham, T.K., Dunn-Rankin, D., and Sirignano, W.A., 2006, "Flame Structure in Small-Scale Liquid Film Combustors," *Proc. Combust. Inst.* **31**, Part 2, 3269–75.

Pimbley, W.T., 1976, "Drop Formation From a Liquid Jet: A Linear One-Dimensional Analysis Considered as a Boundary Value Problem," *IBM J. Res. Develop.* **20**, 148–56.

Pimbley, W.T. and Lee, H.C., 1977, "Satellite Droplet Formation in a Liquid Jet," *IBM J. Res. Develop.* **21**, 21–30.

Piomelli, U., 1999, "Large-Eddy Simulation: Achievements and Challenges," *Prog. Aerospace Sci.* **35**, 335–62.

Poling, B.E., Prausnitz, J.M., and O'Connell, J.P., (2001), *The Properties of Gases and Liquids, 5th ed.*, McGraw-Hill: New York.

Polymeropoulos, C.E., 1974, "Flame Propagation in a One-Dimensional Liquid Fuel Spray," *Combust. Sci. Technol.* **9**, 197–207.

Ponstein, J., 1959, "Instability of Rotating Cylindrical Jets," *Appl. Sci. Res. A* **8**, 425.

Poo, J.Y., and Ashgriz, N., 1998, "Numerical Simulation of Capillary Driven Viscous Flows in Liquid Drops and Films by an Interface Reconstruction Scheme," *AIP Conference Proc.* **197**, 235–45.

Poplow, F., 1994, "Numerical Calculation of the Transition from Subcritical Droplet Evaporation to Supercritical Diffusion," *Int. J. Heat Mass Transfer* **37**, 485–92.

Potter, J.N. and Riley, N., 1980, "Free Convection Over a Burning Sphere," *Combust. Flame* **39**, 83–96.

Prakash, S. and Sirignano, W.A., 1978, "Liquid Fuel Droplet Heating With Internal Circulation," *Int. J. Heat Mass Transfer* **21**, 885–95.

Prakash, S. and Sirignano, W.A., 1980, "Theory of Convective Droplet Vaporization with Unsteady Heat Transfer in the Circulating Liquid Phase," *Int. J. Heat Mass Transfer* **23**, 253–68.

Priem, R.J., 1963, "Theoretical and Experimental Models of Unstable Rocket Combustors," in *Proceedings of the Ninth Symposium (International) on Combustion*, 982–92. The Combustion Institute: Pittsburgh, PA.

Priem, R.J. and Heidmann, M.F., 1960, "Propellant Vaporization as a Design Criterion for Rocket Engine Combustion Chambers," NASA Rep. TR-R67.

Prosperetti, A. and Jones, A.V., 1984, "Pressure Forces in Disperse Two-phase Flow," *International Journal of Multiphase Flow* **10**, 425–40.

Prosperetti, A. and Zhang, D.Z., 1995, "Finite-Particle-Size Effects in Disperse Two-Phase Flows," *Theor. Comput. Fluid Dyn.* **7**, 429–40.

Puri, I.K. and Libby, P.A., 1989, "Droplet Behavior in Counterflowing Streams," *Combust. Sci. Technol.* **66**, 267–292.

Qian, J. and Law, C.K., 1997, "Regimes of Coalescence and Separation in Droplet Collision," *J. Fluid Mech.* **331**, 59–80.

Queiroz, M. and Yao, S.C., 1989, "Parametric Exploration of the Dynamic Behavior of Flame Propagation in Planar Sprays," *Combust. Flame* **76**, 351–68.

Raju, M.S. and Sirignano, W.A., 1989, "Spray Computations in a Centerbody Combustor," *J. Eng. Gas Turbines Power* **1**, 710–18.

Raju, M.S. and Sirignano, W.A., 1990a, "Multicomponent Spray Computations in a Modified Centerbody Combustor," *J. Propul. Power* **6**, 97–105.

Raju, M.S. and Sirignano, W.A., 1990b, "Interaction Between Two Vaporizing Droplets in an Intermediate-Reynolds-Number-Flow," *Phys. Fluids* **2**, 1780–96.

Ramos, J.I., 1992, "Annular Liquid Jets: Formulation and Steady-State Analysis," *Z. Angew. Math. Mech.* **72**, 565–89.

Ramos, J.I., 1996, "One-Dimensional Models of Steady Inviscid, Annular Liquid Jets," *Appl. Math Model.* **20**, 593–607.

Rangel, R.H., 1990, "Spray Entrainment and Mixing in a Temporally-Growing Shear Layer," ASME Winter Annual Meeting.

Rangel, R.H., 1992, "Heat Transfer in Vortically Enhanced Mixing of Vaporizing Droplet Sprays," in *Annual Review of Heat Transfer*, Vol. 4, C.L. Tien, ed., 331–62. Hemisphere: Washington, D.C.

Rangel, R.H. and Continillo, G., 1992, "Theory of Vaporization and Ignition of a Droplet Cloud in the Field of a Vortex," in *Proceedings of the Twenty-Fourth Symposium (International) on Combustion*, 1493–1501. The Combustion Institute: Pittsburgh, PA.

Rangel, R.H. and Fernandez-Pello, A.C., 1984, "Mixed Convective Droplet Combustion with Internal Circulation," *Combust. Sci. Technol.* **42**, 47–65.

Rangel, R.H. and Fernandez-Pello, A.C., 1985, "Droplet Ignition in Mixed Convection," in *Dynamics of Reactive Systems, Part II*, Vol. 105 of Progress in Astronautics and Aeronautics Series, 239–52. AIAA: Washington, D.C.

Rangel, R.H. and Sirignano, W.A., 1986, "Rapid Vaporization and Heating of Two Parallel Fuel Droplet Streams," in *Proceedings of the Twenty-First Symposium (International) on Combustion*, 617–24. The Combustion Institute: Pittsburgh, PA.

Rangel, R.H. and Sirignano, W.A., 1988a, "Unsteady Flame Propagation in a Spray With Transient Droplet Vaporization," in *Proceedings of the Twenty-Second Symposium (International) on Combustion*, 1931–39. The Combustion Institute: Pittsburgh, PA.

Rangel, R.H. and Sirignano, W.A., 1988b, "Two-Dimensional Modeling of Flame Propagation in Fuel Stream Arrangements," in *Dynamics of Reactive Systems Part II: Heterogeneous Combustion and Applications*, Vol. 113 of Progress in Astronautics and Aeronautics Series, A.L. Kuhl, J.R. Bowen, J.S. Leyer, and A. Borisov, eds., 128–50. AIAA: Washington, DC.

Rangel, R.H. and Sirignano, W.A., 1988c, "Nonlinear Growths of Kelvin–Helmholtz Instability: Effect of Surface Tension and Density Ratio," *Phys. Fluids* **31**, 1845–55.

Rangel, R.H. and Sirignano, W.A., 1989a, "Combustion of Parallel Fuel Droplet Streams," *Combust. Flame* **75**, 241–54.

Rangel, R.H. and Sirignano, W.A., 1989b, "An Evaluation of the Point-Source Approximation in Spray Calculations," *Num. Heat Transfer A*, **16**, 37–57.

Rangel, R.H. and Sirignano, W.A., 1991a, "Spray Combustion in Idealized Configurations: Parallel Droplet Streams," in *Numerical Approaches to Combustion Modeling*, Vol. 135 of Progress in Astronautics and Aeronautics Series, E.S. Oran and J.P. Boris, eds., 585–613. AIAA: Washington, DC.

Rangel, R.H. and Sirignano, W.A., 1991b, "The Linear and Nonlinear Shear Instability of a Fluid Sheet." *Phys. Fluids* **3**, 2392–2400.

Ranger, A.A. and Nicholls, J.A., 1969, "Aerodynamic Shattering of Liquid Drops," *AIAA J.* **7**, 285–90.

Ranger, A.A. and Nicholls, J.A., 1972, "Atomization of Liquid Droplets in a Convective Gas Stream," *Int. J. Heat Mass Transfer* **15**, 1203–11.

Ranz, W.E. and Marshall, W.R., 1952, "Evaporation from Drops," *Chem. Eng. Prog.* **48**, 141–46 and 173–80.

Rapaport, D.C., 1995, *The Art of Molecular Dynamics Simulation*. Cambridge University Press: Cambridge.

Rätzsch, M.T. and Kehlen, H., 1983, "Continuous Thermodynamics Model of Complex Mixtures," *Fluid Phase Equilibria* **14**, 225–234.

Ray, I. and Sirignano, W.A., 1992, "Analysis of Shell Formation from a Vaporizing Metal Slurry Droplet," 1992 Fall Meeting of the Western States Section of the Combustion Institute, Livermore, CA.

Rayleigh, Lord (John William Strutt), 1878, "On the Instability of Jets," *Proc. London Math. Soc.* **10**, 4–13.

Redlich, O. and Kwong, J.N.S., 1949, "On the Thermodynamics of Solutions," *Chem. Rev.* **44**, 233–44.

Rein, M., 1993, "Phenomena of Liquid Drop Impact on Solid and Liquid Surfaces," *Fluid Dyn. Res.* **12**, 61–93.

Rein, M., 1995, "Nonlinear Analysis of Two-Dimensional Compressible Liquid–Liquid Impact," *Eur. J. Mech. B* **14**, 301–22.

Rein, M., 1996, "The Transitional Regime Between Coalescing and Splashing Drops," *J. Fluid Mech.* **306**, 145–65.

Reitz, R.D., 1987, "Modeling Atomization Processes in High-Pressure Vaporizing Sprays," *Atomisation Spray Technol.* **3**, 309–37.

Reitz, R.D. and Bracco, F.V., 1982, "Mechanism of Atomization of a Liquid Jet," *Phys. Fluids* **25**, 1730–42.

Renksizbulut, M. and Haywood, R.J., 1988, "Transient Droplet Evaporation With Variable Properties and Internal Circulation at Intermediate Reynolds Numbers," *Int. J. Multiphase Flow* **14**, 189–202.

Renksizbulut, M. and Yuen, M.C., 1983, "Experimental Study of Droplet Evaporation in High-Temperature Air Stream," *J. Heat Transfer* **105**, 384–88.

Rivero, M., Magnaudet, J., and Fabre, J., 1991, "Quelques Resultants Nouveaux Concernant les Forces Exercees Sur Une Inclusion Spherique par en Ecoulement Accelere," *C.R. Acad. Sci. Paris* **312**, II, 1499–1506.

Rivkind, V.Y. and Ryskin, G.M., 1976, "Flow Structure in Motion of a Spherical Drop in a Fluid Medium at Intermediate Reynolds Number," *Proc. Acad. Sci. USSR Mech. Fluids* **1**, 8–15.

Rizk, M.A. and Elghobashi, S.E., 1985, "The Motion of a Spherical Particle Suspended in a Turbulent Flow Near a Plane Wall," *Phys. Fluids* **28**, 806–17.

Rizk, M.A. and Elghobashi, S.E., 1989, "A Two-Equation Turbulence Model for Dispersoid Dilute Confined Two-Phase Flows," *Int. J. Multiphase Flow* **15**, 119–33.

Rizzi, A.W. and Bailey, H.E., 1975, "A Generalized Hyperbolic Marching Method for Chemically Reacting Three-Dimensional Supersonic Flow Using a Splitting Technique," AIAA Second Computational Fluid Dynamics Conference, Hartford, CT.

Roache, P.J., 1976, *Computational Fluid Dynamics*. Hermosa: Albuquerque, NM.

Rohani, M., Iobbi, D.K., Dunn-Ranlin, D., And Jabbari, F., 2009, "Controlling Liquid Jet Breakup with Practical Piezoelectric Devices," *Atomization Sprays* **19**, 135–55.

Rossi, S., Dreizen, E.L., and Law, C.K., 2001, "Combustion of Aluminum Particles in Carbon Dioxide," *Combust. Sci. Technol.* **164**, 209–37.

Rusanov, A.I. and Brodskaya, E.N., 1977, "The Molecular Dynamics Simulation of a Small Drop," *J. Colloid Interface Sci.* **62**, 542–55.

Ryan, W., Annamalai, K., and Caton, J., 1990, "Relation Between Group Combustion and Drop Array Studies," *Combust. Flame* **80**, 313–21.

Rybczynski, W., 1911, "On the Translatory Motion of a Fluid Sphere in a Viscous Medium," *Bull. Int. Acad. Pol. Sci. Lett. A. Sci. Math. Natur. Ser. A.*, 40–46.

Sadhal, S.S. and Ayyaswamy, P.S., 1983, "Flow Past a Liquid Drop with a Large Non-Uniform Radial Velocity," *J. Fluid Mech.* **133**, 65–81.

Sadhal, S.S., Ayyaswamy, P.S., and Chung, J.N., 1997, *Transport Phenomena with Drops and Bubbles*. Springer-Verlag: New York.

Samson, R., Bedeaux, D., Saxton, M.J., and Deutsch, J.M., 1978, "A Simple Model of Fuel Spray Burning. I. Random Sprays. II. Linear Droplet Streams," *Combust. Flame* **31**, 215–29.

Sangiovanni, J.J. and Kesten, A.S., 1976, "Effect of Droplet Interaction on Ignition in Monodispersed Droplet Streams," in *Proceedings of the Sixteenth Symposium (International) on Combustion*, 577–92. The Combustion Institute: Pittsburgh, PA.

Sangiovanni, J.J. and Labowsky, M., 1982, "Burning Times of Linear Fuel Droplet Arrays: A Comparison of Theory and Experiment," *Combust. Flame* **47**, 15.

Sankaran, V. and Menon, S., 2002a, "LES of Spray Combustion in Swirling Flows," *J. Turbulence* **3**, 001.

Sankaran, V. and Menon, S., 2002b, "Vorticity-Scalar Alignments and Small-Scale Structures in Swirling Spray Combustion," *Proc. Combust. Inst.* **29**, 577–84.

Sankaran, V., Porumbel, I., and Menon, S., 2003, "Large-Eddy Simulation of a Single-Cup Gas Turbine Combustor," AIAA Paper 2003-5083.

Savart, F., 1833, "Suite du memoire sur le Cho d'une veine liquide lancee contre un plan circulaire," *Annal. Chim. Phys.* **54**, 113–65.

Schiller, D.N., Bhatia, R., and Sirignano, W.A., 1996a, "Energetic Fuel Droplet Gasification with Liquid-Phase Reaction," *Combust. Sci. Technol.* **113–114**, 471–91.

Schiller, D.N., Li, J., and Sirignano, W.A., 1998, "Transient Heating, Gasification, and Oxidation of an Energetic Liquid Fuel," *Combust. Flame* **114**, 349–58.

Schiller, D.N. and Sirignano, W.A., 1997 "Opposed-Flow Flame Spread Across n-Propanol Pools," in *Proceedings of the Twenty-Sixth International Symposium on Combustion*, 1319–1325. The Combustion Institute: Pittsburgh, PA.

Schiller, D.N., Sirignano, W.A., and Ross, H.D, 1996b, "Computational Analysis of Flame Spread Over Alcohol Pools," *Combust. Sci. Technol.* **118**, 205–58.

Schlotz, D. and Gutheil, E., 2000, "Modeling of Laminar Mono- and Bidisperse Liquid Oxygen/Hydrogen Spray Flames in the Counterflow Configuration," *Combust. Sci. Technol.* **158**, 195–210.

Schulkes, R.M.S.M., 1993, "Dynamics of Liquid Jets Revisited," *J. Fluid Mech.* **250**, 635–50.

Selle, L.C. and Bellan, J., 2007a, "Scalar Dissipation Modeling for Passive and Active Scalars: *A Priori* Study Using Direct Numerical Simulation," *Proc. Combust. Inst.* **31**, 1665–73.

Selle, L.C. and Bellan, J., 2007b, "Evaluation of Assumed-PDF Methods in Two-phase Flows Using Direct Numerical Simulation," *Proc. Combust. Inst.* **31**, 2273–81.

Selle, L.C. and Bellan, J., 2007c, "Characteristics of Transitional Multicomponent Gaseous and Drop-Laden Mixing Layers from Direct Numerical Simulation: Composition Effects," *Phys. Fluids* **19**, doi: 10.1063/1.2734997, 063301- 1–33.

Selle, L.C. and Bellan, J., 2009, "Large Eddy Simulation Composition Equations for Single-Phase and Two-Phase Fully Multicomponent Flows," *Proc. Combust. Inst.* **32**, 2239–46.

Selle, L.C., Okong'o, N., Bellan, J., and Harstad, K.G., 2007, "Modeling of Subgrid Scale Phenomena in Supercritical Transitional Mixing Layers: An *A Priori* Study," *J. Fluid Mech.* **593**, 57–91.

Seminario, J.M., Concha, M.C., Murray, J.S., and Politzer, P., 1994, "Theoretical Analyses of O_2/H_2O Systems Under Normal and Supercritical Conditions," *Chem. Phys. Lett.* **222**, 25–31.

Sengers, J.V., Basu, R.S., and Sengers, J.M.H.L., 1981, "Representative Equations for the Thermodynamic and Transport Properties of Fluids Near the Gas-Liquid Critical Point," NASA Contractor Rep. 3424, NASA, Scientific and Technical Information Branch: Washington, D.C.

Sengers, J.V. and Sengers, J.M.H.L., 1977, "Concepts and Methods for Describing Critical Phenomena in Fluids," NASA Contractor Rep. 149665, NASA: Springfield, VA.

Seth, B., Aggarwal, S.K., and Sirignano, W.A., 1978, "Flame Propagation Through an Air-Fuel Spray With Transient Droplet Vaporization," *Combust. Flame* **32**, 257–70.

Shah, R.K., and London, A.L., 1978, *Laminar Flow Forced Convection in Ducts,* Academic: New York.

Shaw, B.D., Aharon, I.D., Lenhart, D., Dietrich, D.L., and Williams, F.A., 2001a, "Spacelab and Drop-Tower Experiments on Combustion of Methanol/Dodecanol and Ethanol/Dodecanol Mixture Droplets in Reduced Gravity," *Combust. Sci. Technol.* **167**, 29–56.

Shaw, B.D., Clark, B.D., and Wang, D.F., 2001b, "Spacelab Experiments on Combustion of Heptane-Hexadecane Droplets," *AIAA J.* **39**, 2327–35.

Shaw, B.D., Dwyer, H.A., and Wei, J.B., 2002, "Studies on Combustion of Single and Double Streams of Methanol and Methanol/Dodecanol Droplets," *Combust. Sci. Technol.* **174**, 29–50.

Shaw B.D. and Harrison, M.J., 2002, "Influences of Support Fibers on Shapes of Heptane/Hexadecane Mixture Droplets in Reduced Gravity," *Microgravity Sci. Technol.* **13**, 30–40.

Shen, J. and Li, X., 1996, "Instability of an Annular Viscous Liquid Jet," *Acta Mech.* **114**, 167–83.

Sherman, F.S., 1990, *Viscous Flow*. McGraw-Hill: New York.

Shi, X.D., Brenner, M.P., and Nagel, S.R., 1994, "A Cascade of Structure in a Drop Falling From a Faucet," *Science* **265**, 219–22.

Shih, A.T. and Megaridis, C.M., 1996, "Thermocapillary Flow Effects on Convective Droplet Evaporation," *Int. J. Heat Mass Transfer* **39**, 247–57.

Shuen, J.S., Yang, V., and Hsiao, C.C., 1992, "Combustion of Liquid Fuel Droplets in Supercritical Conditions," *Combust. Flame* **89**, 299–319.

Sirignano, W.A., 1972, "Introduction to Analytical Models of High Frequency Combustion Instability," in *Liquid Propellant Rocket Combustion Instability*, D.T. Harrje and F.H. Reardon, eds., NASA SP194. U.S. Government Printing Office: Washington, D.C.

Sirignano, W.A., 1979, "Theory of Multicomponent Fuel Droplet Vaporization," *Arch. Thermodyn. Combust.* **9**, 235–51.

Sirignano, W.A., 1983, "Fuel Droplet Vaporization and Spray Combustion," *Prog. Energy Combust. Sci.* **9**, 291–322.

Sirignano, W.A., 1985a, "Spray Combustion Simulation," in *Numerical Simulation Combustion Phenomena*, R. Glowinski, B. Larrouturou, and R. Teman, eds. Springer-Verlag: Heidelburg.

Sirignano, W.A., 1985b, "Linear Model of Convective Heat Transfer in a Spray," in *Recent Advances in Aerospace Science*, C. Casci, ed. Plenum: New York.

Sirignano, W.A., 1986, "The Formulation of Spray Combustion Models: Resolution Compared to Droplet Spacing," *J. Heat Transfer* **108**, 633–39.

Sirignano, W.A., 1988, "An Integrated Approach to Spray Combustion Model Development," *Combust. Sci. Technol.* **58**, 231–51.

Sirignano, W.A., 1993a, "Computational Spray Combustion," in *Numerical Modeling in Combustion*, T.J. Chung, ed. Hemisphere: Washington, DC.

Sirignano, W.A., 1993b, "Fluid Dynamics of Sprays – 1992 Freeman Scholar Lecture," *J. Fluids Eng.* **115**, 345–78.

Sirignano, W.A., 2002, "A General Super-Scalar for the Combustion of Liquid Fuels" *Proc. Combust. Inst.* **29**, 535–42.

Sirignano, W.A., 2005a, "Volume Averaging for the Analysis of Turbulent Spray Flows" *Int. J. Multiphase Flow* **31**, 675–705.

Sirignano, W.A., 2005b, "Corrigendum to 'Volume Averaging for the Analysis of Turbulent Spray Flows' [International Journal of Multiphase Flow 31 (2005) 675–705]" *Int. J. Multiphase Flow* **31**, 867.

Sirignano, W.A., 2007, "Liquid-Fuel Burning with Non-Unitary Lewis Number," *Combust. Flame* **148**, 177–86.

Sirignano, W.A., 2008, "Recent Theoretical Advances For Liquid-fuel Atomization and Burning," in *Energetic Material Synthesis and Combustion Characterization for Chemical Propulsion,* K. Kuo and K. Hori, eds., 299–319. Begell House: New York.

Sirignano, W.A. and Crocco, L., 1964, "A Shock Wave Model of Unstable Rocket Combustors," *AIAA J.* **3**, 1285–96.

Sirignano, W.A., Dunn-Rankin, D., Strayer, B., and Pham, T., 2001, "Miniature Combustor with Liquid-Fuel Film," *Proceedings of the Western States Section/Combustion Institute Fall Meeting*, October, Salt Lake City, UT.

Sirignano, W.A. and Law, C.K., 1978, "Transient Heating and Liquid Phase Mass Diffusion in Droplet Vaporization," in *Evaporation-Combustion of Fuels*, Vol. 166 of the Advances in Chemistry Series, J.T. Zung, ed., 1–26. American Chemical Society: Washington, D.C.

Sirignano, W.A., and Mehring, C., 2000, "Review of Theory of Distortion and Disintegration of Liquid Streams," *Prog. Energy Combust. Sci.* **26**, 609–55.

Sirignano, W.A. and Mehring, C., 2005, "Distortion and Disintegration of Liquid Streams," *Liquid Rocket Combustion Devices: Aspects of Modeling, Analysis, and Design*, Vol. 200 of AIAA Progress in Astronautics and Aeronautics Series, V. Yang, M. Habiballah, J. Hulba, and M. Popp, eds., 167–249. AIAA: Washington, D.C.

Sirignano, W.A. Pham, T., and Dunn-Rankin, D., 2003, "Miniature Scale Liquid-Fuel Film Combustor," in *Proceedings of the 29th International Combustion Symposium*, 925–31. The Combustion Institute: Pittsburgh, PA.

Sirignano, W.A. and Schiller, D.N., 1996, "Mechanisms of Flame Spread Across Condensed Phase Fuels," in *Physical and Chemical Aspects of Combustion, A Tribute to Irvin Glassman, Combustion Science and Technology* book series, F.L. Dryer and R.F. Sawyer, eds., 353–407. Gordon & Breach: New York.

Sirignano, W.A., Stanchi, S., and Imaoka, R., 2005, "Linear Analysis of Liquid-Film Combustor," *J. Propul. Power* **21**, 1075–91.

Sirignano, W.A. and Wu, G., 2008, "Multicomponent-Liquid-Fuel Vaporization with Complex Configuration," *Int. J. Heat Mass Transfer* **51**, 4759–74.

Sitzki, L., Borer, K., Wussow, S., Schuster, E., Maruta, K., Ronney, P.D., and Cohen, AQ., 2001, "Combustion in Microscale Heat-Recirculating Burners," Paper AIAA 2001–1087, Thirty-Eighth Aerospace Sciences Meeting and Exhibit, Reno, Nevada.

Slattery, J.C., 1967, "Flow of Viscoelastic Fluids Through Porous Media," *AIChE J.* **13**, 1066–71.

Smirnov, N.N., 1985, "Heat and Mass Transfer in a Multi-Component Chemically Reactive Gas Above a Liquid Fuel Layer," *Int. J. Heat Mass Transfer* **28**, 929–38.

Smith, T.M. and Menon, S., 1996, "Model Simulations of Freely Propagating Turbulent Premixed Flames," in *Proceedings of the Twenty-Sixth Symposium (International) on Combustion*, 299–306. The Combustion Institute: Pittsburgh, PA.

Soo, S.L., 1967, *Fluid Dynamics of Multiphase Systems*. Blaisdell: Waltham, MA.

Spalding, D.B., 1951, "Combustion of Fuel Particles," *Fuel* **30**, 121.

Spalding, D.B., 1959, "Theory of Particle Combustion at High Pressure," *ARS J.* **29**, 828–35.

Spalding, D.B., 1980, "Numerical Computation of Multiphase Flow and Heat Transfer," *Recent Advances in Numerical Methods in Fluids*, Vol. 1, C. Taylor and K. Morgan, eds., 139–66. New York: Pineridge Press.

Spangler, C.A., Hilbing, J.H., and Heister, S.D., 1995, "Nonlinear Modeling of Jet Atomization in the Wind-Induced Regime," *Phys. Fluids* **7**, 964–71.

Squire, H.B., 1953, "Investigation of the Instability of a Moving Liquid Film," *Brit. J. Appl. Phys.* **4**, 167–69.

Squires, K.D. and Eaton, J.K., 1990, "Particle Response and Turbulence Modification in Isotropic Turbulence," *Phys. Fluids A* **2**, 1191–1203.

Squires, K.D. and Eaton, J.K., 1991, "Preferential Concentration of Particles by Turbulence," *Phys. Fluids* **3**, 1169–78.

Stapf, P., Maly, R., and Dwyer, H.A., 1998, "A Group Combustion Model for Treating Reactive Sprays in IC Engines," in *Proceedings of the Twenty-Seventh Symposium (International) on Combustion*, 1857–64. The Combustion Institute: Pittsburgh, PA.

Strahle, W.C., 1963, "A Theoretical Study of Unsteady Droplet Burning: Transients and Periodic Solutions," Princeton University, Dept. of Aerospace Engineering. Ph.D. Dissertation.

Strahle, W.C., 1964, "Periodic Solutions to a Convective Droplet Burning Problem: The Stagnation Point," in *Proceedings of the Tenth Symposium (International) on Combustion*, 1315–25. The Combustion Institute: Pittsburgh, PA.

Strahle, W.C., 1965a, "Unsteady Reacting Boundary Layer on a Vaporizing Flat Plate," *AIAA J.* **3**, 1195–98.

Strahle, W.C., 1965b, "Unsteady Laminar Jet Flames at Large Frequencies of Oscillation," *AIAA J.* **3**, 957–60.

Strahle, W.C., 1966, "High Frequency Behavior of the Laminar Jet Flame Subjected to Transverse Sound Waves," in *Proceedings of the Eleventh Symposium (International) on Combustion*, 747–54. The Combustion Institute: Pittsburgh, PA.

Strayer, B.A. and Dunn-Rankin, D., 2001, "Towards a Model for Manipulating Liquid Jet Breakup," *Atomization Sprays* **11**, 415–31.

Street, W.B. and Calado, J.C.G., 1978, "Liquid-Vapor Equilibrium for Hydrogen + Nitrogen at Temperature from 63 to 110 K and Pressures to 57 MPa," *J. Chem. Thermodyn.* **10**, 1089–1100.

Sundararajan, T. and Ayyaswamy, P.S., 1984, "Hydrodynamics and Heat Transfer Associated with Condensation on a Moving Drop: Solutions for Intermediate Reynolds Number," *J. Fluid Mech.* **149**, 33–58.

Suzuki, T. and Chiu, H.H., 1971, "Multi Droplet Combustion on Liquid Propellants," in *Proceedings of the Ninth International Symposium on Space Technology and Science*, 145–54. AGNE: Tokyo, Japan.

Szekely Jr., G.A. and Faeth, G.M., 1982, "Combustion Properties of Carbon Slurry Drops," *AIAA J.* **20**, 422–29.

Takahashi, F., Dryer, F.L., and Heilweil, I.J., 1986, "Combustion Behavior of Free Boron Slurry Droplets," in *Proceedings of the Twenty-First Symposium (International) on Combustion*, 1983–91. The Combustion Institute, Pittsburgh, PA.

Takahashi, F., Dryer, F.L., and Heilweil, I.J., 1989, "Disruptive Burning Mechanism of Free Slurry Droplets," *Combust. Sci. Technol.* **65**, 151–65.

Tal, R., Lee, D.M., and Sirignano, W.A., 1983, "Hydrodynamics and Heat Transfer in Sphere Assemblages-Cylindrical Cell Models," *Int. J. Heat Mass Transfer* **26**, 1265–73.

Tal, R., Lee, D.N., and Sirignano, W.A., 1984a, "Periodic Solutions of Heat Transfer for Flow Through a Periodic Assemblage of Spheres," *Int. J. Heat Transfer* **27**, 1414–17.

Tal, R., Lee, D.A., and Sirignano, W.A., 1984b, "Heat and Momentum Transfer Around a Pair of Spheres in Viscous Flow," *Int. J. Heat Mass Transfer* **27**, 1953–62.

Tal, R. and Sirignano, W.A., 1982, "Cylindrical Cell Model for the Hydrodynamics of Particle Assemblages at Intermediate Reynolds Numbers," *AIChE J.* **28**, 233–37.

Tal, R. and Sirignano, W.A., 1984, private communication of unpublished work.

Talley, D.G. and Yao, S.C., 1986, "A Semi-Empirical Approach to Thermal and Composition Transients Inside Vaporizing Fuel Droplets," in *Proceedings of the Twenty-First Symposium (International) on Combustion*, 609–16. The Combustion Institute: Pittsburgh, PA.

Tamaki, N., Shimizu, M., and Hiroyasu, H., 2001, "Enhancement of the Atomization of a Liquid Jet by Cavitation in a Nozzle Hole," *Atomization Sprays* **11**, 125–38.

Tamaki, N., Shimizu, M., Nishida, K. and Hiroyasu, H., 1998, "Effects of Cavitation and Internal Flow on Atomization of a Liquid Jet," *Atomization Sprays* **8**, 179–97.

Tambour, Y., 1984, "Vaporization of Polydisperse Sprays in a Laminar Boundary Layer Flow: A Sectional Approach," *Combust. Flame* **58**, 103–14.

Tamim, J. and Hallett, W.L.H., 1995, "A Continuous Thermodynamics Model for Multicomponent Droplet Vaporization," *Chem. Eng. Sci.* **50**, 2933–42.

Taylor, G.I., 1959, "The Dynamics of Thin Sheets of Fluid, II., Waves on Fluid Sheets," *Proc. R. Soc. London Ser. A.* **253**, 253–96.

Temkin, S. and Ecker, G.Z., 1989, "Droplet Pair Interactions in a Shock-Wave Flow Field," *J. Fluid Mech.* **202**, 467–97.

Thompson, S.M. and Gubbins, K.E., 1984, "A Molecular Dynamics Study of Liquid Drops," *J. Chem. Phys.* **81**, 530–42.

Ting, L. and Keller, J.B., 1990, "Slender Jets and Thin Sheets with Surface Tension," *SIAM J. Appl. Math.* **50**, 1533–46.

Titchmarsh, E.C., 1962, *Eigenfunction Expansions Associated with Second-Order Differential Equations; Part I.* Clarendon: Oxford.

Tomboulides, A.G., Orszag, S.A., and Karniadakis, G.E., 1991, "Three-Dimensional Simulation of Flow Past a Sphere," *Intl. Soc. Offshore Polar Engineering Proc.*, Edinburgh.

Tong, A.Y. and Chen, S.J., 1988, "Heat Transfer Correlations for Vaporizing Liquid Droplet Arrays in a High Temperature Gas at Intermediate Reynolds Number," *Int. J. Heat Fluid Flow* **9**, 118–30.

Tong, A.Y. and Sirignano, W.A., 1982a, "Transient Thermal Boundary Layer in Heating of Droplet With Internal Circulation: Evaluation of Assumptions," *Combust. Sci. Technol.* **11**, 87–94.

Tong, A.Y. and Sirignano, W.A., 1982b, "Analytical Solution for Diffusion and Circulation in a Vaporizing Droplet," in *Proceedings of the Nineteenth Symposium (International) on Combustion*, 1007–20. The Combustion Institute: Pittsburgh, PA.

Tong, A.Y. and Sirignano, W.A., 1983, "Analysis of Vaporizing Droplet With Slip, Internal Circulation and Unsteady Liquid-Phase and Quasi-Steady Gas-Phase Heat Transfer," ASME–JSME Thermal Joint Engineering Conference.

Tong, A.Y. and Sirignano, W.A., 1986a, "Multicomponent Droplet Vaporization in a High Temperature Gas," *Combust. Flame* **66**, 221–35.

Tong, A.Y. and Sirignano, W.A., 1986b, "Multicomponent Transient Droplet Vaporization with Internal Circulation: Integral Equation Formulation and Approximate Solution," *Num. Heat Transfer* **10**, 253–78.

Tong, A.Y. and Sirignano, W.A., 1989, "Oscillatory Vaporization Fuel Droplets in Unstable Combustor, *J. Propul. Power* **5**, 257–61.

Tsai, J.S. and Sterling, A.M., 1990, "The Application of an Embedded Grid to the Solution of Heat and Momentum Transfer for Spheres in a Linear Array," *Int. J. Heat Mass Transfer* **33**, 2491–502.

Turns, S.R., 2000, *An Introduction to Combustion: Concept and Applications*, 2nd ed., McGraw-Hill, New York.

Twardus, E.M. and Brzustowski, T.A., 1977, "The Interaction Between Two Burning Fuel Droplets," *Arch. Procesow Spalania* **8**, 347–58.

Uhlmann, M., 2005, "An Immersed Boundary Method with Direct Forcing for the Simulation of Particulate Flows," *J. Comput. Phys.* **209**, 448–76.

Uhlmann, M., 2007, "Investigating Turbulent Particulate Channel Flow with Interface-resolved DNS," *International Conference on Multiphase Flow ICMF 2007*, July, Leipzig.

Umemura, A., Ogawa, A., and Oshiwa, N., 1981a, "Analysis of the Interaction Between Two Burning Droplets," *Combust. Flame* **41**, 45–55.

Umemura, A., Ogawa, A., and Oshiwa, N., 1981b, "Analysis of the Interaction Between Two Burning Droplets with Different Sizes," *Combust. Flame* **43**, 111–19.

Waitz, I.A., Gauba, G., and Tzeng, Y.-S., 1998, "Combustors for Micro-Gas Turbine Engines," *J. Fluids Eng.* **120**, 109–117.

Waldman, C.H., 1975, "Theory of Non-Steady State Droplet Combustion," in *Proceedings of the Fifteenth Symposium (International) on Combustion*, 429–35. The Combustion Institute: Pittsburgh, PA.

Wallis, G.B., 1969, *One-Dimensional Two-Phase Flow*. McGraw-Hill: New York.

Walther, D.C. and Fernandez-Pello, A.C., 2001, "Microscale Combustion: Issues and Opportunities," *Eastern States Section Combustion Institute Fall Technical Meeting*, December 3–5, Hilton Head, SC.

Wang, C.H., Hung, W.G., Fu, S.Y., Huang, W.C., and Law, C.K., 2003, "On the Burning and Microexplosion of Collision-Generated Bi-Component Droplets: Miscible Fuels," *Combust. Flame* **134**, 289–300.

Wang, C.H., Lin, C.Z., Hung, W.G., Huang, W.C., and Law, C.K., 2004, "On the Burning Characteristics of Collision-Generated Water/Hexadecane Droplets," *Combust. Sci. Technol.* **176**, 71–93.

Wang, Q. and Squires, K.D., 1996, "Large Eddy Simulation of Particle-Laden Turbulent Channel Flow," *Phys. Fluids* **8**, 1207–23.

Warnatz, J., 1984, "Chemistry of High Temperature Combustion of Alkanes up to Octane," in *Proceedings of the Twentieth Symposium (International) on Combustion*, 845–56. The Combustion Institute: Pittsburgh, PA.

Wei, J.B. and Shaw, B.D., 2006, "Influences of Pressure on Reduced-Gravity Combustion of HAN-Methanol-Water Droplets," *Combust. Flame* **146**, 484–92.

Westbrook, C.K. and Dryer, F.L., 1984, "Chemical Kinetic Modelling of Hydrocarbon Combustion," *Prog. Energy Combust. Sci.* **10**, 1–57.

Whitaker, S., 1966, "The Equations of Motion in Porous Media," *Chem. Eng. Sci.* **21**, 291–300.

Whitaker, S., 1967, "Diffusion and Dispersion in Porous Media," *AIChE Journal* **13**, 420–27.

Whitaker, S., 1973, "The Transport Equations for Multi-Phase Systems," *Chem. Eng. Sci.* **28**, 139–47.

White, F.M., 1991, *Viscous Fluid Flow*. McGraw-Hill: New York.

Wichman, I. and Baum, H.R., 1993, "A Solution Procedure for Low Reynolds Number Combustion Problems Under Microgravity Conditions," in *Heat Transfer in Microgravity*, C.T. Avedisian and V.A. Arpaci, eds. ASME: New York.

Williams, A., 1976, "Fundamentals of Oil Combustion," *Prog. Energy Combust. Sci.* **2**, 167–79.

Williams, F.A., 1962, Introduction to Analytical Models of High Frequency Combustion Instability," in *Proceedings of the Eighth Symposium (International) on Combustion*, 50–63. Williams and Wilkins: Baltimore, MD.

Williams, F.A., 1985, *Combustion Theory; The Fundamental Theory of Chemically Reacting Flow Systems*, 2nd ed. Benjamin/Cummings: Menlo Park, CA.

Williams, F.A., 1997, "Some Aspects of Metal Particle Combustion," in *Physical Aspects of Combustion, A Tribute to Irvin Glassman*, Vol. 4 of Combustion Science and Technology Book Series, F.L. Dryer and R.F. Sawyer, eds., 461–505. Gordon & Breach: New York.

Winer, W.O. and Bair, S., 1987, "The Influence of Ambient Pressure on the Apparent Shear Thinning of Liquid Lubricants – An Overlooked Phenomenon," *Conference Publication Vol. 1 London Inst. Mech. Eng.* C190-87, 395–8.

Wong, S.-C. and Turns, S.R., 1987, "Ignition of Aluminum Slurry Droplets," *Combust. Sci. Technol.* **52**, 221–42.

Wu, X., Law, C.K., and Fernandez-Pello, A.C., 1982, "A Unified Criterion for the Convective Extinction of Fuel Particles," *Combust. Flame* **44**, 113–24.

Xiong, T.Y., Law, C.K., and Miyasaka, K., 1985, "Interactive Vaporization and Combustion of Binary Droplet System," in *Proceedings of the Twentieth Symposium (International) on Combustion*, 1781–87. The Combustion Institute: Pittsburgh, PA.

Yan, C. and Aggarwal, S.K., 2006, "A High-Pressure Quasi-Steady Droplet Vaporization Model for Spray Simulations," *ASME J. Eng. Gas Turbines Power* **128**, 482–92.

Yanenko, N.N., 1971, *The Method of Fractional Steps.* Springer-Verlag: New York.

Yang, C.J., Jackson, S.C., and Avedisian, C.T., 1990, "Combustion of Unsupported Methanol/Dodeconal Mixture Droplets at Low Gravity," *Proc. Combust. Inst.* **23**, 1619–25.

Yang, V., 2008, Private communication of unpublished computations on phase equilibrium of dodecane/oxygen mixture.

Yang, V., Hsieh, K.C., and Shuen, J.S., 1993, "Supercritical Droplet Combustion and Related Transport Phenomena," AIAA 93-0812, 31st Aerospace Sciences Meeting and Exhibit, Reno, NV.

Yang, V. and Lin, N.N., 1994, "Vaporization of Liquid Oxygen (LOX) Droplets at Supercritical Conditions," *Combust. Sci. Technol.* **97**, 247–70.

Yang, V., Lin, N.N., and Shuen, J.S., 1992, "Vaporization of Liquid Oxygen (LOX) Droplets at Supercritical Conditions," AIAA-92-0078, 30th AIAA Aerospace Sciences Meeting and Exhibit, Reno, NV.

Yarin, A.L., 1993, *Free Liquid Jets and Films: Hydrodynamics and Rheology*, Longman, Harlow and Wiley: New York.

Yoshii, N. and Okazaki, S., 1997, "A Large-Scale and Long-Time Molecular Dynamics Study of Supercritical Lennard-Jones Fluid. An Analysis of High Temperature Clusters," *J. Chem. Phys.* **107**, 2020–33.

Young, N.O., Goldstein, J.S., and Block, M.J., 1959, "The Motion of Bubbles in a Vertical Temperature Gradient," *J. Fluid Mech.* **6**, 350–56.

Yuen, M.C., 1968, "Non-Linear Capillary Instability of a Liquid Jet," *J. Fluid Mech.* **33**, 151–63.

Yuen, M.C. and Chen, L.W., 1976, "On Drag of Evaporating Droplets," *Combust. Sci. Technol.* **14**, 147–54.

Zhang, B.L., Card, J.M., and Williams, F.A., 1996, "Application of Rate-Ratio Asymptotics to the Prediction of Extinction for Methanol Droplet Combustion," *Combust. Flame* **105**, 267–90.

Zhang, D.Z. and Prosperetti, A., 1994a, "Ensemble Phase-Averaged Equations for Bubbly Flows," *Phys. Fluids* **6**, 2956–70.

Zhang, D.Z. and Prosperetti, A., 1994b, "Averaged Equations for Inviscid Disperse Two-Phase Flow," *J. Fluid Mech.* **267**, 185–219.

Zhang, D.Z. and Prosperetti, A., 1994c, "Ensemble Phase-Averaged Equations for Bubbly Flows," *Phys. Fluids* **6**, 2956–70.

Zhang, D.Z. and Prosperetti, A., 1994d, "Averaged Equations for Inviscid Disperse Two-Phase Flow," *J. Fluid Mech.* **267**, 185–219.

Zhang, D.Z. and Prosperetti, A., 1997, "Momentum and Energy Equations for Disperse Two-Phase Flows and Their Closure for Dilute Suspensions," *Int. J. Multiphase Flow* **23**, 425–53.

Zhang, D.Z. and Prosperetti, A., 2005, "A Second-Order Method for Three-Dimensional Particle Dispersions," *J. Comput. Phys.* **210**, 292–324.

Zhang, H.Q. and Law, C.K., 2008, "Effects of Temporally Varying Liquid-Phase Mass Diffusivity in Multicomponent Droplet Gasification," *Combust. Flame* **153**, 593–602.

Zhang, P. and Law, C.K. 2007, "Theory of Bouncing and Coalescence in Droplet Collision," Paper No. G16, Fifth U.S. Combustion Meeting, March 25–28, San Diego, CA, 2007.

Zhao, Z., Poulikakis, D., and Fukai, J., 1996, "Heat Transfer and Fluid Dynamics During the Collision of a Liquid Droplet on a Substrate – I. Modeling," *Int. J. Heat Mass Transfer* **39**, 2771–89.

Zhu, G. and Aggarwal, S.K., 2000, "Transient Supercritical Droplet Evaporation With Emphasis on the Effects of Equation of State," *Int. J. Heat Mass Transfer* **43**, 1157–71.

Zhu, G., Reitz, R.D., and Aggarwal, S.K., 2001, "Gas-Phase Unsteadiness and Its Influence on Droplet Vaporization in Sub- and Super-Critical Environments," *Int. J. Heat Mass Transfer* **44/16**, 3081–93.

Index

Printed in the United States
By Bookmasters